STUDIES IN ECONOMIC DESIGN

Series Editor Murat R. Sertel
Turkish Academy of Sciences

Springer

Berlin
Heidelberg
New York
Hong Kong
London
Milan
Paris
Tokyo

Titles in the Series

V. I. Danilov and A. I. Sotskov
Social Choice Mechanisms
VI, 191 pages. 2002. ISBN 3-540-43105-5

T. Ichiishi and T. Marschak (Eds.)
Markets, Games and Organizations
VI, 314 pages. 2003. ISBN 3-540-43897-1

Bhaskar Dutta
Matthew O. Jackson
Editors

Networks and Groups

Models of Strategic Formation

With 71 Figures
and 9 Tables

Springer

Professor Bhaskar Dutta
Indian Statistical Institute
New Delhi 110016
India

Professor Matthew O. Jackson
California Institute of Technology
Division of Humanities and Social Sciences, 228–77
Pasadena, CA 91125
USA

ISBN 3-540-43113-6 Springer-Verlag Berlin Heidelberg New York

Cataloging-in-Publication Data applied for

A catalog record for this bbok is available from the Library of Congress.

Bibliographic information published by Die Deutsche Bibliothek
Die Deutsche Bibliothek lists this publication in the Deutsche Nationalbibliografie; detailed bibliographic data is available in the Internet at *http://dnb.ddb.de*.

Springer-Verlag Berlin Heidelberg New York
a member of BertelsmannSpringer Science + Business Media GmbH

http://www.springer.de

Printed in Germany

Hard cover design: Erich Kirchner, Heidelberg

SPIN 10864286 42/3130 – 5 4 3 2 1 0 – Printed on acid free paper

Preface

When Murat Sertel asked us whether we would be interested in organizing a special issue of the Review of Economic Design on the formation of networks and groups, we were happy to accept because of the growing research on this important topic. We were also pleasantly surprised at the response to our request for submissions to the special issue, receiving a much larger number of submissions than we had anticipated. In the end we were able to put together two special issues of insightful papers on this topic.

Given the growing interest in this topic, we also decided (with encouragement from Murat) to combine the special issues in the form of a book for wider dissemination. However, once we had decided to edit the book, it was natural to move beyond the special issue to include at least some of the papers that have been influential in the literature on the formation of networks. These papers were published in other journals, and we are very grateful to the authors as well as the journals for permission to include these papers in the book.

In collecting these papers, we hope that this book will be a useful base for researchers in the area, as well as a good starting point for those wanting to learn about network and group formation. Having this goal in mind helped us through the difficult task of selecting papers for the volume. Of course some selections were clear simply from following the progression of the early literature. However, as the literature is growing rapidly, there were some fine recent papers that we were forced to exclude. Our selection of the other papers was guided by a desire to strike a balance between broader theoretical issues and models tailored to specific contexts.

The ordering of the papers mainly follows the literature's progression, with an attempt to group papers together by the branches of the literature that they follow. Our introductory chapter provides an overview of the literature's progression and the role of the various papers collected here.

Bhaskar Dutta and Matthew O. Jackson
Pasadena, December 2002

Table of Contents

On the Formation of Networks and Groups

Bhaskar Dutta[1], **Matthew O. Jackson**[2]

[1] Indian Statistical Institute, 7 SJS Sansanwal Marg, New Delhi 110016, India
(e-mail: dutta@isid.ac.in)
[2] Division of Humanities and Social Sciences, California Institute of Technology, Pasadena, CA 91125, USA
(e-mail: jacksonm@hss.caltech.edu and http://www.hss.caltech.edu/~jacksonm/Jackson.html)

Abstract. We provide an introduction to and overview of the volume on *Models of the Strategic Formation of Networks and Groups.*

JEL classification: A14, D20, J00

1 Introduction

The organization of individual agents into networks and groups has an important role in the determination of the outcome of many social and economic interactions. For instance, networks of personal contacts are important in obtaining information about job opportunities (e.g., Boorman (1975) and Montgomery (1991)). Networks also play important roles in the trade and exchange of goods in non-centralized markets (e.g., Tesfatsion (1997, 1998), Weisbuch, Kirman and Herreiner (1995)), and in providing mutual insurance in developing countries (e.g., Fafchamps and Lund (2000)). The partitioning of societies into groups is also important in many contexts, such as the provision of public goods and the formation of alliances, cartels, and federations (e.g., Tiebout (1956) and Guesnerie and Oddou (1981)).

Our understanding of how and why such networks and groups form and the precise way in which these structures affect outcomes of social and economic interaction is the main focus of this volume. Recently there has been concentrated research focused on the formation and design of groups and networks, and their roles in determining the outcomes in a variety of economic and social settings. In this volume, we have gathered together some of the central papers in this recent literature which have made important progress on this topic. These problems are tractable and interesting, and from these works we see that structure matters

We thank Sanjeev Goyal and Anne van den Nouweland for helpful comments on an earlier draft.

and that clear predictions can be made regarding the implications of network and group formation. These works also collectively set a rich agenda for further research.

In this introduction, we provide a brief description of the contributions of each of the papers. We also try to show how these papers fit together, provide some view of the historical progression of the literature, and point to some of the important open questions.

2 A Brief Description of Some Related Literatures

There is an enormous literature on networks in a variety of contexts.

The "social networks" literature in sociology (with some roots in anthropology and social psychology) examines social interactions from theoretical and empirical viewpoints. That literature spans applications from family ties through marriage in 15th century Florence to needle sharing among drug addicts, to networks of friendship and advice among managers. An excellent and broad introductory text to the social networks literature is Wasserman and Faust (1994). One particular area of overlap with economics is the portion of that literature on exchange networks. The Bienenstock and Bonacich (1997) paper in this volume (and discussed more below) is a nice source for some perspective on and references to that literature. The analysis of the incentives to form networks and groups and resulting welfare implications, the focus of most of the papers in this volume, is largely complementary to the social networks literature both in its perspective and techniques.

There are also various studies of networks in economics and operations research of transportation and delivery networks.[1] One example would be the routing chosen by airlines which has been studied by Hendricks, Piccione and Tan (1995) and Starr and Stinchcombe (1992). One major distinguishing feature of the literature that we focus on in this volume is that the parties in the network or group are economic or social actors. A second distinguishing feature is that the focus is on the incentives of individual actors to form networks and groups.

Thus, the focus here is on a series of papers and models that have used formal game theoretic reasoning to study the formation of networks and other social structures.[2]

---

[1] There is also a literature in industrial organization that surrounds network externalities, where, for instance a consumer prefers goods that are compatible with those used by other individuals (see Katz and Shapiro (1994)). There, agents care about who else uses a good, but the larger nuances of a network with links does not play any role. Young (1998) provides some insights into such interactions where network structures provide the fabric for interaction, but are taken to be exogenous.

[2] Also, our focus is primarily on the formation of networks. There is also a continuing literature on incentives in the formation of coalitions that we shall not attempt to survey here, but mention at a few points.

3 Overview of the Papers in the Volume

Cooperation Structures and Networks in Cooperative Games

An important first paper in this literature is by Myerson (1977). Myerson started from cooperative game theory and layered on top of that a network structure that described the possible communication or cooperation that could take place. The idea was that a coalition of individuals could only function as a group if they were connected through links in the network. Thus, this extends standard cooperative game theory where the modeler knows only the value generated by each potential coalition and uses this to make predictions about how the value of the overall society will be split among its members. One perspective on this is that the members of society bargain over how to split the value, and the values of the different coalitions provide threat points in the bargaining.[3] The enrichment from the communication structures added by Myerson is that it provides more insight into which coalitions can generate value and thus what threats are implicit when society is splitting.

More formally, Myerson starts from the familiar notion of a *transferable utility* game (N, v), where N is a set of players and v is a characteristic function denoting the worth $v(S)$ of each coalition $S \subset N$. He defines a *cooperation structure* as an *non-directed* graph g among the individuals. So, a graph represents a list of which individuals are linked to each other, with the interpretation that if individual i is linked to individual j in the graph, then they can communicate and function together. Thus, a network g partitions the set of individuals into different groups who are connected to each other (there is a path from each individual in the group to every other individual in the group). The value of a coalition S under the network g is simply the sum of the values of the sub-coalitions of S across the partition of S induced by g. For instance, consider a cooperative game where the worth of coalition $\{1, 2\}$ is 1, the worth of coalition $\{3, 4\}$ is 1, and the worth of coalition $\{1, 2, 3, 4\}$ is 3. If under the graph g the only links are that individual 1 is linked to 2 and individual 3 is linked to 4, then the worth of the coalition $\{1, 2, 3, 4\}$ under the restriction to the graph g is simply $1 + 1 = 2$, rather than 3, as without any other links this is the only way in which the group can function. So in Myerson's setting, a network or graph g coupled with a characteristic function v results in a *graph-restricted* game v^g.

In Myerson's setting, an *allocation rule*[4] describes the distribution of payoffs amongst individuals for every pair (v, g). This may represent the natural payoff going to each individual, or may represent some additional intervention and transfers. Myerson characterizes a specific allocation rule which eventually became referred to as the *Myerson value*. In particular, Myerson looks at allocation rules that are fair: the gain or loss to two individuals from the addition

[3] From a more normative perspective, these coalitional values may be thought of as providing a method of uncovering how much of the total value that the whole society produces that various individuals and groups are responsible for.

[4] This is somewhat analogous to a solution concept in cooperative game theory.

of a link should be the same for the two individuals involved in the link; and are balanced in that they are spreading exactly the value of a coalition (from the graph-restricted game) among the members of the coalition. Myerson shows that the unique allocation rule satisfying these properties is the *Shapley value* of the graph-restricted game.[5]

While Myerson's focus was on characterizing the allocation rule based on the Shapley value, his extension of cooperative game theory to allow for a network describing the possibilities for cooperation was an important one as it considerably enriches the cooperative game theory model not only to take into account the worth of various coalitions, but also how that worth depends on a structure describing the possibilities of cooperation.

Network Formation More Generally

While Myerson's model provides an important enrichment of a cooperative game, it falls short of providing a general model where value is network dependent. For example, the worth of a coalition $\{1, 2, 3\}$ is the same whether the underlying network is one that only connects 1 to 2 and 2 to 3, or whether it is a complete network that also connects 1 to 3. While this is of interest in some contexts, it is somewhat limited as a model of networks. For instance, it does not permit there to be any cost to a link or any difference between being directly linked to another individual versus only being indirectly linked.

The key departure of Jackson and Wolinsky (1996) from Myerson's approach was to start with a value function that is defined on networks directly, rather than on coalitions. Thus, Jackson and Wolinsky start with a value function v that maps each network into a worth or value. Different networks connecting the same individuals can result in different values, allowing the value of a group to depend not only on who is connected but also how they are connected. This allows for costs and benefits (both direct and indirect) to accrue from connections. In the Jackson-Wolinsky framework, an allocation rule specifies a distribution of payoffs for each pair network and value function. One result of Jackson and Wolinsky is to show that the analysis of Myerson extends to this more general setting, and fairness and component balance again lead to an allocation rule based on the Shapley value.

One of the central issues examined by Jackson and Wolinsky is whether *efficient* (value maximizing) networks will form when self-interested individuals can choose to form links and break links. They define a network to be *pairwise stable* if no pair of individuals wants to form a link that is not present and no individual gains by severing a link that is present.

They start by investigating the question of the compatibility of efficiency and stability in the context of two stylized models. One is the *connections model*,

[5] An interesting feature of Myerson's characterization is that he dispenses with *additivity*, which is one of the key axioms in Shapley's original characterization. This becomes implicit in the balance condition given the network structure.

where individuals get a benefit $\delta \in [0, 1]$ from being linked to another individual and bear a cost c for that link. Individuals also benefit from indirect connections - so a friend of a friend is worth δ^2 and a friend of a friend of a friend is worth δ^3, and so forth. They show that in this connections model efficient networks take one of three forms: an empty network if the cost of links is high, a star-shaped network for middle ranged link costs, and the complete network for low link costs. They demonstrate a conflict between this very weak notion of stability and efficiency - for high and low costs the efficient networks are pairwise stable, but not always for middle costs. This also holds in the second stylized model that they call the *co-author* model, where benefits from links come in the form of synergies between researchers.

Jackson and Wolinsky also examine this conflict between efficiency and stability more generally. They show that there are natural situations (value functions), for which under any allocation rule belonging to a fairly broad class, no efficient network is pairwise stable. This class considers allocation rules which are component balanced (value is allocated to the component of a network which generated it) and are anonymous (do not structure payments based on labels of individuals but instead on their position in the network and role in contributing value in various alternative networks). Thus, even if one is allowed to choose the allocation rule (i.e., transfer wealth across individuals to try to align incentives according to some mild restrictions) it is impossible to guarantee that efficient networks will be pairwise stable. So, the tension between efficiency and stability noted in the connections and co-author models is a much broader problem. Jackson and Wolinsky go on to study various conditions and allocation rules for which efficiency and pairwise stability are compatible.

While Jackson and Wolinsky's work provides a framework for examining the relationship between individual incentives to form networks and overall societal welfare, and suggests that these may be at odds, it leaves open many questions. Under exactly what circumstances (value functions and allocation rules) do individual incentives lead to efficient networks? How does this depend on the specific modeling of the stability of the network as well as the definition of efficiency?

Further Study of the Compatibility of Efficiency and Stability

This conflict between stability and efficiency is explored further in other papers.

Johnson and Gilles (2000) study a variation on the connections model where players are located along a line and the cost of forming a link between individuals i and j depends on the spatial distance between them. This gives a geography to the connections model, and results in some interesting structure to the efficient networks. Stars no longer play a central role and instead chains do. It also has a dramatic impact on the shape of pairwise stable networks, as they have interesting local interaction properties. Johnson and Gilles show that the conflict between efficiency and pairwise stability appears in this geographic version of the connections model, again for an intermediate range of costs to links.

Dutta and Mutuswami (1997) adopt a "mechanism design" approach to reconcile the conflict between efficiency and stability. In their approach, the allocation rule is analogous to a mechanism in the sense that this is an object which can be designed by the planner. The value function is still given exogenously and the network is formed by self-interested players or agents, but properties of the allocation rule such as anonymity are only applied at stable networks. That is, just as a planner may be concerned about the ethical properties of a mechanism only at *equilibrium*, Dutta and Mutuswami assume that one needs to worry only about the ethical properties of an allocation rule on networks which are equilibria of a formation game. That is, the design issue with which Dutta and Mutuswami are concerned is whether it is possible to define allocation rules which are "nice" on the set of equilibrium networks. They construct allocation rules which satisfy some desirable properties on equilibrium graphs. Of course, the construction deliberately uses some ad hoc features "out of equilibrium".

The network formation game that is considered by Dutta and Mutuswami is discussed more fully below, and offers an alternative to the notion of pairwise stability.

The paper by Jackson (2001) examines the tension between efficiency and stability in further detail. He considers three different definitions of efficiency, which consider the degree to which transfers are permitted among individuals. The strong efficiency criterion of Jackson and Wolinsky is only appropriate to the extent that value is freely transferable among individuals. If more limited transfers are possible (for instance, when one considers component balance and anonymity), then a constrained efficiency notion or even Pareto efficiency become appropriate. Thus the notion of efficiency is tailored to whether the allocation rule is arising naturally, or to what extent one is considering some further intervention and reallocation of value. Jackson studies how the tension between efficiency and stability depends on this perspective and the corresponding notion of efficiency used. He shows how this applies in several models including the Kranton and Minehart (1998) model, and a network based bargaining model due to Corominas-Bosch (1999). He also shows that the Myerson allocation rule generally has difficulties guaranteeing even Pareto efficiency, especially for low costs to links. Individuals have incentives to form links to better their bargaining position and thus their resulting Myerson allocation. When taking these incentives together, this can result in over-connection to a point where all individuals suffer.

Stability and Efficiency in Directed Networks Models

Non-directed networks capture many applications, especially those where mutual consent or effort is required to form a link between two individuals. However, there are also some applications where links are directed or unilateral. That is, there are contexts where one individual may form a link with a second individual without the second individual's consent, as would happen in sending a paper to another individual. Other examples include web links and one-sided compatibility of products (such as software). Such settings lead to different incentives in the formation of networks, as mutual consent is not needed to form a link. Hence the analysis of such directed networks differs from that of non-directed networks.

Bala and Goyal (2000a) analyze a communication model which falls in this directed network setting. In their model, value flows costlessly through the network along directed links. This is similar to the connections model, but with δ close to 1 and with directional flow of communication or information. Bala and Goyal focus on the dynamic formation of networks in this directed communications model. The network formation game is played repeatedly, with individuals deciding on link formation in each period. Bala and Goyal use a version of the best response dynamics, where agents choose links in response to what happened in the previous period, and with some randomization when indifferent. In this setting, for low enough costs to links, the process leads naturally to a limiting network which has the efficient structure of a wheel.

Dutta and Jackson (2000) show that while efficiency is obtained in the Bala and Goyal communication model, the tension between efficiency and stability reemerges in the directed network setting if one looks more generally. As one might expect, the nature of the conflict between stability and efficiency in directed networks differs from that in non-directed networks. For instance, the typical (implicit) assumption in the directed networks framework is that an agent can *unilaterally* form a link with any other agent, with the cost if any of link formation being borne by the agent who forms the link. It therefore makes sense to say that a directed network is stable if no agent has an incentive to either break an existing link or create a new one.[6] Using this definition of stability, Dutta and Jackson show that efficiency can be reconciled with stability either by distributing benefits to *outsiders* who do not contribute to the productive value of the network or by violating equity; but that the tension between stability and efficiency persists if one satisfies anonymity and does not distribute value to such outsiders.

Bala and Goyal (2000) also analyze the efficiency-stability conflict for a hybrid model of information flow, where the cost of link formation is borne by the agent setting up the link, but where both agents can access each other's information regardless of who initiated the link. Bala and Goyal give the example of a telephone call, where the person who makes the call bears the cost of the call, but both persons are able to exchange information. In their model, however, each link is *unreliable* in the sense that there is a positive probability that the link will fail to transmit the information. Bala and Goyal find that if the cost of forming links is low or if the network is highly reliable, then there is no conflict between efficiency and stability. Bala and Goyal also analyze the structure of stable networks in this setting.

[6] In contrast, the implicit assumption in the undirected networks framework is that both i and j have to agree in order for the link ij to form.

Modeling the Formation of Networks

Notice that pairwise stability used by Jackson and Wolinsky is a very weak concept of stability - it only considers the addition or deletion of a single link at a time. It is possible that under a pairwise stable network some individual or group would benefit by making a dramatic change to the network. Thus, pairwise stability might be thought of as a necessary condition for a network to be considered stable, as a network which is *not* pairwise stable may not be formed irrespective of the actual process by which agents form links. However, it is not a sufficient condition for stability. In many settings pairwise stability already dramatically narrows the class of networks, and noting a tension between efficiency and pairwise stability implies that such a tension will also exist if one strengthens the demands on stability. Nevertheless, one might wish to look beyond pairwise stability to explicitly model the formation process as a game. This has the disadvantage of having to specify an ad hoc game, but has the advantage of permitting the consideration of richer forms of deviations and threats of deviations. The volume contains several papers devoted to this issue.

This literature owes its origin to Aumann and Myerson (1988), who modeled network formation in terms of the following extensive form game.[7] The extensive form presupposes an exogenous ranking of pairs of players. Let this ranking be $(i_1j_1, \ldots, i_nj_n)$. The game is such that the pair i_kj_k decide on whether or not to form a link knowing the decisions of all pairs coming before them. A decision to form a link is binding and cannot be undone. So, in equilibrium such decisions are made with the knowledge of which links have already formed (or not), and with predictions as to which links will form as a result of the given pair's decision. Aumann and Myerson assume that after all pairs have either formed links or decided not to, then allocations come from the Myerson value of the resulting network g and some graph restricted cooperative game v^g. They are interested in the subgame perfect equilibrium of this network formation game.

To get a feeling for this, consider a symmetric 3-person game where $v(S) = 0$ if $\#S = 1$, $v(S) = 40$ if $\#S = 2$ and $v(N) = 48$. An efficient graph would be one where at least two links form so that the grand coalition can realize the full worth of 48. Suppose the ranking of the pairs is 12, 13, 23. Then, if 1 and 2 decide to form the link 12 and refrain from forming links with 3, then they each get 20. If all links form, then each player gets 16. The unique subgame perfect equilibrium in the Aumann-Myerson extensive form is that only the link 12 will form, which is inefficient.

A crucial feature of the game form is that if pair i_kj_k decide not to form a link, but some other pair coming after them does form a link, then i_kj_k are allowed to reconsider their decision.[8] It is this feature which allows player 1 to make a credible threat to 2 of the form "I will not form a link with 3 if you do not. But if you do form a link with 3, then I will also do so." This is what

[7] A precursor of the network formation literature can be found in Boorman (1975).

[8] As Aumann and Myerson remark, this procedure is like bidding in bridge since a player is allowed to make a fresh bid if some player bids after her.

sustains $g = \{12\}$ as the equilibrium link. Notice that after the link 12 has been formed, if 1 refuses to form a link with 3, then 2 has an incentive to form the link with 3 - this gives her a payoff of $29\frac{1}{3}$ *provided* 1 cannot come back and form the complete graph. So, it is the possibility of 1 and 3 coming back into the game which deters 2 from forming the link with 3.

Notice that such threats cannot be levied when the network formation is simultaneous. Myerson (1991) suggested the following simultaneous process of link formation. Players simultaneously announce the set of players with whom they want to form links. A link between i and j forms if both i and j have announced that they want a link with the other. Dutta, van den Nouweland, and Tijs (1998)[9] model link formation in this way in the context of the Myerson model of cooperation structures. Moreover, they assume that once the network is formed, the eventual distribution of payoffs is determined by some allocation rule within a class containing the Myerson value. The entire process (formation of links as well as distribution of payoffs) is a normal form game. Their principal result is that for all superadditive games, a complete graph (connecting the grand coalition) or graphs that are payoff equivalent will be the undominated equilibrium or coalition-proof Nash equilibrium.

The paper by Slikker and van den Nouweland (2000) considers a variant on the above analysis, where they introduce an explicit cost of forming links. This makes the analysis much more complicated, but they are still able to obtain solutions at least for the case of three individuals. With costs to links, they find the surprising result that link formation may not be monotone in link costs: it is possible that as link costs increase more links are formed. This depends in interesting ways on the Myerson value, the way that individual payoffs vary with the network structure, and also on the modeling of network formation via the Aumann and Myerson extensive form.

Dutta and Mutuswami (1997) (discussed above) use the same normal form game for link formation in the context of the network model of Jackson and Wolinsky. They note the relationship between various solution concepts such as strong equilibrium and coalition proof Nash equilibrium to pairwise stability.[10] They (as well as Dutta, van den Nouweland and Tijs (1998)) also discuss the importance of considering only undominated strategies and/or deviations by at least two individuals in this sort of game, so as to avoid degenerate Nash equilibria where no agent offers to form any links knowing that nobody else will.

Bargaining and Network Formation

One aspect that is present in all of the above mentioned analyses is that the network formation process and the way in which value is allocated to members of a network are separated. Currarini and Morelli (2000) take the interesting view

[9] See also Qin(1996).

[10] See also Jackson and van den Nouweland (2001) for a detailed analysis of a strong equilibrium based stability concept where arbitrary coalitions can modify their links.

that the allocation of value among individuals may take place simultaneously with the link formation, as players may bargain over their shares of value as they negotiate whether or not to add a link.[11] The game that Currarini and Morelli analyze is one where players are ranked exogenously. Each player sequentially announces the set of players with whom he wants to form a link as well as a payoff demand, as a function of the history of actions chosen by preceding players. Both players involved in a link must agree to form the link. In addition, payoff demands within each component of the resulting graph must be consistent. Currarini and Morelli show that for a large class of value functions, all subgame perfect equilibria are efficient. This differs from what happens under the Aumann and Myerson game. Also, as it applies for a broad class of value functions, it shows that the tension between stability and efficiency found by Jackson and Wolinsky may be overcome if bargaining over value is tied to link formation.

Gerber (2000) looks at somewhat similar issues in the context of coalition formation. With a few exceptions, the literatures on bargaining and on coalition formation, either look at how worth is distributed taking as given that the grand coalition will form, or look at how coalitions form taking as given how coalitions distribute worth. Gerber stresses the simultaneous determination of the payoff distribution and the coalition structure, and defines a new solution for general NTU games. This solution views coalitions as various interrelated bargaining games which provide threat points for the bargaining with any given coalition, and ultimately the incentives for individuals to form coalitions. Gerber's solution concept is based on a consistency condition which ties these games together. Gerber shows how this applies in some special cases (including the marriage market which ties in with the networks models) as well as several examples, and illustrates the important differences between her solution and others.

Dynamic Network Formation

The papers discussed above have largely analyzed network formation in static settings (taking an extensive form to be essentially static). The main exception is that of best response dynamics in the directed communications model of Bala and Goyal (2000a).

Watts (2001) also departs from the static modeling tradition.[12] In the context of the connections model of Jackson and Wolinsky, she considers a framework where pairs of agents meet over time, and decide whether or not to form or sever links with each other. Agents are myopic and so base their decision on how the decision on the given link affects their payoffs, given the current network in place. The network formation process is said to reach a *stable state* if no additional links are formed or broken in subsequent periods. A principal result

[11] See also Slikker and van den Nouweland (2001) and Mutuswami and Winter (2000).

[12] The literature on the dynamic formation of networks has grown following Watts' work, and there are a number of recent papers that study various stochastic models of network formation. These include Jackson and Watts (1998, 1999), Goyal and Vega-Redondo (1999), Skyrms and Pemantle (2000), and Droste, Gilles and Johnson (2000).

is that a stable state is often inefficient, although this depends on the precise cost and benefit parameters. A particularly interesting result applies to a cost range where a star network is both pairwise stable and efficient, but where there are also some inefficient networks that are stable states. Watts shows that as the number of individuals increases, the probability[13] that a star forms goes to 0. Thus as the population increases the particular ordering which is needed to form a star (the efficient network) becomes less and less likely relative to orderings leading to some other stable states.

Networks for the Trade and Exchange of Goods

There has also been study of network models in some other specific contexts. For instance, the two papers by Kranton and Minehart (1998, 2000) focus on networks of buyers and sellers, where goods are to be exchanged between connected individuals, but the terms of trade can depend on the overall set of opportunities that the connected individuals have. The first paper considers buyers with private values who can bid in auctions of sellers to whom they are connected. Buyers gain from being involved in more auctions as they then have a better chance of obtaining an object and at a lower expected cost. Sellers gain from having more buyers participate in their auction as it increases the expected highest valuation and thus willingness to bid, and also increases the competition among buyers. Kranton and Minehart show the striking result that the change in expected utility that any buyer sees from adding a link to some seller is precisely the overall social gain from adding that link. Thus, if only buyers face costs to links, then they have incentives to form a socially efficient network. They also show that if sellers face costs to invest in providing the good for sale, then inefficiency can arise.[14]

In the second paper, Kranton and Minehart (2000) develop a theory of competition in networks which intends to look more generally at how the connection structure among buyers and sellers affects terms of trades. The new concept that they introduce is that of "opportunity paths" which describe the various ways in which individuals can effect trades. The pattern of opportunity paths is central in determining the trades that occur, and Kranton and Minehart provide a series of results indicating how the opportunity paths determine the resulting prices and utilities to the various agents in the network.

Bloch and Ghosal (2000) also analyze how interrelationships among buyers and sellers affect terms of trade. Their analysis is not so network dependent, but focuses more directly on the issue of cartel formation amongst buyers and sellers. In particular, they are concerned with how collusion on one side of the market affects cartel formation on the other side of the market. They build on the bargaining model of Rubinstein and Wolinsky (1990), where a random matching

[13] Links are identified randomly and then agents decide whether to add or delete them.

[14] Jackson (2001) points out that a similar inefficiency result holds in Kranton and Minehart's model if sellers face any costs to links and pairwise stability is required.

process is a central determinant of the terms of trade. They find that there is at most one stable cartel structure, which if it exists consists of equal size structures and cartels each remove one trader from the market. This suggests the emergence of a balance between the two sides of the market.

The paper by Bienenstock and Bonacich (1997) provides discussion of how cooperative game theory concepts can be useful in modeling network exchange. In discussing the way in which notions of transferable utility cooperative game theory can be applied to study exchange of goods in networks, Bienenstock and Bonacich provide a nice overview of the network exchange literature, and some pointed discussion about the alternative behavioral assumptions that can be made, and how utility theory and viewing things as a cooperative game can be a useful lens. An important point that they make is that using game theoretic tools allows for an understanding of how structural outcomes depend on underlying characteristic function and how this relates to the structure itself. That network structure is important in determining power and distribution of resources is a fundamental understanding in most of the work on the subject. Bienenstock and Bonacich (1997) outline why cooperative game theory can be a useful tool in studying this relationship. They discuss the possible use of the kernel in some detail.

Networks in Other Specific Contexts

The remaining papers in the volume, Barbera and Dutta (2000) and Gehrig, Regibeau, and Rocket (2000), are concerned with different issues connected with the organizational structure of firms.

Barbera and Dutta consider a labor-managed firm where there are two types of tasks and two types of workers (skilled and unskilled), with the skilled workers being more productive in the first type of task. Their interest is in the possibility of constructing payment or reward schemes which will induce workers to reveal types correctly, and thereby sort themselves correctly into tasks. They show that using various hierarchical structures in rewards, can provide strong incentives for workers to properly sort themselves into tasks.

Gehrig, Regibeau, and Rocket study the internal organization of firms in a context where the firm has to evaluate cost-reducing R & D projects. They compare hierarchies versus polyarchies. In hierarchies unanimous approval by a number of reviewers is required in order for the project to be approved, while the approval of any one reviewer is sufficient in a polyarchy. They allow for different reviewers to have different hurdles for approval, and then study conditions under which the polyarchy can dominate the hierarchy.

4 Some Important Open Questions

We end our introduction to this volume by briefly describing some important issues regarding network and group formation which deserve closer attention, and are suggested through the collection of papers here.

Even a cursory look at the papers in this volume indicates that the conflict between stability and efficiency is of significant interest. Nevertheless, much remains to be known about the conditions under which the conflict will arise. Some of the papers have examined this conflict in the abstract, and others in the context of very pointed and specific models. While we see some themes emerging, we do not yet have an overarching characterization of exactly what leads to an alignment between individual incentives and societal welfare, and what leads these to diverge. Jackson (2001) suggests that at least some of the tension can be categorized as coming from two separate sources: one source is that of a classic externality where individuals do not internalize the full societal effects of their forming or severing a link; and another source is the incentives of individuals to form or sever links in response to how the network structure affects bargaining power and the allocation of value, rather than in response to how it affects overall value. Whether inefficiency can always be traced to one (or both) of these sources and more basically whether this is a useful taxonomy, are open questions.

There are also several issues that need to be addressed in the general area of the formation of networks. It becomes clear from comparisons within and across some of the papers, that the specific way in which one models network stability or formation can matter. This is clearly borne out by comparing the Aumann–Myerson prediction that inefficient networks might form with that of Dutta et al. who find efficiency at least in superadditive games. We need to develop a fuller understanding of how the specifics of the process matter, and tie this to different sorts of applications to get some sense of what modeling techniques fit different sorts of problems.

Perhaps the most important (and possibly the hardest) issue regarding modeling the formation of networks is to develop fuller models of networks forming over *time*, and in particular allowing for players who are *farsighted*. Farsightedness would imply that players' decisions on whether to form a network are not based solely on current payoffs, but also on where they expect the process to go and possibly from emerging steady states or cycles in network formation. We see some of this in the Aumann and Myerson (1988) extensive form, but it is artificially cut by the finiteness of the game. It is conceivable that, at least in some contexts, farsightedness may help in ensuring efficiency of the stable state. For instance, if there are increasing returns to network formation, then myopic considerations may result in the null network being formed since no one (or pair) may want to incur the initial cost. However, the initial costs may well be recouped in the long-run, thereby facilitating the formation of efficient networks. This is only one small aspect of what farsighted models might bring.

More work can also be undertaken in constructing, analyzing, and characterizing "nice" allocation rules, as well as ones that might arise naturally under certain conditions. There are essentially two prominent single-valued solution concepts in cooperative game theory - the Shapley value and the nucleolus. While there is a close connection between characteristic functions and value functions, the special structure of networks may allow for the construction of allocation rules which do not have any obvious correspondence with solution concepts in cooperative game theory.

Also, the papers collected in this volume are all theoretical in nature. Many of them provide very pointed predictions regarding various aspects of network formation, albeit in highly stylized environments. Some of these predictions can be tested both in experiments,[15] as well as being brought directly to the data. The models can also be applied to some areas of particular interest, for example to examine whether decentralized labor markets, which depend a great deal on connections and network structure, function efficiently.

References

Aumann, R., Myerson, R. (1988) Endogenous Formation of Links Between Players and Coalitions: An Application of the Shapley Value. In: A. Roth *The Shapley Value*. Cambridge University Press pp. 175–191.

Bala, V., Goyal, S. (2000) A Strategic Analysis of Network Reliability *Review of Economic Design* 5: 205–228.

Bala, V., Goyal, S. (2000a) Self-Organization in Communication Networks, *Econometrica* 68: 1181–1230.

Barbera, S., Dutta, B. (2000) Incentive Compatible Reward Schemes for Labour-Managed Firms, *Review of Economic Design* 5: 111–128.

Bienenstock, E., Bonacich, P. (1997) Network Exchange as a Cooperative Game, *Rationality and Society* 9: 37–65.

Bloch, F., Ghosal, S. (2000) Buyers' and Sellers' Cartels on Markets with Indivisible Goods, *Review of Economic Design* 5: 129–148.

Boorman, S. (1975) A Combinatorial Optimization Model for Transmission of Job Information through Contact Networks, *Bell Journal of Economics* 6: 216–249.

Corominas-Bosch, M. (1999) On Two-Sided Network Markets, Ph.D. dissertation: Universitat Pompeu Fabra.

Corbae, D., Duffy, J. (2000) Experiments with Network Economies, mimeo: University of Pittsburgh.

Currarini, S., Morelli, M. (2000) Network Formation with Sequential Demands, *Review of Economic Design* 5: 229–250.

Droste, E., Gilles, R., Johnson, C. (2000) Evolution of Conventions in Endogenous Social Networks, mimeo: Virginia Tech.

Dutta, B., Jackson, M.O. (2000) The Stability and Efficiency of Directed Communication Networks, *Review of Economic Design* 5: 251–272.

Dutta, B., Mutuswami, S. (1997) Stable Networks, *Journal of Economic Theory* 76: 322–344.

Dutta, B., van den Nouweland, A., Tijs, S. (1998) Link Formation in Cooperative Situations, *International Journal of Game Theory* 27: 245–256.

Fafchamps, M., Lund, S. (2000) Risk-Sharing Networks in Rural Philippines, mimeo: Stanford University.

Gerber, A. (2000) Coalition Formation in General NTU Games, *Review of Economic Design* 5: 149–177.

Gehrig, T., Regibeau, P., Rockett, K. (2000) Project Evaluation and Organizational Form, *Review of Economic Design* 5: 177-200

Goyal, S., Vega-Redondo, F. (1999) Learning, Network Formation and Coordination, mimeo: Erasmus University.

Guesnerie, R., Oddou, C. (1981) Second best taxation as a game, *Journal of Economic Theory* 25: 67–91.

Hendricks, K., Piccione, M., Tan G. (1995) The Economics of Hubs: The Case of Monopoly, *Review of Economic Studies* 62: 83–100.

[15] See Charness and Corominas-Bosch (1999) and Corbae and Duffy (2000) for recent examples of testing such predictions.

Jackson, M.O. (2001) The Stability and Efficiency of Economic and Social Networks, mimeo. (2002) in *Advances in Economic Design*, Edited by Murat Sertel, Springer Verlag.

Jackson, M.O., van den Nouweland, A. (2001) Strongly Stable Networks, mimeo Caltech and University of Oregon.

Jackson, M.O., Watts, A. (1998) The Evolution of Social and Economic Networks, Caltech WP # 1044. Forthcoming: *Journal of Economic Theory*.

Jackson, M.O., Watts, A. (1999) On the Formation of Interaction Networks in Social Coordination Games mimeo. Forthcoming: *Games and Economic Behavior*.

Jackson, M.O., Wolinsky, A. (1996) A Strategic Model of Social and Economic Networks, *Journal of Economic Theory* 71: 44–74.

Johnson, C., Gilles, R.P. (2000) Spatial Social Networks, *Review of Economic Design* 5: 273–300.

Katz, M., Shapiro, C. (1994) Systems Competition and Networks Effects, *Journal of Economic Perspectives* 8: 93–115.

Kranton, R., Minehart, D. (1998) A Theory of Buyer-Seller Networks, forthcoming: *American Economic Review*.

Kranton, R., Minehart, D. (2000) Competition for Goods in Buyer-Seller Networks, *Review of Economic Design* 5: 301–332.

Liebowitz, S., Margolis, S. (1994) Network Externality: An Uncommon Tragedy, *Journal of Economic Perspectives* 8: 133–150.

Montgomery, I. (1991) Social Networks and Labor Market Outcomes, *The American Economic Review* 81: 1408–1418.

Mutuswami, S., Winter, E. (2000) Subscription Mechanisms for Network Formation, mimeo: CORE and Hebrew University in Jerusalem.

Myerson, R. (1977) Graphs and Cooperation in Games, *Math. Operations Research* 2: 225–229.

Myerson, R. (1991) *Game Theory: Analysis of Conflict*, Harvard University Press: Cambridge, MA.

Qin, C-Z. (1996) Endogenous Formation of Cooperation Structures, *Journal of Economic Theory* 69: 218–226.

Skyrms, B., Pemantle, R. (2000) A Dynamic Model of Social Network Formation, *Proceedings of the National Academy of Sciences* 97: 9340–9346.

Slikker, M., van den Nouweland, A. (2000) Network Formation Models with Costs for Establishing Links, *Review of Economic Design* 5: 333–362.

Slikker, M., van den Nouweland, A. (2001) A One-Stage Model of Link Formation and Payoff Division, *Games and Economic Behavior* 34: 153-175.

Starr, R.M., Stinchcombe, M.B. (1992) Efficient Transportation Routing and Natural Monopoly in the Airline Industry: An Economic Analysis of Hub-Spoke and Related Systems, UCSD dp 92-25.

Tesfatsion, L. (1997) A Trade Network Game with Endogenous Partner Selection. In: H. Amman et al. (eds.), *Computational Approaches to Economic Problems*, Kluwer Academic Publishers, 249–269.

Tesfatsion, L. (1998) Gale-Shapley matching in an Evolutionary Trade Network Game, Iowa State University Economic Report no. 43.

Tiebout, C.M. (1956) A Pure Theory of Local Expenditures, *Journal of Political Economy* 64: 416–424.

Wasserman, S., Faust, K. (1994) *Social Network Analysis: Methods and Applications*, Cambridge University Press.

Watts, A. (2001) A Dynamic Model of Network Formation, *Games and Economic Behavior* 34: 331–341.

Weisbuch, G., Kirman, A., Herreiner (1995) Market Organization, mimeo École Normal Superieure.

Young, H.P. (1998) Individual Strategy and Social Structure, Princeton University Press, Princeton.

Graphs and Cooperation in Games

Roger B. Myerson

Graduate School of Management, Nathaniel Leverone Hall, Northwestern University, Evanston, Illinois 60201, USA

Abstract. Graph-theoretic ideas are used to analyze cooperation structures in games. Allocation rules, selecting a payoff for every possible cooperation structure, are studied for games in characteristic function form. Fair allocation rules are defined, and these are proven to be unique, closely related to the Shapley value, and stable for a wide class of games.

1 Introduction

In the study of games, one often assumes either that all players will cooperate with each other, or, else that the game will be played noncooperatively. However, there are many intermediate possibilities between universal cooperation and no cooperation. (See Aumann and Drèze [1974] for one systematic study of the implications of partial cooperation.) In this paper we use ideas from graph theory to provide a framework within which we can discuss a broad class of partial cooperation structures and study the question of how the outcome of a game should depend on which players cooperate with each other.

Let N be a nonempty finite set, to be interpreted as the set of players. A *graph* on N is a set of unordered pairs of distinct members of N. We will refer to these unordered pairs as *links*, and we will denote the link between n and m by $n : m$. (So $n : m = m : n$, since the link is an unordered pair.) Let g^N be the complete graph of all links:

$$g^N = \{n : m \mid n \in N, m \in N, n \neq m\} . \tag{1}$$

Then let GR be the set of all graphs on N, so that

$$GR = \{g \mid g \subseteq g^N\} . \tag{2}$$

Our basic idea is that players may cooperate in a game by forming a series of bilateral agreements among themselves. These bilateral cooperative agreements can be represented by links between the agreeing players, so any cooperation structure can be represented by a set of agreement links. In this way, we can identify the set of all possible *cooperation structures* with *GR*, the set of all graphs on the set of players.

2 Coalitions and Connectedness

A *coalition* is a nonempty subset of N. We will need a few basic concepts of connectedness, to relate coalitions and cooperation graphs.

Suppose $S \subseteq N$, $g \in GR$, $n \in S$ and $m \in S$ are given. Then we say that n and m are *connected in* S *by* g iff there is a path in g which goes from n to m and stays within S. That is, n and m are connected in S by g if $n = m$ or if there is some $k \geq 1$ and a sequence $(n^0, n^1, \ldots, n^k)$ such that $n^0 = n$, $n^k = m$, and $n^{i-1} : n^i \in g$ and $n^i \in S$ for all i from 1 to k.

Given $g \in GR$ and $S \subseteq N$, there is a unique partition of S which groups players together iff they are connected in S by g. We will denote this partition by S/g (read "S divided by g"), so:

$$S/g = \{\{(i \mid i \text{ and } j \text{ are connected in } S \text{ by } g\} \mid j \in S\} \,. \qquad (3)$$

We can interpret S/g as the collection of smaller coalitions into which S would break up, if players could only coordinate along the links in g.

For example, if $N = \{1, 2, 3, 4, 5\}$ and $g = \{1:2, 1:4, 2:4, 3:4\}$ then $\{1, 2, 3\}/g = \{\{1, 2\}, \{3\}\}$ and $N/g = \{\{1, 2, 3, 4\}, \{5\}\}$.

When we speak of *connectedness* without reference to any specific coalition, we will always mean *connectedness in* N. Given a cooperation graph g, the connectedness-partition N/g is the natural coalition structure to associate with graph g. The idea is that, even if two players do not have a direct agreement link between themselves, they may still effectively cooperate if they both have an agreement with the same mutual friend or if they are otherwise connected by the cooperation graph.

3 Allocation Rules

We can now turn to the question posed in the first paragraph: how will the outcome of a given game depend on the cooperation structure?

Let v be a game in characteristic function form. That is, v is a function which maps each coalition S to a real number $v(S)$. Each number $v(S)$ is interpreted as the wealth of transferrable utility which the members of S would have to divide among themselves if they were to cooperate together and with no one outside S. We can let *GR* be the set of all possible cooperation structures for the game v, and the outcomes of v can be represented by payoff allocation vectors in $\mathbb{R}^N$. So we can describe how the outcome of v might depend on the cooperation structure by a function $Y : GR \to \mathbb{R}^N$, mapping cooperation graphs to allocation vectors.

The idea is that $Y_n(g)$ (the n-component of $Y(g)$) should be the utility payoff which player n would expect in game v if g represented pattern of cooperative agreements between the players.

Formally, we define an *allocation rule* for v to be any function $Y : GR \to \mathbb{R}^N$ such that

$$\forall g \in GR, \forall S \in N/g, \quad \sum_{n \in S} Y_n(g) = v(S)\,. \tag{4}$$

This condition (4) asserts that, if S is a connected component of g, then the members of S ought to allocate to themselves the total wealth $v(S)$ available to them. This expresses our idea that N/g is the natural coalition structure to associate with a cooperation graph g. Notice that the allocation within a connected coalition S still depends on the actual graph g. For example, an allocation rule might give higher payoff to player 1 in $g_1 = \{1 : 2, 1 : 3, 3 : 4\}$ than in $g_2 = \{1 : 2, 2 : 3, 3 : 4\}$, because 1's position is more essential to coordinating the others in g_1. In each case, however, condition (4) requires that $\sum_{n=1}^{4} Y_n(g_1) = v(\{1, 2, 3, 4\}) = \sum_{n=1}^{4} Y_n(g_2)$.

We use the symbol $\backslash$ to denote removal of a member from a set. Thus $g\backslash n : m = \{i : j \mid i : j \in g, i : j \neq n : m\}$.

An allocation rule $Y : GR \to \mathbb{R}^N$ is *stable* iff:

$$\begin{aligned}&\forall g \in GR, \forall n : m \in g,\\ &Y_n(g) \geq Y_n(g\backslash n : m) \quad \text{and } Y_m(g) \geq Y_m(g\backslash n : m)\,.\end{aligned} \tag{5}$$

A stable allocation rule has the property that two players always benefit from reaching a bilateral agreement. So if the allocation rule were stable, then all players would want to be linked to as many others as possible, and we could expect the complete cooperation graph g^N to be the cooperation structure of the game.

In general, a characteristic function game can have many stable allocation rules. For example, consider the two-player "Divide the Dollar" game: $N = \{1, 2\}$, $v(\{1\}) = v(\{2\}) = 0$ and $v(\{1, 2\}) = 1$. To be an allocation rule for v, Y must satisfy $Y_1(\varnothing) = 0$, $Y_2(\varnothing) = 0$, and $Y_1(\{1 : 2\}) - Y_2(\{1 : 2\}) = 1$. ($\varnothing$ is the empty graph, with no links.) Stability then requires $Y_1(\{1 : 2\}) \geq 0$ and $Y_2(\{1 : 2\}) \geq 0$.

To narrow the range of allocation rules under consideration, we may seek allocation rules which are equitable in some sense. One equity principle we may apply is the equal-gains principle: that two players should gain equally from their bilateral agreement.

We define an allocation rule $Y : GR \to \mathbb{R}^N$ to be *fair* iff:

$$\forall g \in GR, \forall n : m \in g, Y_n(g) - Y_n(g\backslash n : m) = Y_m(g) - Y_m(g\backslash n : m)\,. \tag{6}$$

For example, in the Divide the Dollar game, the only fair allocation rule has $Y_1(\{1 : 2\}) = 0.5$ and $Y_2(\{1 : 2\}) = 0.5$, so that both players gain 0.5 units of transferable utility from their agreement link.

To state our main result, we need one more definition. Given a characteristic function game v and a graph g, define v/g to be a characteristic function game so that

$$\forall S \subseteq N, \quad (v/g)(S) = \sum_{T \in S/g} v(T) \, . \tag{7}$$

(Recall the definition of S/g in (3).) One may interpret v/g as the characteristic function game which would result if we altered the situation represented by v, requiring that players can only communicate along links in g.

Theorem. *Given a characteristic function game v, there is a unique fair allocation rule $Y : GR \to \mathbb{R}^N$ (satisfying (4) and (6)). This fair allocation rule also satisfies*

$$Y(g) = \varphi(v/g), \forall g \in GR \, ,$$

where $\varphi(\cdot)$ is the Shapley value operator. Furthermore, if v is superadditive then the fair allocation rule is stable.

(Recall that a game v is *superadditive* iff: $\forall S \subseteq N$, $\forall T \subseteq N$, if $S \cap T = \varnothing$ then $v(S \cup T) \geq v(S) + v(T)$.)'

(For proof, see Sect. 5.)

Since $v/g^N = v$ (where g^N is the complete graph on N), we get $Y(g^N) = \varphi(v)$ for the fair allocation rule Y. Thus our notions of cooperation graphs and fair allocation rules provide a new derivation of the Shapley value. (See Shapley [1953] and Harsanyi [1963] for other approaches.)

4 Example

Let $N = \{1, 2, 3\}$, and consider the characteristic function game v where:

$$v(\{1\}) = v(\{2\}) \;=\; v(\{3\}) = 0, \quad v(\{1,3\}) = v(\{2,3\}) = 6, \quad \text{and}$$
$$v(\{1,2\}) \;=\; v(\{1,2,3\}) = 12 \, .$$

The fair allocation rule for this game is as follows:

$$Y(\varnothing) = (0,0,0), \qquad Y(\{1:2, 1:3\}) = (7,4,1),$$
$$Y(\{1:2\}) = (6,6,0), \qquad Y(\{1:2, 2:3\}) = (4,7,1),$$
$$Y(\{1:3\}) = (3,0,3), \qquad Y(\{1:3, 2:3\}) = (3,3,6),$$
$$Y(\{2:3\}) = (0,3,3), \qquad Y(\{1:2, 1:3, 2:3\}) = (5,5,2) \, .$$

The Shapley value of v is $\varphi(v) = Y(g^N) = (5,5,2)$.

This example was chosen because most other well-known solution concepts, the core and the nucleolus and the bargaining set, all select the single allocation $(6,6,0)$ for this game. These solution concepts are all based on ideas about what it means for the universal coalition N to be stable against objections. According to the argument for the core, $(5,5,2)$ should be an unstable allocation because players 1 and 2 could earn 12 units wealth for themselves, which exceeds the wealth 5+5 = 10 given to them. But when we shift our perspective from coalitions to cooperation gaphs, this argument evaporates, and the value $(5, 5, 2)$ actually is part of a stable fair allocation rule. If any one player were to break either or both of his cooperation links, then his fair allocation would decrease. To be sure, if both players 1 and 2 were to simultaneously break their links with 3, then both would benefit; but each would benefit even more if he continued to cooperate with player 3 while the other alone broke his link to player 3.

5 Proof of the Theorem

We show first that there can be at most one fair allocation rule for a given game v. Indeed, suppose $Y^1 : GR \to \mathbb{R}^N$ and $Y^2 : GR \to \mathbb{R}^N$ both satisfy (4) and (6) and are different. Let g be a graph with a minimum number of links such that $Y^1(g) \neq Y^2(g)$; set $y^1 = Y^1(g)$ and $y^2 = Y^2(g)$, so that $y^1 \neq y^2$. By the minimality of g, if $n : m$ is any link of g, then $Y^1(g\backslash n : m) = Y^2(g\backslash n : m)$. Hence (6) yields

$$y_n^1 - y_m^1 = Y_n^1(g\backslash n : m) - Y_m^1(g\backslash n : m) = Y_n^2(g\backslash n : m) - Y_m^2(g\backslash n : m) = y_n^2 - y_m^2 .$$

Transposing, we deduce

$$y_n^1 - y_n^2 = y_m^1 - y_m^2$$

whenever n and m are linked, and so also when they are in the same connected component S of g. Thus we may write $y_n^1 - y_n^2 = d_S(g)$, where $d_S(g)$ depends on S and g only, but not on n. But by (4) we have $\sum_{n\in S} y_n^1 = \sum_{n\in S} y_n^2$. Hence $0 = \sum_{n\in S}(y_n^1 - y_n^2) = |S|d_S(g)$, and so $d_S(g) = 0$. Hence $y^1 = y^2$ after all a contradiction. That is, there can be at most one fair allocation rule for v.

It now remains only to show that $Y(g) = \varphi(v/g)$ implies that (4) and (6) are satisfied, along with (5) if v is superadditive.

We show (4) first. Select any $g \in GR$. For each $S \in N/g$, define u^S to be a characteristic function game such that:

$$u^S(T) = \sum_{R\in(T\cap S)/g} v(R), \ \forall T \subseteq V .$$

Now, any two players connected in T by g are also connected in N by g, so

$$T/g = \bigcup_{S\in N/g} (T \cap S)/g .$$

Therefore $v/g = \sum_{S\in N/g} u^S$. But S is a carrier of u^S, because $u^S(T) = u^S(T\cap S)$. So, using the carrier axiom of Shapley [1953] for any $S \in N/g$ and any $T \in N/g$:

$$\sum_{n\in S} \varphi_n(u^T) = \begin{cases} u^T(N), & \text{if } S = T; \\ 0, & \text{if } S \cap T = \varnothing . \end{cases}$$

Thus, by linearity of φ, if $S \in N/g$ then

$$\sum_{n\in S}\varphi_n(v/g) = \sum_{T\in N/g}\sum_{n\in S} v_n(u^T) = u^S(N) = \sum_{R\in S/g} v(R) = v(S)\,.$$

To show (6) holds, select any $g \in GR$ and any $n : m \in g$. Let $w = v/g - v/(g\backslash n : m)$. Observe that $S/g = S(g\backslash n : m)$ if $\{n,m\} \not\subseteq S$. So if $n \notin S$ or $m \in S$ we get:

$$w(S) = \sum_{T\in S/g} v(T) = \sum_{T\in S(g\backslash n:m)} v(T) = 0\,.$$

So the only coalitions with nonzero wealth in w are coalitions containing both n and m. So by the symmetry axiom of Shapley [1953] it follows that $\varphi_n(w) = \varphi_m(w)$. By linearity of φ, $\varphi_n(v/g) - \varphi_n(v/g\backslash n : m) = \varphi_n(w) = \varphi_m(w) = \varphi_m(v/g) - \varphi_m(v/g\backslash n : m)$.

Finally we show (5). Observe that $S/(g\backslash n : m)$ always refines S/g as a partition of S, and if $n \notin S$ then $S/(g\backslash n : m) = S/g$. So, if v is superadditive:

$$(v/g)(S) = \sum_{T\in S/g} v(T) \geq \sum_{T\in S/(g\backslash n:m)} v(T) = (v/(g\backslash n : m))(S)$$

and the inequality becomes an equality if $n \notin S$. Thus, if $w = v/g - v/(g\backslash n : m)$, then $w(S) \geq 0$ for all S and $w(S) = 0$ if $n \notin S$. Hence $w(S \cup \{n\}) \geq w(S)$ for all S, and so $\varphi_n(w) \geq 0$, by the representation of the Shapley value as an expected marginal contribution (see Shapley [1953]). Thus $\varphi_n(v/g) - \varphi_n(v/(g\backslash n : m)) = \varphi_n(w) \geq 0$, proving stability.

Acknowledgements. The author is indebted to Kenneth Arrow and Truman Bewley for many conversations on this subject, and to Robert Aumann for detailed and useful comments.

References

[1] Aumann, R. J. and Dreze, J. H. (1974). Cooperative Games with Coalition Structures. *International Journal of Game Theory* III: 217–237.

[2] Harsanyi, J. C. (1963). A Simplified Bargaining Model for the n-Person Cooperative Game. *International Economic Review* IV: 194–220.

[3] Shapley, L. S. (1953). A Value for n-Person Games. In *Contributions to the Theory of Games II*. H. W. Kuhn and A. W. Tucker (eds.) Princeton, Princeton University Press, pp. 307–317.

A Strategic Model of Social and Economic Networks

Matthew O. Jackson[1], **Asher Wolinsky**[2]

[1] Division of the Humanities and Social Sciences, California Institue of Technology, Pasadena, CA 91125, USA
[2] Department of Economics, Northwestern University, Evanston IL 60208, USA

Abstract. We study the stability and efficiency of social and economic networks, when self-interested individuals can form or sever links. First, for two stylized models, we characterize the stable and efficient networks. There does not always exist a stable network that is efficient. Next, we show that this tension persists generally: to assure that there exists a stable network that is efficient, one is forced to allocate resources to nodes that are not responsible for any of the production. We characterize one such allocation rule: the equal split rule, and another rule that arises naturally from bargaining of the players.

JEL classification: A14, D20, J00

1 Introduction

Network structures play an important role in the organization of some significant economic relationships. Informal social networks are often the means for communicating information and for the allocation of goods and services which are not traded in markets. Among such goods one can mention not only invitations to parties and other forms of exchanging friendship, but also information about job openings, business opportunities and the like. In the context of a firm, the formal network through which relevant information is shared among the employees may have an important effect on the firm's productivity. In both contexts, the place of an agent in the network may affect not only his or her productivity, but also his or her bargaining position relative to others and this might be reflected in the design of such organizations.

We thank Kyle Bagwell, Dawn Iacobucci, Ehud Kalai, Bart Lipman, James Montgomery, Roger Myerson, Anne van den Nouweland, Jeroen Swinkels, Sang-Seung Yi, and an anonymous referee, for helpful written comments and/or conversations. Our understanding and presentation of the results have benefited from the interaction in seminar presentations, whose participants we thank without listing. Support from U.S.-Israel BSF Grant 90-00165 is gratefully acknowledged

The main goal of this paper is to begin to understand which networks are stable, when self-interested individuals choose to form new links or to sever existing links. This analysis is designed to give us some predictions concerning which networks are likely to form, and how this depends on productive and redistributive structures. In particular, we will examine the relationship between the set of networks which are productively efficient, and those which are stable. The two sets do not always intersect. Our analysis begins in the context of several stylized models, and then continues in the context of a general model.

This work is related to a number of literatures which study networks in a social science context. First, there is an extensive literature on social networks from a sociological perspective (see Wellman and Berkowitz [28] for one recent survey) covering issues ranging from the inter-family marriage structure in 15th century Florence to the communication patterns among consumers (see Iacobucci and Hopkins [11]). Second, occasional contributions to microeconomic theory have used network structures for such diverse issues as the internal organization of firms (e.g., Boorman [2], Keren and Levhari [16]), employment search (Montgomery [18]), systems compatibility (see Katz and Shapiro [15]), information transmission (Goyal [5]), and the structure of airline routes (Hendricks et al. [7,8], Starr and Stinchcombe [26]). Third, there is a formal game theoretic literature which includes the marriage problem and its extensions (Gale and Shapley [4], Roth and Sotomayor [24]), games of flow (Kalai and Zemel [14]), and games with communication structures (Aumann and Myerson [1], Kalai et al. [13], Kirman et al. [17] and Myerson [19]). Finally, the operations research literature has examined the optimization of transportation and communications networks. One area of that research studies the allocation of costs on minimal cost spanning trees, and makes explicit use of cooperative game theory. (See Sharkey [25] for a recent survey.)

The main contribution of this paper to these existing literatures is the modeling and analysis of the stability of networks when the nodes themselves (as individuals) choose to form or maintain links. The issue of graph endogeneity has been studied in specific contexts including cooperative games under the Shapley value (see Aumann and Myerson [1]) and the marriage problem (see Roth and Sotomayor [24]). The contribution here lies in the diversity and generality of our analysis, as well as in the focus on the tension between stability and efficiency.

Of the literatures we mentioned before, the one dealing with cooperative games that have communication structures is probably the closest in methodology to our analysis. This direction was first studied by Myerson [19], and then by Owen [22], van den Nouweland and Borm [21], and others (see van den Nouweland [20] for a detailed survey). Broadly speaking, the contribution of that literature is to model restrictions on coalition formation in cooperative games. Much of the analysis is devoted to some of the basic issues of cooperative game theory such as the characterization of value allocations with communication structures. Our work differs from that literature in some important respects. First, in our framework the value of a network can depend on exactly how agents are interconnected, not just who they are directly or indirectly connected to. Unlike games with communication, different forms of organization might generate different levels of profit or utility, even if they encompass (interconnect) exactly the same players. Second, we focus on network stability and formation and its relationship to efficiency. Third, an important aspect of our work is the application of this approach to some specific models of the organization of firms and network allocation mechanisms of non-market goods.

The paper proceeds as follows. In Sect. 2 we provide the definitions comprising the general model. In Sect. 3 we examine several specific versions of the model with stylized value functions. For each of these models we describe the efficient networks and the networks which are stable. We note several instances of incompatibility between efficiency and stability. In Sect. 4, we return to the general model to study means of allocating the total production or utility of a network. We examine in detail which types of allocation rules allow for stability of efficient networks. We conclude with a result characterizing the implications of equal bargaining power for allocation rules.

2 Definitions

Let $\mathscr{N} = \{1, \ldots, N\}$ be the finite set of players. The network relations among these players are formally represented by graphs whose *nodes* are identified with the players and whose arcs capture pairwise relations.

2.1 Graphs

The complete graph, denoted g^N, is the set of all subsets of $\mathscr{N}$ of size 2. The set of all possible graphs on $\mathscr{N}$ is then $\{g \mid g \subset g^N\}$. Let ij denote the subset of $\mathscr{N}$ containing i and j and is referred to as the *link* ij. The interpretation is that if $ij \in g$, then nodes i and j are directly connected (sometimes referred to as adjacent), while if $ij \notin g$, then nodes i and j are not directly connected.[1]

Let $g + ij$ denote the graph obtained by adding link ij to the existing graph g and $g - ij$ denote the graph obtained by deleting link ij from the existing graph g (i.e., $g + ij = g \cup \{ij\}$ and $g - ij = g \backslash \{ij\}$).

Let $N(g) = \{i \mid \exists j \text{ s.t. } ij \in g\}$ and $n(g)$ be the cardinality of $N(g)$. A *path in g connecting* i_1 and i_n is a set of distinct nodes $\{i_1, i_2, \ldots, i_n\} \subset N(g)$ such that $\{i_1 i_2, i_2 i_3, \ldots, i_{n-1} i_n\} \subset g$.

The graph $g' \subset g$ is a *component of* g, if for all $i \in N(g')$ and $j \in N(g')$, $i \neq j$, there exists a path in g' connecting i and j, and for any $i \in N(g')$ and $j \in N(g)$, $ij \in g$ implies $ij \in g'$.

[1] The graphs analyzed here are nondirected. That is, it is not possible for one individual to link to another, without having the second individual also linked to the first. (Graphs where unidirectional links are possible are sometimes called digraphs.) Furthermore, links are either present or not, as opposed to having connections with variable intensities (a valued graph). See Iacobucci [10] for a detailed set of definitions for a general analysis of social networks. Such alternatives are important, but are beyond the scope of our analysis.

2.2 Values and Allocations

Our interest will be in the total productivity of a graph and how this is allocated among the individual nodes. These notions are captured by a value function and an allocation function.

The *value of a graph* is represented by $v : \{g \mid g \subset g^N\} \to I\!R$. The set of all such functions is V. In some applications the value will be an aggregate of individual utilities or productions, $v(g) = \sum_i u_i(g)$, where $u_i : \{g \mid g \subset g^N\} \to I\!R$.

A graph $g \subset g^N$ is *strongly efficient* if $v(g) \geq v(g')$ for all $g' \subset g^N$. The term strong efficiency indicates maximal total value, rather than a Paretian notion. Of course, these are equivalent if value is transferable across players.

An *allocation rule* $Y : \{g \mid g \subset g^N\} \times V \to I\!R^N$ describes how the value associated with each network is distributed to the individual players. $Y_i(g, v)$ is the payoff to player i from graph g under the value function v.

2.3 Stability

As our interest is in understanding which networks are likely to arise in various contexts, we need to define a notion which captures the stability of a network. The definition of a stable graph embodies the idea that players have the discretion to form or sever links. The formation of a link requires the consent of both parties involved, but severance can be done unilaterally.

The graph g is *pairwise stable* with respect to v and Y if

(i) for all $ij \in g$, $Y_i(g, v) \geq Y_i(g - ij, v)$ and $Y_j(g, v) \geq Y_j(g - ij, v)$, and

(ii) for all $ij \notin g$, if $Y_i(g, v) < Y_i(g + ij, v)$ then $Y_j(g, v) > Y_j(g + ij, v)$.

We shall say that g is *defeated* by g' if $g' = g - ij$ and (i) is violated for ij, or if $g' = g + ij$ and (ii) is violated for ij.

Condition (ii) embodies the assumption that, if i strictly prefers to form the link ij and j is just indifferent about it, then it will be formed.

The notion of pairwise stability is not dependent on any particular formation process. That is, we have not formally modeled the procedure through which a graph is formed. Pairwise stability is a relatively weak notion among those which account for link formation and as such it admits a relatively larger set of stable allocations than might a more restrictive definition or an explicit formation procedure. (See Sect. 5 for more discussion of this). For our purposes, such a weak definition provides strong results, since in many instances it already narrows the set of graphs substantially.

There are many obvious modifications of the above definition which one might consider. One obvious strengthening would be to allow changes to be made by coalitions which include more than two players. To keep the presentation uncluttered, we will go through the analysis using only the stability notion defined above and relegate all the remarks on other variations to Sect. 5.

3 Two Specific Models

We begin by analyzing several stylized versions of the general model outlined in the last section. There are innumerable versions which one can think of. The examples presented in this section are meant to capture some basic and diverse issues arising in social and economic networks. In particular, we illustrate what the application of pairwise stability predicts concerning which graphs might form and whether or not these are strongly efficient.

3.1 The Connections Model

This first example models social communication among individuals.[2] Individuals directly communicate with those to whom they are linked. Through these links they also benefit from indirect communication from those to whom their adjacent nodes are linked, and so on. The value of communication obtained from other nodes depends on the distance to those nodes. Also, communication is costly so that individuals must weigh the benefits of a link against its cost.

Let $w_{ij} \geq 0$ denote the "intrinsic value" of individual j to individual i and c_{ij} denote the cost to i of of maintaining the link ij. The utility of each player i from graph g is then

$$u_i(g) = w_{ii} + \sum_{j \neq i} \delta^{t_{ij}} w_{ij} - \sum_{j:ij \in g} c_{ij},$$

where t_{ij} is the number of links in the shortest path between i and j (setting $t_{ij} = \infty$ if there is no path between i and j), and $0 < \delta < 1$ captures the idea that the value that i derives from being connected to j is proportional to the proximity of j to i.[3] Less distant connections are more valuable than more distant ones, but direct connections are costly. Here

$$v(g) = \sum_{i \in \mathcal{N}} u_i(g).$$

3.1.1 Strong Efficiency in the Connections Model

In what follows we focus on the symmetric version of this model, where $c_{ij} = c$ for all ij and $w_{ij} = 1$ for all $j \neq i$ and $w_{ii} = 0$. The term *star* describes a component in which all players are linked to one central player and there are no other links: $g \subset g^N$ is a star if $g \neq \emptyset$ and there exists $i \in \mathcal{N}$ such that if $jk \in g$, then either $j = i$ or $k = i$. Individual i is the center of the star.

Proposition 1. *The unique strongly efficient network in the symmetric connections model is*

[2] Goyal [5] considers a related model. His is a non-cooperative game of one sided link formation and it differs in some of the specifications as well, but it is close in terms of its flavor and motivation.

[3] The shortest path is sometimes called the geodesic, and t_{ij} the geodesic distance.

(i) the complete graph g^N if $c < \delta - \delta^2$,

(ii) a star encompassing everyone if $\delta - \delta^2 < c < \delta + \frac{(N-2)}{2}\delta^2$, and

(iii) no links if $\delta + \frac{(N-2)}{2}\delta^2 < c$.

Proof. (i) Given that $\delta^2 < \delta - c$, any two agents who are not directly connected will improve their utilities, and thus the total value, by forming a link.

(ii) and (iii). Consider g', a component of g containing m individuals. Let $k \geq m - 1$ be the number of links in this component. The value of these direct links is $k(2\delta - 2c)$. This leaves at most $m(m-1)/2 - k$ indirect links. The value of each indirect link is at most $2\delta^2$. Therefore, the overall value of the component is at most

$$k(2\delta - 2c) + (m(m-1) - 2k)\delta^2. \tag{1}$$

If this component is a star then its value would be

$$(m-1)(2\delta - 2c) + (m-1)(m-2)\delta^2. \tag{2}$$

Notice that (1) – (2) = $(k - (m-1))(2\delta - 2c - 2\delta^2)$, which is at most 0 since $k \geq m - 1$ and $c > \delta - \delta^2$, and less than 0 if $k > m - 1$. The value of this component can equal the value of the star only when $k = m - 1$. Any graph with $k = m - 1$, which is not a star, must have an indirect connection which has a path longer than 2, getting value less than $2\delta^2$. Therefore, the value of the indirect links will be below $(m-1)(m-2)\delta^2$, which is what we get with star.

We have shown that if $c > \delta - \delta^2$, then any component of a strongly efficient graph must be a star. Note that any component of an strongly efficient graph must have nonnegative value. In that case, a direct calculation using (2) shows that a single star of $m + n$ individuals is greater in value than separate stars of m and n individuals. Thus if the strongly efficient graph is nonempty, it must consist of a single star. Again, it follows from (2) that if a star of n individuals has nonnegative value, then a star of $n + 1$ individuals has higher value. Finally, to complete (ii) and (iii) notice that a star encompassing everyone has positive value only when $\delta + \frac{(N-2)}{2}\delta^2 > c$. □

This result has some of the same basic intuition as the hub and spoke analysis of Hendricks, Piccione, and Tan [8] and Starr and Stinchcombe [26], except that the values of graphs are generated in different manners.

3.1.2 Stability in the Connections Model Without Side Payments

Next, we examine some implications of stability for the allocation rule $Y_i(g) = u_i(g)$. This specification might correspond best to a social network in which by convention no payments are exchanged for "friendship."

Proposition 2. *In the symmetric connections model with $Y_i(g) = u_i(g)$:*

(i) A pairwise stable network has at most one (non-empty) component.

(ii) For $c < \delta - \delta^2$, the unique pairwise stable network is the the complete graph, g^N.

(iii) For $\delta - \delta^2 < c < \delta$, a star encompassing all players is pairwise stable, but not necessarily the unique pairwise stable graph.

(iv) For $\delta < c$, any pairwise stable network which is nonempty is such that each player has at least two links and thus is inefficient.[4]

Proof. (i) Suppose that g is pairwise stable and has two or more non-trivial components. Let u^{ij} denote the utility which accrues to i from the link ij, given the rest of g: so $u^{ij} = u_i(g+ij)-u_i(g)$ if $ij \notin g$ and $u^{ij} = u_i(g)-u_i(g-ij)$ if $ij \in g$. Consider $ij \in g$. Then $u^{ij} \geq 0$. Let kl belong to a different component. Since i is already in a component with j, but k is not, it follows that $u^{kj} > u^{ij} \geq 0$, since k will also receive δ^2 in value for the indirect connection to i, which which is not included in u^{ij}. For similar reasons, $u^{jk} > u^{lk} \geq 0$. This contradicts pairwise stability, since $jk \notin g$.

(ii) It follows from the fact that in this cost range, any two agents who are not directly connected benefit from forming a link.

(iii) It is straightforward to verify that the star is stable. It is the unique stable graph in this cost range if $N = 3$. It is never the unique stable graph if N=4. (If $\delta - \delta^3 < c < \delta$, then a line is also stable, and if $c < \delta - \delta^3$, then a circle [5] is also stable.)

(iv) In this range, pairwise stability precludes "loose ends" so that every connected agent has at least two links. This means that the star is not stable, and so by Proposition 1, any non-empty pairwise stable graph must be inefficient. □

Remark. The results of Proposition 2 would clearly still hold if one strengthens pairwise stability to allow for deviations by groups of individuals instead of just pairs. This would lean even more heavily on the symmetry assumption.

Remark. Part (iv) implies that in the high cost range (where $\delta < c$) the only non-degenerate networks which are stable are those which are over-connected from an efficiency perspective. (We will return to this tension between strong efficiency and stability later, in the analysis of the general model.) Since $\delta < c$, no individual is willing to maintain a link with another individual who does not bring additional new value from indirect connections. Thus, each node must have at least two links, or none. This means that the star cannot be stable: the center will not wish to maintain links with any of the end nodes.

The following example features an over-connected pairwise stable graph. The example is more complex than necessary (a circle with $N = 5$ will illustrate the same point), but it illustrates that pairwise stable graphs can be more intricate than the simple stars and circles.

Example 1. Consider the "tetrahedron" in Fig. 1. Here $N = 16$. A star would involve 15 links and a total value of $30\delta + 210\delta^2 - 30c$. The tetrahedron has 18

[4] If $\delta + \frac{(N-2)}{2}\delta^2 > c$, then all pairwise stable networks are inefficient since then the empty graph is also inefficient.

[5] $g \subset g^N$ is a *circle* if $g \neq \emptyset$ and there exists $\{i_1, i_2, \ldots, i_n\} \subset \mathcal{N}$ such that $g = \{i_1 i_2, i_2 i_3, \ldots, i_{n-1} i_n, i_n i_1\}$.

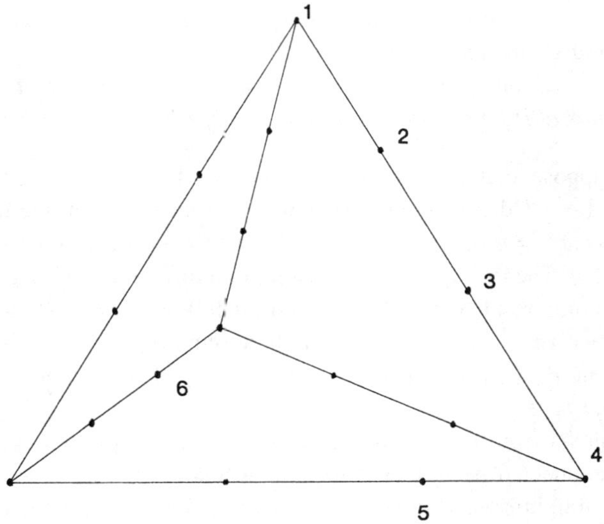

Fig. 1.

links and a total value of $36\delta + 48\delta^2 + 60\delta^3 + 72\delta^4 + 24\delta^5 - 36c$, which (since $c > \delta$ and $\delta < 1$) is less than that of the star.

Let us verify that the tetrahedron is pairwise stable. (Recall that u^{ij} denotes the utility which accrues to i from the link ij, given the rest of g: so $u^{ij} = u_i(g + ij) - u_i(g)$ if $ij \notin g$ and $u^{ij} = u_i(g) - u_i(g - ij)$ if $ij \in g$.) Given the symmetry of the graph, the following inequalities assure pairwise stability of the graph: $u^{12} \geq 0$, $u^{21} \geq 0$, $u^{23} \geq 0$, $u^{13} \leq 0$, $u^{14} \leq 0$, $u^{15} \leq 0$, and $u^{26} \leq 0$. The first three inequalities assure that no one wants to sever a link. The next three inequalities assure that no new link can be improving to two agents if one of those agents is a "corner" agents. The last inequality assures that no new link can be improving to two agents if both of those agents are not "corner" agents. It is easy to check that $u^{21} > u^{12}$, $u^{23} > u^{12}$, $u^{13} < u^{14}$, $u^{15} < u^{14}$, and $u^{14} < u^{26}$. Thus we verify that $u^{12} \geq 0$ and $u^{26} \leq 0$.

$u^{12} = \delta - \delta^8 + \delta^2 - \delta^7 + \delta^3 - \delta^6 + 2(\delta^4 - \delta^5) - c,$

$u^{26} = \delta - \delta^5 + \delta^2 - \delta^4 + \delta^2 - \delta^5 + 2(\delta^3 - \delta^4) - c,$

If $c = 1$ and $\delta = .9$, then (approximately) $u^{12} = .13$ and $u^{26} = -.17$.

In this example, the graph is stable since each link connects an individual indirectly to other valuable individuals. The graph cannot be too dense, since it then becomes too costly to maintain links relative to their limited benefit. The graph cannot be too sparse as nodes will have incentives to add additional links to those links which are currently far away and/or sever current links which are not providing much value.

Before proceeding, we remark that the results presented for the connections model are easily adapted to replace $\delta^{t_{ij}}$ by any nonincreasing function $f(t_{ij})$, by simply substituting $f(t_{ij})$ whenever $\delta^{t_{ij}}$ appears. One such alternative specification is a truncated connections model where players benefit only from connections which are not more distant than some bound D. The case of $D = 2$, for example, has the natural interpretation that i benefits from j only if they are directly connected or if they have a "mutual friend" to whom both are directly connected. It is immediate to verify that Propositions 1 and 2 continue to hold for the truncated connections models. In addition we have the following observations.

Proposition 3. *In the truncated connections model with bound D*

(i) $t_{ij} \leq 2D - 1$ *for all i and j which belong to a pairwise stable component.*

(ii) For $D = 2$ *and* $\delta < c$ *no member in a pairwise stable component is in a position to disconnect all the paths connecting any two other players by unilaterally severing links.*

Proof. (i) Suppose $t_{ij} > 2D - 1$. Consider one of the shortest paths between i and j. Let m belong to this path and $t_{mj} = 1$. Note that $t_{ik} > D$, for any k such that j belongs to the shortest path between m and k and such that $t_{mk} \leq D$. This is because $t_{jk} \leq D - 1$ and $t_{ij} > 2D - 1$. Therefore, $u^{ij} > u^{mj}$ (the inequality is strict since u^{ij} includes the value to i of the connection to m which is not present in u^{mj}) so i wants to link directly to j. (Recall the notation u^{ij} from the proof of Proposition 2.) An analogous argument establishes that j wants to link directly to i.

(ii) Suppose that player i occupies such a position. Let j and k be such that i can unilaterally disconnect them and such that t_{jk} is the longest minimal path among all such pairs. Since by (i), $t_{jk} \leq 3$, there is at least one of them, say j, such that $t_{ij} = 1$. But then i prefers to sever the link to j, since the maximality of t_{jk} implies that there is no h to whom i's only indirect connection passes through j (otherwise $t_{hk} > t_{jk}$). □

There are obvious extensions to the connections model which seem quite interesting. For instance, one might have a decreasing value for each connection (direct or indirect) as the total amount of connectedness increases. Also, if communication is modeled as occuring with some probability across each link, then one cares not only about the shortest path, but also about other paths in the event that communication fails across some link in the shortest path.[6]

3.1.3 Stability in the Connections Model with Side Payments

In the connections model with side payments, players are able to exchange money in addition to receiving the direct benefits from belonging to the network. The allocation rule will reflect these side payments which might be agreed upon in bilateral negotiations or otherwise. This version exposes another source of discrepancy between the strongly efficient and stable networks. Networks which produce high values might place certain players in key positions that will allow them to "claim" a disproportionate share of the total value. This is particularly true for the strongly efficient star-shaped network. This induces other players to

[6] Two such alternative models are discussed briefly in the appendix of Jackson and Wolinsky [12]

form additional links that mitigate this power at the expense of reducing the total value. This consideration is illustrated by the following example.

Example 2. Let $N = 3$ and v be as in the basic connections model. The graph $g = \{12, 23\}$ is strongly efficient for $\delta - \delta^2 < c < \delta$. Suppose that the allocation rule Y allocates the whole value of any graph to the players having links in the graph and reflects equal bargaining power in the sense that $Y_i(g, v) - Y_i(g - ij, v) = Y_j(g, v) - Y_j(g - ij, v)$ for all g, i and j (we characterize this equal bargaining power rule in Theorem 4). Then $Y_1(g, v) = Y_3(g, v) = \delta + \frac{2}{3}\delta^2 - c$ and $Y_2(g, v) = 2\delta + \frac{2}{3}\delta^2 - 2c$.[7] That is, each of the peripheral players pays the center $\frac{1}{3}\delta^2$. In the alternative network $g' = \{12, 23, 31\}$ (the circle), $Y_1(g', v) = Y_2(g', v) = Y_3(g', v) = 2\delta - 2c$, and no side payments are exchanged. In the range $\delta - \frac{2}{3}\delta^2 < c < \delta$ the strongly efficient network g is uniquely stable, but in the range $\delta - \delta^2 < c < \delta - \frac{2}{3}\delta^2$ the inefficient network g' is the only stable one.

As mentioned above, the reason for the tension between efficiency and stability is the strong bargaining position of the center in g: when c is not too large, g is destabilized by the link between the peripheral players who increase their share at the expense of the center.

This version of the connections model can be adapted to discuss issues in the internal organization of firms. Consider a firm whose output depends on the organization of the employees as a network. The network would capture here the structure of communication and hence coordination between workers. The nodes of the graph correspond to the workers. (For simplicity we exclude the owner from the graph, although it is not necessary for the result). The total value of the firm's output, v, is as above. The allocation rule, Y, specifies the distribution of the total value between the workers (wages) and the firm (profit). It captures the outcome of wage bargaining within the firm, where labor contracts are not binding, and where the bargained wage of a worker is half the surplus associated with that worker's employment. The assumption built into this rule is that the position of a worker who quits cannot be filled immediately, so $Y_i(g - i, v)$ and $v(g - i) - \sum_{j \neq i} Y_j(g - i, v)$ are identified as the bargaining disagreement points of the worker and firm respectively (where $g - i$ denotes the graph which remains when all links including i are deleted). Thus

$$Y_i(g, v) = Y_i(g - i, v) + \frac{1}{2}\Big[v(g) - \sum_{j \neq i} Y_j(g, v) - Y_i(g - i, v) - \Big(v(g - i) - \sum_{j \neq i} Y_j(g - i, v)\Big)\Big].$$

[7] To see this, notice that $Y_1(g - 23, v) = Y_2(g - 23, v) = \delta - c$, $Y_3(g - 23, v) = 0$, and $Y_1(g - 12, v) = 0$, $Y_2(g - 12, v) = Y_3(g - 12, v) = \delta - c$. Then from equal bargaining power, we have that $Y_2(g, v) - (\delta - c) = Y_1(g, v) - 0 = Y_3(g, v) - 0$. Then using the fact that $Y_1(g, v) + Y_2(g, v) + Y_3(g, v) = 4\delta + 2\delta^2 - 4c$, one can solve for $Y(g, v)$.

If we think of the owner as external to the network, this Y is not balanced, as the firm's profit is $v - \sum_i Y_i$.[8]

Example 3. Let $N = 3$ and v be as above. Assume $Y_i(g - i, v) = 0$ which means that a worker who quit is not paid. The graph $g = \{12, 23\}$ is strongly efficient for $\delta - \delta^2 < c < \delta$. Note that $Y_1(g, v) = Y_3(g, v) = \frac{2}{3}\delta + \frac{1}{2}\delta^2 - \frac{2}{3}c$ and $Y_2(g, v) = \frac{4}{3}\delta + \frac{1}{2}\delta^2 - \frac{4}{3}c$, leaving a profit of $\frac{4}{3}\delta + \frac{1}{2}\delta^2 - \frac{4}{3}c$ for the firm. Consider $g' = \{12, 23, 31\}$. Here $Y_1(g', v) = Y_2(g', v) = Y_3(g', v) = \frac{4}{3}\delta - \frac{4}{3}c$, leaving a profit of $2(\delta - c)$ for the firm.

In the range $\delta - \delta^2 < c < \delta - \frac{3}{4}\delta^2$ the network g is the strongly efficient form, but the network g' is more profitable to the firm, since it weakens the bargaining position of the worker occupying the center position in the graph g. This point complements existing work on internal wage bargaining and its consequences for the structure of firms. Stole and Zweibel (1993) investigate how internal wage bargaining distorts employment decisions, the extent of investment in capital, and the division of the workforce among activities (see also Grout [6] and Horn and Wolinsky [9]). The current framework adds explicitly the network organization of the firm.

3.2 The Co-Author Model

Here nodes are interpreted as researchers who spend time writing papers. Each node's productivity is a function of its links. A link represents a collaboration between two researchers. The amount of time a researcher spends on any given project is inversely related to the number of projects that researcher is involved in. Thus, in contrast to the connections model, here indirect connections will enter utility in a negative way as they detract from one's co-author's time.

The fundamental utility or productivity of player i given the network g is

$$u_i(g) = \sum_{j:ij \in g} w_i(n_i, j, n_j) - c(n_i),$$

where $w_i(n_i, j, n_j)$ is the utility derived by i from a direct contact with j when i and j are involved in n_i and n_j projects, respectively, and $c(n_i)$ is the cost to i of maintaining n_i links.

We analyze a more specific version of this model where utility is given by the following expression. For $n_i > 0$,

$$u_i(g) = \sum_{j:ij \in g} \left[\frac{1}{n_i} + \frac{1}{n_j} + \frac{1}{n_i n_j}\right] = 1 + \left(1 + \frac{1}{n_i}\right) \sum_{j:ij \in g} \frac{1}{n_j},$$

and for $n_i = 0$, $u_i(g) = 0$. This form assumes that each researcher has a unit of time which they allocate equally across their projects. The output of each project

[8] If the owner is included explicitly as a player, then Y coincides with the equal bargaining power rule examined in Sect. 4.

depends on the total time invested in it by the two collaborators, $\frac{1}{n_i}+\frac{1}{n_j}$, and on some synergy in the production process captured by the interactive term $\frac{1}{n_i n_j}$.

The interactive term is inversely proportional to the number of projects each author is involved with. Here there are no direct costs of connection. The cost of connecting with a new author is that the new link decreases the strength of the interaction term with existing links.[9]

Proposition 4. *In this co-author model: (i) if N is even, then the strongly efficient network is a graph consisting of $N/2$ separate pairs, and (ii) a pairwise stable network can be partitioned into fully intraconnected components, each of which has a different number of members. (If m is the number of members of one such component and n is the next largest in size, then $m > n^2$).*

Proof. To see (i), notice that

$$\sum_{i\in N} u_i(g) = \sum_{i:n_i>0}\sum_{j:ij\in g}\left[\frac{1}{n_i}+\frac{1}{n_j}+\frac{1}{n_i n_j}\right],$$

so that

$$\sum_{i\in N} u_i(g) \leq 2N + \sum_{i:n_i>0}\sum_{j:ij\in g}\frac{1}{n_i n_j},$$

and equality can only hold if $n_i > 0$ for all i. To finish the proof of (i), notice that $\sum_{i:n_i>0}\sum_{j:ij\in g}\frac{1}{n_i n_j} \leq N$, with equality only if $n_i = 1 = n_j$ for all i and j, and $3N$ is the value of $N/2$ separate pairs.

To see (ii), consider i and j who are not linked. It follows directly from the formula for $u_i(g)$ that i will strictly want to link to j if and only if

$$\frac{1}{n_j+1}\left(1+\frac{1}{n_i+1}\right) > \left[\frac{1}{n_i}-\frac{1}{n_i+1}\right]\sum_{k:k\neq j,ik\in g}\frac{1}{n_k},$$

(substitute 0 on the right hand side if $n_i = 0$) which simplifies to

$$\frac{n_i+2}{n_j+1} > \frac{1}{n_i}\sum_{k:k\neq j,ik\in g}\frac{1}{n_k}. \qquad (*)$$

The following facts are then true of a pairwise stable network.

1. If $n_i = n_j$, then $ij \in g$.

We show that if $n_j \leq n_i$, then i would like to link to j. Note that $\frac{n_i+2}{n_j+1} > 1$ while the right hand side of $(*)$ is at most 1 (the average of n_i fractions). Therefore, i would like to link to j.

2. If $n_h \leq \mathrm{Max}\{n_k | ik \in g\}$, then i wants to link to h.

Let j be such that $ij \in g$ and $n_j = \mathrm{Max}\{n_k | ik \in g\}$. If $n_i \geq n_j - 1$ then $\frac{n_i+2}{n_h+1} \geq 1$. If $\frac{n_i+2}{n_h+1} > 1$ then $(*)$ clearly holds for i's link to h. If $\frac{n_i+2}{n_h+1} = 1$, then it must be that $n_h \geq 2$ and so $n_j \geq 2$. This means that the right hand side of $(*)$

[9] An alternative version of the co-author model appears in the appendix of Jackson and Wolinsky [12].

when calculated for adding the link h will be strictly less than 1. Thus $(*)$ will hold. If $n_i < n_j - 1$, then $\frac{n_i+1}{n_j} < \frac{n_i+2}{n_j+1} < \frac{n_i+2}{n_h+1}$. Since $ij \in g$, it follows from $(*)$ that

$$\frac{n_i+1}{n_j} \geq \frac{1}{n_i - 1} \sum_{k:k \neq j, ik \in g} \frac{1}{n_k}.$$

Also

$$\frac{1}{n_i - 1} \sum_{k:k \neq j, ik \in g} \frac{1}{n_k} \geq \frac{1}{n_i} \sum_{k:ik \in g} \frac{1}{n_k}$$

since the extra element on the right hand side is $1/n_j$ which is smaller than (or equal to) all terms in the sum. Thus $\frac{n_i+2}{n_h+1} > \frac{1}{n_i} \sum_{k:ik \in g} \frac{1}{n_k}$.

Facts 1 and 2 imply that all players with the maximal number of links are connected to each other and nobody else. [By 1 they must all be connected to each other. By 2, anyone connected to a player with a maximal number of links would like to connect to all players with no more than that number of links, and hence all those with that number of links.] Similarly, all players with the next to maximal number of links are connected to each other and to nobody else, and so on.

The only thing which remains to be shown is that if m is the number of members of one (fully intraconnected) component and n is the next largest in size, then $m > n^2$. Notice that for i in the next largest component not to be willing to hook to j in the largest component it must be that $\frac{n_i+2}{n_j} + 1 \leq \frac{1}{n_i}$ (using $(*)$, since all nodes to which i is connected also have n_i connections). Thus $n_j + 1 \geq n_i(n_i + 2)$. It follows that $n_j > n_i^2$. □

The combination of the efficiency and stability results indicates that stable networks will tend to be over-connected from an efficiency perspective. This happens because authors only partly consider the negative effect their new links have on the productivity of links with existing co-authors.

4 The General Model

We now turn to analyzing the general model.

As we saw in Propositions 1 and 2, as well as in some of the examples in the previous section, efficiency and pairwise stability are not always compatible. That is, there are situations in which no strongly efficient graphs are pairwise stable. Does this persist in general? In other words, if we are free to structure the allocation rule in any way we like, is it possible to find one such that there is always at least one strongly efficient graph which is pairwise stable? The answer, provided in Theorem 1 below, depends on whether the allocation rule is balanced across components or is free to allocate resources to nodes which are not productive.

Definition. *Given a permutation* $\pi : \mathcal{N} \to \mathcal{N}$, *let* $g^\pi = \{ij | i = \pi(k), j = \pi(l), kl \in g\}$. *Let* v^π *be defined by* $v^\pi(g^\pi) = v(g)$.[10]

[10] In the language of social networks, g^π and g are said to be isomorphic.

Definition. *The allocation rule Y is* anonymous *if, for any permutation π, $Y_{\pi(i)}(g^\pi, v^\pi) = Y_i(g, v)$.*

Anonymity states that if all that has changed is the names of the agents (and not anything concerning their relative positions or production values in some network), then the allocations they receive should not change. In other words, the anonymity of Y requires that the information used to decide on allocations be obtained from the function v and the particular g, and not from the label of an individual.

Definition. *An allocation rule Y is* balanced *if $\sum_i Y_i(g, v) = v(g)$ for all v and g.*

A stronger notion of balance, component balance, requires Y to allocate resources generated by any component to that component. Let $C(g)$ denote the set of components of g. Recall that a component of g is a maximal connected subgraph of g.

Definition. *A value function v is* component additive *if $v(g) = \sum_{h \in C(g)} v(h)$.* [11]

Definition. *The rule Y is* component balanced *if $\sum_{i \in N(h)} Y_i(g, v) = v(h)$ for every g and $h \in C(g)$ and component additive v.*

Note that the definition of component balance only applies when v is component additive. Requiring it otherwise would necessarily contradict balance.

Theorem 1. *If $N \geq 3$, then there is no Y which is anonymous and component balanced and such that for each v at least one strongly efficient graph is pairwise stable.*

Proof. Let $N = 3$ and consider (the component additive) v such that, for all i, j, and k, $v(\{ij\}) = 1$, $v(\{ij, jk\}) = 1 + \epsilon$ and $v(\{ij, jk, ik\}) = 1$. Thus the strongly efficient networks are of the form $\{ij, jk\}$. By anonymity and component balance, $Y_i(\{ij\}, v) = 1/2$ and

$$Y_i(\{ij, jk, ik\}, v) = Y_k(\{ij, jk, ik\}, v) = 1/3. \qquad (*)$$

Then pairwise stability of the strongly efficient network requires that $Y_j(\{ij, jk\}, v) \geq 1/2$, since $Y_j(\{ij\}, v) = 1/2$. This, together with component balance and anonymity, implies that $Y_i(\{ij, jk\}, v) = Y_k(\{ij, jk\}, v) \leq 1/4 + \epsilon/2$. But this and $(*)$ contradict stability of the strongly efficient network when ϵ is sufficiently small ($< 1/6$), since then i and k would both gain from forming a link. This example is easily extended to $N > 3$, by assigning $v(g) = 0$ to any g which has a link involving a player other than players 1, 2 or 3. □

Theorem 1 says that there are value functions for which there is no anonymous and component balanced rule which supports strongly efficient networks as pairwise stable, even though anonymity and component balance are reasonable in

[11] This definition implicitly requires that the value of disconnected players is 0. This is not necessary. One can redefine components to allow a disconnected node to be a component. One has also to extend the definition of v so that it assigns values to such components.

many scenarios. It is important to note that the value function used in the proof is not at all implausible, and is easily perturbed without upsetting the result.[12] Thus one can make the simple observation that this conflict holds for an open set of value functions.

Theorem 1 does not reflect a simple nonexistence problem. We can find an anonymous and component balanced Y for which there always exists a pairwise stable network. To see a rule which is both component balanced and anonymous, and for which there always exists a pairwise stable network, consider $\overline{Y}$ which splits equally each component's value among its members. More formally, if v is component additive let $\overline{Y}_i(g, v) = v(h)/n(h)$ (recalling that $n(h)$ indicates the number of nodes in the component h) where $i \in N(h)$ and $h \in C(g)$,[13] and for any v that is not component additive let $\overline{Y}_i(g, v) = v(g)/N$ for all i. A pairwise stable graph for $\overline{Y}$ can be constructed as follows. For any component additive v find g by constructing components $h_1, \ldots, h_n$ sequentially, choosing h_i to maximize $v(h)/n(h)$ over all nonempty components which use only nodes not in $\cup_{j=1}^{i-1} N(h_j)$ (and setting $h_i = \emptyset$ if this value is always negative). The implication of Theorem 1 is that such a rule will necessarily have the property that, for some value functions, all of the networks which are stable relative to it are also inefficient.

The conflict between efficiency and stability highlighted by Theorem 1 depends both on the particular nature of the value function and on the conditions imposed on the allocation rule. This conflict is avoided if attention is restricted to certain classes of value functions, or if conditions on the allocation rule are relaxed. The following discussion will address each of these in turn. First, we describe a family of value functions for which this conflict is avoided. Then, we discuss the implications of relaxing the anonymity and component balance conditions.

Definition. *A link ij is* critical *to the graph g if $g - ij$ has more components than g or if i is linked only to j under g.*

A critical link is one such that if it is severed, then the component that it was a part of will become two components (or one of the nodes will become disconnected). Let h denote a component which contains a critical link and let h_1 and h_2 denote the components obtained from h by severing that link (where it may be that $h_1 = \emptyset$ or $h_2 = \emptyset$).

Definition. *The pair (g, v) satisfies* critical link monotonicity *if, for any critical link in g and its associated components h, h_1, and h_2, we have that $v(h) \geq v(h_1) + v(h_2)$ implies that $v(h)/n(h) \geq \max[v(h_1)/n(h_1), v(h_2)/n(h_2)]$.*

Consider again $\overline{Y}$ as defined above. The following is true.

---

[12] One might hope to rely on group stability to try to retrieve efficiency. However, group stability will simply refine the set of pairwise stable allocations. The result will still be true, and in fact sometimes there will exist no group stable graph.

[13] Use the convention that $n(\emptyset) = 1$ and $i \in N(\emptyset)$ if i is not linked to any other node.

Claim. *If g is strongly efficient relative to a component additive v, then g is pairwise stable for $\overline{Y}$ relative to v if and only if (g, v) satisfies critical link monotonicity.*

Proof. Suppose that g is strongly efficient relative to v and is pairwise stable for $\overline{Y}$ relative to v. Then for any critical link ij, it must be that i and j both do not wish to sever the link. This implies that $v(h)/n(h) \geq \max[v(h_1)/n(h_1), v(h_2)/n(h_2)]$. Next, suppose that g is strongly efficient relative to a component additive v and that the critical link condition is satisfied. We show that g is pairwise stable for $\overline{Y}$ relative to v. Adding or severing a non-critical link will only change the value of the component in question without changing the number of nodes in that component. By strong efficiency and component additivity, the value of this component is already maximal and so there can be no gain. Next consider adding or severing a critical link. Severing a critical link leads to no benefit for either node, since by strong efficiency and component additivity $v(h) \geq v(h_1) + v(h_2)$, which by the critical link condition implies that $v(h)/n(h) \geq \max[v(h_1)/n(h_1), v(h_2)/n(h_2)]$. By strong efficiency and component additivity, adding a critical link implies that $v(h) \leq v(h_1) + v(h_2)$ (where h_1 and h_2 are existing components and h is the new component formed by adding the critical link). Suppose to the contrary that g is not stable to the addition of the critical link. Then, without loss of generality it is the case that $v(h)/n(h) > v(h_1)/n(h_1)$ and $v(h)/n(h) \geq v(h_2)/n(h_2)$. Taking a convex combination of these inequalities (with weights $n(h_1)/n(h)$ and $n(h_2)/n(h)$) we find that $v(h) > v(h_1) + v(h_2)$, contradicting the fact that $v(h) \leq v(h_1) + v(h_2)$. □

To get some feeling for the applicability of the critical link condition, notice that if a strongly efficient graph has no critical links, then the condition is trivially satisfied. This is true in Proposition 1, parts (i) and (iii), for instance. Note, also, that the strongly efficient graphs described in Proposition 1 (ii) and Proposition 4 (i) satisfy the critical link condition, even though they consist entirely of critical links. Clearly, the value function described in the proof of Theorem 1 does not satisfy the critical link condition.

Consider next the role of the anonymity and component balance conditions in the result of Theorem 1. The proof of Theorem 1 uses anonymity, but it can be argued that the role of anonymity is not central in that a weaker version of Theorem 1 holds if anonymity is dropped. A detailed statement of this result appears in Sect. 5. The component balance condition, however, is essential for the result of Theorem 1.

To see that if we drop the component balance condition the conflict between efficiency and stability can be avoided, consider the equal split rule ($Y_i(g, v) = v(g)/N$). This is not component balanced as all agents always share the value of a network equally, regardless of their position. This rule aligns the objectives of all players with value maximization and, hence, it results in strongly efficient graphs being pairwise stable. In what follows, we identify conditions under which the equal split rule is the only allocation rule for which strongly efficient graphs are pairwise stable. This is made precise as follows.

Definition. *The value function* v *is* anonymous *if* $v(g^\pi) = v(g)$ *for all permutations* π *and graphs* g.

Anonymity of v requires that v depends only on the shape of g.

Definition. Y *is* independent of potential links *if* $Y(g, v) = Y(g, w)$ *for all graphs* g *and value functions* v *and* w *such that there exists* $j \neq i$ *so that* v *and* w *agree on every graph except* $g + ij$.

Such an independence condition is very strong. It requires that the allocation rule ignore some potential links. However, many allocation rules, such as the equal split and the one based on equal bargaining power (Theorem 4 below), satisfy independence of potential links.

Theorem 2. *Suppose that* Y *is anonymous, balanced, and independent of potential links. If* v *is anonymous and all strongly efficient graphs are stable, then* $Y_i(g, v) = v(g)/N$, *for all* i *and strongly efficient* g*'s.*

Proof. If g^N is strongly efficient the result follows from the anonymity of v and Y. The rest of the proof proceeds by induction. Suppose that $Y_i(g, v) = v(g)/N$, for all i and strongly efficient g's which have k or more links. Consider a strongly efficient g with $k - 1$ links. We must show that $Y_i(g, v) = v(g)/N$ for all i.

First, suppose that i is not fully connected under g and $Y_i(g, v) > v(g)/N$. Find j such that $ij \notin g$. Let w coincide with v everywhere except on $g + ij$ (and all its permutations) and let $w(g + ij) > v(g)$. Now, $g + ij$ is strongly efficient for w and so by the inductive assumption, $Y_i(g + ij, w) = w(g + ij)/N > v(g)/N$. By the independence of potential links (applied iteratively, first changing v only on $g + ij$, then on a permutation of $g + ij$, etc.), $Y_i(g, w) = Y_i(g, v) > v(g)/N$. Therefore, for $w(g + ij) - v(g)$ sufficiently small, $g + ij$ is defeated by g under w (since i profits from severing the link ij), although $g + ij$ is strongly efficient while g is not – a contradiction.

Next, suppose that i is not fully connected under g and that $Y_i(g, v) < v(g)/N$. Find j such that $ij \notin g$. If $Y_j(g, v) > v(g)/N$ we reach a contradiction as above. So $Y_j(g, v) \leq v(g)/N$. Let w coincide with v everywhere except on $g + ij$ (and all its permutations) where $w(g + ij) = v(g)$ Now, $g + ij$ is strongly efficient for w and hence, by the inductive assumption, $Y_i(g + ij, w) = Y_j(g + ij, w) = v(g)/N$. This and the independence of potential links imply that $Y_i(g + ij, w) = v(g)/N > Y_i(g, v) = Y_i(g, w)$ and $Y_j(g + ij, w) = v(g)/N \geq Y_j(g, v) = Y_j(g, w)$. But this is a contradiction, since g is strongly efficient for w but is unstable. Thus we have shown that for any strongly efficient g, $Y_i(g, v) = v(g)/N$ for all i which are not fully connected under g. By anonymity of v and Y (and total balance of Y), this is also true for i's which are fully connected. □

Remark. The proof of Theorem 2 uses anonymity of v and Y only through their implication that any two fully connected players get the same allocation. We can weaken the anonymity of v and Y and get a stronger version of Theorem 2. The allocation rule Y satisfies *proportionality* if for each i and j there exists a constant k_{ij} such that $Y_i(g, v)/Y_j(g, v) = k_{ij}$ for any g in which both i and j are fully connected and for any v. The new Theorem 2 would read: *Suppose Y satisfies proportionality and is independent of potential links. If all strongly efficient graphs are pairwise stable, then $Y_i(g, v) = s^i v(g)$, for all i, v, and g's which are strongly efficient relative to v, where $s^i = Y_i(g^N, v)/v(g^N)$.* The proof proceeds like that of Theorem 2 with s^i taking the place of $1/N$.

Theorem 2 only characterizes Y at strongly efficient graphs. If we require the right incentives holding at all graphs then the characterization is made complete.

Definition. *Y is* pairwise monotonic *if g' defeats g implies that $v(g') > v(g)$.*

Pairwise monotonicity is more demanding than the stability of strongly efficient networks, and in fact it is sufficiently strong (coupled with anonymity, balance, and independence of potential links) to result in a unique allocation rule for anonymous v. That is, the result that $Y_i(g, v) = v(g)/N$ is obtained for all g, not just strongly efficient ones, providing the following characterization of the equal split rule.

Theorem 3. *If Y is anonymous, balanced, is independent of potential links, and is pairwise monotonic, then $Y_i(g, v) = v(g)/N$, for all i, and g, and anonymous v.*

Proof. The theorem is proven by induction. By the anonymity of v and Y and $Y_i(g^N, v) = v(g^N)/N$. We show that if $Y_i(g, v) = v(g)/N$ for all g where g has at least k links, then this is true when g has at least $k - 1$ links.

First, suppose that i is not fully connected under g and $Y_i(g, v) > v(g)/N$. Find j such that $ij \notin g$. Let w coincide with v everywhere except that $w(g + ij) > v(g)$. By the inductive assumption, $Y_i(g + ij, w) = w(g + ij)/N$. By the independence of potential links, $Y_i(g, w) = Y_i(g, v) > v(g)/N$. Therefore, for $w(g + ij) - v(g)$ sufficiently small $g + ij$ is defeated by g under w (since i profits from severing ij), while $w(g + ij) > w(g)$, contradicting pairwise monotonicity.

Next, suppose that i is not fully connected under g and that $Y_i(g, v) < v(g)/N$. Find j such that $ij \notin g$. If $Y_j(g, v) > v(g)/N$ we reach a contradiction as above. So $Y_j(g, v) \leq v(g)/N$. Let w coincide with v everywhere except on $g + ij$ where $w(g + ij) = v(g)$. By the inductive assumption, $Y_i(g + ij, w) = Y_j(g + ij, w) = w(g + ij)/N$. This and the independence of potential links imply that $Y_i(g+ij, w) = w(g+ij)/N = v(g)/N > Y_i(g, v) = Y_i(g, w)$ and $Y_j(g+ij, w) = w(g + ij)/N = v(g)/N \geq Y_j(g, v) = Y_j(g, w)$. This is a contradiction, since $w(g) = w(g + ij)$ but g is defeated by $g + ij$.

Thus we have shown that $Y_i(g, v) = v(g)/N$ for all i which are not fully connected under g. By anonymity of v and Y (and total balance of Y), this is also true for i's which are fully connected. □

Note that the equal split rule, $Y_i(g, v) = v(g)/N$, for all i and g, satisfies anonymity, balance, pairwise monotonicity, and is independent of potential links. Thus a converse of the theorem also holds.

Theorem 1 documented a tension between pairwise stability and efficiency. If one wants to guarantee that efficient graphs are stable, then one has to violate component balance (as the equal split rule does). In some circumstances, the rule by which resources are allocated may not be subject to choice, but may instead be determined by some process, such as bargaining among the individuals in the network. We conclude with a characterization of allocation rules satisfying equal bargaining power.

Definition. *An allocation rule Y satisfies* equal bargaining power[14] *(EBP) if for all v, g, and $ij \in g$*

$$Y_i(g, v) - Y_i(g - ij, v) = Y_j(g, v) - Y_j(g - ij, v).$$

Under such a rule every i and j gain equally from the existence of their link relative to their respective "threats" of severing this link.

The following theorem is an easy extension of a result by Myerson [19].

Theorem 4. *If v is component additive, then the unique allocation rule Y which satisfies component balance and equal bargaining power (EBP) is the Shapley value of the following game $U_{v,g}$ in characteristic function form.*[15] *For each S, $U_{v,g}(S) = \sum_{h \in C(g|_S)} v(h)$, where $g|_S = \{ij \in g \ : \ i \in S$ and $j \in S\}$.*

Although Theorem 4 is easily proven by extending Myerson's [19] proof to our setting (see the appendix for details), it is an important strengthening of his result. In his formulation a graph represents a communication structure which is used to determine the value of coalitions. The value of a coalition is the sum over the value of the subcoalitions which are those which are intraconnected via the graph. For example, the value of coalition $\{1, 2, 3\}$ is the same under graph $\{12, 23\}$ as it is under graph $\{12, 13, 23\}$. In our formulation the value depends explicitly on the graph itself, and thus the value of any set of agents depends not only on the fact that they are connected, but on exactly how they are connected.[16] In all of the examples we have considered so far, the shape of the graph has played an essential role in the productivity.

The potential usefulness of Theorem 4 for understanding the implications of equal bargaining power, is that it provides a formula which can be used to study the stability properties of different organizational forms under various value functions. For example, the following corollary brings two implications.

Corollary. *Let Y be the equal bargaining power rule from Theorem 4, and consider a component balanced v and any g and $ij \in g$.*

---

[14] Such an allocation rule, in a different setting, is called the "fair allocation rule" by Myerson [19].

[15] $Y_i(g, v) = SV_i(U_{v,g})$, where the Shapley value of a game U in characteristic function form is $SV_i(U) = \sum_{S \subset \mathcal{N} - i}(U(S + i) - U(S))\frac{\#S!(N - \#S - 1)!}{N!}$.

[16] The graph structure is still essential to Myerson's formulation. For instance, the value of the coalition $\{1, 3\}$ is not the same under graph $\{12, 23\}$ as it is under graph $\{12, 13, 23\}$, since agents 1 and 3 cannot communicate under the graph $\{12, 23\}$ when agent 2 is not present.

$$\textit{If, for all } g' \subset g, \quad v(g') \geq v(g' - ij), \quad \textit{then } Y_i(g, v) \geq Y_i(g - ij, v).$$

$$\textit{If, for all } g' \subset g, \quad v(g') \geq v(g' + ij), \quad \textit{then } Y_i(g, v) \geq Y_i(g + ij, v).$$

This follows directly from inspection of the Shapley value formula.

The first line of the Corollary means, for example, that if v is such that links are of diminishing marginal contribution, then stable networks will not be too sparse in the sense that a subgraph of the strongly efficient graph won't be stable. Thus, in some circumstances, the equal bargaining power rule will guarantee that strongly efficient graphs are pairwise stable. However, as we saw in Theorem 1 this will not always be the case.

5 Discussion of the Stability Notion

The notion of stability that we have employed throughout this paper is one of many possible notions. We have selected this notion, not because it is necessarily more compelling than others, but rather because it is a relatively weak notion that still takes into account both link severance and link formation (and provides sharp results for most of our analysis). The purpose of the following discussion is to consider the implications of modifying this notion. At the outset, it is clear that stronger stability notions (admitting fewer stable graphs) will just strengthen Theorems 1, 2, and 3 (as well as Propositions 2, 3, and 4). That is, stronger notions would allow the conclusions to hold under the same or even weaker assumptions. Some of the observations derived in the examples change, however, depending on how the stability notion is strengthened.

Let us now consider a few specific variations on the stability notion and comment on how the analysis is affected. First, let us consider a stronger stability notion that still allows only link severance by individuals and link formation by pairs, but implicitly allows for side payments to be made between two agents who deviate to form a new link.

The graph g' defeats g under Y and v (allowing for side payments) if either

(i) $g' = g - ij$ and $Y_i(g, v) < Y_i(g', v)$ or $Y_j(g, v) < Y_j(g', v)$, or

(ii) $g' = g + ij$ and $Y_i(g', v) + Y_j(g', v) > Y_i(g, v) + Y_j(g, v)$.

We then say that g is pairwise stable allowing for side payments under Y and v, if it is not defeated by any g' according to the above definition.

Note that in a pairwise stable network allowing for side payments payoffs are still described by Y rather than Y plus transfers. This reflects the interpretation that Y is the allocation to each agent when one includes the side payments that have already been made. The network, however, still has to be immune against deviations which could involve additional side payments. This interpretation introduces an asymmetry in the consideration of side payments since severing a link, (i), can be done unilaterally, and so the introduction of additional side payments will not change the incentives, while adding a link, (ii), requires the consent of two agents and additional side payments relative to the new graph may play a role.[17]

Under this notion of stability allowing for side payments, a version of Theorem 1 holds without the anonymity requirement.

Theorem 1′. *If $N \geq 3$, then there is no Y which is component balanced and such that for each v no strongly efficient graph is defeated (when allowing for side payments) by an inefficient one.*

The proof is in the appendix. As this version reproduces the impossibility result of Theorem 1 without the anonymity restriction on Y, it supports our earlier assertion that this result was not driven by the anonymity of Y, but rather by the component balance condition.

Stability with side payments also results in stronger versions of Theorems 2 and 3 which are included in the appendix.

Another possible strengthening of the stability notion would allow for richer combinations of moves to threaten the stability of a network. Note that the basic stability notion we have considered requires only that a network be immune to one deviating action at a time. It is not required that a network be immune to more complicated deviations, such as a simultaneous severance of some existing links and an introduction of a new link by two players (which is along the lines of the stability notion used in studying the marriage problem). It is also not required that a network be immune to deviations by more than two players simultaneously. Actually, the notion of pairwise stability that we have employed does not even contemplate the severance of more than one link by a single player.

The general impact of such stronger stability notions would be to strengthen our results, with the possible complication that in some cases there may exist no stable network. As an example, reconsider the co-author model and allow any pair of players to simultaneously sever any set of their existing links. Based on Proposition 4 part (ii), we know that any graph that could be stable under such a new definition must have fully intraconnected components. However, now a pair of players can improve for themselves by simultaneously severing all their links, except the one joining them. It follows that no graph is stable.

A weaker version of the stability notion can be obtained by alterring (ii) to require that both deviating players who add a link be strictly better off in order for a new graph to defeat an old one. The notion we have used requires that one player be strictly better off and the other be weakly better off. Most of our discussion is not sensitive to this distinction; however, the conclusions of Theorems 2 and 3 are, as illustrated in the following example. Let $N = \{1, 2, 3, 4\}$, $g = \{14, 23, 24, 34\}$, and consider v with $v(g) = 1$, $v(g') = 1$ if g' is a permutation of g, and $v(g') = 0$ for any other g'. Consider Y such that $Y_1(g', v) = 1/8$ $Y_2(g', v) = Y_3(g', v) = 1/4$ and $Y_4(g', v) = 3/8$ if g' is a permutation of g, and $Y_i(g', v) = 0$ otherwise. Specify $Y_i(g', w) = w(g')/N$ for $w \neq v$, except if g' is a permutation of g and w agrees with v on g and all its

[17] The results still hold if (i) is also altered to allow for side payments.

subgraphs, in which case set $Y_i(g', w) = Y_i(g', v)$. This Y is anonymous, balanced, and independent of potential links. However, it is clear that $Y_1(g, v) \neq v(g)/N$. To understand where Theorems 2 and 3 fail consider $g' = g + 12$ and w which agrees with v on all subgraphs of g but gives $w(g+12) = 1$. Under the definition of stability that we have used in this paper, $g+12$ defeats g since player 1 is made better off and 2 is unchanged ($Y_1(g + 12, w) = 1/4 = Y_2(g + 12, w)$), however, under this weakened notion of stability $g + 12$ does not defeat g.

One way to sort out the different notions of stability would be to look more closely at the non-cooperative foundations of this model. Specifications of different procedures for graph formation (e.g., an explicit non-cooperative game) and equilibria of those procedures, would lead to notions of stability. Some of the literature on communication structures has taken this approach to graph formation (see, e.g., Aumann and Myerson [1], Qin [23], and Dutta, van den Nouweland, and Tijs [3]). Let us make only one observation in this direction. Central to our notion of stability is the idea that a deviation can include two players who come together to form a new link. The concept of Nash equilibrium does not admit such considerations. Incorporating deviations by pairs (or larger groups) of agents might most naturally involve a refinement of Nash equilibrium which explicitly allows for such deviations, such as strong equilibrium, coalition-proof Nash equilibrium,[18] or some other notion which allows only for certain coalitions to form. This constitutes a large project which we do not pursue here.

Appendix

Theorem 1′. *If $N \geq 3$, then there is no Y which is component balanced and such that for each v no strongly efficient graph is defeated (allowing for side payments) by an inefficient one.*

Remark. In fact, it is not required that no strongly efficient graph is defeated by an inefficient one, but rather that there is some strongly efficient graph which is not defeated by any inefficient one and such that any permutation of that graph which is also strongly efficient is not defeated by any inefficient one. This is clear from the following proof.

Proof. Let $N = 3$ and consider the same v given in the Proof of Theorem 1. (For all i, j, and k, $v(\{ij\}) = 1$, $v(\{ij, jk\}) = 1 + \epsilon$ and $v(\{ij, jk, ik\}) = 1$, where the strongly efficient networks are of the form $\{ij, jk\}$.) Without loss of generality, assume that $Y_1(\{12\}, v) \geq 1/2$ and $Y_2(\{23\}, v) \geq 1/2$. (Given the component balance, there always exists such a graph with some relabelling of players.) Since $\{12, 13\}$ cannot be defeated by $\{12\}$, it must be that $Y_1(\{12, 13\}, v) \geq 1/2$. It follows from component balance that $1/2 + \epsilon \geq Y_2(\{12, 13\}, v) + Y_3(\{12, 13\}, v)$. Since $\{12, 13\}$ cannot be defeated by $\{12, 13, 23\}$, it must be that

[18] One can try to account for the incentives of pairs by considering an extensive form game which sequentially considers the addition of each link and uses a solution such as subgame perfection (as in Aumann and Myerson [1]). See Dutta, van den Nouweland, and Tijs [3] for a discussion of this approach and an alternative approach based on coalition-proof Nash equilibrium.

$$1/2+\epsilon \geq Y_2(\{12,13\},v)+Y_3(\{12,13\},v)$$
$$\geq Y_2(\{12,13,23\},v)+Y_3(\{12,13,23\},v). \qquad (*)$$

Similarly

$$1/2+\epsilon \geq Y_1(\{12,23\},v)+Y_3(\{12,23\},v)$$
$$\geq Y_1(\{12,13,23\},v)+Y_3(\{12,13,23\},v). \qquad (**)$$

Now note that adding $(*)$ and $(**)$ we get

$$Y_2(\{12,13\},v)+Y_3(\{12,13\},v)+Y_1(\{12,23\},v)+Y_3(\{12,23\},v)$$
$$\geq Y_1(\{12,13,23\},v)+Y_2(\{12,13,23\},v)+2Y_3(\{12,13,23\},v).$$

Note that $Y_3(\{12,13,23\},v) \geq 0$. This is shown as follows:[19] Let $Y_3(\{12,13,23\}) = a$. By balance, $Y_1(\{12,13,23\})+Y_2(\{12,13,23\}) = 1-a$. Since $\{13,23\}$ is not defeated by $\{12,13,23\}$, this implies that $Y_1(\{13,23\})+Y_2(\{13,23\}) \geq 1-a$. Then balance implies that $Y_3(\{13,23\}) \leq \varepsilon + a$. Since $\{13,23\}$ is not defeated by $\{13\}$ or $\{23\}$, this implies that $Y_3(\{13\}) \leq \varepsilon+a$ and $Y_3(\{23\}) \leq \varepsilon+a$. Component balance then implies that $Y_1(\{13\}) \geq 1-\varepsilon-a$ and $Y_2(\{23\}) \geq 1-\varepsilon-a$. The facts that $\{13,12\}$ is not defeated by $\{13\}$ and $\{12,23\}$ is not defeated by $\{23\}$ imply that $Y_1(\{13,12\}) \geq 1-\varepsilon-a$ and $Y_2(\{12,23\}) \geq 1-\varepsilon-a$. Balance then implies that $Y_2(\{13,12\})+Y_3(\{13,12\}) \leq 2\varepsilon+a$ and $Y_1(\{12,23\})+Y_3(\{12,23\}) \leq 2\varepsilon+a$. Then, since neither $\{13,12\}$ nor $\{12,23\}$ is defeated by $\{12,13,23\}$, it follows that $Y_2(\{13,12,23\})+Y_3(\{13,12,23\}) \leq 2\varepsilon+a$ and $Y_1(\{12,13,23\})+Y_3(\{12,13,23\}) \leq 2\varepsilon+a$. Given that $Y_3(\{12,13,23\}) = a$ this implies that $Y_2(\{13,12,23\}) \leq 2\varepsilon$ and $Y_1(\{12,13,23\}) \leq 2\varepsilon$. So, $Y_1(\{13,12,23\})+Y_2(\{13,12,23\})+Y_3(\{12,13,23\}) \leq 4\varepsilon+a$. By balance these sum to 1, so if $\varepsilon \leq 1/4$ then it must be that $a \geq 0$.

By component balance, we rewrite the inequality from before as

$$2+2\epsilon - Y_1(\{12,13\},v) - Y_2(\{12,23\},v) \geq 1+Y_3(\{12,13,23\},v).$$

Thus

$$Y_1(\{12,13\},v)+Y_2(\{12,23\},v) \leq 1+2\epsilon.$$

Then since no strongly efficient graph is defeated by an inefficient one, we know that $Y_1(\{12,13\},v) \geq Y_1(\{12\},v)$ and $Y_2(\{12,23\},v) \geq Y_2(\{23\},v)$, and so

$$Y_1(\{12\},v)+Y_2(\{23\},v) \leq 1+2\epsilon.$$

Since $Y_1(\{12\},v) \geq 1/2$, we know that $Y_2(\{23\},v) \leq 1/2+2\epsilon$. Thus, by component balance

$$Y_3(\{23\},v) \geq 1/2-2\epsilon.$$

Since $\{13,23\}$ cannot be defeated by $\{23\}$, it must be that $Y_3(\{13,23\},v) \geq 1/2-2\epsilon$. It follows from component balance that $1/2+3\epsilon \geq Y_1(\{13,23\},v)+$

[19] We thank Juan D. Moreno Ternero for suggesting that we show this, as it was not shown in earlier versions of the paper and does take a few lines to verify.

$Y_2(\{13,23\},v)$. Since $\{13,23\}$ cannot be defeated by $\{12,13,23\}$, it must be that

$$\begin{aligned} 1/2+3\epsilon &\geq Y_1(\{13,23\},v)+Y_2(\{13,23\},v) \\ &\geq Y_1(\{12,13,23\},v)+Y_2(\{12,13,23\},v). \end{aligned} \qquad (***)$$

Adding $(*)$, $(**)$, and $(***)$, we find

$$3/2+5\epsilon \geq 2[Y_1(\{12,13,23\},v)+Y_2(\{12,13,23\},v)+Y_3(\{12,13,23\},v)] = 2,$$

which is impossible for $\epsilon < 1/10$.

Again, this is easily extended to $N > 3$, by assigning $v(g) = 0$ to any g which has a link involving a player other than players 1, 2 or 3. □

Definition. *The allocation rule Y is* continuous, *if for any g, and v and w that differ only on g and for any ϵ, there exists δ such that $|v(g) - w(g)| < \delta$ implies $|Y_i(g,v) - Y_i(g,w)| < \epsilon$ for all $i \in N(g)$.*

Theorem 2′. *Suppose that Y is anonymous, balanced, continuous, and is independent of potential links. If v is anonymous and no strongly efficient graph is defeated (allowing for side payments) by an inefficient one, then, $Y_i(g,v) = v(g)/N$, for all i and strongly efficient g's.*

Proof. If g^N is strongly efficient the result follows from the anonymity of v and Y. The rest of the proof proceeds by induction. Suppose that $Y_i(g,v) = v(g)/N$, for all i and strongly efficient g's which have k or more links. Consider a strongly efficient g with $k-1$ links. We must show that $Y_i(g,v) = v(g)/N$ for all i.

First, suppose that i is not fully connected under g and $Y_i(g,v) > v(g)/N$. Find j such that $ij \notin g$. Let w coincide with v everywhere except on $g+ij$ (and all its permutations) and let $w(g+ij) > v(g)$. Now, $g+ij$ is strongly efficient for w and so by the inductive assumption, $Y_i(g+ij,w) = w(g+ij)/N > v(g)/N$. By the independence of potential links (applied iteratively, first changing v only on $g+ij$, then on a permutation of $g+ij$, etc.), $Y_i(g,w) = Y_i(g,v) > v(g)/N$. Therefore, for $w(g+ij) - v(g)$ sufficiently small, $g+ij$ is defeated by g under w (since i profits from severing the link ij), although $g+ij$ is strongly efficient while g is not – a contradiction.

Next, suppose that i is not fully connected under g and that $Y_i(g,v) < v(g)/N$. Find j such that $ij \notin g$. If $Y_j(g,v) > v(g)/N$ we reach a contradiction as above. So $Y_j(g,v) \leq v(g)/N$. Let $\epsilon < [v(g)/N - Y_i(g,v)]/2$ and let w coincide with v everywhere except on $g+ij$ (and all its permutations) and let $w(g+ij) = v(g)+\delta/2$ where δ is the appropriate $\delta(\epsilon)$ from the continuity definition. Now, $g+ij$ is strongly efficient for w and hence, by the inductive assumption, $Y_i(g+ij,w) = Y_j(g+ij,w) = [v(g)+\delta/2]/N$. Define u which coincides with v and w everywhere except on $g+ij$ (and all its permutations) and let $u(g+ij) = w(g) - \delta/2$. By the continuity of Y, $Y_i(g+ij,u) \geq v(g)/N - \epsilon$ and $Y_j(g+ij,u) \geq v(g)/N - \epsilon$. Thus, we have reached a contradiction, since g is strongly efficient for u but defeated by $g+ij$ since $Y_i(g+ij,u)+Y_j(g+ij,u) \geq 2v(g)/N - 2\epsilon > 2v(g)/N - [v(g)/N - Y_i(g,v)] \geq Y_i(g,u)+Y_j(g,u)$. Thus we have shown that for a strongly efficient g, $Y_i(g, v) = v(g)/N$ for all i which are not fully connected under g. By anonymity of v and Y (and total balance of Y), this is also true for i's which are fully connected. □

Remark. The definition of "defeats" allows for side payments in (ii), but not in (i). To be consistent, (i) could be altered to read $Y_i(g', v) + Y_j(g', v) > Y_i(g, v) + Y_j(g, v)$, as side payments can be made to stop an agent from severing a link. Theorem 2 is still true. The proof would have to be altered as follows. Under the new definition (i) the cases $ij \notin g$ and $Y_i(g, v) + Y_j(g, v) > 2v(g)/N$ or $Y_i(g, v) + Y_j(g, v) < 2v(g)/N$ would follow roughly the same lines as currently is used for the case where $ij \notin g$, and $Y_i(g, v) < v(g)/N$ and $Y_j(g, v) \leq v(g)/N$. (For $Y_i(g, v) + Y_j(g, v) > 2v(g)/N$ the argument would be that ij would want to sever ij from $g + ij$ when $g + ij$ is strongly efficient.) Then notice that it is not possible that for all $ij \notin g$, $Y_i(g, v) + Y_j(g, v) = 2v(g)/N$, without having only two agents ij who are not fully connected, in which case anonymity requires that they get the same allocation, or by having $Y_i = v(g)/N$ for all i which are not fully connected.

Theorem 2 only characterizes Y at strongly efficient graphs. If we require the right incentives holding at all graphs then the characterization is made complete:

Definition. Y *is* pairwise monotonic allowing for side payments *if g' defeats (allowing for side payments) g implies that $v(g') \geq v(g)$.*

Theorem 3′. *If Y is anonymous, balanced, is independent of potential links, and is pairwise monotonic allowing for side payments, then $Y_i(g, v) = v(g)/N$, for all i, and g, and anonymous v.*

Proof. The theorem is proven by induction. By the anonymity of v and Y and $Y_i(g^N, v) = v(g^N)/N$. We show that if $Y_i(g, v) = v(g)/N$ for all g where g has at least k links, then this is true when g has at least $k - 1$ links.

First, suppose that i is not fully connected under g and $Y_i(g, v) > v(g)/N$. Find j such that $ij \notin g$. Let w coincide with v everywhere except that $w(g + ij) > v(g)$. By the inductive assumption, $Y_i(g + ij, w) = w(g + ij)/N$. By the independence of potential links, $Y_i(g, w) = Y_i(g, v) > v(g)/N$. Therefore, for $w(g+ij, w) - v(g)$ sufficiently small $g+ij$ is defeated by g under w (since i profits from severing ij), while $w(g + ij) > w(g)$, contradicting pairwise monotonicity.

Next, suppose that i is not fully connected under g and that $Y_i(g, v) < v(g)/N$. Find j such that $ij \notin g$. If $Y_j(g, v) > v(g)/N$ we reach a contradiction as above. So $Y_j(g, v) \leq v(g)/N$. Let w coincide with v everywhere except that $w(g + ij) < v(g)$ and $v(g)/N - w(g + ij)/N < \frac{1}{2}(v(g)/N - Y_i(g, v))$. Thus $2w(g + ij)/N > v(g)/N + Y_i(g, v)) \geq Y_j(g, v)) + Y_i(g, v))$. By the inductive assumption, $Y_i(g + ij, w) = Y_j(g + ij, w) = w(g + ij)/N$. Thus, we have reached a contradiction, since $w(g) > w(g + ij)$ but g is defeated by $g + ij$ since $Y_i(g + ij, w) + Y_j(g + ij, w) > Y_i(g, w) + Y_j(g, w)$.

Thus we have shown that $Y_i(g, v) = v(g)/N$ for all i which are not fully connected under g. By anonymity of v and Y (and total balance of Y), this is also true for i's which are fully connected. □

Proof of Theorem 4. Myerson's [19] proof shows that there is a unique Y which satisfies equal bargaining power (what he calls fair, having fixed our v) and such that $\sum Y_i$ is a constant across i's in any connected component when other components are varied (which is guaranteed by our component balance condition).

We therefore have only to show that $Y_i(g,v) = SV_i(U_{v,g})$ (as defined in the footnote below Theorem 4) satisfies component balance and equal bargaining power.

Fix g and define Y^g by $Y^g(g') = SV(U_{v,g\cap g'})$. (Notice that $U_{v,g\cap g'}$ substitutes for what Myerson calls v/g'. With this in mind, it follows from Myerson's proof that Y^g satisfies equal bargaining power and that for any connected component h of g $\sum_{i\in h} Y_i^g(g) = U_{v,g}(N(h))$. Since $Y^g(g) = Y(g)$, this implies that $\sum_{i\in h} Y_i^g(g) = U_{v,g}(N(h)) = v(h)$, so that Y satisfies component balance. Also, since Y^g satisfies equal bargaining power, we have that $Y_i^g(g) - Y_i^g(g-ij) = Y_j^g(g) - Y_j^g(g-ij)$. Now, $Y_i^g(g-ij) = SV_i(U_{v,g\cap g-ij}) = SV_i(U_{v,g-ij}) = Y_i(g-ij)$. Therefore, $Y_i(g) - Y_i(g-ij) = Y_j(g) - Y_j(g-ij)$, so that Y satisfies equal bargaining power as well.

References

1. Aumann and Myerson (1988) Endogenous Formation of Links Between Players and Coalitions: An Application of the Shapley Value. In: A. Roth (ed.) *The Shapley Value*, Cambridge University Press, Cambridge, pp 175–191
2. Boorman, S. (1975) A Combinatorial Optimization Model for Transmission of Job Information through Contact Networks *Bell J. Econ.* 6: 216–249
3. Dutta, B., van den Nouweland, A., Tijs, S. (1998) Link Formation in Cooperative Situations. *International Journal of Game Theory* 27: 245–256
4. Gale, D., Shapley, L. (1962) College Admissions and the Stability of Marriage. *Amer. Math. Monthly* 69: 9–15
5. Goyal, S. (1993) Sustainable Communication Networks, Discussion Paper TI 93-250, Tinbergen Institute, Amsterdam-Rotterdam
6. Grout, P. (1984) Investment and Wages in the Absence of Binding Contracts, *Econometrica* 52: 449–460
7. Hendricks, K., Piccione, M., Tan, G. (1994) *Entry and Exit in Hub–Spoke Networks*. mimeo, University of British Columbia
8. Hendricks, K., Piccione, M., Tan, G. (1995) The Economics of Hubs: The Case of Monopoly. *Rev. Econ. Stud.* 62: 83–100
9. Horn, H., Wolinsky, A. (1988) Worker Substitutability and Patterns of Unionisation, *Econ. J.* 98: 484–497
10. Iacobucci, D. (1994) Chapter 4: Graph Theory. In: S. Wasserman, Faust, K. (eds.) *Social Networks: Analyses, Applications and Methods*, Cambridge University Press, Cambridge
11. Iacobucci D., Hopkins, N. (1992) Modeling Dyadic Interactions and Networks in Marketing. *J. Marketing Research* 29: 5–17
12. Jackson, M., Wolinsky, A. (1994) A Strategic Model of Social and Economic Networks CMSEMS Discussion paper 1098, Northwestern University, revised May 1995, dp 1098R
13. Kalai, E., Postlewaite, A., Roberts, J. (1978) Barriers to Trade and Disadvantageous Middlemen: Nonmonotonicity of the Core. *J. Econ. Theory* 19: 200–209
14. Kalai, E., Zemel, E. (1982) Totally Balanced Games and Games of Flow. *Math. Operations Research* 7: 476–478
15. Katz, M., Shapiro, C. (1994) Systems Competition and Network Effects. *J. Econ. Perspectives* 8: 93–115

16. Keren, M., Levhari, D. (1983) The internal Organization of the Firm and the Shape of Average Costs. *Bell J. Econ.* 14: 474–486
17. Kirman, A., Oddou, C., Weber, S. (1986) Stochastic Communication and Coalition Formation. *Econometrica* 54: 129–138
18. Montgomery, J. (1991) Social Networks and Labor Market Outcomes: Toward an Economic Analysis. *Amer. Econ. Rev.* 81: 1408–1418
19. Myerson, R. (1977) Graphs and Cooperation in Games. *Math. Operations Research* 2: 225–229
20. Nouweland, A. van den (1993) Games and Graphs in Economic Situations. Ph.D. dissertation, Tilburg University
21. Nouweland, A. van den, Borm, P. (1991) On the Convexity of Communication Games. *Int. J. Game Theory* 19: 421–430
22. Owen, G. (1986) Values of Graph Restricted Games. *SIAM J. Algebraic and Discrete Methods* 7: 210–220
23. Qin, C. (1994) Endogenous Formation of Cooperation Structures. University of California at Santa Barbara
24. Roth, A. Sotomayor, M. (1989) Two Sided Matching Econometric Society Monographs No. 18: Cambridge University Press
25. Sharkey, W. (1993) Network Models in Economics. Forthcoming in *The Handbook of Operations Research and Management Science*
26. Starr, R., Stinchcombe, M. (1992) An Economic Analysis of the Hub and Spoke System. mimeo: UC San Diego
27. Stole, L., Zweibel, J. (1993) *Organizational Design and Technology Choice with Nonbinding Contracts*. mimeo
28. Wellman, B., Berkowitz, S. (1988) *Social Structure: A Network Approach*. Cambridge University Press, Cambridge

Spatial Social Networks

Cathleen Johnson[1], **Robert P. Gilles**[2]

[1] Research Associate, Social Research and Demonstration Corp. (SRDC), 50 O'Connor St., Ottawa, Ontario K1P 6L2, Canada (e-mail: johnson@srdc.org)
[2] Department of Economics (0316), Virginia Tech, Blacksburg, VA 24061, USA (e-mail: rgilles@vt.edu)

Abstract. We introduce a spatial cost topology in the network formation model analyzed by Jackson and Wolinsky, *Journal of Economic Theory* (1996), 71: 44–74. This cost topology might represent geographical, social, or individual differences. It describes variable costs of establishing social network connections. Participants form links based on a cost-benefit analysis. We examine the pairwise stable networks within this spatial environment. Incentives vary enough to show a rich pattern of emerging behavior. We also investigate the subgame perfect implementation of pairwise stable and efficient networks. We construct a multistage extensive form game that describes the formation of links in our spatial environment. Finally, we identify the conditions under which the subgame perfect Nash equilibria of these network formation games are stable.

JEL classification: A14, C70, D20

Key Words: Social networks, implementation, spatial cost topologies

1 Introduction

Increasing evidence shows that *social capital* is an important determinant in trade, crime, education, health care and rural development. Broadly defined, social capital refers to the institutions and relationships that shape a society's social interactions (see Woolcock [27]). Anecdotal evidence for the importance of social

Corresponding author: Cathleen Johnson.

We are very grateful for the constructive comments of Matt Jackson and an anonymous referee. We also like to thank Vince Crawford, Marco Slikker, Edward Droste, Hans Haller, Dimitrios Diamantaras, and Sudipta Sarangi for comments on previous drafts of this paper. We acknowledge Jay Hogan for his programming support.

Part of this research was done while visiting the CentER for Economic Research, Tilburg University, Tilburg, The Netherlands. Financial support from the Netherlands Organization for Scientific Resrarch (NWO), grant B46-390, is gratefully acknowledged.

capital formation for the well-functioning of our society is provided by Jacobs [17] on page 180: "These [neighborhood] networks are a city's irreplaceable social capital. When the capital is lost, from whatever cause, the income from it disappears, never to return until and unless new capital is slowly and chancily accumulated." Knack and Keefer [19] recently explored the link between social capital and economic performance. They found that trust and civic cooperation have significant impacts on aggregate economic activity. Social networks, especially those networks that take into account the social differences among persons, are the media through which social capital is created, maintained and used. In short, spatial social networks convey social capital. It is our objective to study the formation and the structure of such spatial social networks.

Social networks form as individuals establish and maintain relationships.[1] Being "connected" greatly benefits an individual. Yet, maintaining relationships is costly. As a consequence individuals limit the number of their active relationships. These social-relationship networks develop from the participants' comparison of costs versus benefits of connecting.

To study spatial social networks we extend the Jackson-Wolinsky [16] framework by introducing a *spatial* cost topology. Thus, we incorporate the main hypotheses from Debreu [7] that players located closer to one another incur less cost to establish communication. We limit our analysis to the simplest possible implementation of this spatial cost topology within the Jackson-Wolinsky framework. Individuals are located along the real line as in Akerlof's [1] model of social distance, and the distance between two individuals determines the cost of establishing a direct link between them. The consequences of this simple extension are profound. A rich structure of social networks emerges, showing the relative strength of the specificity of the model.

First, we identify the *pairwise stable networks* introduced by Jackson and Wolinsky [16]. We find an extensive typology of such networks. We mainly distinguish two classes: If costs are *high* in relation to the potential benefits, only the empty network is stable. If costs are *low* in relation to the potential benefits, an array of stable network architectures emerges. However, we derive that *locally complete networks* are the most prominent stable network architecture in this spatial setting. In these networks, localities are completely connected. This represents a situation frequently studied and applied in spatial games, as exemplified in the literature on local interaction, e.g., Ellison [10] and Goyal and Janssen [13]. This result also confirms the anecdotal evidence from Jacobs [17] on city life. Furthermore, we note that the networks analyzed by Watts and Strogetz [26] and the notion of the *closure* of a social network investigated by Coleman [6] also fall within this category of locally complete networks.

Next, we turn to the consideration of Pareto optimal and efficient spatial social networks. A network is *efficient* if the total utility generated is maximal.

[1] Watts and Strogetz [26] recently showed with computer simulations using deterministic as well as stochastic elements one can generate social networks that are highly efficient in establishing connections between individuals. This refers to the "six degrees of separation" property as perceived in real life networks.

Pareto optimality leads to an altogether different collection of networks. We show that efficient networks exist that are not pairwise stable. This is comparable to the conflict demonstrated by Jackson and Wolinsky [16].

Finally, we present an analysis of the subgame perfect implementation of stable networks by creating an appropriate network formation game. We introduce a class of defined, multi-stage *link formation games* in which all pairs of players sequentially have the potential to form links. The order in which pairs take action is given exogenously.[2] We show that subgame perfect Nash equilibria of such link formation games may consist of pairwise-stable networks only.

Related Literature

In the literature on network formation, economists have developed cost-benefit theories to study the processes of link formation and the resulting networks. One approach in the literature is the formation of social and economic relationships based on cost considerations only, thus neglecting the benefit side of such relationships. Debreu [7], Haller [14], and Gilles and Ruys [12] theorized that costs are described by a topological structure on the set of individuals, being a *cost topology*. Debreu [7] and Gilles and Ruys [12] base the cost topology explicitly on characteristics of the individual agents. Hence, the space in which the agents are located is a topological space expressing individual characteristics. We use the term "neighbors" to describe agents who have similar individual characteristics. The *more similar* the agents, with regard to their individual characteristics, the *less costly* it is for them to establish relationships with each other. Haller [14] studies more general cost topologies. The papers cited investigate the coalitional cooperation structures that are formed based on these cost topologies. Thus, cost topologies are translated into constraints on coalition formation. Neglecting the benefits from network formation prevents these theories from dealing with the hypothesis that the more *dissimilar* the agents, the more beneficial their interactions might be.

A second approach in the literature emphasizes the benefits resulting from social interaction. The cost topology is a priori given and reduced to a set of constraints on coalition formation or to a given network. Given these constraints on social interaction, the allocation problem is investigated. For an analysis of constraints on coalition formation and the core of an economy, we refer to, e.g., Kalai et al. [18] and Gilles et al. [11]. Myerson [21] initiated a cooperative game theoretic analysis of the allocation problem under such constraints. For a survey of the resulting literature, we also refer to van den Nouweland [22] and Borm, van den Nouweland and Tijs [5].

More recently the focus has turned to a full cost-benefit analysis of network formation. In 1988, Aumann and Myerson [2] presented an outline of such a

[2] Our link formation game differs from the network formation game considered by Aumann and Myerson [2] in that each pair of players takes action only once. In the formation game considered by Aumann and Myerson, all pairs that did not form links are asked repeatedly whether they want to form a link or not. See also Slikker and van den Nouweland [24].

research program. However, not until recently has this type of program been initiated. Within the resulting literature we can distinguish three strands: a purely cooperative approach, a purely noncooperative approach, and an approach based on both considerations, in particular the equilibrium notion of pairwise stability.

The cooperative approach was initiated by Myerson [21] and Aumann and Myerson [2]. Subsequently Qin [23] formalized a non-cooperative link formation game based on these considerations. In particular, Qin showed this link formation game to be a potential game as per Monderer and Shapley [20]. Slikker and van den Nouweland [24] have further extended this line of research. Whereas Qin only considers costless link formation, Slikker and van den Nouweland introduce positive link formation costs. They conclude that due to the complicated character of the model, results beyond the three-player case seem difficult to obtain.

Bala and Goyal [3] and [4] use a purely non-cooperative approach to network formation resulting into so-called *Nash networks*. They assume that each individual player can create a one-sided link with any other player. This concept deviates from the notion of pairwise stability at a fundamental level: a player cannot refuse a connection created by another player, while under pairwise stability both players have to consent explicitly to the creation of a link. Bala and Goyal show that the set of Nash networks is significantly different from the ones obtained by Jackson and Wolinsky [16] and Dutta and Mutuswami [9].

Jackson and Wolinsky [16] introduced the notion of a *pairwise stable network* and thereby initiated an approach based on cooperative as well as non-cooperative considerations. Pairwise stability relies on a cost-benefit analysis of network formation, allows for both link severance and link formation, and gives some striking results. Jackson and Wolinsky prominently feature two network types: the star network and the complete network. Dutta and Mutuswami [9] and Watts [25] refined the Jackson-Wolinsky framework further by introducing other stability concepts and derived implementation results for those different stability concepts.

2 Social Networks

We let $N = \{1, 2, ..., n\}$ be the set of players, where $n \geqq 3$. We introduce a spatial component to our analysis. As remarked in the introduction, the spatial dispersion of the players could be interpreted to represent the social distance between the players. We require players to have a *fixed* location on the real line $\mathbb{R}$. Player $i \in N$ is located at x_i. Thus, the set $X = \{x_1, ..., x_n\} \subset [0, 1]$ with $x_1 = 0$ and $x_n = 1$ represents the spatial distribution of the players. Throughout the paper we assume that $x_i < x_j$ if $i < j$ and the players are located on the unit interval. This implies that for all $i, j \in N$ the *distance* between i and j is given by $d_{ij} := |x_i - x_j| \leqq 1$.

Network relations among players are formally represented by graphs where the nodes are identified with the players and in which the edges capture the pairwise relations between these players. These relationships are interpreted as *social* links that lead to benefits for the communicating parties, but on the other hand are costly to establish and to maintain.

We first discuss some standard definitions from graph theory. Formally, a *link* ij is the subset $\{i,j\}$ of N containing i and j. We define $g^N := \{ij \mid i,j \in N\}$ as the collection of all links on N. An arbitrary collection of links $g \subset g^N$ is called an (undirected) *network* on N. The set g^N itself is called the *complete network* on N. Obviously, the family of all possible networks on N is given by $\{g \mid g \subset g^N\}$. The number of possible networks is $\sum_{k=1}^{c(n,2)} c(c(n,2),k) + 1$, where for every $k \leqq n$ we define $c(n,k) := \frac{n!}{k!(n-k)!}$.

Two networks $g, g' \subset g^N$ are said to be of the same *architecture* whenever it holds that $ij \in g$ if and only if $n - i + 1, n - j + 1 \in g'$. It is clear that this defines an equivalence relation on the family of all networks. Each equivalence class consists exactly of two mirrored networks and will be denoted as an "architecture."[3]

Let $g{+}ij$ denote the network obtained by adding link ij to the existing network g and $g - ij$ denote the network obtained by deleting link if from the existing network g, i.e., $g + ij = g \cup \{ij\}$ and $g - ij = g \setminus \{ij\}$.

Let $N(g) = \{i \mid ij \in g \text{ for some } j\} \subset N$ be the set of players involved in at least one link and let $n(g)$ be the cardinality of $N(g)$. A *path* in g connecting i and j is a set of distinct players $\{i_1, i_2, ..., i_k\} \subset N(g)$ such that $i_1 = i$, $i_k = j$, and $\{i_1 i_2, i_2 i_3, ..., i_{k-1} i_k\} \subset g$. We call a network *connected* if between any two nodes there is a path. A *cycle* in g is a path $\{i_1, i_2, ..., i_k\} \subset N(g)$ such that $i_1 = i_k$. We call a network *acyclic* if it does not contain any cycles. We define t_{ij} as the number of links in the shortest path between i and j. A *chain* is a connected network composed of exactly one path with a spatial requirement.

Definition 1. *A network $g \subset g^N$ is called a* **chain** *when (i) for every $ij \in g$ there is no h such that $i < h < j$ and (ii) g is connected.*

Since $i < j$ if and only if $x_i < x_j$, there exists exactly one chain on N and it is given by $g = \{12, 23, \ldots, (n-1)n\}$.

Let $i,j \in N$ with $i < j$. We define $i \leftrightarrow j := \{h \in N \mid i \leqq h \leqq j\} \subset N$ as the set of all players that are spatially located between i and j and including i and j. We let $n(ij)$ denote the cardinality of the set $i \leftrightarrow j$. Furthermore, we introduce $\ell(ij) := n(ij) - 1$ as the *length* of the set $i \leftrightarrow j$. The set $i \leftrightarrow j$ is a *clique* in g if $g^{i \leftrightarrow j} \subset g$ where $g^{i \leftrightarrow j}$ is the complete network on $i \leftrightarrow j$.

Definition 2. *A network g is called* **locally complete** *when for every $i < j : ij \in g$ implies $i \leftrightarrow j$ is a clique in g.*

Locally complete networks are networks that consist of spatially located cliques. These networks can range in complication from any subnetwork of the chain to the complete network. In a locally complete network, a connected agent will always be connected to at least one of his direct neighbors and belong to a complete subnetwork.

To illustrate the social relevance of locally complete networks we refer to Jacobs [17], who keenly observes the intricacy of social networks that turn city

[3] Bala and Goyal [4] define an architecture as a set of networks that are equivalent for arbitrary permutations. We only allow for mirror permutations to preserve the cost topology.

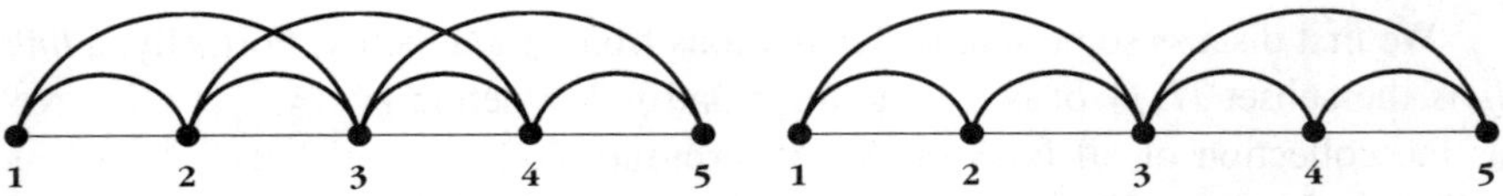

Fig. 1. Examples of locally complete networks

streets, blocks and sidewalk areas into a city neighborhood. Using the physical space of a city street or sidewalk as an example of the space for the players, the concept of local completeness could be interpreted as each player knowing everyone on his block or section of the sidewalk.

Definition 3. *Let $i, j \in N$. The set $i \leftrightarrow j \subset N$ is called a* **maximal clique** *in the network $g \subset g^N$ if it is a clique in g and for every player $h < i$, $h \leftrightarrow j$ is not a clique in g and for every player $h > j$, $i \leftrightarrow h$ is not a clique in g.*

A maximal clique in a certain network is a subset of players that represent a maximal complete subnetwork of that network. For some results in this paper a particular type of locally complete network is relevant.

Definition 4. *Let $k \leqq n$. A network g is called* **regular of order k** *when for every $i, j \in N$ with $\ell(ij) = k$, the set $i \leftrightarrow j$ is a maximal clique.*

Examples of regular networks are the empty network and the chain; the empty network is regular of order zero, while the chain is regular of order one. The complete network is regular of order $n - 1$.

Finally, we introduce the concept of a *star* in which one player is directly connected to all other players and these connections are the only links in the network. Formally, the star with player $i \in N$ as its center is given by $g_i^s = \{ij \mid j \neq i\} \subset g^N$.

To illustrate the concepts defined we refer to Fig. 1. The left network is the second order regular network for $n = 5$. The right network is locally complete, but not regular.

3 A Spatial Connections Model

A network creates benefits for the players, but also imposes costs on those players who form links. Throughout we base benefits of a player $i \in N$ on the connectedness of that player in the network: For each player $i \in N$ her individual payoffs are described by a utility function $u_i : \{g \mid g \subset g^N\} \to \mathbb{R}$ that assigns to every network a (net) benefit for that player.

Following Jackson and Wolinsky [16] and Watts [25] we model the total *value* of a certain network $g \subset g^N$ as

$$v(g) = \sum_{i \in N} u_i(g). \tag{1}$$

This formulation implies that we assume a transferable utility formulation.

We modify the Jackson-Wolinsky *connections model*[4] by incorporating the spatial dispersion of the players into a non-trivial cost topology. This is pursued by replacing the cost concept used by Jackson and Wolinsky with a cost function that varies with the spatial distance between the different players.

Let $c : g^N \to \mathbb{R}_+$ be a general cost function with $c(ij) \geqq 0$ being the cost to create or maintain the link $ij \in g^N$. We simplify our notation to $c_{ij} = c(ij)$. In the Jackson-Wolinsky connections model the resulting utility function of each player i from network $g \subset g^N$ is now given by

$$u_i(g) = w_{ii} + \sum_{j \neq i} w_{ij} \delta^{t_{ij}} - \sum_{j : ij \in g} c_{ij} , \tag{2}$$

where t_{ij} is the number of links in the shortest path in g between i and j, $w_{ij} \geqq 0$ denotes the intrinsic value of individual i to individual j, and $0 < \delta < 1$ is a communication depreciation rate. In this model the parameter δ is a depreciation rate based on network connectedness, not a spatial depreciation rate.

Using the Jackson-Wolinsky connections model and a linear cost topology we are now able to re-formulate the utility function for each individual player to arrive at a *spatial connections model*. We assume that the n individuals are uniformly distributed along the real line segment $[0, 1]$. We define the cost of establishing a link between individuals i and j as $c_{ij} = c \cdot \ell(ij)$ where $c \geqq 0$ is the spatial unit cost of connecting. Finally, we simplify our analysis further by setting for each $i \in N$: $w_{ii} = 0$ and $w_{ij} = 1$ if $i \neq j$. This implies that the utility function for $i \in N$ in the Jackson-Wolinsky connections model — given in (2) — reduces to

$$u_i(g) = \sum_{j \neq i} \delta^{t_{ij}} - c \sum_{j : ij \in g} \ell(ij) . \tag{3}$$

The formulation of the individual benefit functions given in Eq. (3) will be used throughout the remainder of this paper. For several of our results and examples we make an additional simplifying assumption that $c = \frac{1}{n-1}$.

The concept of *pairwise stability* (Jackson and Wolinsky [16]) represents a natural state of equilibrium for certain network formation processes; The formation of a link requires the consent of both parties involved, but severance can be done unilaterally.

Definition 5. *A network $g \subset g^N$ is* **pairwise stable** *if*

1. *for all $ij \in g$, $u_i(g) \geqq u_i(g - ij)$ and $u_j(g) \geqq u_j(g - ij)$, and*
2. *for all $ij \notin g$, $u_i(g) < u_i(g + ij)$ implies that $u_j(g) > u_j(g + ij)$.*

Overall efficiency of a network is in the literature usually expressed by the total utility generated by that network. Consequently, a network $g \subset g^N$ is *efficient* if

[4] Jackson and Wolinsky discuss two specific models, the connections model and the co-author model, and a general model. The connectons model and the co-author model are completely characterized by a specific formulation of the individual utility functions based on the assumptions underlying the sources of the benefits of a social network. Here we only consider the connections model.

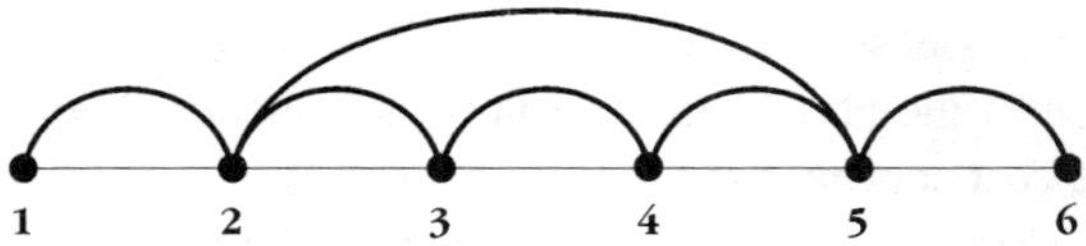

Fig. 2. Pairwise stable network for $n = 6$, $c = \frac{1}{5}$, $\delta = \frac{7}{10}$

g maximizes the value function $v = \sum_N u_i$ over the set of all potential networks $\{g \mid g \subset g^N\}$, i.e., $v(g) \geqq v(g')$ for all $g' \subset g^N$.

3.1 Pairwise Stability in the Spatial Connections Model

The spatial aspect of the cost topology enables us to identify pairwise stable networks with spatially discriminating features. For example, individuals may attempt to maintain a locally complete network but refuse to connect to more distant neighbors. Conversely, it may benefit individuals who are locally connected to maintain a connection with a player who is far away and also well-connected locally. Such a link would have a large spatial cost but it could have an even larger benefit. The example depicted in Fig. 2 illustrates a relatively simple non-locally complete network in which players 2 and 5 enjoy the benefits of close connections as well as the indirect benefits of a distant, costly connection. (Here, we call a network *non-locally complete* if it is not locally complete.) A star is a highly organized non-locally complete network.

Example 1. *Let $n = 6$, $c = \frac{1}{n-1} = \frac{1}{5}$, and $\delta = \frac{7}{10}$. Consider the network depicted in Fig. 2. This non-locally complete network is pairwise stable for the given values of c and δ. We observe that players 2 and 5 maintain a link 50% more expensive than a potential link to player 4 or 3 respectively. The pairwise stability of this network hinges on the fact that the direct and indirect benefits, δ and δ^2, are high relative to the cost of connecting. In this example $u_2(g) = 3\delta + 2\delta^2 - 5c$. If player 2 severed her long link then her utility, $u_2(g - 25)$, would be $\delta + \sum_{k=1}^{4} \delta^k - 2c$. $u_2(g) - u_2(g - 25) = \delta + \delta^2 - \delta^3 - \delta^4 - 3c = 0.0069 > 0$. Each players is willing to incur higher costs to maintain relationships with distant players in order to reap the high benefits from more valuable indirect connections.* ♦

We investigate which networks are pairwise stable in the spatial connections model. We distinguish two major mutually exclusive cases: $\delta > c$ and $\delta \leqq c$. For $\delta > c$ there is a complex array of possibilities. We highlight the locally complete and non-locally complete insights below and leave the remaining results for the appendix. For a proof to Proposition 1 we refer to Appendix A.

For all δ and c we define

$$\hat{n}(c, \delta) := \left\lfloor \frac{\delta}{c} \right\rfloor \tag{4}$$

where $\left\lfloor \frac{\delta}{c} \right\rfloor$ indicates the smallest integer greater than or equal to $\frac{\delta}{c}$.

Proposition 1. *Let $\delta > c > 0$.*

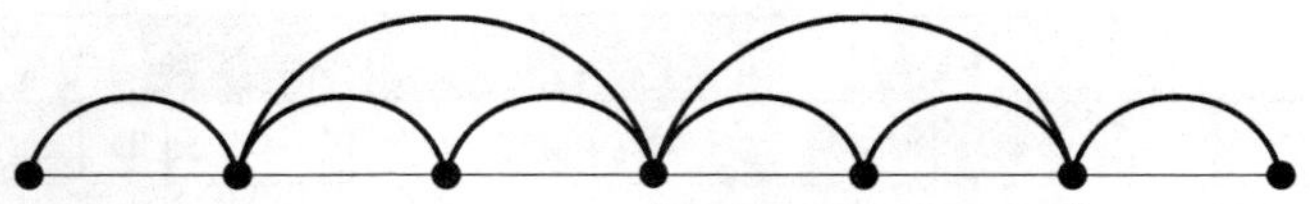

Fig. 3. Pairwise stable network for $n = 7$, $c = \frac{1}{6}$, $\delta = \frac{1}{2}$

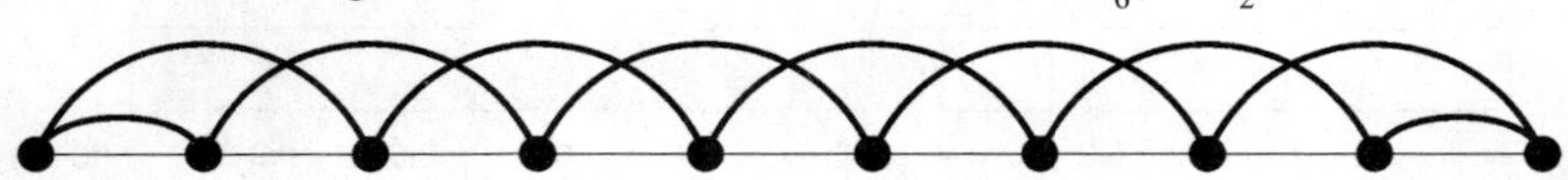

Fig. 4. Example of a cyclic pairwise stable network

(a) *For $[\hat{n}(c,\delta) - 1] \cdot c < \delta - \delta^2$ and $\hat{n}(c,\delta) \geqq 3$, there exists a pairwise stable network which is regular of order $\hat{n}(c,\delta) - 1$.*

(b) *For $c > \delta - \delta^2$, there is no pairwise stable network which contains a clique of a size of at least three players.*

(c) *For $c > \delta - \delta^2$, $\delta > \frac{1}{2}$ and $c = \frac{1}{n-1}$, for $n \leqq 5$ the chain is the only regular pairwise stable network, for $n = 6$ there are certain values of δ for which the chain is pairwise stable, and for $n \geqq 7$ the chain is never pairwise stable.*

To illustrate why the restrictions of $\hat{n}(c,\delta) \geqq 3$ and $[\hat{n}(c,\delta) - 1] \cdot c < \delta - \delta^2$ are placed in the formulation of Proposition 1(a) we refer the following example.

Example 2. *Let $n = 7$, $c = \frac{1}{n-1} = \frac{1}{6}$, and $\delta = \frac{1}{2}$. Consider the network depicted in Fig. 3. This network is pairwise stable for the given values of c and δ.*
We identify two maximal cliques of size 2, $\{1,2\}$ and $\{6,7\}$, and two maximal cliques of the size 3, $\{2,3,4\}$ and $\{4,5,6\}$. Thus, this pairwise stable network is locally complete, but it is not regular of any order. With reference to Proposition 1(a) we note that $\hat{n}(c,\delta) = 3$. However, $[\hat{n}(c,\delta) - 1] \cdot c = \frac{1}{3} > \delta - \delta^2 = \frac{1}{4}$. ♦

For $\delta < c$ the analysis becomes involved in particular due to the possibility of cyclic pairwise stable networks. A proof of the next proposition on acyclic networks only can be found in Appendix A.

Proposition 2. *Let $0 < \delta \leqq c = \frac{1}{n-1}$.*

(a) *For $\delta < c$ there exists exactly one acyclic pairwise stable network, the empty network.*

(b) *For $\delta = c$ there exist exactly two acyclic pairwise stable networks, the empty network and the chain.*

The following example illustrates the possibilities if we allow for cycles.

Example 3. *Consider a network g_c for n even, i.e., we can write $n = 2k$. The network g_c is defined as the unique cycle given by*

$$g_c = \{12, (n-1)n\} \cup \{i\,(i+2) \mid i = 1, \ldots, n-2\}.$$

For $k = 5$ the resulting network is depicted in Fig. 4. This cyclic network is pairwise stable for $\delta = \sqrt[4]{\frac{1}{5}} \sim 0.66874$ and $c = \delta \frac{1-2\delta^k}{2-2\delta} \sim 0.739$

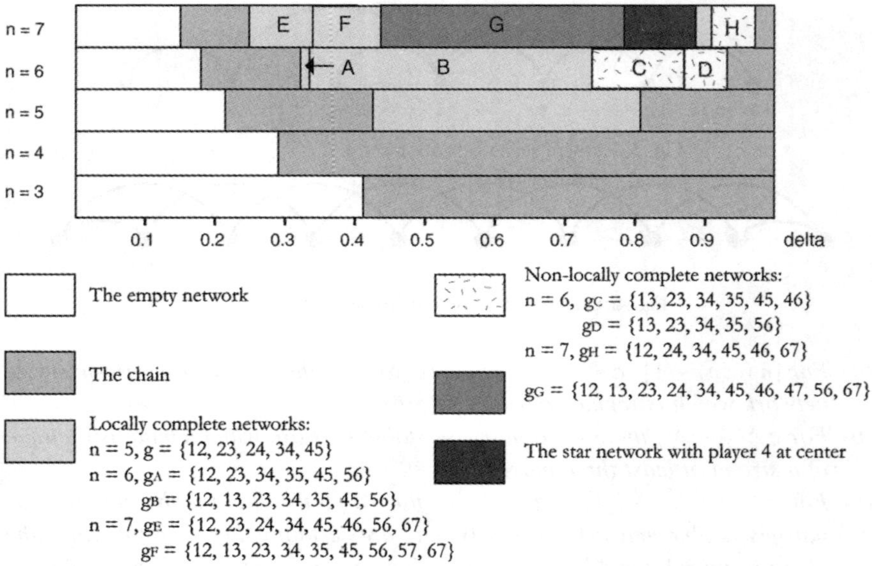

Fig. 5. Typology of the efficient networks for $n \leqq 7$

3.2 Efficiency in the Spatial Connections Model

Recall that a network $g \subset g^N$ is *efficient* if g maximizes the value function $v = \sum_N u_i$ over the set of all potential networks $\{g \mid g \subset g^N\}$, i.e., $v(g) \geqq v(g')$ for all $g' \subset g^N$. We show that efficient networks exist that are not pairwise stable. This is consistent with the insight derived by Jackson and Wolinsky [16] regarding efficient networks.

Our main result shows that for $c > \delta$ any efficient network is either the chain or the empty network. This is mainly due to the fact that the chain is the least expensive connected graph.

Theorem 1. *Let* $0 < \delta < c = \frac{1}{n-1}$.

(a) For $c > \delta + \frac{1}{n-1} \sum_{k=2}^{n-1} (n-k)\delta^k$*, the only efficient network is the empty network.*

(b) For $c < \delta + \frac{1}{n-1} \sum_{k=2}^{n-1} (n-k)\delta^k$*, the only efficient network is the chain.*

For a proof of Theorem 1 we refer to Appendix A.

Next we turn our attention numerical computations of highest valued networks. Even for relatively small numbers of players the number of possible networks can be very large, requiring us to use a computer program to calculate the value of all social networks for each n. We limit our computations to $n \leqq 7$ as the number of possible networks for $n = 8$ exceeds 250 million. Given n, $c = \frac{1}{n-1}$, and δ, Fig. 5 summarizes our results. Figure 6 identifies the ranges of δ for which the social networks are both pairwise stable and efficient. Numerical values corresponding to Figs. 5 and 6 can be found in Appendix A.

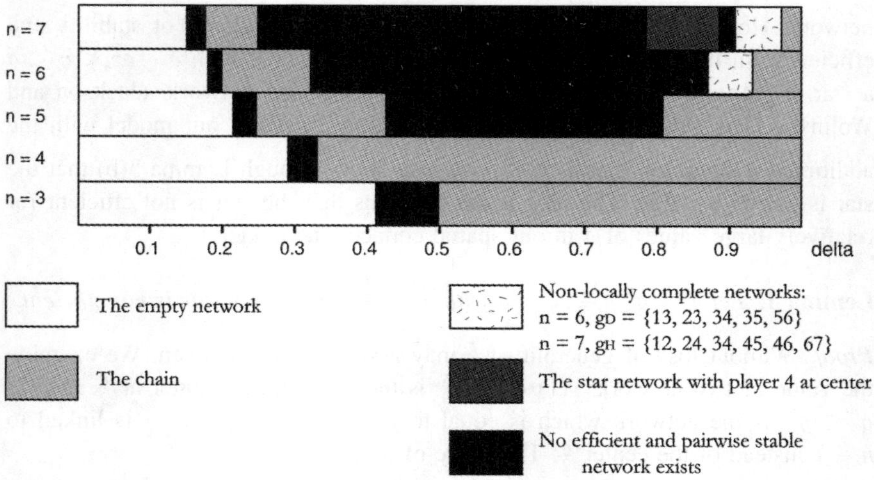

Fig. 6. Efficient and pairwise stable networks for $n \leqq 7$

We highlight some simple observations on pairwise stability and efficiency by comparing Figs. 5 and 6. We focus on one non-locally complete network with 6 players and four networks with 7 players to illustrate some of the conflict and coincidence that occurs between efficiency and pairwise stability.

For $n = 6$ and the range of δ labelled C the non-locally complete network g_C is efficient. The three links $\{34, 35, 45\}$ give this network a locally complete aspect that renders it unstable. The intuition for this instability is found in the proof to Proposition 1.

For $n = 7$, the locally complete network g_E is efficient in range E. This network is described in Example 3.1 and depicted in Fig. 3. Similarly the locally complete network g_F is efficient in range F. Neither network is pairwise stable. The star g_4^s is efficient for $\delta \in [0.7887, 0.8811]$ as well as pairwise stable by Lemma 3(b) found in Appendix A. Finally, for range H, the network g_H is efficient and pairwise stable. This network architecture is discussed in the proof of Lemma 1 below.

We conclude that the empty network is always pairwise stable if it is efficient. For $n \leqq 7$ and $\delta \geqq c = \frac{1}{n-1}$, if the chain is efficient it is also pairwise stable. For relatively high δ, the chain is always efficient because the relative difference between direct and indirect connections is quite small.[5] Proposition 1(c) reminds us that for $n \geqq 7$ and $\delta > \frac{1}{2}$ the chain is never pairwise stable. Finally, for $n \leqq 7$, a locally complete network with a clique of three or more players is never efficient and pairwise stable.

Bala and Goyal [3] demonstrated that for relatively high and low connection costs, pairwise stability and efficiency coincide. We arrive at a different insight. For relatively high cost the empty network is both the unique pairwise stable and efficient network. For relatively low cost we see the chain emerge as an efficient

[5] The chain is efficient for $n = 5$ if $\delta \in [0.215, 0.4287] \cup [0.8129, 1)$; $n = 6$ if $\delta \in [0.1727, 0.3141] \cup [0.9307, 1)$; and $n = 7$ if $\delta \in [0.1465, 0.2467] \cup [0.9695, 1)$.

network. However, Proposition 1(c) rules out any coincidence of stability and efficiency. In the standard non-spatial connections model with $\delta - \delta^2 < c < \delta$ a star is pairwise stable as well as the unique efficient network. (Jackson and Wolinsky [16], Proposition 1(ii) and Proposition 2(iii).) In our model with the additional assumption that $\delta > \frac{\lfloor \frac{n}{2} \rfloor}{n-1}$ we also show through Lemma 3(b) that the star is pairwise stable. The next result confirms that the star is not efficient for relatively large values of δ in our spatial connections model.

Lemma 1. *Let $\delta^2 - \delta^3 < \frac{c}{2}$, $c < \delta$ with $\delta > \frac{\lfloor \frac{n}{2} \rfloor}{n-1}$. Then any star is not efficient.*

Proof. Without loss of generality we may assume that n is even. We examine the value of two networks: (1) $g^s \subset g^N$ is the star with its center at $\frac{n}{2}$ and (2) $g' \subset g^N$ is the network which is equal to g^s except that player n is linked to $n-1$ instead of the center $\frac{n}{2}$. The value of g^s is

$$v\left(g^s\right) = 2(n-1)\delta + (n-1)(n-2)\delta^2 - 4\sum_{k=1}^{\frac{n-1}{2}} kc - 2\binom{n}{2}c. \tag{5}$$

The value of the network g' is

$$v\left(g'\right) = 2(n-1)\delta + (n-2)(n-3)\delta^2 + 2\delta^2 + 2(n-3)\delta^3 - 4\sum_{k=1}^{\frac{n-1}{2}} kc - 2c. \tag{6}$$

The difference between Eq. (5) and Eq. (6) is

$$2(n-3)\delta^2 - 2(n-3)\delta^3 - (n-2)c$$

This is negative when $\delta^2 < \delta^3 + \frac{(n-2)}{2(n-3)}c$. We conclude that the star g^s may not be the network with the highest value.

4 Implementation of Pairwise Stable Networks

Implementation of pairwise stable networks has been explored in the literature for the Jackson-Wolinsky framework with a binary cost topology. Watts [25] explicitly models the connections model of Jackson and Wolinsky [16] as an extensive form game. She bases her analysis on the myopic players playing the Grim Strategy[6] to illustrate the resulting equilibria of such a game. Dutta and Mutuswami [9] look at the relationship between stability and efficiency, but they use a static, strategic form framework. In the spatial connections model we look at a natural extensive form game in which all pairs meet exactly once to form

[6] Watts [25] defines the Grim Strategy as follows: Each player agrees to link with the fist two players he meets. Secondly, each player never severs a link as long as all the other players cooperate. However, if player i deviates, then every player $j \neq i$ severs all ties with i and refuses to form any links with i for the rest of the game. Thus, if player i deviates, his payoff will be 0 in all future periods.

a link. We investigate the subgame perfect Nash equilibria of this game and show that for certain orders in which the pairs meet we can implement specific pairwise stable networks. A full analysis of this game with random order of play is deferred to future research.

Initially, in our game none of the players are connected. Over multiple playing rounds, players make contact with the other players and determine whether to form a link with each other or not. Exactly one pair of players meets each round — or "stage." Each pair of players meets once and only once in the course of the game. The resulting extensive form game is called the *link formation game*.

We remark that our link formation game differs considerably from the one formulated in Aumann and Myerson [2]. There the pairs that did not link in previous stages of the game, meet again to reconsider their decision. The game continues until a stable state has been reached in which no remaining unlinked pairs of players are willing to reconsider. Obviously our structure implies that the "order of play" is crucial, while the Aumann-Myerson structure this is not the case. On the other hand the analysis of our game is more convenient and rather strong results can be derived.

Formally, an "order of play" in the link formation game is represented by a bijection $O : g^N \to \{1, \ldots, c(n,2)\}$ that assigns to every potential pair of players $\{i,j\} \subset N$ a unique index $O_{ij} \in \{1, \ldots, c(n,2)\}$. The set of all orders is denoted by $\mathbb{O}$.

The link formation game has therefore $c(n,2)$ stages. In stage k of the game the pair $\{i,j\} \subset N$ such that $O_{ij} = k$ play a subgame. For any two players, i and j with $i < j$, the choice set facing each player is $A_i(ij) = \{C_{ij}, R_{ij}\}$ and $A_j(ij) = \{C_{ij}, R_{ij}\}$, where C_{ij} represents the offer to establish the link ij and R_{ij} represents the refusal to establish the link ij. Players will form a link when it is mutually agreed upon, i.e., link ij is established if and only if both players i and j select action C_{ij}. No link will be formed if either player refuses to form the link, i.e., when either one of the players i or j selects R_{ij}. Link formation is permanent; no player can sever the links that were formed during earlier stages of the game. The sequence of actions, recorded as the history of the game, determines in a straightforward fashion the resulting network. We emphasize that all players have complete information in this game.

To complete the description of strategies in the link formation game with order of play $O \in \mathbb{O}$ we introduce the notion of a *(feasible) history*. A history is a listing $h \in H(O) := \cup_{k=1}^{c(n,2)} H_k(O)$, where

$$\begin{aligned} H_k(O) &= X_1(O) \times \cdots \times X_k(O) \text{ with for every } 1 \leqq p \leqq k \\ X_p(O) &= A_i(ij) \times A_j(ij) \text{ for } \{i,j\} \subset N \text{ with } O_{ij} = p. \end{aligned}$$

The history $h = (h_1, \ldots, h_k) \in H_k(O)$ is said to have a length of k, where $h_p \in X_p(O)$ for every $1 \leqq p \leqq k$. A history describes all actions undertaken by the players in the link formation game up till a certain moment in that game. The network $g(h) \in g^N$ corresponding to history $h = (h_1, \ldots, h_k) \in H_k(O)$ is defined as the network that has been formed up till stage k of the link formation game with order O, i.e., $ij \in g(h)$ if and only if $O_{ij} \leqq k$ and $x_{O_{ij}} = \left(C_{ij}, C_{ij}\right)$. Now we are able to introduce for each player $i \in N$ the *strategy set*

$$S_i = \prod_{ij \in g^N} \prod_{h \in H_{O_{ij}}(O)} A_i(ij). \tag{7}$$

A strategy for player i assigns to every potential link ij of which i is a member, and every possible history of the link formation game up till stage O_{ij} an action. A strategy tuple in the link formation game is now given by $a \equiv (a_1, \dots, a_n) \in S := \prod_{i \in N} S_i$. With each strategy $a \in A$ we can define the resulting network as $g_a \subset g^N$. Furthermore, player i receives a payoff $u_i(g_a)$ for every strategy tuple $a \in S$.

Formally, for any order of play $O \in \mathbb{O}$ the above describes a game tree $\mathscr{G}_O$. This implies that for order $O \in \mathbb{O}$ the *link formation game* Γ_O may be described by the $(2n+2)$-tuple

$$\Gamma_O = (N, \mathscr{G}_O, S_1, \dots, S_n, u_1, \dots, u_n). \tag{8}$$

Since the link formation game is a well-defined extensive form game, we can use the concept of subgame perfection to analyze the formation of networks. Next we investigate the nature of the subgame perfect Nash equilibria of the link formation game developed above.

Our analysis mainly considers the case that $c < \delta$. As shown in Proposition 1 there is a wide range of non-trivial pairwise stable networks in this situation. It can be shown that there is a set of efficient and pairwise stable networks can be implemented as subgame perfect equilibria of link formation games. First we address the conditions under which regular networks can be implemented as subgame perfect equilibria of the link formation game.

Theorem 2. *Let $m \in \{1, \dots, n-1\}$. Then for (c, δ) satisfying*

$$\frac{1}{m+1}\delta + \left(\frac{n-1}{m+1} + 1\right)\delta^2 < c < \frac{1}{m}\delta - \frac{1}{m}\delta^2 \tag{9}$$

there exists an order of play $O \in \mathbb{O}$ such that the regular network of order m can be supported as a subgame perfect Nash equilibrium of the link formation game with order O.

A proof of this theorem is given in Appendix B.

We remark that the chain is the unique regular network of order one on N. By substituting $m = 1$ into the condition (9) for the implementation of the chain as a subgame perfect Nash equilibrium of a link formation game.

From this main implementation result above we are able to derive some further conclusions. Our first conclusion concerns the support of the complete network as a subgame perfect Nash equilibrium in the link formation game. Such a complete network can be supported for high enough benefits in relation to the link costs:

Corollary 1. *For $(n-1)c < \delta - \delta^2$ and for any order of play $O \in \mathbb{O}$, the complete network g^N can be supported as a subgame perfect Nash equilibrium of the link formation game with order O.*

Proof. The assertion follows from a slight modification of part (1) in the proof of Theorem 2 for $m = n - 1$. (Remark that the complete network on N is the unique regular network of order $n - 1$.) Here the order of the game is irrelevant, thus showing that any order of play leads to the establishment of the strategy $\widehat{a}$ as given in the proof of Theorem 2 as a subgame perfect Nash equilibrium.

Finally we consider under which conditions the identified subgame perfect Nash equilibria generate a pairwise stable network. The following corollary of Proposition 1 and Theorem 2 summarizes some insights:

Corollary 2. *The following properties hold:*

(a) Suppose that $\frac{1}{2}\delta + \frac{n+1}{2}\delta^2 < c = \frac{1}{n-1} < \delta$. Then there exists an order of play $O \in \mathbb{O}$ such that at least one subgame perfect Nash equilibrium of the link formation game with order O is pairwise stable.

(b) Suppose that $\hat{n}(c,\delta) \geqq 3$. If

$$\frac{1}{\hat{n}(c,\delta)+1}\delta + \frac{n+\hat{n}(c,\delta)}{\hat{n}(c,\delta)+1}\delta^2 < c < \frac{1}{\hat{n}(c,\delta)}\delta - \frac{1}{\hat{n}(c,\delta)}\delta^2 \qquad (10)$$

then there exists an order of play $O \in \mathbb{O}$ such that at least one subgame perfect Nash equilibrium of the link formation game with order O is pairwise stable.

Proof. **(a)** First we remark that $\delta > \frac{1}{n-1} > \frac{1}{n+3}$ implies that

$$c > \frac{1}{2}\delta + \frac{n+1}{2}\delta^2 > \delta - \delta^2. \qquad (11)$$

Now condition (11) implies that Lemma 3(a) holds. Hence, the chain is pairwise stable. From (11) it follows that Theorem 2 holds, implying that there is an order of play O such that the chain can be supported as a SPNE of that link formation game.

(b) First we remark that from (10) it follows immediately that $[\hat{n}(c,\delta)-1]\cdot c < \hat{n}(c,\delta)\cdot c < \delta - \delta^2$, and so Proposition 1(a) is satisfied. Hence, the regular network of order $\hat{n}(c,\delta)-1$ is pairwise stable. Furthermore, from (10) it follows through Theorem 2 that the regular network of order $\hat{n}(c,\delta)-1$ can be supported as a subgame perfect Nash equilibrium for some order of play $O \in \mathbb{O}$ in the link formation game.

We use an example to illustrate the tension between the order of play, efficiency and pairwise stability when $c < \delta$.

Example 4. *Consider the case where* $n = 5$, $c = \frac{1}{n-1} = \frac{1}{4}$, *and* $\delta = \frac{5}{8}$. *The star* $g_3^s = \{13, 23, 34, 35\}$ *is pairwise stable but not efficient. The chain,* $g^c = \{12, 23, 34, 45\}$, *is also pairwise stable and has a higher total value than the star. The locally complete network* $g^l = \{12, 23, 24, 34, 45\}$ *is efficient but not pairwise stable; maintaining link 24 decreases utility for both player 2 and player 4.*

g	$u_1(g) = u_5(g)$	$u_2(g) = u_4(g)$	$u_3(g)$	$v(g) = \sum_i u_i(g)$
g_3^s	$\delta + 3\delta^2 - 2c$	$\delta + 3\delta^2 - c$	$4\delta - 6c$	$8\delta + 12\delta^2 - 12c$
g^c	$\delta + \delta^2 + \delta^3 + \delta^4 - c$	$2\delta + \delta^2 + \delta^3 - 2c$	$2\delta + 2\delta^2 - 2c$	$8\delta + 6\delta^2 + 4\delta^3 + 2\delta^4 - 8c$
g^l	$\delta + 2\delta^2 + \delta^3 - c$	$3\delta + \delta^2 - 4c$	$2\delta + 2\delta^2 - 2c$	$10\delta + 8\delta^2 + 2\delta^3 - 12c$

Player 3 prefers the chain or the locally complete network over the star; all other players prefer the star to the chain. Players 2 and 4 also prefer the star over the locally complete network. Depending on the order of play, we can generate the star or the chain; yet never both from the same ordering. For the star to form, we must allow pairs $\{12, 45\}$ *to refuse to connect before player 3 has an opportunity to refuse any connection to the furthest star points. The order of play* $\{\mathbf{12}, \mathbf{45}, 23, 34, \mathbf{15}, \mathbf{14}, \mathbf{25}, \mathbf{24}, 13, 35\}$ *guarantees that the star with the center at 3 forms. The pairs bold-faced in the ordering will not form a link because both players will refuse to make the connection to guarantee the that the network that each of them prefers to form will indeed form. For the chain to form, we must allow player 3 to refuse the links* $\{13, 35\}$ *before other players have the opportunity to refuse the links* $\{12, 45\}$. *An ordering that would result in the chain is* $\{\mathbf{13}, \mathbf{35}, 23, 34, \mathbf{15}, \mathbf{14}, \mathbf{25}, \mathbf{24}, 12, 45\}$.
If there was a strategy available to encourage players 2 and 4 into enduring the link 24, the players could create the efficient locally complete graph $g^l = \{12, 23, 24, 34, 45\}$. *This is because there are two pairwise stable graphs with one link lower than* g^l*: the chain and the non-locally complete graph* $\{12, 24, 34, 45\}$. ♦

Next we turn to an example for the case $c > \delta$.

Example 5. *Consider* $n = 5$, $c = \frac{1}{n-1} = \frac{1}{4}$, *and* $\delta = 0.22$. *As shown in Theorem 1 the chain is the unique efficient network. However, it is not pairwise stable for these parameter values. In the link formation game the chain can be generated by an order similar to that of the order described in proof of Theorem 2. The difference is that the pairs ij where* $n(ij) = 2$ *must meet in a specific order: the players located at the end points must meet their direct neighbors first, before any interior pairs meet. For example, the order of play* $\{12, 45, 23, 34, \mathbf{13}, \mathbf{24}, \mathbf{35}, \mathbf{14}, \mathbf{25}, \mathbf{15}\}$ *guarantees that the chain forms. The first four pairs are ordered so the pairs located nearest to the endpoints have the option of linking. By backward induction, players 2 and 4 realize that if they do not offer to connect to players 1 and 5 respectively, they will not form an attractive potential link for player 3.* ♦

The two examples above capture how for a given order of play players can strategically influence the creation of a network. We continue with showing that

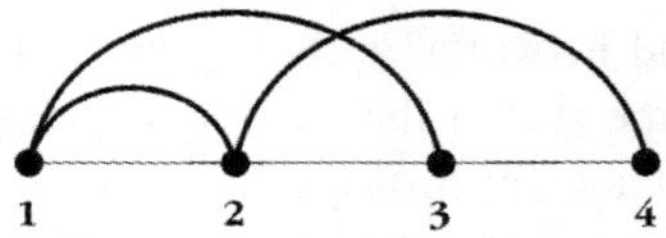

Fig. 7. Generated network for $n = 4$, $c = \frac{1}{3}$, $\delta = \frac{7}{10}$

for a given order we find an outcome where players create a network that is neither efficient nor pairwise stable.

Example 6. *Given* $n = 4$, $c = \frac{1}{n-1}$ *and* $\delta = 0.7$. *The order* $\{\mathbf{34}, \mathbf{23}, 12, 13, 24, \mathbf{14}\}$ *generates the network* $g = \{12, 13, 24\}$ *which is neither efficient nor pairwise stable. (See Fig. 7.) Both the chain and non-locally complete graph* $\{12, 13, 34\}$ *are Pareto Superior to* g. *Furthermore, given the opportunity, players 3 and 4 would benefit from forming a link. This order creates this graph because players use their linking strategies as votes against the graphs that they earn the least in. The central players, 2 and 3, will refuse link* 23 *so as to not become the center of the star. The players located at the end points, 1 and 4, refuse link* 14 *so as to veto the graph* $\{12, 14, 34\}$ *in which they have very little positive utility. Player 3 refuses the first link merely to flip the resulting network architecture forcing player 2 to maintain two links in the network.* ♦

We observe that specific structures can emerge from specific orders for certain parameter values. Our results differ from other models of sequential network formation. With myopic players as implemented by Watts [25] sequential play would result in a pairwise stable network; one would not obtain an efficient network as in Example 4. In addition, Aumann and Myerson [2] introduce a sequential game with foresight, but allow unlinked players a last chance to form a link. This would eliminate the possibility of a network as in Example 4 to form. These results suggest that future research should investigate how to model network formation.

5 Concluding Remarks

In this paper we introduced a spatial cost topology in a specific network formation model analyzed by Jackson and Wolinsky [16]. There are four main assumptions that determine our results: (i) links are undirected in the sense that both players incur the cost of maintaining the link, (ii) we apply a linear cost topology, (iii) we assume a polynomial benefit function, and (iv) the benefit function is founded on a uniform, constant δ. Extensions of this model can address changes to any of these four hypotheses.

With regard to (i), Bala and Goyal [3] and [4] develop a non-cooperative model of social communication where links are directional. Furthermore, concerning assumption (iii), Bala and Goyal [3] introduce the possibility that links are not fully reliable, thereby changing the benefit function. Adding both of these changes to our model may make some types of spatial social networks unstable.

Since Bala and Goyal find networks in their setup to be 'super-connected', we would not expect to see the star or the chain emerge as a stable network when links that are not fully reliable, even if the links are undirected.

As for assumption (ii) we mention replacing the linear cost topology with arbitrary cost functions would unnecessarily complicate the analysis and most likely not lead to richer insights.

Benefits could also be generated by playing games on the network. The benefit from forming a direct link would then be the payoff from the game played by two linked players. Droste, Gilles and Johnson [8] explore such a model. They use an evolutionary framework inspired by Jackson and Watts [15] and Ellison [10]. Their agents are spatially located on a circle and play a coordination game in an endogenously formed network. In this framework there are no benefits from indirect connections. Their main result is that the emerging network is always locally complete. One could also include indirect benefits in the payoff structure of the game. We hypothesize in such a setting there emerge other pairwise stable networks.

Appendices

A Proofs From Sect. 3

We first show some intermediate results.

Lemma 2. *For $\delta > c > 0$, every pairwise stable network is connected.*

Proof. Assume there exists a pairwise stable network $g \subset g^N$ that is not connected. Since the network is not connected there will be two direct neighbors, i and j, with the following characteristics: player i is in one connected component of g and player j is in another connected component of g. Since i and j are direct neighbors, the cost to i and j to connect is equal to c. The benefit of connection to each will always be at least the direct benefit of δ. Therefore, both i and j will always want to form a connection since $\delta > c$.

Lemma 3. *Let $\delta > c > 0$ and $c = \frac{1}{n-1}$.*

(a) For $c > \delta - \delta^2$ and $\delta < \frac{1}{2}$, the chain is pairwise stable.
(b) For $c > \delta - \delta^2$ and $\delta > \frac{\lfloor \frac{n}{2} \rfloor}{n-1}$, there exists a star which is pairwise stable.

Proof. Let $g \subset g^N$ be pairwise stable.

(a) Suppose g is the chain on N. The net benefit to any player severing a link with their nearest neighbor would be at most $c - \delta < 0$. Therefore no player will sever a link.
A player $i \in N$ will connect to a player j with $\ell(ij) = 2$ only if $2c \leqq \delta - \delta^{\bar{n}}$. Because $\delta^2 < \frac{\delta}{2}$ and $\delta > c > \delta - \delta^2$, we know $2c > \delta \geqq \delta - \delta^{\bar{n}}$. Thus, player i will not make such a connection.

Next consider j with $\ell(ij) \geqq 3$. Player i will make a link with j if the net benefit of such a connection is positive. Let $\ell(ij) = k$. For k odd, the net benefit for player i connecting to player j is

$$\delta + 2\sum_{l=2}^{\frac{k-1}{2}} \delta^l + \delta^{\frac{k-1}{2}+1} \quad - \sum_{m=\tilde{n}-(k-2)}^{\tilde{n}} \delta^m - kc.$$

For k even, the net benefit for player i connecting to player j is

$$\delta + 2\sum_{l=2}^{\frac{k}{2}} \delta^l \quad - \sum_{m=\tilde{n}-(k-2)}^{\tilde{n}} \delta^m - kc.$$

We proceed with a proof by induction with regard to the parameter $k \geqq 3$. When $k = 3$, the net benefit expression above simplifies to $\delta + \delta^2 - \delta^{\tilde{n}} - \delta^{\tilde{n}-1} - 3c$. If $3c + \delta^{\tilde{n}} + \delta^{\tilde{n}-1} < \delta + \delta^2$, player i would consider making a link with player j. This expression is never true for $\delta < \frac{1}{2}$ and $c > \delta - \delta^2$. For higher values of k the positive elements of the net benefit value increase by less than δ^2 and the negative elements increase by c. As $c > \delta^2$, the net benefit function decreases with respect to k. Thus for any $k \geqq 3$, player i will not consider creating a link with player j.

Thus, we have shown that no player will sever or add a link when g is chain on N and, therefore, the chain is pairwise stable.

(b) Let g be a star on N with the central player located at $\lfloor \frac{n}{2} \rfloor$. Refer to all players except the center as "points." The benefit of maintaining a connection to the center for all points is $\delta + (n-2)\delta^2$. The maximal cost of any connection in this star is $\lfloor \frac{n}{2} \rfloor \cdot c = \frac{\lfloor \frac{n}{2} \rfloor}{n-1} < \delta$. Thus, no player will sever a connection, not even the center. The net benefit of adding an additional connection for a player is $\delta - \delta^2 < c$. Thus, the star with the central player located at $\lfloor \frac{n}{2} \rfloor$ is pairwise stable.

This completes the proof of Lemma 3

A.1 Proof of Proposition 1

Let $g \subset g^N$ be pairwise stable.

(a) Consider $g \subset g^N$ on N to be regular of order $\hat{n}(c, \delta) - 1$. Then the maximal net benefit of severing a link $ij \in g$ within a clique in g would be $c_{ij} + \delta^2$. Since $c_{ij} \leqq [\hat{n}(c, \delta) - 1] \cdot c < \delta - \delta^2$, it holds that $\delta > c_{ij} + \delta^2$ and, so, no player would be willing to sever a link. An additional link would form if $c_{ij} \leqq \delta - \delta^{\tilde{n}}$, where $\delta^{\tilde{n}}$ represents the value of an indirect connection lost due to a shorter path being created when a new link is created in a connected network. Since by Lemma 2 the network is connected, if a player were to add a link, his net benefit would be composed of three parts: The benefit of the new link and possibly higher indirect connections, the loss of indirect connections replaced by a shorter path created by the new link, and the cost of maintaining the link. We let $\delta^{\tilde{n}}$ represent the value of an indirect connection lost due to a shorter path being created when a new link is created. If more than one indirect connection is replaced by a shorter path, we use the convention of ranking the benefits $\delta^{\tilde{n}}$ by decreasing $\tilde{n}$. We know that $c_{ij} \geqq \hat{n}(c,\delta) \cdot c > [\hat{n}(c,\delta) - 1] \cdot c$ because the location for any player that i could form an additional link with would lie beyond the maximal clique. Using the definition of $\hat{n}(c,\delta)$, we know that $c_{ij} \geqq \delta$. Therefore no player will try to form an additional link outside the maximal clique. Hence, g is pairwise stable.

(b) Suppose $g^{i \leftrightarrow j} \subset g \subset g^N$ with $\ell(ij) \geqq 2$. If player i severs one of his links to a player within the clique $i \leftrightarrow j$, the resulting benefits from replacing a direct with an indirect connection are $\delta^2 + c - \delta > 0$. Therefore, player i will sever one of his connections. This shows that networks with a cliques of at least 3 members are not pairwise stable, thus showing the assertion.

(c) From assertion (b) shown above, it follows that any pairwise stable network $g \subset g^N$ does not contain a clique of at least three players. This implies that the chain is the only regular pairwise stable network to be investigated. Let g be the chain on N. First note that since $c < \delta$ no player has an incentive to sever a link in g. We will discuss three subcases, $n \geqq 7$, $n = 6$, and $n \leqq 5$.

(1) Assume $n \geqq 7$. Select two players i and j, $i < j$, who are neither located at the end locations of the network nor direct neighbors. Also assume that $\ell(ij) = 3$. If i were to connect to a player j the minimum net benefit of such a connection to either i or j would be $\delta + \delta^2 - \delta^3 - \delta^4 - 3c$. The maximal cost of connection c_{ij} when $\ell(ij) = 3$ is $\frac{1}{2}$ since $c = \frac{1}{n-1} \leqq \frac{1}{6}$. Since $\delta > \frac{1}{2}$, the minimum benefit, $\delta + \delta^2 - \delta^3 - \delta^4$, of such a connection is greater than the maximal cost. Thus, the additional connection will be made.[7] Also, note that player i is not connected to j's neighbor to the left. This player has essentially been skipped over by player i. Nor does player i have any incentive to form a link with the player that was skipped over. A connection to this player would cost $2c$, and the benefit would only be $\delta - \delta^2$. Thus, the chain is not pairwise stable.

(2) Assume $n = 6$. From assertion (b) shown above, we need only to examine two situations of link addition for two players i and j: a) $\ell(ij) = 3$ and $1 \neq i \neq n$, and b) $\ell(ij) \geqq 3$, $i = 1$ or $i = n$.

a) Select two players i and j with i not located at the end of the network, i.e., $1 \neq i \neq n$, and $\ell(ij) = 3$. If i were to connect to a player j the cost of such a connection would be $3c = \frac{3}{5}$ and the net benefit of this connection would be $\delta + \delta^2 - \delta^3 - \delta^4$. Because $c > \delta - \delta^2$, and $\delta > \frac{1}{2}$, we know that

[7] Because $\frac{1}{6} \geqq c > \delta - \delta^2$, and $\delta > \frac{1}{2}$, we know that $\delta + \delta^2 - \delta^3 - \delta^4$ has a minimum value of $\left(\frac{1}{2} + \frac{1}{6}\sqrt{3}\right) + \left(\frac{1}{2} + \frac{1}{6}\sqrt{3}\right)^2 - \left(\frac{1}{2} + \frac{1}{6}\sqrt{3}\right)^3 - \left(\frac{1}{2} + \frac{1}{6}\sqrt{3}\right)^4$ which is approximately equal to 0.53. Here we note that this minimum is attained in a corner solution determined by the constrained $\delta - \delta^2 < c$.

$\delta + \delta^2 - \delta^3 - \delta^4$ has a minimum value which is less than $\frac{3}{5}$.[8]
b) Select two players i and j with i, $\ell(ij) \geqq 3$, $i = 1$ or $i = n$. If player j were to connect to player i the minimum cost of a connection would be $3c$ or $\frac{3}{5}$. The minimal net benefit of such connection would be $\delta - \delta^{\tilde{n}}$ where $\tilde{n} \in \{3, 4, 5\}$. Since $c > \delta - \delta^2$, we know that $3c > \delta - \delta^{\tilde{n}}$. We can conclude that a link to an end agent will never be stable from such a distance.
We thus conclude that for δ such that $\delta + \delta^2 - \delta^3 - \delta^4 \leqq \frac{3}{5}$ the chain is pairwise stable and for some values of δ a non-locally complete network is stable.
(3) Assume $n \leqq 5$. Select three players i, j and k, where $i < j < k$ and j, k with $\ell(ij) \geqq 2$. We know $ij \notin g$ and $ik \notin g$. Suppose that $i = 1$. If player j were to make a new connection with player i, the maximum net benefit of such a connection to player j would be $\delta - \delta^2 - 2c < 0$. For player k we have that $\ell(ik) \geqq 3$, so, the net benefit of such a connection for player k would be at most $\delta - \delta^3 - 3c < 0$. If the player at the opposite end of the network linked with player i the net benefit would always be negative.[9]
We conclude that no player would decide to connect with a player at either end points of the chain. From this it can easily be concluded that a similar argument can be applied to the other players for the case $n = 5$. (Note that the cases $n \leqq 4$ are trivially excluded.) Therefore, no player will form an additional link, and we conclude that the chain is pairwise stable.

This completes the proof of Proposition 1

A.2 Proof of Proposition 2

In this proof we first introduce some auxiliary notions. We define a path $\{i_1, \ldots, i_m\} \subset N(g)$ in the network $g \subset g^N$ to be *terminal* if $\#\{i_m j \in g \mid j \in N(g)\} = 1$ and for every $k = 2, \ldots, m-1$ it holds that $\#\{i_k j \in g \mid j \in N(g)\} = 2$. We also say that player i_1 *anchors* this terminal path.

(a) Let $\emptyset \subset g^N$ represent the empty network on N. For any two players the cost of connecting is at least c and the benefit of connection to each is equal toδ. Since $\delta < c$, no two players would like to add a link. So, the empty network $\emptyset$ is pairwise stable.
We now consider a network $g \subset g^N$ that is non-empty, pairwise stable, as well as acyclic. Hence, in the network $g \subset g^N$ there is at least one player $i \in N(g) \neq \emptyset$ such that $\#\{ij \in g \mid j \in N(g) \setminus \{i\}\} = 1$. Clearly since $\delta < c$, player $j \neq i$ with $ij \in g$ is better off by severing the link with i. Thus, g

[8] Because $\frac{1}{5} \geqq c > \delta - \delta^2$, and $\delta > \frac{1}{2}$, we know that the polynomial $\delta + \delta^2 - \delta^3 - \delta^4$ has a minimum value given by $\left(\frac{1}{2} + \frac{1}{10}\sqrt{5}\right) + \left(\frac{1}{2} + \frac{1}{10}\sqrt{5}\right)^2 - \left(\frac{1}{2} + \frac{1}{10}\sqrt{5}\right)^3 - \left(\frac{1}{2} + \frac{1}{10}\sqrt{5}\right)^4$ which is approximately equal to 0.594. Again this minimum is determined by the constraint $\delta - \delta^2 < c$.

[9] For $n = 3$, $\delta - \delta^2 - 2c < 0$. For $n = 4$, $\delta - \delta^3 - 3c < 0$. For $n = 5$, $\delta + \delta^2 - \delta^3 - \delta^4 - 4c < 0$. ($\delta + \delta^2 - \delta^3 - \delta^4$ is maximized at $\delta = \frac{1}{8} + \frac{1}{8}\sqrt{17}$ at a value of approximately 0.62 and $4c = 1$).

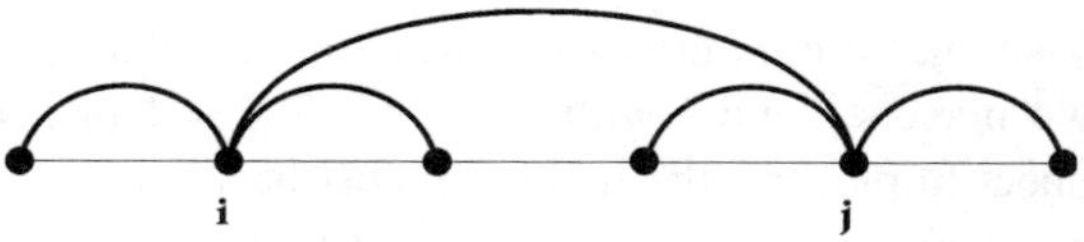

Fig. 8. Case for $\ell(ij) = 3$

cannot be pairwise stable. Therefore we conclude that any acyclic pairwise stable network has to be empty.

(b) It is obvious that both the empty network and the chain on N are pairwise stable given that $\delta = c$. Next let $g \subset g^N$ be pairwise stable, non-empty, as well as acyclic. We first show that g is connected.
Suppose to the contrary that g is not connected. Then there will be two direct neighbors i and j with: player i is in a non-empty connected component of g of size at least 2 and player j is in another connected component of g. (Here we remark that $\{j\}$ is a trivially connected component of any network in which j is not connected to any other individual.) Since i and j are direct neighbors, the cost to i and j to connect is c. The net benefit for i of making a connection to j is then at least $\delta - c = 0$. The net benefit for j for making a connection to i is at least $\delta + \delta^2 - c = \delta^2 > 0$. Therefore, g is not pairwise stable. This contradicts our hypothesis and therefore g has to be connected.
Next we show that g is the chain. Suppose to the contrary that g is not the chain. From the assumptions it can easily be derived that there exists a player $i \in N$ with $\#\{ij \in g \mid j \in N(g)\} \geqq 3$.
First, we show that there is no player $j \in N$ with $ij \in g$, $\ell(ij) \geqq 2$, and the link ij is the initial link in a terminal path in g that is anchored by player i. Suppose to the contrary that such a player j exists and that the length of this terminal path is m. Then the net benefit for player i to sever ij is at least

$$2c - \sum_{k=1}^{m} \delta^k = 2\delta - \frac{\delta - \delta^{m+1}}{1-\delta} = \frac{\delta - 2\delta^2 + \delta^{m+1}}{1-\delta} > \delta \frac{1-2\delta}{1-\delta} \geqq 0$$

since $\delta = c = \frac{1}{n-1} \leqq \frac{1}{2}$. Thus, we conclude that player i is better off by severing the link ij. Hence, there is no player $j \in N$ with $ij \in g$, $\ell(ij) \geqq 2$, and the link ij is the initial link in a terminal path in g that is anchored by player i.
From this property it follows that the *only* case not covered is that $n \geqq 6$ and there exists a player j with $ij \in g$, $\ell(ij) \geqq 3$, $\#\{jh \in g \mid h \in N(g)\} = 3$, and that the two other links at j have length 1 that are connected to terminal paths, respectively of length m_1 and m_2. (The smallest network satisfying this case is depicted in Fig. 8 and is the situation with $n = 6$ and $\ell(ij) = 3$.)
The maximal net benefit of agent i to sever ij is

$$\begin{aligned} 2c - \delta - \sum_{k=2}^{m_1} \delta^k - \sum_{k=2}^{m_2} \delta^k &= \delta - \frac{\delta^2 - \delta^{m_1+1}}{1-\delta} - \frac{\delta^2 - \delta^{m_2+1}}{1-\delta} \\ &= \delta \frac{1 - 3\delta + \delta^{m_1+1} + \delta^{m_2+1}}{1-\delta} \end{aligned}$$

$$> \delta \frac{1-3\delta}{1-\delta}$$

Since $n \geqq 6$ it follows immediately that $\delta \leqq \frac{1}{5}$, and thus the term above is positive. This shows that g cannot be pairwise stable. Thus, every non-empty acyclic pairwise stable network has to be the chain.

This completes the proof of Proposition 2.

A.3 Proof of Theorem 1

(a) We partition the collection of all potential networks $\{g \mid g \subset g^N\}$ into four relevant classes: (a) $\emptyset \subset g^N$ the empty network, (b) $g^c \subset g^N$ the chain , (c) all acyclic networks, and (d) any network with a clique of at least three players. For each of these four classes we consider the value of the networks in that subset: The value of $\emptyset$ is zero. $v(g^c) = 2\sum_{k=2}^{n-1}(n-k)\delta^k - 2(n-1)c < 0$ from the condition on c and δ.

We partition acyclic networks into two groups: (i) all partial networks of the chain and (ii) all other acyclic networks.

(i) Take $\emptyset \neq g \subset g^c \subset g^N$ with $g \neq g^c$. Then g is not connected and there exists $ij \in g$ with $n(ij) = 2$. Since $c > \delta + \frac{1}{n-1}\sum_{k=2}^{n-1}(n-k)\delta^k$ deleting ij increases the total value of the network. By repeated application we conclude that $v(g) < 0$.

(ii) Assume $g \neq g^c$ is acyclic and not a subset of the chain. We define with g the partial chain $\emptyset \neq g^p \subseteq g^c \subset g^N$ given by $ij \in g$ if and only if $i \leftrightarrow j \in N(g^p)$. There are two situations: (A) the total cost of g is identical to the total cost of the corresponding g^p, (B) The total cost of g is higher than the total cost of the corresponding g^p.

Situation (A) could only occur if there is a player k with $ij \in g$ and $k \in i \leftrightarrow j$. Now $v(g^p) > v(g)$ due to more direct and possibly indirect connections.

Next consider situation (B). Assume g has one link ij with $n(ij) \geqq 3$. The cost of g is at least $2c$ higher than the cost of g^p. The gross benefit of g is at most $2\delta^2$ higher than that of of g^p.[10]

Next consider $g \subset g^N$ with $K \geqq 2$ links where $n(ij) \geqq 3$. As compared to g^p, the value of g is decreased at least by $K \cdot 2c$. The maximum gross benefit of g is thus at most $2K\delta^2$ higher than the corresponding g^p.[11] In either subcase as $c > \delta > \delta^2$ we conclude that $v(g) < v(g^p)$.

Finally we consider $g \subset g^N$ containing $ij \in g$ with $n(ij) = 3$. We can quickly rule out any network with a clique greater than 3 as a candidate for higher utility than the chain. (Indeed, given the conditions for c and δ, the sum, of any extra benefits generated by forming a longer link on the chain could not compensate for the minimum additional cost of $2c$.) Next, we examine

---

[10] This value would be lessened by at least $-2\delta^{n-1}$ if g was connected.

[11] This value would be lessened by at least $-\sum_{m=1}^{k}\delta^{(n-m)}$ if g was connected.

the possibility of a cycle having a higher value than the chain. Two links of length 2 must be present to have a cycle other than a trivial cycle of a neighborhood of three players or a clique of 3 players. These two links would cost at least $6c$ more than the total cost of the chain. A cycle that is nowhere locally complete has a gross maximal value of $2n\sum_{k=1}^{\frac{n}{2}-1}\delta^k + n\delta^{\frac{n}{2}}$. Recall that the chain has a gross value of $2\sum_{k=1}^{n-1}(n-k)\,\delta^k$. The gross value of the cycle exceeds that of the chain by $\frac{-n\delta^{\frac{1}{2}n}+n\delta^{\frac{1}{2}n+2}-2\delta^{n+1}+2\delta}{(\delta-1)^2} < 6c$. Thus, g is not efficient.

(b) The value of the chain network is $2\sum_{k=1}^{n-1}(n-k)\,\delta^k - 2(n-1)c$. For any value of n, given the condition $c < \delta + \frac{1}{n-1}\sum_{k=2}^{n-1}(n-k)\delta^k$, $v(g^c) > 0$ and $v(g^c) > v(g)$ for every $g \subsetneq g^c$. We refer to the preceding proof to verify that the value of any other network formation is less than the chain for $c > \delta$. The chain is the efficient network formation.

This completes the proof of Theorem 1.

A.4 Calculations for Example 3

We investigate for which values of k and (δ, c) with $0 < \delta < c < 1$ the described cyclic network g_c is pairwise stable. It is clear that there is only one condition to be considered, namely whether the severance of one of the links of length 2 in g_c is beneficial for one of the players. The net benefit of severing a link of length 2 is

$$\Delta = 2c - \sum_{m=1}^{k-1}\delta^m + \sum_{m=k+1}^{n-1}\delta^m = 2c - \frac{\delta - \delta^k - \delta^{k+1} + \delta^n}{1-\delta} \tag{12}$$

We analyze when $\Delta \leqq 0$. Remark that $\delta - \delta^k - \delta^{k+1} + \delta^n > \delta\left(1 - 2\delta^k\right)$. Now we consider values of (δ, c) such that

$$\delta\frac{1-2\delta^k}{1-\delta} = 2c > 2\delta \tag{13}$$

We note that for high enough values of k condition (13) is indeed feasible. As an example we consider $k = 5$ and $\delta = \sqrt[k-1]{\frac{1}{k}} = \sqrt[4]{\frac{1}{5}} \sim 0.66874$. Then

$$\frac{1-2\delta^k}{1-\delta} = \frac{1-2\left(\sqrt[4]{\frac{1}{5}}\right)^5}{1-\sqrt[4]{\frac{1}{5}}} \sim 2.211$$

and we conclude that condition (13) is indeed satisfied for

$$c = \delta\frac{1-2\delta^k}{2-2\delta} = \sqrt[4]{\frac{1}{5}}\,\frac{1-2\left(\sqrt[4]{\frac{1}{5}}\right)^5}{2-2\sqrt[4]{\frac{1}{5}}} \sim 0.739$$

For further details we refer to Example 3.1 and Fig. 4.

A.5 Numerical Values of δ for Figs. 5 and 6

Fig. 5

n = 7 [0, 0.1464], [0.1465, 0.2467], [0.2468, 0.3480], [0.3481, 0.4299], [0.4300, 0.7886], [0.7887, 0.8811], [0.8812, 0.9030], [0.9031, 0.9694], [0.9695, 1)

n = 6 [0, 0.1726], [0.1727, 0.3141], [0.3142, 0.3375], [0.3376, 0.7236], [0.7237, 0.8788], [0.8789, 0.9306], [0.9307, 1)

n = 5 [0, 0.2149], [0.2150, 0.4287], [0.4288, 0.8128], [0.8129, 1)
n = 4 [0, 0.2799], [0.2800, 1)
n = 3 [0, 0.4142], [0.4143, 1)

Fig. 6

n = 7 [0, 0.1464], [0.1465, 0.1666], [0.1667, 0.2467], [0.2468, 0.3480], [0.3481, 0.4299], [0.4300, 0.7886], [0.7887, 0.8811], [0.8812, 0.9030], [0.9031, 0.9694], [0.9695, 1)

n = 6 [0, 0.1726], [0.1727, 0.1999], [0.2, 0.3141], [0.3142, 0.3375], [.3376, 0.7236], [0.7237, 0.8788], [0.8789, 0.9306], [0.9307, 1)

n = 5 [0, 0.2149], [0.2150, 0.2499], [0.2500, 0.4287], [0.4288, 0.8128], [0.8129, 1)
n = 4 [0, 0.2799], [0.2800, 0.3333], [0.3334, 1)
n = 3 [0, 0.4142], [0.4143, 0.4999], [0.5000, 1);

B Proofs From Sect. 4

B.1 Proof of Theorem 2

Let $m \in \{1, \ldots, n-1\}$. First we remark that (9) stated in Theorem 2 is indeed a feasible condition on the parameters c and δ. Namely, this holds for low enough values of δ; to be exact $\delta < (m(n+m)+m+1)^{-1}$.
Now we partition the set of potential links g^N into $n-m$ subsets $\{G_0, G_{m+1}, \ldots, G_n\}$ where we define

$$\begin{aligned} G_0 &= \left\{ij \in g^N \mid n(ij) \leqq m+1\right\} \\ G_k &= \left\{ij \in g^N \mid n(ij) = k\right\} \text{ where } k \in \{m+2, \ldots, n\} \end{aligned}$$

We now consider the order $\widetilde{O} := \left(\widetilde{G}_0, \widetilde{G}_n, \widetilde{G}_{n-1}, \ldots, \widetilde{G}_{m+2}\right) \in \mathbb{O}$, where $\widetilde{G}_k$ is an enumeration of G_k, $k = 0, m+2, \ldots, n$. We now show that the regular network of order m is a subgame perfect Nash equilibrium of the link formation game corresponding to the order $\widetilde{O}$. For that purpose we apply backward induction to this link formation game.

We define the strategy tuple $\widehat{a}$ by $\widehat{a}_i\,(ij,h) = C_{ij}$ $\left(\text{where } h \in H\left(\widetilde{O}\right)\right)$ if and only if $ij \in G_0$.[12] From this definition it is clear that the resulting network $g_{\widehat{a}}$ is the unique network on N that is regular of order m. We proceed to show that the strategy described by $\widehat{a}$ is indeed a best response to any history in the link formation game, following the backward induction method.

(1) When any player i is paired with a player j where $ij \in G_0$, i.e., $n\,(ij) \leqq m+1$, both players will choose to make a connection because those connections will always have a positive net benefit because a lower bound for the net benefit of such a link is given by $\delta - \delta^2 - \left[n\,(ij) - 1\right] \cdot c \geqq \delta - \delta^2 - m \cdot c > 0$ from the right-hand side of condition (9). This is independent of the number of links made in the previous or later stages of the game. Hence, we conclude that if $n\,(ij) \leqq m + 1$, the history in the link formation game with order $\widetilde{O}$ does not affect the willingness to make the connection ij.

Next, we proceed by checking the remaining pairs:

(2) Let $k \in \{m+1, \ldots, n-1\}$ and $i,j \in N$ with $i < j$ be such that $n\,(ij) = k+1 \geqq m+2$ and let $h \in H_{\widetilde{O}_{ij}}\left(\widetilde{O}\right)$ be an arbitrary history of the link formation game up till stage $\widetilde{O}_{ij}$. Then given the backward induction hypothesis that in later stages no links will be formed, the network $g\,(h)$ only consists of links of length less than $m+1$ and links of lengths k and higher. This implies that player j can be connected to at most $2m$ players with links of length m or less and to at most with $(n - k + 1)$ players with links of length k and higher. So, an upper bound for the net benefits $U_i\,(ij)$ for player i of creating a direct link with player j can be constructed to be

$$\begin{aligned} U_i\,(ij) &\leqq \delta + (n - k + 2m + 1)\,\delta^2 - kc \\ &\leqq \delta + (n+m)\,\delta^2 - (m+1)c \\ &< \delta + (n+m)\,\delta^2 - (m+1)\left(\frac{1}{m+1}\delta + \left(\frac{n-1}{m+1} + 1\right)\delta^2\right) \\ &= 0 \end{aligned}$$

We conclude that player i will not have any incentives to create a link with player j in the link formation game with order $\widetilde{O}$.

Thus we conclude from (1) and (2) above that the strategy $\widehat{a}$ is indeed a subgame perfect Nash equilibrium of the link formation game with order $\widetilde{O}$. This shows that the regular network of order m can be supported as such for the parameter values described in the assertion.

[12] Hence, this strategy prescribes that all links are formed in the first $|G_0|$ stages of the game corresponding to all pairs in G_0. Furthermore, irrespective of the history in the link formation game up till that moment there are no links formed in the final $C\,(n,2) - |G_0|$ stages of the link formation game corresponding to the pairs in $G_{m+1}, \ldots, G_n$. Obviously the outcome of this strategy is that $ij \in g_{\widehat{a}}$ if and only if $n\,(ij) \leqq m$.

References

[1] Akerlof, G. (1997) Social distance and social decisions. *Econometrica* 65: 1005–1027
[2] Aumann, R.J., Myerson, R.B. (1988) Endogenous formation of links between coalitions and players: An application of the Shapley value. In: Roth, A.E. (ed.) *The Shapley Value*. Cambridge University Press, Cambridge
[3] Bala, V., Goyal, S. (1998) A strategic analysis of network reliability. Mimeo, Econometric Institute, Erasmus University, Rotterdam, the Netherlands, December
[4] Bala, V., Goyal, S. (2000) A non-cooperative theory of network formation. Discussion Paper TI 99-025/1, Tinbergen Institute, Rotterdam, the Netherlands. *Econometrica* 68: 1181–1230
[5] Borm, P., van den Nouweland, A., Tijs, S. (1994) Cooperation and communication restrictions: A survey. In: Gilles, R.P., Ruys, P.H.M. (eds.) *Imperfections and Behavior in Economic Organizations*. Kluwer Academic Publishers, Boston
[6] Coleman, J.S. (1990) *Foundations of Social Theory*. The Belknap Press of Harvard University Press, Cambridge, Massachusetts, and London, England
[7] Debreu, G. (1969) Neighboring economic agents. *La Décision* 171: 85–90
[8] Droste, E.J.R., Gilles, R.P., Johnson, C. (1999) Evolution of conventions in endogenous social networks. mimeo, Virginia Polytechnic Institute and State University, Blacksburg, VA, November
[9] Dutta, B., Mutuswami, S. (1997) Stable networks. *Journal of Economic Theory* 76: 322–344
[10] Ellison, G. (1993) Learning, local interaction, and coordination. *Econometrica* 61: 1047–1071
[11] Gilles, R.P., Haller, H.H., Ruys, P.H.M. (1994) The modelling of economies with relational constraints on coalition formation. In: Gilles, R.P., Ruys, P.H.M. (eds.) *Imperfections and Behavior in Economic Organizations*. Kluwer Academic Publishers, Boston
[12] Gilles, R.P., Ruys, P.H.M. (1990) Characterization of economic agents in arbitrary communication structures. *Nieuw Archief voor Wiskunde* 8: 325–345
[13] Goyal, S., Janssen, M.C.W. (1997) Non-exclusive conventions and social coordination. *Journal of Economic Theory* 77: 34–57
[14] Haller, H. (1994) Topologies as infrastructures. In: Gilles, R.P., Ruys, P.H.M. (eds.) *Imperfections and Behavior in Economic Organizations*. Kluwer Academic Publishers, Boston
[15] Jackson, M.O., Watts, A. (1999) The evolution of social and economic networks. mimeo, Caltech, Pasadena, CA, March
[16] Jackson, M.O., Wolinsky, A. (1996) A strategic model of social and economic networks. *Journal of Economic Theory* 71: 44–74
[17] Jacobs, J. (1961) *The Death and Life of Great American Cities*. Random House, New York
[18] Kalai, E., Postlewaite, A., Roberts, J. (1978) Barriers to trade and disadvantageous middlemen: Nonmonotonicity of the core. *Journal of Economic Theory* 19: 200–209
[19] Knack, S., Keefer, P. (1997) Does social capital have an economic payoff? A cross-country investigation. *Quarterly Journal of Economics* 112: 1251–1288
[20] Monderer, D., Shapley, L. (1996) Potential games. *Games and Economic Behavior* 14: 124–143
[21] Myerson, R.B., (1997) Graphs and cooperation in games. *Mathematics of Operations Research* 2: 225–229
[22] Nouweland, A. van den (1993) *Games and Graphs in Economic Situations*. Dissertation, Tilburg University, Tilburg, The Netherlands
[23] Qin, C-Z. (1996) Endogenous formation of cooperation structures. *Journal of Economic Theory* 69: 218–226
[24] Slikker, M., van den Nouweland, A. (1999) network formation models with costs for establishing links. *FEW Research Memorandum* 771, Faculty of Economics and Business Administration, Tilburg University, Tilburg, The Netherlands
[25] Watts, A. (1997) A dynamic model of network formation. mimeo, Vanderbilt University, Nashville, TN, September
[26] Watts, D.J., Strogetz, S.H. (1998) Collective dynamics of 'small-world' networks. *Nature* 393: 440–442
[27] Woolcock, M. (1998) Social capital and economic development: Toward a theoretical synthesis and policy framework. *Theory and Society* 27: 151–208

Stable Networks

Bhaskar Dutta, Suresh Mutuswami

Indian Statistical Institute, 7, S.J.S. Sansanwal Marg, New Delhi 110016, India

Abstract. A network is a graph where the nodes represent agents and an arc exists between two nodes if the corresponding agents interact bilaterally. An exogeneous value function gives the value of each network, while an allocation rule describes how the value is distributed amongst the agents. M. Jackson and A. Wolinsky (1996, *J. Econ. Theory* **71**, 44–74) have recently demonstrated a potential conflict between stability and efficiency in this framework. In this paper, we use an implementation approach to see whether the tension between stability and efficiency can be resolved.

JEL classification: C72, D20

1 Introduction

The interaction between agents can often be fruitfully described by a *network structure* or *graph*, where the nodes represent the agents and an arc exists between two nodes if the corresponding agents interact bilaterally. Network structures have been used in a wide variety of contexts ranging from social networks, (Wellman and Berkowitz [16]), information transmission (Goyal [4]), internal organization of firms (Marschak and Reichelstein [10]), cost allocation schemes (Henriet and Moulin [7]), to the structure of airline routes (Hendricks et al. [6])[1].

In a recent paper, Jackson and Wolinsky [8] focus on the *stability* of networks. Their analysis is designed to give predictions concerning which networks are likely to form when self-interested agents can choose to form new links or severe existing links. They use a specification where a *value function* gives the value

We are most grateful to Matt Jackson for several helpful discussions and suggestions. Thanks are also due to Sudipto Bhattacharyya, an anonymous referee, and an Associate Editor for comments on earlier versions of the paper. An earlier version of the paper was written when the first author was visiting Caltech. Their hospitality is gratefully acknowledged.

[1] See van den Nouweland [13] and Sharkey [15] for detailed surveys and additional references.

(or total product) of each graph or network, while an allocation rule gives the distribution of value amongst the agents forming the network. A principal result of their analysis shows that efficient graphs (that is, graphs of maximum value) may not be stable when the allocation rule treats individuals symmetrically.

The main purpose of this paper is to subject the potential conflict between stability and efficiency of graphs to further scrutiny. In order to do this, we follow Dutta et al. [3] and assume that agents' decisions on whether or not to form a link with other agents can be represented as a game in strategic form.[2] In this "link formation" game, each player announces a set of players with whom he or she wants to form a link. A link between two players is formed if both players want the link. This rule determines the graph corresponding to any n-tuple of announcements. The value function and the allocation rule then give the payoff function of the strategic form game.

Since the link formation game is a well-defined strategic-form game, one can use any equilibrium concept to analyze the formation of networks. In this paper, we will define a graph to be *strongly stable* (respectively *weakly stable*) if it corresponds to a *strong Nash* equilibrium (respectively *coalition-proof Nash equilibrium*) of the link formation game. Although Jackson and Wolinsky [8] did not use the link formation game, their specification assumed that only *two*-person coalitions can form; their notion of *pairwise stability* is implied by our concept of strong stability. Hence, it follows straightaway from their analysis that there is a conflict between strong stability and efficiency if the allocation rule is symmetric.

How can we ensure that efficient graphs will form? One possibility is to use allocation rules which are *not* symmetric. For instance, fix a vector of weights $w = (w_1, w_2, \ldots, w_n)$. Call an allocation rule *w-fair* if the gains or losses to players i and j from the formation of the *new* link (ij) is proportional to w_i/w_j. *w-fair* rules are symmetric only if $w_i = w_j$ for all i and j. However, the vector of weights w can be chosen so that there is only a "slight" departure from symmetry. We first show that the class of w-fair rules coincides with the class of *weighted Shapley values* of an appropriately defined transferable utility game. We then go on to construct a value function under which *no* efficient graph is strongly stable for any w-fair allocation rule. Thus, the relaxation of symmetry in this direction does not help.

A second possibility is to use weak stability instead of strong stability. However, again we demonstrate a conflict between efficiency, symmetry and (weak) stability.

We then go on to adopt an *implementation* or *mechanism design* approach. Suppose the implicit assumption or prediction is that only those graphs which correspond to strong Nash equilibria of the link formation game will form. Then, our interest in the *ethical* properties of the allocation rule should be restricted only to how the rule behaves *on the class of these graphs*. Hence, if we want

[2] This game was originally suggested by Myerson [12] and subsequently used by Qin [14]. See also Hart and Kurz [5] who discuss a similar strategic form game in the context of the endogeneous formation of coalition structures.

symmetry of the allocation rule, we should be satisfied if the allocation rule is symmetric on the *subdomain* of strongly stable graphs.

We analyse *two* specific problems within this general approach. In the first *design* problem, we construct an allocation rule which ensures that (i) the class of strongly stable graphs is a *nonempty* subset of the set of efficient graphs, and (ii) satisfies the restriction that the rule is symmetric on the class of strongly stable graphs. This result is proved under a very mild restriction on the class of value functions. The second result is much stronger, but is proved for a *more* restrictive class of value functions. More specifically, we construct an allocation rule which (given the restrictions on the class of value functions) guarantees that (i) there is at least *one* strongly stable graph, (ii) all *weakly stable* graphs are efficient, and (iii) the allocation rule is symmetric on the class of weakly stable graphs. Thus, this achieves a kind of "double" implementation in strong Nash and coalition-proof Nash equilibrium.

A common feature of the allocation rules constructed by us is that these distribute the value of stable graphs *equally* amongst all agents. Obviously, this ensures symmetry of the allocation rules on the class of stable graphs. Of course, the rules do not treat agents symmetrically on some graphs which are not stable. Indeed, the asymmetries are carefully constructed so as to ensure that the other requirements of the design problem(s) are satisfied.

The plan of this paper is as follows. In Sect. 2, we provide definitions of some key concepts. Section 3 describes the link formation game, while Sect. 4 contains the results. We conclude in Sect. 5.

2 Some Definitions

Let $N = \{1, 2, \ldots, n\}$ be a finite set of agents with $n \geq 3$. Interactions between agents are represented by graphs whose vertices represent the players, and whose arcs depict the pairwise relations. The *complete* graph, denoted g^N, is the set of *all* subsets of N of size 2. G is the set of *all* possible graphs on N, so that $G = \{g \mid g \subset g^N\}$.

Given any $g \in G$, let $N(g) = \{i \in N \mid \exists j \text{ such that } (ij) \in g\}$.

The link (ij) is the subset of N containing i, j, $g + (ij)$ and $g - (ij)$ are the graphs obtained from g by adding and subtracting the link (ij) respectively.

i and j are *connected* in g if there is a sequence $\{i_0, i_1, \ldots, i_K\}$ such that $i_0 = i$, $i_K = j$ and $(i_k i_{k+1}) \in g$ for all $k = 0, 1, \ldots, K-1$. We will use $C(g)$ to denote the set of connected components of g. g is said to be *fully connected* (respectively *connected* on S) if all pairs of agents in N (respectively in S) are connected. g is *totally disconnected* if $g = \{\varnothing\}$. If h is a component of g, then $N(h) = \{i \mid (ij) \in h \text{ for some } j \in N \setminus \{i\}\}$, and n_h denotes the cardinality of $N(h)$.

The value of a graph is represented by a function $v : G \to \mathbb{R}$. We will only be interested in the set V of such functions satisfying *Component Additivity*.

Definition 2.1. *A value function is component additive if* $v(g) = \sum_{h \in C(g)} v(h)$.

We interpret the value function to indicate the total "output" produced by agents in N when they are "organized" according to a particular graph. For instance, the members of N may be workers in a firm. The graph g then represents the internal organization of the firm, that is the structure of communication amongst the workers. Alternatively, N could be a set of (tax) auditors and supervisors, and g could represent a particular hierarchical structure of auditors and supervisors. In this case, $v(g)$ is the (expected) tax revenue realized from a population of tax payers when g is in "operation".

It is worth emphasizing at this point that the value function is a very general concept. In particular, it is more general than Myerson's [11] games with cooperation structure. A cooperation structure is a graph in our terminology. Given any exogeneously specified transferable utility game (N, u) and a graph g, we define for each $S \subset N$, the restricted graph on S as $g \mid S \equiv \{(ij) \in g \mid i, j \in S\}$. The *graph-restricted* game (N, u^g) specifies the *worth* of a coalition as follows.

$$\text{For all } S \subset N, u^g(S) = \sum_{h \in C(g|S)} u(N(h)) . \tag{2.1}$$

As (2.1) makes clear, the value or worth of a given set of agents in Myerson's formulation depends on whether they are connected or not, whereas in the Jackson–Wolinsky approach, the value of a coalition can in principle depend on *how* they are connected.

Given v, g is *strongly efficient* if $v(g) \geq v(g')$ for all $g' \in G$. Let $E(v)$ denote the set of strongly efficient graphs.

Finally, an *allocation rule* $Y : V \times G \to \mathbb{R}^N$ describes how the value associated with each network is distributed to the individual players. $Y_1(v, g)$ will denote the payoff to player i from graph g under the value function v. Clearly, an allocation rule corresponds to the concept of a *solution* in cooperative game theory.

Given a permutation $\pi : N \to N$, let $g^\pi = \{(ij) \mid i = \pi(k), j = \pi(l), (kl) \in g\}$. Let v^π be defined by $v^\pi(g^\pi) = v(g)$.

The following condition imposes the restriction that all agents should be treated symmetrically by the allocation rule. In particular, *names* of the agents should not determine their allocation.

Definition 2.2 *Y is* anonymous *on $G' \subseteq G$ if for* all *pairs* $(v, g) \in V \times G'$, and for all permutations π, $Y_{\pi(i)}(v^\pi, g^\pi) = Y_i(v, g)$.

Remark 2.3. If Y is *anonymous* on G, we say that Y is *fully* anonymous.

Definiton 2.4. *Y is* component balanced *if $\sum_{i \in N(h)} Y_i(v, g) = v(h)$ for every $g \in G$, $h \in C(g)$.*

Component balance implies that cross-subsidization is ruled out. We will restrict attention to component balanced allocation rules throughout the paper.[3]

[3] Jackson and Wolinsky [8] point out that the conflict between anonymity, stability, and efficiency disappears if the rule is not component balanced.

3 The Link Formation Game

In this section, we describe the strategic form game which will be used to model the endogeneous formation of networks or graphs.[4] The following description of the link formation game assumes a specific value function v and an allocation rule Y. Let $\gamma \equiv (v, Y)$.

The linking game $\Gamma(\gamma)$ is given by the $(n+2)$-tuple $(N; S_1, \ldots, S_n, f^\gamma)$, where for each $i \in N$, S_i is player i's *strategy* set with $S_i = 2^{N \setminus \{i\}}$ and the *payoff function* is the mapping $f^\gamma : S \equiv \prod_{i \in N} S_i \Rightarrow \mathbb{R}^N$ given by

$$f_i^\gamma(s) = Y_i(v, g(s)) \tag{3.1}$$

for all $s \in S$, with

$$g(s) = \{(ij) \mid j \in s_i, i \in s_j\} . \tag{3.2}$$

So, a typical strategy of player i in $\Gamma(\gamma)$ consists of the set of players with whom i wants to form a link. Then, (3.2) states that a link between i and j forms if and only if they both want to form this link. Hence, each strategy vector gives rise to a unique graph $g(s)$. Finally, the payoff to player i associated with s is simply $Y_1(v, g(s))$, the payoff that, is given by the allocation rule for the graph *induced by* s.[5]

We now define some equilibrium concepts for $\Gamma(\gamma)$.

Definition 3.1. *A strategy vector* $s^* \in S$ *is a* strong Nash equilibrium *(SNE) of* $\Gamma(\gamma)$ *if there is* no $T \subseteq N$ *and* $s \in S$ *such that*

(i) $s_i = s_i^*$ *for all* $i \notin T$.

(ii) $f_i^\gamma(s) > f_i^\gamma(s^*)$ *for all* $i \in T$.

The second equilibrium concept that will be used in this paper is that of coalition-proof Nash equilibrium (CPNE). In order to define the concept of CPNE of $\Gamma(\gamma)$, we need some more notation. For any $T \subset N$, and $s^*_{N|T} \in S_{N|T} \equiv \prod_{i \in N|T} S_i$, let $\Gamma(\gamma, s^*_{N|T})$ denote the game induced on T by $s^*_{N|T}$. So,

$$\Gamma(\gamma, s^*_{N|T}) = \langle T, \{S_i\}_{i \in T}, \overline{f}^\gamma \rangle . \tag{3.3}$$

where for all $j \in T$, for all $s_T \in S_T, \overline{f}_j^\gamma(s_T) = f_j^\gamma(s_T, s^*_{N|T})$.

The, set of CPNE of $\Gamma(\gamma)$ is defined inductively on the set of players.

Definition 3.2. *In a single-player game,* s^* *is a CPNE of* $\Gamma(\gamma)$ *iff* f_i^* *maximises* $f_i^\gamma(s)$ *over* S. *Let* $\Gamma(\gamma)$ *be a game with n players, where* $n > 1$. *Suppose CPNE have been defined for all games with less than n players. Then, (i)* $s^* \in S$ *is* self-enforcing *if for all* $T \subset N$, s_T^* *is a CPNE of* $\Gamma(\gamma, s^*_{N|T})$*; and (ii)* $s^* \in S$ *is a CPNE of* $\Gamma(\gamma)$ *if it is self-enforcing and moreover there does not exist another self-enforcing strategy vector* $s \in S$ *such that* $f_i^\gamma(s) > f_i^\gamma(s^*)$ *for all* $i \in N$.

[4] Aumann and Myerson [1] use an *extensive* form approach in modeling the endogeneous formation of cooperation structures.

[5] We will say that g is induced by s if $g \equiv g(s)$, where $g(s)$ satisfies (3.2).

Our interest lies not in the strategy vectors which are SNE or CPNE of $\Gamma(\gamma)$, but in the *graphs* which are induced by these equilibria. This motivates the following definition.

Definition 3.3. *g^* is* strongly stable *[respectively* weakly stable*] for $\gamma = (v, Y)$ if g^* is induced by some s which is a SNE [respectively CPNE] of $\Gamma(\gamma)$.*

Hence, a strongly stable graph is induced or supported by a strategy vector which is a strong Nash equilibrium of the linking game. Of course, a strongly stable graph must also be weakly stable.

Finally, in order to compare the Jackson–Wolinsky notion of *pairwise stability*, suppose the following constraints are imposed on the set of possible deviations in Definition 3.1. First, the deviating coalition can contain at most *two* agents. Second, the deviation can consist of severing just *one* existing link or forming *one additional* link. Then, the set of graphs which are immune to such deviations is called pairwise stable. Obviously, if g^* is strongly stable, then it must be pairwise stable.

4 The Results

Notice that strong stability (as well as weak stability) has been defined for a *specific* value function v and allocation rule Y. Of course, which network structure is likely to form must depend upon both the value function as well as on the allocation rule. Here, we adopt the approach that the value function is given *exogeneously*, while the allocation rule itself can be "chosen" or "designed".

Within this general approach, it is natural to seek to construct allocation rules which are (ethically) attractive and which also lead to the formation of stable network structures which maximize output, no matter what the exogeneously specified value function. This is presumably the underlying motive behind Jackson and Wolinsky's search for a symmetric allocation rule under which at least *one* strongly efficient graph would be pairwise stable for every value function.

Given their negative result, we initially impose weaker requirements. First, instead of *full anonymity*, we only ask that the allocation rule be *w-fair*, a condition which is defined presently. However, we show that there can be value functions under which *no* strongly efficient graph is strongly stable.[6] Second, we retain full anonymity but replace strong stability by weak stability. Again, we construct a value function under which the unique strongly efficient graph is *not* weakly stable.

Our final results, which are the main results of the paper, explicitly adopt an implementation approach to the problem. Assuming that strong Nash equilibrium is the "appropriate" concept of equilibrium and that the individual agents decide to form network relations through the link formation game is equivalent to predicting that only strongly stable graphs will form. Let $S(\gamma)$ be the set of strongly stable graphs corresponding to $\gamma \equiv (v, Y)$. Instead of imposing full anonymity,

[6] We point out below that strong stability can be replaced by pairwise stability.

we only require that the allocation rule be anonymous on the *restricted* domain $S(\gamma)$. However, we now require that for all permissible value functions, $S(\gamma)$ is contained in the set of strongly efficient graphs, instead of merely *intersecting* with it, which was the "target" sought to be achieved in the earlier results. We are able to construct an allocation rule which satisfies these requirements.

Suppose, however, that the designer has some doubt whether strong Nash equilibrium is really the "appropriate" notion of equilibrium. In particular, she apprehends that weakly stable graphs may also form. Then, she would want to ensure anonymity of the allocation rule over the larger class of weakly stable graphs, as well as efficiency of these graphs. Assuming a stronger restriction on the class of permissible value functions, we are able to construct an allocation rule which satisfies these requirements. In addition the allocation rule also guarantees that the set of strongly stable graphs is nonempty.

Our first result uses w-fairness. Fix a vector $w = (w_1, \ldots, w_n) \gg 0$.

Definition 4.1. *An allocation rule Y is* w-fair *if for all $v \in V$, for all $g \in G$, for all $i, j \in N$,*

$$\frac{1}{w_i}[Y_i(v,g) - Y_i(v, g-(ij))] = \frac{1}{w_j}[Y_j(v,g) - Y_j(v, g-(ij))] \,.$$

In Proposition 4.1 below, we show that *the* unique allocation rule which satisfies w-fairness and component balance is the *weighted Shapley value* of the following characteristic function game.

Take any $(v, g) \in V \times G$. Recall that for any $S \subseteq N$, the restricted graph on S is denoted $g \mid S$. Then, the *TU* game $u_{v,g}$ is given by:

For all $S \subseteq N, u_{v,g}(S) = \sum_{h \in C(g|S)} v(h)$.

Proposition 4.2 *For all $v \in V$, the unique w-fair allocation rule Y which satisfies components balance is the weighted Shapley value of $u_{v,g}$.*

Proof. The proof is omitted since it is a straightforward extension of the corresponding result in Dutta et al. [3]. □

Remark 4.3. This proposition is similar to corresponding results of Dutta et al. [3] and Jackson and Wolinsky [8]. The former proved that w-fair allocation rules satisfying component balance are the weighted Shapley values (also called weighted Myerson values) of the *graph-restricted* game given by any exogeneous TU game and any graph g. Of course, the set of graph-restricted games is a strict subset of V, and hence Proposition 4.2 is a formal generalization of their result. Jackson and Wolinsky show that where $w_1 = w_j$ for all i, j, then the unique w-fair allocation rule satisfying component balance is the Shapley value of $u_{v,g}$.

Our first result on stability follows. The motivation for proving this result is the following. Since the weight vector w can be chosen to make the allocation rule "approximately" anonymous (by choosing w to be very close to the unit vector $(1, \ldots, 1)$), we may "almost" resolve the tension between stability, efficiency, and symmetry unearthed by Jackson–Wolinsky by using such a w-fair allocation rule. However, the next result rules out this possibility.

Theorem 4.4. *Suppose $w \gg 0$. Then, there is no w-fair allocation rule Y satisfying component balance and such that for each $v \in V$, at least one strongly efficient graph is strongly stable.*

Proof. Let $N = \{1,2,3\}$, and choose any w such that $w_1 \geq w_2 \geq w_3 > 0$ and $\sum_{i=1}^{3} w_i = 1$.

Now, consider the (component additive) v such that $v(\{(ij)\}) = 1$, $v(\{(ij),(jk)\}) = 1+\varepsilon$, and $v(g^N) = 1+2\varepsilon$, where $\varepsilon \in \left(0, \frac{1}{2}(1-(w_2/(w_1+w_3)))\right)$.

Using Proposition 4.2, the unique w-fair allocation rule Y satisfying component balance is the weighted Shapley value of $u_{v,g}$. Routine calculation yields

$$Y_i(v, g^N) = 2\varepsilon w_i + w_i \left(\frac{w_k}{w_i + w_j} + \frac{w_j}{w_i + w_k} \right) \quad \text{for all } i,j,k \in N \tag{4.1}$$

$$Y_i(v, \{(ij)\}) = \frac{w_i}{w_i + w_j} \quad \text{for all } i,j \in N . \tag{4.2}$$

From (4.1) and (4.2), and using $\sum_{i=1}^{3} w_i = 1$

$$Y_i(v, \{(ij)\}) - Y_i(v, g^N) = w_i \left(1 - 2\varepsilon - \frac{w_j}{w_i + w_k} \right) . \tag{4.3}$$

Remembering that $w_1 \geq w_2 \geq w_3$ and that $\varepsilon < \frac{1}{2}(1-(w_2/(w_1+w_3))))$, (4.3) yields

$$Y_i(v, \{(ij)\}) - Y_i(v, g^N) > 0 \quad \text{for } i \in \{2,3\} \tag{4.4}$$

which implies that g^N is not strongly stable since $\{2,3\}$ will break links with 1 to move to the graph $\{(2,3)\}$. Since g^N is the unique strongly efficient graph, the theorem follows. □

Remark 4.5. Note that since only a pair of agents need form a coalition to "block" g^N, the result strengthens the intuitive content of the Jackson–Wolinsky result.

Our next result uses weak stability instead of strong stability.

Theorem 4.6. *There is no fully anonymous allocation rule Y satisfying component balance such that for each $v \in V$, at least one strongly efficient graph is weakly stable.*

Proof. Let $N = \{1,2,3\}$, and consider v such that $v(g^N) = 1 = v(\{(ij)\})$ and $v(\{(ij),(jk)\}) = 1+2\varepsilon$. Assume that $0 < \varepsilon < \frac{1}{12}$.

Since Y is fully anonymous and component balanced, $Y_i(v, \{(ij)\}) = Y_j$, $(v, \{(ij)\}) = \frac{1}{2}$. Let $g^j \equiv \{(ij),(jk)\}$. Note that $\{g^j \mid j \in N\}$ is the set of strongly efficient graphs. Choose any $j \in N$. Then, $Y_j(v, g^j) \geq \frac{1}{2}$. For, suppose $Y_j(v, g^j) < \frac{1}{2}$. Then, j can deviate *unilaterally* to change g^j to $\{(ij)\}$ or $\{(jk)\}$ by breaking the link with i or k respectively. So, if $Y_j(v, g^j) < \frac{1}{2}$ and g^j is induced by s, then s is not a Nash equilibrium, and hence not a CPNE.

So, $Y_j(v, g^j) \geq \frac{1}{2}$. Since Y is fully anonymous and component balanced, $Y_i(v, g^j) = Y_k(v, g^j) \leq \frac{1}{4} + \varepsilon$. Again, full anonymity of Y ensures that $Y_i(v, g^N) = \frac{1}{3}$ for all $i \in N$.

Hence, $\{i, k\}$ can deviate from g^j and form the additional link (ik). This will precipitate the complete graph. From preceding arguments, the deviation is profitable if $\varepsilon < \frac{1}{12}$.

Letting s^N denote the strategy n-tuple which induces g^N one notes that s^N is a Nash equilibrium. Hence, the deviation of $\{i, k\}$ to s^N is not deterred by the possibility of a further deviation by either i or k. So, g^j is *not* weakly stable. This completes the proof of the theorem. □

Remark 4.7. Again note that only a 2-person coalition has to form to block g^j. So, the result could have been proved in terms of "pairwise weak stability", which is strictly weaker than pairwise stability. Hence, this generalizes Jackson and Wolinsky's basic result.

Definition 4.8. *v satisfies* monotonicity *if for all $g \in G$, for all $i, j \in N$, $v(g + (ij)) \geq v(g)$.*

Thus, a monotonic value function has the property that additional links never decrease the value of the graph.[7] A special class of monotonic value functions, the class considered by Dutta et al. [3], is the set of graph-restricted games derived from superadditive TU games. Of course, there are also other contexts which might give rise to monotonic value functions. Dutta et al. proved that for the class of graph-restricted games derived from superadditive TU games, a large class of component balanced allocation rules (including all w-fair rules with $w \gg 0$) has the property that the set of weakly stable graphs is a *subset* of the set of graphs which are *payoff-equivalent* to g^N. Moreover, g^N itself is weakly stable.[8] Their proof can be easily extended to cover *all* monotonic value functions. We state the following result.

Theorem 4.9. *Suppose v is monotonic. Let Y be any w-fair allocation rule with $w \gg 0$, and satisfying component balance. Then, g^N is weakly stable for (v, Y). Moreover, if g is weakly stable for (v, Y), then g is payoff-equivalent to g^N.* [9]

The result is true for a larger class of allocation rules, which is not being defined here to save space.

Proof. The proof is omitted since it is almost identical to that of Dutta et al. [3]. □

Remark 4.10. Note that Theorem 4.9 ensures that *only* strongly efficient graphs are weakly stable. Thus, if our prediction is that only weakly stable graphs will form, then this result guarantees that there will be no loss in efficiency. This guarantee

[7] Hence, g^N is strongly efficient.

[8] g and g' are payoff-equivalent under (v, Y) if $Y(v, g) = Y(v, g')$. Also, note that if v is monotonic, then g^N and hence graphs which are payoff-equivalent, to g^N are strongly efficient.

[9] The result is true for a larger class of allocation rules, which is not being defined here to save space.

is obviously stronger than that provided if *some* stable graphs is strongly efficient. In the latter case, there is the possibility that *other* stable graphs are inefficient, and since there is no reason to predict that *only* the efficient stable graph will form, inefficiency can still occur.

Unfortunately, monotonicity of the value function is a stringent requirement. There are a variety of problems in which the optimum network is a *tree* or a *ring*. For example, cost allocation problems give rise to the minimum-cost spanning tree. Efficient airline routing or optimal trading arrangements may also imply that the *star* or *ring* is the efficient network.[10] Indeed, in cases where there is a (physical) cost involved in setting up an additional link, g^N will seldom be the optimal network.

This provides the motivation to follow the "implementation approach" and prove results similar to that of Theorem 4.9, but covering *nonmonotonic* value functions. First, we construct a component balanced allocation rule which is anonymous on the set of *strongly stable graphs* and which ensures that *all* strongly stable graphs are strongly efficient.

In order to prove this result, we impose a restriction on the class of value functions.

Definition 4.11. *The set of* admissible *value functions is* $V^* = \{v \in V \mid v(g) > 0$ *iff* g *is not totally disconnected*$\}$.

So, a value function is admissible if *all* graphs (except the trivial one in which no pair of agents is connected) have positive value.[11]

Before we formally define the particular allocation rule which will be used in the proof of the next theorem, we discuss briefly the key properties which will have to be satisfied by the rule.

Choose some efficient $g^* \in G$. Suppose s^* induces g^*, and we want to ensure that g^* is strongly stable. Now, consider any g which is different from g^*, and let s induce g. Then, the allocation rule must *punish* at least one agent who has deviated from s^* to s. This is possible only if a deviant can be *identified*. This is trivial if there is some $(ij) \in g \backslash g^*$, because then both i and j must concur in forming the extra link (ij). However, if $g \subset g^*$, say $(ij) \in g^* \backslash g$, then *either* i *or* j can unilaterally break the link. The only way to ensure that the deviant is punished, is to punish *both* i and j.

Several simple punishment schemes can be devised to ensure that at least two agents who have deviated from s^* are punished sufficiently to make the deviation unprofitable. However, since the allocation rule has to be component balanced, these punishment schemes may result in some other agent being given more than the agent gets in g^*. This possibility creates a complication because the punishment scheme has to *satisfy an additional property*. Since we also want to

[10] See Hendricks et al. [6] on the "hub" and "spokes" model of airline routing. See also Landa [9] for an interesting account of why the ring is the efficient institutional arrangement for organization of exchange amongst tribal communities in East Papua New Guinea.

[11] In common with Jackson and Wolinsky, we are implicitly assuming that the value of a disconnected player is zero. This assumption can be dropped at the cost of some complicated notation.

ensure that inefficient graphs are *not* strongly stable, agents have to be provided with an incentive to deviate from any graph which is not strongly efficient. Hence, the punishment scheme has to be relatively more sophisticated.

Choose some strongly efficient g^* with $C(g^*) = \{h_1^*, \ldots, h_j^*\}$, and let $\succ$ be a *strict* ordering on arcs of g^*. Consider any other graph g, and let $C(g) = \{h_,, \ldots, h_K\}$.

The first step in the construction of the allocation rule is to choose agents who will be punished in some components $h_k \in C(g)$. For reasons which will become clear later on, we only need to worry about components h_k such that $D(h_k) = \{i \in N(h_k) \mid (ij) \in g^* \text{ for some } j \notin N(h_k)\}$ is nonempty. For such components, choose $i(h_k) \equiv i_k$ such that $\forall j \in N(h_k)\backslash\{i_k\}, \forall m \notin N(h_k)$:

$$(jm) \in g^* \Rightarrow (i_k l) \succ (jm) \quad \text{for some } (i_k l) \in g^*, l \notin N(h_k)\,. \tag{4.5}$$

We will say that g is a $*$*-supergraph* of g^* if for each $h^* \in C(g^*)$, there is $h \in C(g)$ such that $N(h^*) \subseteq N(h)$. Note that the fully connected graph is a $*$*-supergraph* of every graph.

Lemma 4.12. *Suppose $g = (h_1, \ldots, h_K)$ is not a $*$-supergraph of g^*. Then, $\exists k, l \in \{1, \ldots, K\}$ such that $(i_k i_l) \in g^*$.*

Proof. Since g is not a $*$*-supergraph*, it follows that g is not fully connected, and that there exists a component h and players i, j such that $i \in N(h)$, $j \notin N(h)$ and $(ij) \in g^*$. Indeed, assume that for each $h_k \in C(g)$, the set $D(h_k)$ is nonempty.[12] For every $k = 1, 2, \ldots, K$, there is $j_k \notin N(h_k)$ such that $(i_k j_k) \in g^*$ and $(i_k j_k) \succ (ij)$ for all $i \in D(h_k)\backslash\{ik\}$ and for all $j \notin N(h_k)$ with $(ij) \in g^*$. Let the $\succ$-maximal element within the set $\{(i_1, j_1), \ldots, (i_K j_K)\}$ be $(i_{k^*} j_{k^*})$. Let $j_{k^*} \in N(h_l)$. Note that from the definition of the pair $(i_{k^*} j_{k^*})$, it follows that $l \neq k^*$. Also, $(i_{k^*} j_{k^*}) \in g^*$. It therefore follows that $i_l = j_{k^*}$ and $j_l = i_{k^*}$. Hence, $(i_{k^*} i_l) \in g^*$. This completes the proof of the lemma.

The implication of Lemma 4.12 is the following. Suppose one or more agents deviate from g^* to some $g \in G$ with components $\{h_1, \ldots, h_K\}$. Then, the set of agents $\{i(h_1), \ldots, i(h_K)\}$ must contain a deviator. This property will be used intensively in the proof of the next theorem.

Theorem 4.13. *Let $v \in V^*$. Then, there is a component balanced allocation rule Y^* such that the set of strongly stable graphs is nonempty and contained in $E(v)$. Moreover, Y^* is anonymous on the set of strongly stable graphs.*

Proof. Choose any $v \in V^*$. Fix $g^* \in E(v)$. Let $C(g^*) = \{h_1^*, \ldots, h_K^*\}$. An allocation rule Y^* satisfying the required properties will be constructed which ensures that g^* is strongly stable. Moreover, no g will be strongly stable unless it is in $E(v)$.

For any $S \subseteq N$ with $|S| \geq 2$, let G_S be the set of graphs which are connected on S, and have no arcs outside S. So,

[12] Otherwise, we can restrict attention to those components for which $D(h_k)$ is nonempty.

$$G_S = \{g \in G \mid g \text{ is connected on } S \text{ and } N(g) = S\}\ .$$

Let

$$a_S = \min_{g \in G_S} \frac{v(g)}{|S|(n-2)}\ .$$

Choose any ε such that

$$0 < \varepsilon < \min_{S \subseteq N} a_S\ . \tag{4.6}$$

The allocation rule Y^* is defined by the following rules. Choose any g.

(Rule 1) For any $h \in C(g)$, suppose $N(h) = \bigcup_{i \in I} N(h_i^*)$ for some (non-empty) $I \subseteq \{1, \ldots, K\}$. Then,

$$Y_i^*(v, g) = \frac{v(h)}{n_h} \quad \text{for all } i \in N(h)\ .$$

(Rule 2) Suppose $N(h) \neq \bigcup_{i \in I} N(h_i^*) \forall I \subseteq \{1, \ldots, K\}$. Then, g is not a $*$-*supergraph* of g^*. Choose $j_h \in N(h)$ such that $j_h \neq i_h$. Then,

$$Y_i^*(v, g) = \begin{cases} (n_h - 1)\varepsilon & \text{if } i \neq j_h \\ v(h) - (n_h - 1)^2 \varepsilon & \text{otherwise}\ . \end{cases}$$

Clearly, the rule defined above is component balanced. We will show later that Y^* is anonymous on the set of strongly stable graphs. We first show that the efficient graph g^* is strongly stable under the above allocation rule.

Let s^* be the strategy profile defined as follows: For all $i \in N$, $s_i^* \{j \in N \mid (ij) \in g^*\}$. Clearly, s^* induces g^* in $\gamma = (v, Y^*)$. We need to show that s^* is a SNE of $\Gamma(\gamma)$.

Consider any $s \neq s^*$, and let g be induced by s. Also, let $T = \{i \in N \mid s_i \neq s_i^*\}$. Suppose $h \in C(g)$. If $N(h) = \bigcup_{i \in I} N(h_i^*)$ for some nonempty subset I of $\{1, \ldots, K\}$, then $Y_i^*(v, g) = v(h)/n_h$ for all $i \in N(h)$. However, since g^* is efficient, there exists some $i \in I$ such that $v(h_i^*)/n_{h_i^*} \geq v(h)/n_h$. So, no member of h_i^* is better-off as a result of the deviation. Also, note that $T \cap N(h_i^*) \neq \varnothing$. So, T does not have a profitable deviation in this case.

Suppose there is $h \in C(g)$ such that $N(h) \neq \bigcup_{i \in I} N(h^*)$ for any nonempty subset I of $\{1, \ldots, K\}$. Then, g is not a $*$-*supergraph* of g^*, and let $C(g) = \{h_1, \ldots, h_L\}$. From the above lemma, there exists $(i_k i_l) \in g^*$ where i_k and i_l are the players who are punished in h_k and h_l respectively. Obviously, $T \cap \{i_k, i_l\} \neq \phi$. But from Rule (2), it follows that $Y_{i_k}^*(v, g) = (n_{h_k} - 1)\varepsilon$ and $Y_{i_l}^*(v, g) = (n_{h_l} - 1)\varepsilon$. Given the value of ε, it follows that both i_k and i_h are worse-off from the deviation.

We now show that if g is strongly stable, then $g \in E(v)$. So suppose that g is an inefficient graph.

(i) If g is an inefficient graph which is a $*$-*supergraph* of g^*, then there exist $h \in C(g), h^* \in C(g^*)$ such that $N(h^*) \subseteq N(h)$ and

$$Y_i^*(v,g) = \frac{v(h)}{n_h} < Y_i^*(v,g^*) = \frac{v(h^*)}{n_{h^*}} \quad \text{for all } i \in N(h^*) .$$

So, each $i \in N(h^*)$ can deviate to the strategy s_i^*. This will induce the component h^* where they are all strictly better off.

(ii) Suppose that g is not a $*$-*supergraph* of g^*. Let $C(g) = (h_1, \ldots, h_K)$. Without loss of generality, let $n_{h_1} \leq \ldots \leq n_{h_K}$. Since g is not a $*$-*supergraph* of g^*, Rule (2) of the allocation rule applies and we know that there exist $h_k, h_l \in C(g)$, and $i_{h_k} \in N(h_k), i_{h_l} \in N(h_i)$ such that $Y_{i_{h_k}}^*(v,g) = (n_{h_k} - 1)\varepsilon$ and $Y_{i_{h_l}}^*(v,g) = (n_{h_l} - 1)\varepsilon$. Let $\overline{s}$ be such that

(i) $\forall j \notin \{i_{h_k}, i_{h_l}\}, \overline{s}_j = s_j$.

(ii) $\overline{s}_{i_{h_k}} = \{j \mid j \in s_{i_{h_k}} \text{ or } j = i_{h_l}\}$.

(iii) $\overline{s}_{i_{h_l}} = \{j \mid j \in s_{i_{h_l}} \text{ or } j = i_{h_k}\}$.

Let $\overline{g}$ be the graph induced by $\overline{s}$. Notice that $\overline{g} = g + (i_{h_k} i_{h_l})$. We claim that

$$Y_j^*(v,\overline{g}) > Y_j^*(v,g) \forall j \in \{i_{h_k}, i_{h_l}\} . \tag{4.7}$$

Let $\overline{h} \in C(\overline{g})$ be the component containing players i_{h_k} and i_{h_l}. Notice that $n_{\overline{h}} > \max(n_{h_k}, n_{h_l})$. Given the value of ε, it follows that

$$Y_j^*(v,\overline{g}) \geq (n_{\overline{h}} - 1)\varepsilon \quad \forall j \in \{i_{h_k}, i_{h_l}\} .$$

This shows that the coalition $\{i_{h_k}, i_{h_l}\}$ has a deviation which makes both players better off.

The second half of the proof also shows that g is strongly stable only if g is a $*$-*supergraph* of g^*. From Rule (1), it is clear that Y^* is anonymous on all such graphs. This observation completes the proof of the theorem. □

We have remarked earlier that we need to restrict the class of permissible value functions in order to prove the analogue of a "double implementation" in strong Nash and coalition-proof Nash equilibrium. In order to explain the motivation underlying the restricted class of value functions, we first show in Example 4.14 below that the allocation rule used in Theorem 4.13 *cannot* be used to prove the double implementation result. In particular, this allocation rule does not ensure that weakly stable graphs are efficient.

Example 4.14. Let $N = \{1, 2, 3, 4\}$. Consider a value function such that $v(g^*) = 4$, $v(g_1) = 3.6$, $v(g_2) = v(g_3) = 2.9$, where $g^* = \{(14), (13), (23), (12)\}$, $g_1 = \{(12), (13), (34)\}$, $g_2 = \{(12), (13)\}$ and $g_3 = \{(13), (34)\}$. Also, $v(\{(ij)\}) = 1.6$. Finally, the value of other graphs is such that $\varepsilon = 0.4$ satisfies (4.6). Note that g^* is the unique efficient graph. Let the strict order on links (used in the construction of the allocation rule in Theorem 4.13) be given by

$$(13) \succ (23) \succ (14) \succ (12) .$$

Consider the graph $g = \{(12), (34)\}$. Then, from (Rule 2) and the specification of $\succ$, we have $Y_2^*(v,g) = Y_4^*(v,g) = 1.2$, $Y_1^*(v,g) = Y_3^*(v,g) = 0.4$. Now, g is weakly stable, but not efficient.

To see that g is weakly stable, notice first that agents 2 and 4 have no profitable deviation. Second, check using the particular specification of $\succ$ that $Y_3^*(v,g_2) = 1.3 > Y_3^*(v,g_1) = 0.9$, $Y_1^*(v,\{(13)\}) > Y_1^*(v,g_2)$ and $Y_3^*(v,g_3) = 0.8 > Y_3^*(v,\{(13)\}) = 0.4$.

Finally, consider the 2-person link formation game T with player set $\{1,3\}$ generated from the original game by fixing the strategies of players 2 and 4 at $s_2 = \{1\}$, $s_4 = \{3\}$. Routine inspection yields that there is *no* Nash equilibrium in this game. This shows that g is weakly stable.

In order to rule out inefficient graphs from being stable, we need to give some coalition the ability to deviate credibly. However, the allocation rule constructed earlier fails to ensure this essentially because agents can severe links and become the "residual claimant" in the new graph. For instance, notice that in the previous example, if 3 "deviates" from g_1 to g_2 by breaking ties with 4, then 3 becomes the residual claimant in g_2. Similarly, given g_2, 1 breaks links with 2 to establish $\{(13)\}$, where she is the residual claimant.

To prevent this jockeying for the position of the residual claimant, one can impose the condition that on all inefficient graphs, players are punished according to a *fixed* order. Unfortunately, while this does take care of the problem mentioned above, it gives rise to a new problem. It turns out that in some cases the *efficient* graph itself may not be (strongly) stable. The following example illustrates this.

Example 4.15. Let $N = \{1,2,3,4\}$. Let g^*, the unique efficient graph be $\{(12),(23),(34),(41)\}$, let $g = \{(12),(34)\}$. Assume tat $v(g^*) = 4$ and $v(\{(ij)\}) = 1.5$ for all $\{i,j\} \subset N$. The values of the other graphs are chosen so that

$$\min_{S \subseteq N} \min_{g \in G_S} \frac{v(g)}{(|N|-2)|S|} = 0.25 \ .$$

Choose $\varepsilon = 0.25$ and let $\succ_p$ be an ordering on N such that $1 \succ_p 2 \succ_p 3 \succ_p 4$. Applying the allocation rule specified above, it follows that

$$Y_1(v,g^*) = 1 \quad \text{for all } i \in N$$
$$Y_2(v,g) = Y_4(v,g) = 1.25$$

and

$$Y_1(v,g) = Y_3(v,g) = 0.25 \ .$$

One easily checks that the coalition $\{2,4\}$ can deviate from the graph g^* to induce the graph g. This deviation makes both deviators better off. The symmetry of the value function on graphs of the form $\{(ij)\}$ now implies that *no* fixed order will work here.

This explains why we need to impose a restriction on the class of value functions. We impose a restriction which ensures that for some efficient graph

g^*, there is a "central" agent within each component, that is, an agent who is connected to every other agent in the component. This restriction is defined formally below.

Definition 4.16. *A graph g is* focussed *if for each $h \in C(g)$, there is $i_h \in N(h)$ such that $(i_h j) \in h$ for all $j \in N(h)\backslash\{i_h\}$.*

Let $\overline{V}$ be the set or all value functions v such that

(i) $v(g) = 0$ only if g is completely disconnected.

(ii) There exists $g^* \in E(v)$ such that g^* is focussed.

We now assume that the class of permissible value functions is $\overline{V}$. This is a much stronger restriction than the assumption used in the previous theorem. However, there are several interesting problems which give rise to such value functions. Indeed, the two special models discussed by Jackson and Wolinsky (the *symmetric connections* and *coauthor* models) both give rise to value functions in $\overline{V}$.

Choose some $v \in \overline{V}$, and let $g^* \in E(v)$ be focussed. Assume that $(h_1^*, \ldots, h_K^*)$ are the components of g^*, and let i_k be the player who is connected to all other players in $N(h_k)$.[13]

Let $\succ_p$ be a strict order on the player set N satisfying the following conditions:

$$\text{(i)} \quad \forall i, j \in N, \quad \text{if } i \in N(h_k), j \in N(h_l) \quad \text{and } k < l, \quad \text{then } i \succ_p j\ .$$

$$\text{(ii)} \quad i_k \succ_p j \quad \text{for all } j \in N(h_k^*)\backslash\{i_k\}, \quad k = 1, \ldots, K\ .$$

So, $\succ_p$ satisfies two properties. First, all agents in $N(h_k^*)$ are ranked above agents in $N(h_{k+1}^*)$. Second, *within* each component, the player who is connected to all other players is ranked first. Finally, choose any ε satisfying (4.6).

The allocation rule Y^* is defined by the following rules. Choose any g and $h \in C(g)$.

(Rule 1) Suppose $N(h) = N(h^*)$ for some $h^* \in C(g^*)$. Then,

$$Y_i^*(v, g) = \frac{v(h)}{n_h} \quad \text{for all } i \in N(h)\ .$$

(Rule 2) Suppose $N(h) \subset N(h^*)$ for some $h^* \in C(g^*)$. Let j_h be the "minimal" element of $N(h)$ under the order $\succ_p$. Then, for all $i \in N(h)$,

$$Y_i^*(v, g) = \begin{cases} (n_h - 1)\varepsilon & \text{if } i \neq j_h \\ v(h) - (n_h - 1)^2\varepsilon & \text{if } i = j_h\ . \end{cases}$$

(Rule 3) Suppose $N(h) \nsubseteq N(h^*)$ for any $h^* \in C(g^*)$. Let j_h be the "minimal" element of $N(h)$ under the order $\succ_p$. Then, for all $i \in N(h)$,

[13] If more than one such player exists, then any selection rule can be employed.

$$Y_i^*(v,g) = \begin{cases} \dfrac{\varepsilon}{2} & \text{if } i \neq j_h \\ v(h) - \dfrac{(n_h-1)\varepsilon}{2} & \text{if } i = j_h \ . \end{cases}$$

The allocation rule has the following features. First, provided a component consists of the same set of players as some component in g^*, the value of the component is divided equally amongst all the agents in the component. Second, punishments are levied in all other cases. The punishment is more severe if players form links *across* components in g^*.

Let s^* be the strategy profile given by $s_i^* = \{j \in N \mid (ij) \in g^*\}$ for all $i \in N$, and let $C(g^*) = \{h_1^*, \ldots, h_K^*\}$. We first show that if agents in components $h_1^*, \ldots, h_k^*$ are using the strategies s_i^*, then no group of agents in h_k^* will find it profitable to deviate. Moreover, this is independent of the strategies chosen by agents in components coming "after" h_k^*.

Lemma 4.17. *Let $v \in \overline{V}$. Suppose s is the strategy profile given by $s_i = s_i^* \forall i \in N(h_k^*)$, $\forall k = 1, \ldots, \overline{K}$ where $\overline{K} \leq K$. Then, there is no s' such that $f_i^\gamma(s') > f_i^\gamma(s)$ for all $i \in T$ where $s_i' = s_i^*$ for all $i \in N(h_k^*)$, $k < \overline{K}$ and $T = \{i \in N(h_{\overline{K}}^*) \mid s_i' \neq s_i\}$.*

Proof. Consider the case $\overline{K} = 1$. Let g be the graph induced by s. Note that $h_i^* \in C(g)$.

Consider any s', and let g' be the graph induced by s'. Suppose $T = \{i \in N(h_1^*) \mid s_i \neq s_i'\} \neq \varnothing$.

Case (1): There exists $h \in C(g')$ such that $N(h) = N(h_1^*)$.
In this case, Rule (1) applies, and we have

$$Y_i^*(v,g') = \frac{v(h)}{|N(h)|} \leq \frac{v(h_1^*)}{|N(h_1^*)|} = Y_i^*(v,g) \quad \forall i \in N(h_1^*) \ .$$

So no $i \in N(h_1^*)$ benefits from the deviation.

Case (2): There exists $h \in C(g')$ such that $N(h) \cap N(h_1^*) \neq \phi$, and $N(h) \not\subseteq N(h_1^*)$.

In this case, Rule (3) applies, and we have

$$Y_i^*(v,g') = \frac{\varepsilon}{2} < Y_i^*(v,g) \forall i \in N(h_1^*) \cap N(h) \ .$$

Noting that $N(h_1^*) \cap N(h) \cap T \neq \varnothing$, we must have $f_i^\gamma(s) > f_i^\gamma(s')$ for some $i \in T$.

Case (3): There exists $h \in C(g')$ such that $N(h) \subset N(h_1^*)$.
Noting that there is i_1 who is connected to everyone in $N(h_1^*)$, either $i_1 \in T$ or $T = N(h)$. If $i_1 \in T$, then since $f_{i_1}^\gamma(s') \leq (n_h - 1)\varepsilon < f_{i_1}^\gamma(s)$, the lemma is true. Suppose is $i_1 \notin T$. Ruling out the trivial case where a single agent breaks away,[14] we have $|T| \geq 2$. From Rule 2 or Rule 3, at least one of the agents must be worse off.

[14] The agent then gets 0.

Hence, in all possible cases, there is some $i \in T$ who does not benefit from the deviation.

The proof can be extended in an obvious way for all values of $\overline{K}$. □

Lemma 4.18. *Let $v \in \overline{V}$. Let g be the graph induced by a strategy profile s. Suppose there exists $h \in C(g)$ such that $N(h) \subset N(h_1^*)$. Then, g is not weakly stable.*

Proof. If s is not a Nash equilibrium of $\Gamma(\gamma)$, then there is nothing to prove. So, assume that s constitutes a Nash equilibrium.

We will prove the lemma by showing that there is a credible deviation from s for a coalition $D \subset N(h_1^*)$, $|D| = 2$. The game induced on the coalition D is defined as $\Gamma(\gamma, s_{N\setminus D}) = \langle D, \{S_i\}_{i \in D}, \overline{f}^{\gamma} \rangle$ where $\overline{f}_j^{\gamma}(s'_D) = Y_j^*(v, g(s'_D, s_{N\setminus D}))$ for all $j \in D$. We show that there is a Nash equilibrium in this two-person game which Pareto-dominates the payoff corresponding to s.

Suppose there is $i \in N(h_1^*)\setminus N(h)$, $j \notin N(h_1^*)$ such that $(ij) \in g$. Then, $Y_i^*(v, g) = \varepsilon/2$. Since s is a Nash equilibrium, this implies that i by a unilateral deviation cannot induce a graph g' in which i will be in some component such that $N(h') \subseteq N(h_1^*)$.

Now, let j be the $\succ_p$-maximal agent in $N(h)$. Consider the coalition $D = \{i, j\}$. Choose $s'_i = \{j\}$, and let s'_j be the best response to s'_i in the game $\Gamma(\gamma, s_{N\setminus D})$. Then, (s'_i, s'_j) must be a Nash equilibrium in $\Gamma(\gamma, s_{N\setminus D})$.[15] Using Rule (2), it is trivial to check that both i and j gain by deviating to s' from s.

Hence, we can now assume that if $N(h) \subset N(h_1^*)$, then there exist $\{h_1, \ldots, h_L\} \subseteq C(g)$ such that $N(h_1^*) = \bigcup_{i=1,\ldots,L} N(h_i)$.[16] Note that $L \geq 2$.

W.l.o.g, let 1 be the $\succ_p$-maximal agent in $N(h_1^*)$, and $1 \in N(h_1)$. Let i be the $\succ_p$-maximal agent in $N(h_2)$, and let $D = \{1, i\}$.

Suppose $L > 2$. Then, consider $\overline{s}_1 = s_1 \cup \{i\}$, and let $\overline{s}_i$ be the best response to $\overline{s}_1$ in the game $\Gamma(\gamma, s_{N\setminus D})$. Note that 1 can never be the residual claimant in any component, and that $1 \in \overline{s}_i$. It then follows that $(\overline{s}_1, \overline{s}_i)$ is a Nash equilibrium in $\Gamma(\gamma, s_{N\setminus D})$ which Pareto-dominates the payoffs (of D) corresponding to the original strategy profile s.

Suppose $L = 2$. Let $\overline{S} = \{\overline{s} \mid \overline{s} = (\overline{s}_1, \overline{s}_i, \overline{s}_{-D})$ for some $(\overline{s}, \overline{s}_i) \in S_1 \times S_i$. Let $\bar{G}$ be the set of graphs which can be induced by D subject to the restriction that both 1 and i belong to a component which is connected on $N(h_1^*)$. Let $\bar{g}$ be such that $v(\bar{g}) = \max_{g \in G} v(g)$, and suppose that $\bar{s}$ induces $\bar{g}$. Then, note that $i \in \bar{s}_1$ and $i \in \bar{s}_i$.

Now, $Y_1^*(v, \bar{g}) = Y_i^*(v, \bar{g}) = v(\bar{g})/n_h$. Clearly, $Y_j^*(v, \bar{g}) > Y_j^*(v, g)$ for $j \in D$. If $(\bar{s}_1, \bar{s}_i)$ is a Nash equilibrium in $\Gamma(\gamma, s_{N\setminus D})$, then this completes the proof of the lemma. Suppose $(\bar{s}_1, \bar{s}_i)$ is not a Nash equilibrium of $\Gamma(\gamma, s_{N\setminus D})$. Then, the only possibility is that i has a profitable deviation since 1 can never become the residual claimant. Let $\tilde{s}_i$ be the best responpse to $\bar{s}_1$ in $\Gamma(\gamma, s_{N\setminus D})$. Note that $1 \in \tilde{s}_i$. Let $\tilde{g}$ denote the induced graph. We must therefore have $Y_1^*(v, \tilde{g}) >$

[15] The fact that i has no profitable deviation from s'_i follows from the assumption that the original strategy profile is a Nash equilibrium.

[16] Again, we are ignoring the possible existence of isolated individuals.

$Y_1^*(v, g)$.[17] Obviously, $Y_i^*(v, \tilde{g}) > Y_i^*(v, g)$. Since $\bar{s}_1$, is also a best response to $\tilde{s}_i$ in $\Gamma(\gamma, s_{N \setminus D})$, this completes the proof of the lemma.

We can now prove the following.

Theorem 4.19. *Let $v \in \bar{V}$. Then, there exists a component balanced allocation rule Y satisfying the following*

(i) The Set of strongly stable graphs is nonempty.

(ii) If g is weakly stable, then $g \in E(v)$.

(iii) Y is anonymous over the set of weakly stable graphs.

Proof. Clearly, the allocation rule Y defined above is component balanced. We first show that the efficient graph g^* is strongly stable by showing that s^* is a strong Nash equilibrium of $\Gamma(\gamma)$.

Let $C(g^*) = \{h_1^*, \ldots, h_K^*\}$.

Let $s \neq s^*$, g be the graph induced by s, and $T = \{i \in N \mid s_i \neq s_i^*\}$. Let $t^* = \arg\min_{1 \leq j \leq K} s_i \neq s_i^*$ for some $i \in N(h_j^*)$.

By Lemma 4.17, it follows that at least one member in $N(h_{t^*}) \cap T$ does not profit by deviating from the strategy s^*. This shows that the graph g^* is strongly stable.

We now show that if g is not efficient, then it cannot be weakly stable. Let s be a strategy profile which induces the graph g. We have the following cases.

Case (1a): There exists $h \in C(g)$ such that $N(h_1^*) = N(h)$ and $v(h) < v(h_1^*)$.

Suppose all individuals i in $N(h_i^*)$ deviate to s_i^*. Clearly, all individuals in $N(h_1^*)$ gain from this deviation. Moreover Lemma 4.17 shows that no subcoalition of $N(h_1^*)$ has any further profitable deviation. Hence, s cannot be a CPNE of $\Gamma(\gamma)$ in this case.

Case (1b): There does not exist $h \in C(g)$ such that $N(h) \subseteq N(h_1^*)$.

In this case all players in $N(h_1^*)$ are either isolated (in which case they get zero) or they are in (possibly different) components which contain players not in $N(h_1^*)$. Using Rule (3) of the allocation rule, it follows that

$$Y_1(v, g) \leq \frac{\varepsilon}{2} \quad \forall i \in N(h_1^*) .$$

So all players in $N(h_1^*)$ can deviate to the strategy s_i^*. Obviously, this will make them strictly better off. That this is a credible deviation follows from Lemma 4.17.

Case (1c): There exists $h \in C(g)$, such that $N(h) \subset N(h_1^*)$.

In this case, it follows from Lemma 4.18 that there is a credible deviation for a coalition $D \subset N(h_1^*)$.

Case (2): If there exists $h \in C(g)$ such that $N(h) = N(h_1^*)$ and $v(h) = v(h_1^*)$, then apply the arguments of Case 1 to h_2^* and so on.

[17] This follows since 1 is now in a component containing more agents.

The preceding arguments show that if g is weakly stable, then:

(i) $N(h_i) = N(h_i^*)$ for each $i \in \{1, \dots, K\}$.

(ii) $v(h_i) = v(h_i^*)$ for each $i \in \{1, \dots, K\}$.

These show that all weakly stable graphs must be efficient. Furthertnore, it follows from Rule (1) that Y is anonymous on all such graphs. This completes the proof of the theorem. □

Notice that in both Theorems 4.13 and 4.19, we have imposed the requirement that the allocation rule satisfy component balance on *all* graphs, and not just on the set of stable (or weakly stable) graphs. This raises the obvious question as to why the two properties of component balance and anonymity have been treated asymmetrically in the paper.

The answer lies in the fact that component balance has a *strategic* role, while anonymity is a purely ethical property. Consider, for instance, the "equal division" allocation rule which specifies that each agent gets $v(g)/n$ on all graphs g. This rule violates component balance.[18] Let the value function be such that $(v(\{12\})/2) > (v(g^*)/n)$ where g^* is some efficient graph. Then, given the equal division rule, agents i and j both do strictly better by breaking away from the rest of the society since the total reward given to them by this allocation rule is less than what they can get by themselves. On the other hand, Theorems 4.13 and 4.19 show that *some* allocation rules which are component balanced ensure that no set of agents wants to break away.

Readers will notice the obvious analogy with the literature on implementation. There, mechanisms which waste resources "out of equilibrium" will not be renegotiation-proof since all agents can move away to a Pareto-superior outcome. Here, the violation of component balance implies that *all* agents in some *component* can agree on a jointly better outcome.

There is also another logical motivation which can be provided for this asymmetric treatment of component balance and anonymity.[19] In view of the Jackson–Wolinsky result, one or both the conditions must be relaxed in order to resolve the tension between stability and efficiency. This paper shows that simply relaxing anonymity out of equilibrium is sufficient. Since we have also argued that the violation of ethical conditions such as anonymity on graphs which are not likely to be observed is not a matter for concern, our results suggest an interesting avenue for avoiding the potential conflict between stability and efficiency in the context of this framework.

5 Conclusion

The central theme in this paper has been to examine the possibility of constructing allocation rules which will ensure that efficient networks of agents form when the

[18] The referee rightly points out that this rule implements the set of efficient graphs.

[19] We are grateful to the Associate Editor for this suggestion.

individual agents decide to form or severe links amongst themselves. Exploiting the insights provided by Jackson and Wolinsky [8], it is shown that in general it may not be possible to reconcile efficiency with stability if the allocation rule is required to be anonymous on all graphs.

However, we go on to argue that if our prediction is that *only* efficient graphs will form, then the requirement that the allocation rule be anonymous on all graphs is unnecessarily stringent. We suggest that a "mechanism design" approach is more appropriate and show that under almost all value functions, the nonempty set of (strongly) stable graphs will be a subset of the efficient graphs under an allocation rule which is anonymous on the domain of strongly stable graphs. A stronger domain restriction allows us to prove that the above result also holds when strong stability is replaced by weak stability. Since these allocation rules will treat agents symmetrically on the graphs which are "likely to be observed", it seems that stability can be reconciled with efficiency after all.

References

1. R. Aumann and R. Myerson (1988) Endogenous formation of links between players and coalitions: An application of the Shapley value. In: (A. Roth (Ed.) *The Shapley Value*, Cambridge Univ. Press, Cambridge, UK.
2. B. D. Bernheim, B. Peleg, M. Whinston (1987) Coalition-proof Nash equilibria I. Concepts, *J. Econ. Theory* 42: 1–12.
3. B. Dutta, A. van den Nouweland, and S. Tijs (1995) Link Formation in Cooperative Situations, Discussion Paper 95-02, Indian Statistical Institute, New Delhi.
4. S. Goyal (1993) Sustainable Communication Networks, Tinbergen Institute Discussion Paper TI 93-250.
5. S. Hart and M. Kurz (1983) Endogenous formation of cooperative structures, *Econometrica* 51: 1047–1064.
6. K. Hendricks, M. Piccione, G. Tan (1995) The economics of hubs: The case of monopoly, *Rev. Econ. Stud.* 62: 83–100.
7. D. Henriet and H. Moulin (1995) Traffic-based cost allocation in a network, *Rund J. Econ.* 27: 332–345.
8. B. M. Jackson and A. Wolinsky (1996) A strategic model of economic and social networks, *J. Econ. Theory* 71: 44–74.
9. J. Landa (1983) The enigma of the Kula ring: Gift exchanges and primitive law and order, *Int. Rev. Luiv Econ.* 3: 137–160.
10. T. Marschak, S. Reichelstein (1993) Communication requirements for individual agents in network mechanisms and hierarchies. In: J. Ledyard (Ed.) *The Economics of Information Decentralization: Complexity, Efficiency and Stability*, Kluwer Academic Press, Boston.
11. R. Myerson (1977) Graphs and cooperation in games, *Math. Oper. Res.* 2: 225–229.
12. R. Myerson (1991) Game Theory: Analysis of Conflict, Harvard Univ. Press, Cambridge.
13. A. van den Nouweland (1993) Games and Graphs in Economic Situations, Ph. D. thesis, Tilburg University, The Netherlands.
14. C. Qin (1996) Endogenous formation of cooperative structures, *J. Econ. Theory* 69: 218–226.
15. W. Sharkey (1993) Network models in economics. In: *The Handbook of Operations Research and Management Science*.
16. B. Wellman and S. Berkowitz (1988) *Social Structure: A Network Approach*, Cambridge Univ. Press, Cambridge, UK.

The Stability and Efficiency of Economic and Social Networks

Matthew O. Jackson

HSS 228-77, California Institute of Technology, Pasadena, California 91125, USA
e-mail: jacksonm@hss.caltech.edu and http://www.hss.caltech.edu/~jacksonm/Jackson.html.

Abstract. This paper studies the formation of networks among individuals. The focus is on the compatibility of overall societal welfare with individual incentives to form and sever links. The paper reviews and synthesizes some previous results on the subject, and also provides new results on the existence of pairwise-stable networks and the relationship between pairwise stable and efficient networks in a variety of contexts and under several definitions of efficiency.

1 Introduction

Many interactions, both economic and social, involve network relationships. Most importantly, in many interactions the specifics of the network structure are important in determining the outcome. The most basic example is the exchange of information. For instance, personal contacts play critical roles in obtaining information about job opportunities (e.g., Boorman (1975), Montgomery (1991), Topa (1996), Arrow and Berkowitz (2000), and Calvo-Armengol (2000)). Networks also play important roles in the trade and exchange of goods in non-centralized markets (e.g., Tesfatsion (1997, 1998), Weisbuch, Kirman and Herreiner (1995)), and in providing mutual insurance in developing countries (e.g., Fafchamps and Lund (1997)).

Although it is clear that network structures are of fundamental importance in determining outcomes of a wide variety of social and economic interactions, far beyond those mentioned above, we are only beginning to develop theoretical models that are useful in a systematic analysis of how such network structures

This paper is partly based on a lecture given at the first meeting of the Society for Economic Design in Istanbul in June 2000. I thank Murat Sertel for affording me that opportunity, and Semih Koray for making the meeting a reality. I also thank the participants of SITE 2000 for feedback on some of the results presented here. I am grateful to Gabrielle Demange, Bhaskar Dutta, Alison Watts, and Asher Wolinsky for helpful conversations.

form and what their characteristics are likely to be. This paper outlines such an area of research on network formation. The aim is to develop a systematic analysis of how incentives of individuals to form networks align with social efficiency. That is, when do the private incentives of individuals to form ties with one another lead to network structures that maximize some appropriate measure of social efficiency?

This paper synthesizes and reviews some results from the previous literature on this issue,[1] and also presents some new results and insights into circumstances under private incentives to form networks align with social efficiency.

The paper is structured as follows. The next section provides some basic definitions and a few simple stylized examples of network settings that have been explored in the recent literature. Next, three definitions of efficiency of networks are presented. These correspond to three perspectives on societal welfare which differ based on the degree to which intervention and transfers of value are possible. The first is the usual notion of Pareto efficiency, where a network is Pareto efficient if no other network leads to better payoffs for all individuals of the society. The second is the much stronger notion of efficiency, where a network is efficient if it maximizes the sum of payoffs of the individuals of the society. This stronger notion is essentially one that considers value to be arbitrarily transferable across individuals in the society. The third is an intermediate notion of efficiency that allows for a natural, but limited class of transfers to be made across individuals of the society. With these definitions of efficiency in hand, the paper turns its focus on the existence and properties of pairwise stable networks, i.e., those where individuals have no incentives to form any new links or sever any existing links. Finally, the compatibility of the different efficiency notions and pairwise stability is studied from a series of different angles.

2 Definitions

Networks[2]

A set $N = \{1, \ldots, n\}$ of individuals are connected in a network relationship. These may be people, firms, or other entities depending on the application.

[1] There is a large and growing literature on network interactions, and this paper does not attempt to survey it. Instead, the focus here is on a strand of the economics literature that uses game theoretic models to study the formation and efficiency of networks. Let me offer just a few tips on where to start discovering the other portions of the literature on social and economic networks. There is an enormous "social networks" literature in sociology that is almost entirely complementary to the literature that has developed in economics. An excellent and broad introductory text to the social networks literature is Wasserman and Faust (1994). Within that literature there is a branch which has used game theoretic tools (e.g., studying exchange through cooperative game theoretic concepts). A good starting reference for that branch is Bienenstock and Bonacich (1997). There is also a game theory literature that studies communication structures in cooperative games. That literature is a bit closer to that covered here, and the seminal reference is Myerson (1977) which is discussed in various pieces here. A nice overview of that literature is provided by Slikker (2000).

[2] The notation and basic definitions follow Jackson and Wolinsky (1996) when convenient.

The network relationships are reciprocal and the network is thus modeled as a non-directed graph. Individuals are the nodes in the graph and links indicate bilateral relationships between the individuals.[3] Thus, a network g is simply a list of which pairs of individuals are linked to each other. If we are considering a pair of individuals i and j, then $\{i,j\} \in g$ indicates that i and j are linked under the network g.

There are many variations on networks which can be considered and are appropriate for different sorts of applications.[4] Here it is important that links are bilateral. This is appropriate, for instance, in modeling many social ties such as marriage, friendship, as well as a variety of economic relationships such as alliances, exchange, and insurance, among others. The key important feature is that it takes the consent of *both* parties in order for a link to form. If one party does not consent, then the relationship cannot exist. There are other situations where the relationships may be unilateral: for instance advertising or links to web sites. Those relationships are more appropriately modeled by directed networks.[5] As some degree of mutual consent is the more commonly applicable case, I focus attention here on non-directed networks.

An important restriction of such a simple graph model of networks is that links are either present or not, and there is no variation in intensity. This does not distinguish, for instance, between strong and weak ties which has been an important area of research.[6] Nevertheless, the simple graph model of networks is a good first approximation to many economic and social interactions and a remarkably rich one.

For simplicity, write ij to represent the link $\{i,j\}$, and so $ij \in g$ indicates that i and j are linked under the network g.

More formally, let g^N be the set of all subsets of N of size 2. $G = \{g \subset g^N\}$ denotes the set of all possible networks or graphs on N, with g^N being the full or complete network.

For instance, if $N = \{1,2,3\}$ then $g = \{12, 23\}$ is the network where there is a link between individuals 1 and 2, a link between individuals 2 and 3, but no link between individuals 1 and 3.

The network obtained by adding link ij to an existing network g is denoted by $g + ij$ and the network obtained by deleting link ij from an existing network g is denoted $g - ij$.

For any network g, let $N(g)$ be the set of individuals who have at least one link in the network g. That is, $N(g) = \{i \mid \exists j \text{ s.t. } ij \in g\}$.

---

[3] The word "link" follows Myerson's (1977) usage. The literature in economics and game theory has largely followed that terminology. In the social networks literature in sociology, the term "tie" is standard. Of course, in the graph theory literature the terms vertices and edges (or arcs) are standard. I will try to keep a uniform usage of individual and link in this paper, with the appropriate translations applying.

[4] A nice overview appears in Wasserman and Faust (1994).

[5] For some analysis of the formation and efficiency of such networks see Bala and Goyal (2000) and Dutta and Jackson (2000).

[6] For some early references in that literature, see Granovetter (1973) and Boorman (1975).

Paths and Components

Given a network $g \in G$, a *path* in g between i an j is a sequence of individuals $i_1, \ldots, i_K$ such that $i_k i_{k+1} \in g$ for each $k \in \{1, \ldots, K-1\}$, with $i_1 = i$ and $i_K = j$.

A (connected) *component* of a network g, is a nonempty subnetwork $g' \subset g$, such that

- if $i \in N(g')$ and $j \in N(g')$ where $j \neq i$, then there exists a path in g' between i and j, and
- if $i \in N(g')$ and $j \notin N(g')$ then there does not exist a path in g between i and j.

Thus, the components of a network are the distinct connected subgraphs of a network.

The set of components of g is denoted $C(g)$. Note that $g = \cup_{g' \in C(g)} g'$.

Value Functions

Different network configurations lead to different values of overall production or overall utility to a society. These various possible valuations are represented via a value function.

A *value function* is a function $v : G \to \mathbb{R}$.

I maintain the normalization that $v(\emptyset) = 0$.

The set of all possible value functions is denoted $\mathscr{V}$.

Note that different networks that connect the same individuals may lead to different values. This makes a value function a much richer object than a characteristic function used in cooperative game theory. For instance, a soceity $N = \{1, 2, 3\}$ may have a different value depending on whether it is connected via the network $g = \{12, 23\}$ or the network $g^N = \{12, 23, 13\}$.

The special case where the value function depends only on the groups of agents that are connected, but not how they are connected, corresponds to the communication networks considered by Myerson (1977).[7] In most applications, however, there may be some cost to links and thus some difference in total value across networks even if they connect the same sets of players, and so this more general and flexible formulation is more powerful and encompasses many more applications.

[7] To be precise, Myerson started with a transferable utility cooperative game in characteristic function form, and layered on top of that network structures that indicated which agents could communicate. A coalition could only generate value if its members were connected via paths in the network. But, the particular structure of the network did not matter, as long as the coalition's members were connected somehow. In the approach taken here (following Jackson and Wolinsky (1996)), the value is a function that is allowed to depend on the specific network structure. A special case is where $v(g)$ only depends on the coalitions induced by the component structure of g, which corresponds to the communication games.

It is also important to note that the value function can incorporate costs to links as well as benefits. It allows for arbitrary ways in which costs and benefits may vary across networks. This means that a value function allows for externalities both within and across components of a network.

Allocation Rules

A value function only keeps track of how the total societal value varies across different networks. We also wish to keep track of how that value is allocated or distributed among the individuals forming a network.

An *allocation rule* is a function $Y : G \times \mathcal{V} \to \mathbb{R}^N$ such that $\sum_i Y_i(g, v) = v(g)$ for all v and g.[8]

It is important to note that an allocation rule depends on both g and v. This allows an allocation rule to take full account of an individual i's role in the network. This includes not only what the network configuration is, but also and how the value generated depends on the overall network structure. For instance, consider a network $g = \{12, 23\}$ in a situation where $v(g) = 1$. Individual 2's allocation might be very different on what the value of other networks are. For instance, if $v(\{12, 23, 13\}) = 0 = v(\{13\})$, then in a sense 2 is essential to the network and may receive a large allocation. If on the other hand $v(g') = 1$ for all networks, then 2's role is not particularly special. This information can be relevant, which is why the allocation rule is allowed (but not required) to depend on it.

There are two different perspectives on allocation rules that will be important in different contexts. First, an allocation rule may simply represent the natural payoff to different individuals depending on their role in the network. This could include bargaining among the individuals, or any form of interaction. This might be viewed as the "naturally arising allocation rule" and is illustrated in the examples below. Second, an allocation rule can be an object of economic design, i.e., representing net payoffs resulting from natural payoffs coupled with some intervention via transfers, taxes, or subsidies. In what follows we will be interested in when the natural underlying payoffs lead individuals to form efficient networks, as well as when intervention can help lead to efficient networks.

Before turning to that analysis, let us consider some examples of models of social and economic networks and the corresponding value functions and allocation rules that describe them.

Some Illustrative Examples

Example 1. *The Connections Model* (Jackson and Wolinsky (1996))

---

[8] This definition builds balance $\left(\sum_i Y_i(g, v) = v(g)\right)$ into the definition of allocation rule. This is without loss of generality for the discussion in this paper, but there may be contexts in which imbalanced allocation rules are of interest.

The basic connections model is described as follows. Links represent social relationships between individuals; for instance friendships. These relationships offer benefits in terms of favors, information, etc., and also involve some costs. Moreover, individuals also benefit from indirect relationships. A "friend of a friend" also results in some benefits, although of a lesser value than a "friend," as do "friends of a friend of a friend" and so forth. The benefit deteriorates in the "distance" of the relationship. For instance, in the network $g = \{12, 23, 34\}$ individual 1 gets a benefit δ from the direct connection with individual 2, an indirect benefit δ^2 from the indirect connection with individual 3, and an indirect benefit δ^3 from the indirect connection with individual 4. For $\delta < 1$ this leads to a lower benefit from an indirect connection than a direct one. Individuals only pay costs, however, for maintaining their direct relationships. These payoffs and benefits may be relation specific, and so are indexed by ij.

Formally, the payoff player i receives from network g is

$$Y_i(g) = \sum_{j \neq i} \delta_{ij}^{t(ij)} - \sum_{j:ij \in g} c_{ij},$$

where $t(ij)$ is the number of links in the shortest path between i and j (setting $t(ij) = \infty$ if there is no path between i and j).[9] The value function in the connections model of a network g is simply $v(g) = \sum_i Y_i(g)$.

Some special cases are of particular interest. The first is the "symmetric connections model" where there are common δ and c such that $\delta_{ij} = \delta$ and $c_{ij} = c$ for all i and j. This case is studied extensively in Jackson and Wolinsky (1996).

The second is one with spatial costs, where there is a geography to locations and c_{ij} is related to distance (for instance, if individuals are spaced equally on a line then costs are proportional to $|i - j|$). This is studied extensively in Johnson and Gilles (2000).

Example 2. *The Co-Author Model* (Jackson and Wolinsky (1996))

The co-author model is described as follows. Each individual is a researcher who spends time working on research projects. If two researchers are connected, then they are working on a project together. The amount of time researcher i spends on a given project is inversely related to the number of projects, n_i, that he is involved in. Formally, i's payoff is represented by

$$Y_i(g) = \sum_{j:ij \in g} \frac{1}{n_i} + \frac{1}{n_j} + \frac{1}{n_i n_j}$$

for $n_i > 0$, and $Y_i(g) = 0$ if $n_i = 0$.[10] The total value is $v(g) = \sum_i Y_i(g)$.

[9] $t(ij)$ is sometimes referred to as the geodesic.

[10] It might also make sense to set $Y_i(g) = 1$ when an individual has no links, as the person can still produce reseach! This is not in keeping with the normalization of $v(\emptyset) = 0$, but it is easy to simply subtract 1 from all payoffs and then view Y as the extra benefits above working alone.

Note that in the co-author model there are no directly modeled costs to links. Costs come indirectly in terms of diluted synergy in interaction with co-authors.

Example 3. *A Bilateral Bargaining Model* (Corominas-Bosch (1999))

Corominas-Bosch (1999) considers a bargaining model where buyers and sellers bargain over prices for trade. A link is necessary between a buyer and seller for a transaction to occur, but if an individual has several links then there are several possibilities as to whom they might transact with. Thus, the network structure essentially determines bargaining power of various buyers and sellers.

More specifically, each seller has a single unit of an indivisible good to sell which has no value to the seller. Buyers have a valuation of 1 for a single unit of the good. If a buyer and seller exchange at a price p, then the buyer receives a payoff of $1-p$ and the seller a payoff of p. A link in the network represents the opportunity for a buyer and seller to bargain and potentially exchange a good.[11]

Corominas-Bosch models bargaining via the following variation on a Rubinstein bargaining protocol. In the first period sellers simultaneously each call out a price. A buyer can only select from the prices that she has heard called out by the sellers to whom she is linked. Buyers simultaneously respond by either choosing to accept some single price offer they received, or to reject all price offers they received.[12] If there are several sellers who have called out the same price and/or several buyers who have accepted the same price, and there is any discretion under the given network connections as to which trades should occur, then there is a careful protocol for determining which trades occur (which is essentially designed to maximize the number of eventual transactions).

At the end of the period, trades are made and buyers and sellers who have traded are cleared from the market. In the next period the situation reverses and buyers call out prices. These are then either accepted or rejected by the sellers connected to them in the same way as described above. Each period the role of proposer and responder switches and this process repeats itself indefinitely, until all remaining buyers and sellers are not linked to each other.

Buyers and sellers are impatient and discount according to a common discount factor $0 < \delta < 1$. So a transaction at price p in period t is worth only $\delta^t p$ to a seller and $\delta^t(1-p)$ to a buyer.

Corominas-Bosch outlines a subgame perfect equilibrium of the above game, and this equilibrium has a very nice interpretation as the discount factor approaches 1.

Some easy special cases are as follows. First, consider a seller linked to each of two buyers, who are only linked to that seller. Competition between the buyers to accept the price will lead to an equilibrium price of 1. So the payoff to the

[11] In the Corominas-Bosch framework links can only form between buyers and sellers. One can fit this into the more general setting where links can form between any individuals, by having the value function and allocation rule ignore any links except those between buyers and sellers.

[12] So buyers accept or reject price offers, rather than accepting or rejecting the offer of some specific seller.

seller in such a network will be 1 while the payoff to the buyers will be 0. This is reversed for a single buyer linked to two sellers. Next, consider a single seller linked to a single buyer. That corresponds to Rubinstein bargaining, and so the price (in the limit as $\delta \to 1$) is 1/2, as are the payoffs to the buyer and seller.

More generally, which side of the market outnumbers the other is a bit tricky to determine as it depends on the overall link structure which can be much more complicated than that described above. Quite cleverly, Corominas-Bosch describes an algorithm[13] for subdividing any network into three types of subnetworks: those where a set of sellers are collectively linked to a larger set of buyers and sellers get payoffs of 1 and buyers 0, those where the collective set of sellers is linked to a same-sized collective set of buyers and each get payoff of 1/2, and those where sellers outnumber buyers and sellers get payoffs of 0 and buyers 1. This is illustrated in Fig. 1 for a few networks.

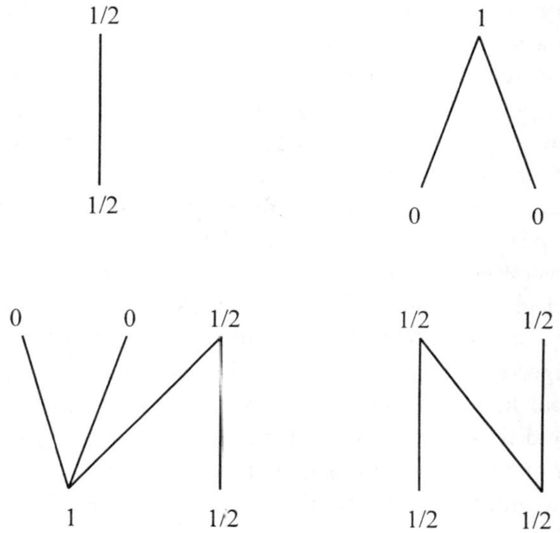

Fig. 1.

While the algorithm prevents us from providing a simple formula for the allocation rule in this model, the important characteristics of the allocation rule for our purposes can be summarized as follows.

(i) if a buyer gets a payoff of 1, then some seller linked to that buyer must get a payoff of 0, and similarly if the roles are reversed,

[13] The decomposition is based on Hall's (marriage) Theorem, and works roughly as follows. Start by identifying groups of two or more sellers who are all linked only to the same buyer. Regardless of that buyer's other connections, take that set of sellers and buyer out as a subgraph where that buyer gets a payoff of 1 and the sellers all get payoffs of 0. Proceed, inductively in k, to identify subnetworks where some collection of more than k sellers are collectively linked to k or fewer buyers. Next reverse the process and progressively in k look for at least k buyers collectively linked to fewer than k sellers, removing such subgraphs and assigning those sellers payoffs of 1 and buyers payoffs of 0. When all such subgraphs are removed, the remaining subgraphs all have "even" connections and earn payoffs of 1/2.

(ii) a buyer and seller who are only linked to each other get payoffs of 1/2, and
(iii) a connected component is such that all buyers and all sellers get payoffs of 1/2 if and only if any subgroup of k buyers in the component can be matched with at least k distinct sellers and vice versa.

In what follows, I will augment the Corominas-Bosch model to consider a cost to each link of c_s for sellers and c_b for buyers. So the payoff to an individual is their payoff from any trade via the bargaining on the network, less the cost of maintaining any links that they are involved with.

Example 4. *A Model of Buyer-Seller Networks* (Kranton and Minehart (1998))

The Kranton and Minehart model of buyer-seller networks is similar to the Corominas-Bosch model described above except that the valuations of the buyers for a good are random and the determination of prices is made through an auction rather than alternating offers bargaining.

The Kranton and Minehart model is described as follows. Again, each seller has an indivisible object for sale. Buyers have independently and identically distributed utilities for the object, denoted u_i. Each buyer knows her own valuation, but only the distribution over other buyers' valuations, and similarly sellers know only the distribution of buyers' valuations.

Again, link patterns represent the potential transactions, however, the transactions and prices are determined by an auction rather than bargaining. In particular, prices rise simultaneously across all sellers. Buyers drop out when the price exceeds their valuation (as they would in an English or ascending oral auction). As buyers drop out, there emerge sets of sellers for whom the remaining buyers still linked to those sellers is no larger than the set of sellers. Those sellers transact with the buyers still linked to them.[14] The exact matching of whom trades with whom given the link pattern is done carefully to maximize the number of transactions. Those sellers and buyers are cleared from the market, and the prices continue to rise among remaining sellers, and the process repeats itself.

For each link pattern every individual has a well-defined expected payoff from the above described process (from an ex-ante perspective before buyers know their u_i's). From this expected payoff can be deducted costs of links to both buyers and sellers.[15]

This leads to well-defined allocation rules Y_i's and a well-defined value function v. The main intuitions behind the Kranton and Minehart model are easily seen in a simple case, as follows.

Consider a situation with one seller and n buyers. Let the u_i's be uniformly and independently distributed on $[0, 1]$. In this case the auction simplifies to a

[14] It is possible, that several buyers drop out at once and so one or more of the buyers dropping out will be selected to transact at that price.

[15] Kranton and Minehart (1998) only consider costs of links to buyers. They also consider potential investment costs to sellers of producing a good for sale, but sellers do not incur any cost per link. Here, I will consider links as being costly to sellers as well as buyers.

standard second-price auction. If k is the number of buyers linked to the seller, the expected revenue to the seller is the second order statistic out of k, which is $\frac{k-1}{k+1}$ for a uniform distribution. The corresponding expected payoff to a bidder is $\frac{1}{k(k+1)}$.[16]

For a cost per link of c_s to the seller and c_b to the buyer, the allocation rule for any network g with $k \geq 1$ links between the buyers and seller is[17]

$$Y_i(g) = \begin{cases} \frac{1}{k(k+1)} - c_b & \text{if } i \text{ is a linked buyer} \\ \frac{k-1}{k+1} - kc_s & \text{if } i \text{ is the seller} \\ 0 & \text{if } i \text{ is a buyer without any links.} \end{cases} \tag{1}$$

The value function is then

$$v(g) = \sum_i Y_i(g) = \frac{k}{k+1} - k(c_s + c_b).$$

Thus, the total value of the network is simply the expected value of the good to the highest valued buyer less the cost of links.

Similar calculations can be done for larger numbers of sellers and more general network structures.

Some Basic Properties of Value and Allocation Functions

Component Additivity

A value function is *component additive* if $v(g) = \sum_{g' \in C(g)} v(g')$ for all $g \in G$.

Component additive value functions are ones for which the value of a network is simply the sum of the value of its components. This implies that the value of one component does not depend on the structure of other components. This condition is satisfied in Examples 1–4, and is satisfied in many economic and social situations. It still allows for arbitrary ways in value can depend on the network configuration within a component. Thus, it allows for externalities among individuals within a component.

An example where component additivity is violated is that of alliances among competing firms (e.g., see Goyal and Joshi (2000)), where the payoff to one set of interconnected firms may depend on how other competing firms are interconnected. So, what component additivity rules out is externalities *across* components of a network, but it still permits them within components.

[16] Each bidder has a $\frac{1}{k}$ chance of being the highest valued bidder. The expected valuation of the highest bidder for k draws from a uniform distribution on [0,1] is $\frac{k}{k+1}$, and the expected price is the expected second highest valuation which is $\frac{k-1}{k+1}$. Putting these together, the ex-ante expected payoff to any single bidder is $\frac{1}{k}\left(\frac{k}{k+1} - \frac{k-1}{k+1}\right) = \frac{1}{k(k+1)}$.

[17] For larger numbers of sellers, the Y_i's correspond to the V_i^b and V_i^s's in Kranton and Minehart (1999) (despite their footnote 16) with the subtraction here of a cost per link for sellers.

Component Balance

When a value function is component additive, the value generated by any component will often naturally be allocated to the individuals among that component. This is captured in the following definition.

An allocation rule Y is component balanced if for any component additive v, $g \in G$, and $g' \in C(g)$

$$\sum_{i \in N(g')} Y_i(g', v) = v(g').$$

Note that component balance only makes requirements on Y for v's that are component additive, and Y can be arbitrary otherwise. If v is not component additive, then requiring component balance of an allocation rule $Y(\cdot, v)$ would necessarily violate balance.

Component balance is satisfied in situations where Y represents the value naturally accruing in terms of utility or production, as the members of a given component have no incentive to distribute productive value to members outside of their component, given that there are no externalities across components (i.e., a component balanced v). This is the case in Examples 1–4, as in many other contexts.

Component balance may also be thought of as a normative property that one wishes to respect if Y includes some intervention by a government or outside authority - as it requires that that value generated by a given component be allocated among the members of that component. An important thing to note is that if Y violates component balance, then there will be some component receiving less than its net productive value. That component could improve the standing of all its members by seceding. Thus, one justification for the condition is as a component based participation constraint.[18]

Anonymity and Equal Treatment

Given a permutation of individuals π (a bijection from N to N) and any $g \in G$, let $g^\pi = \{\pi(i)\pi(j) | ij \in g\}$. Thus, g^π is a network that shares the same architecture as g but with the specific individuals permuted.

A value function is *anonymous* if for any permutation π and any $g \in G$, $v(g^\pi) = v(g)$.

Anonymous value functions are those such that the architecture of a network matters, but not the labels of individuals.

Given a permutation π, let v^π be defined by $v^\pi(g) = v(g^{\pi^{-1}})$ for each $g \in G$.

[18] This is a bit different from a standard individual rationality type of constraint given some outside option, as it may be that the value generated by a component is negative.

An allocation rule Y is *anonymous* if for any v, $g \in G$, and permutation π, $Y_{\pi(i)}(g^{\pi}, v^{\pi}) = Y_i(g, v)$.

Anonymity of an allocation rule requires that if all that has changed is the labels of the agents and the value generated by networks has changed in an exactly corresponding fashion, then the allocation only change according to the relabeling. Of course, anonymity is a type of fairness condition that has a rich axiomatic history, and also naturally arises situations where Y represents the utility or productive value coming directly from some social network.

Note that anonymity allows for asymmetries in the ways that allocation rules operate even in completely symmetric networks. For instance, anonymity does *not* require that each individual in a complete network get the same allocation. That would be true only in the case where v was in fact anonymous. Generally, an allocation rule can respond to different roles or powers of individuals and still be anonymous.

An allocation rule Y satisfies *equal treatment of equals* if for any anonymous $v \in \mathscr{V}$, $g \in G$, $i \in N$, and permutation π such that $g^{\pi} = g$, $Y_{\pi(i)}(g, v) = Y_i(g, v)$.

Equal treatment of equals says that all allocation rule should give the same payoff to individuals who play exactly the same role in terms of symmetric position in a network under a value function that depends only on the structure of a network. This is implied by anonymity, which is seen by noting that $(g^{\pi}, v^{\pi}) = (g, v)$ for any anonymous v and a π as described in the definition of equal treatment of equals. Equal treatment of equals is more of a symmetry condition that anonymity, and again is a condition that has a rich background in the axiomatic literature.

Some Prominent Allocation Rules

There are several allocation rules that are of particular interest that I now discuss. The first naturally arises in situations where the allocation rule comes from some bargaining (or other process) where the benefits that accrue to the individuals involved in a link are split equally among those two individuals.

Equal Bargaining Power and the Myerson Value

An allocation rule satisfies equal bargaining power if for any component additive v and $g \in G$

$$Y_i(g) - Y_i(g - ij) = Y_j(g) - Y_j(g - ij).$$

Note that equal bargaining power does *not* require that individuals split the marginal value of a link. It just requires that they equally benefit or suffer from its addition. It is possible (and generally the case) that $Y_i(g) - Y_i(g - ij) + Y_j(g) - Y_j(g - ij) \neq v(g) - v(g - ij)$.

It was first shown by Myerson (1977), in the context of communication games, that such a condition leads to an allocation that is a variation on the Shapley value. This rule was subsequently referred to as the Myerson value (e.g., see Aumann and Myerson (1988)).

The Myerson value also has a corresponding allocation rule in the context of networks beyond communication games, as shown by Jackson and Wolinsky (1996). That allocation rule is expressed as follows.

Let

$$g|_S = \{ij \ : \ ij \in g \text{ and } i \in S, j \in S\}.$$

Thus $g|_S$ is the network found deleting all links except those that are between individuals in S.

$$Y_i^{MV}(g, v) = \sum_{S \subset N \setminus \{i\}} (v(g|_{S \cup i}) - v(g|_S)) \left(\frac{\#S!(n - \#S - 1)!}{n!} \right) \quad (2)$$

The following proposition from Jackson and Wolinsky (1996) is an extension of Myerson's (1977) result from the communication game setting to the network setting.

Proposition 1. (Myerson (1977), Jackson and Wolinsky (1996))[19] *Y satisfies component balance and equal bargaining power if and only if $Y(g, v) = Y^{MV}(g, v)$ for all $g \in G$ and any component additive v.*

The surprising aspect of equal bargaining power is that it has such strong implications for the structuring of the allocation rule.

Egalitarian Rules

Two other allocation rules that are of particular interest are the egalitarian and component-wise egalitarian rule.

The *egalitarian* allocation rule Y^e is defined by

$$Y_i^e(g, v) = \frac{v(g)}{n}$$

for all i and g.

The egalitarian allocation rule splits the value of a network equally among all members of a society regardless of what their role in the network is. It is clear that the egalitarian allocation rule will have very nice properties in terms of aligning individual incentives with efficiency.

However, the egalitarian rule violates component balance. The following modification of the egalitarian rule respects component balance.

[19] Dutta and Mutuswami (1997) extend the characterization to allow for weighted bargaining power, and show that one obtains a version of a weighted Shapley (Myerson) value.

The *component-wise egalitarian* allocation rule Y^{ce} is defined as follows for component additive v's and any g.

$$Y_i^{ce}(g,v) = \begin{cases} \frac{v(h)}{|N(h)|} & \text{if there exists } h \in C(g) \text{ such that } i \in h, \\ 0 & \text{otherwise.} \end{cases}$$

For any v that is not component additive, set $Y^{ce}(\cdot, v) = Y^{e}(\cdot, v)$.

The component-wise egalitarian splits the value of a component network equally among all members of that component, but makes no transfers across components.

The component-wise egalitarian rule has some nice properties in terms of aligning individual incentives with efficiency, although not quite to the extent that the egalitarian rule does.[20]

3 Defining Efficiency

In evaluating societal welfare, we may take various perspectives. The basic notion used is that of Pareto efficiency - so that a network is inefficient if there is some other network that leads to higher payoffs for all members of the society. The differences in perspective derive from the degree to which transfers can be made between individuals in determining what the payoffs are.

One perspective is to see how well society functions on its own with no outside intervention (i.e., where Y arises naturally from the network interactions). We may also consider how the society fares when some intervention in the forms of redistribution takes place (i.e., where Y also incorporates some transfers). Depending on whether we allow arbitrary transfers or we require that such intervention satisfy conditions like anonymity and component balance, we end up with different degrees to which value can be redistributed. Thus, considering these various alternatives, we are led to several different definitions of efficiency of a network, depending on the perspective taken. Let us examine these in detail. I begin with the weakest notion.

Pareto Efficiency

A network g is *Pareto efficient* relative to v and Y if there does not exist any $g' \in G$ such that $Y_i(g',v) \geq Y_i(g,v)$ for all i with strict inequality for some i.

This definition of efficiency of a network takes Y as fixed, and hence can be thought of as applying to situations where no intervention is possible.

Next, let us consider the strongest notion of efficiency.[21]

---

[20] See Jackson and Wolinsky (1996) Section 4 for some detailed analysis of the properties of the egalitarian and component-wise egalitarian rules.

[21] This notion of efficiency was called strong efficiency in Jackson and Wolinsky (1996).

Efficiency

A network g is *efficient* relative to v if $v(g) \geq v(g')$ for all $g' \in G$.

This is a strong notion of efficiency as it takes the perspective that value is fully transferable. This applies in situations where unlimited intervention is possible, so that any naturally arising Y can be redistributed in arbitrary ways.

Another way to express efficiency is to say that g is efficient relative to v if it is Pareto efficient relative to v and Y *for all* Y. Thus, we see directly that this notion is appropriate in situations where one believes that arbitrary reallocations of value are possible.

Constrained Efficiency

The third notion of efficiency falls between the other two notions. Rather than allowing for arbitrary reallocations of value as in efficiency, or no reallocations of value as in Pareto efficiency, it allows for reallocations that are anonymous and component balanced.

A network g is *constrained efficient* relative to v if there does not exist any $g' \in G$ and a component balanced and anonymous Y such that $Y_i(g', v) \geq Y_i(g, v)$ for all i with strict inequality for some i.

Note that g is constrained efficient relative to v if and only if it is Pareto efficient relative to v and Y for every component balanced and anonymous Y.

There exist definitions of constrained efficiency for any class of allocation rules that one wishes to consider. For instance, one might also consider that class of component balanced allocation rules satisfying equal treatment of equals, or any other class that is appropriate in some context.

The relationship between the three definitions of efficiency we consider here is as follows. Let $PE(v, Y)$ denote the Pareto efficient networks relative to v and Y, and similarly let $CE(v)$ and $E(v)$ denote the constrained efficient and efficient networks relative to v, respective.

Remark: If Y is component balanced and anonymous, then $E(v) \subset CE(v) \subset PE(v, Y)$.

Given that there always exists an efficient network (any network that maximizes v, and such a network exists as G is finite), it follows that there also exist constrained efficient and Pareto efficient networks.

Let us also check that these definitions are distinct.

Example 5. $E(v) \neq CE(v)$

Let $n = 5$ and consider an anonymous and component additive v such that the complete network g^N has value 10, a component consisting of pair of individuals with one link between them has value 2, and a completely connected component among three individuals has value 9. All other networks have value 0.

The only efficient networks are those consisting of two components: one component consisting of a pair of individuals with one link and the other component consisting of a completely connected triad (set of three individuals). However, the completely connected network is constrained efficient.

To see that the completely connected network is constrained efficient even though it is not efficient, first note that any anonymous allocation rule must give each individual a payoff of 2 in the complete network. Next, note that the only network that could possibly give a higher allocation to all individuals is an efficient one consisting of two components: one dyad and one completely connected triad. Any component balanced and anonymous allocation rule must allocate payoffs of 3 to each individual in the triad, and 1 to each individual in the dyad. So, the individuals in the dyad are worse off than they were under the complete network. Thus, the fully connected network is Pareto efficient under every Y that is anonymous and component balanced. This implies that the fully connected network is constrained efficient even though it is not efficient. This is pictured in Fig. 2.

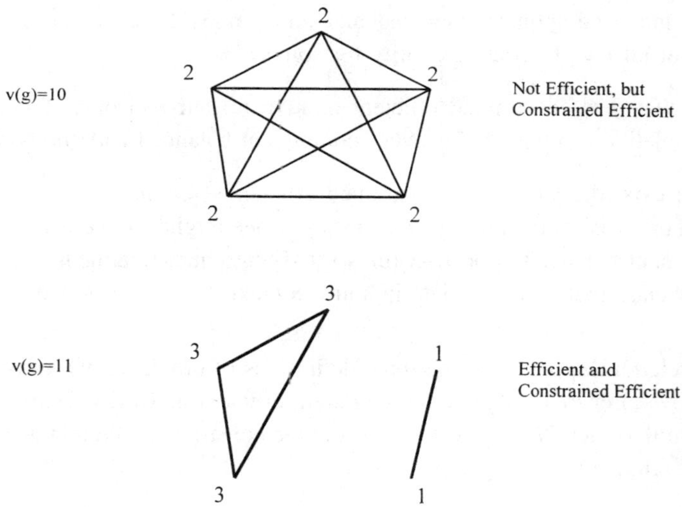

Fig. 2.

Example 6. $CE(v) \neq PE(v, Y)$

Let $n = 3$. Consider an anonymous v where the complete network has a value of 9, a network with two links has a value of 8, and a network of a single link network has any value.

Consider a component balanced and anonymous Y that allocates 3 to each individual in the complete network, and in any network with two links allocates

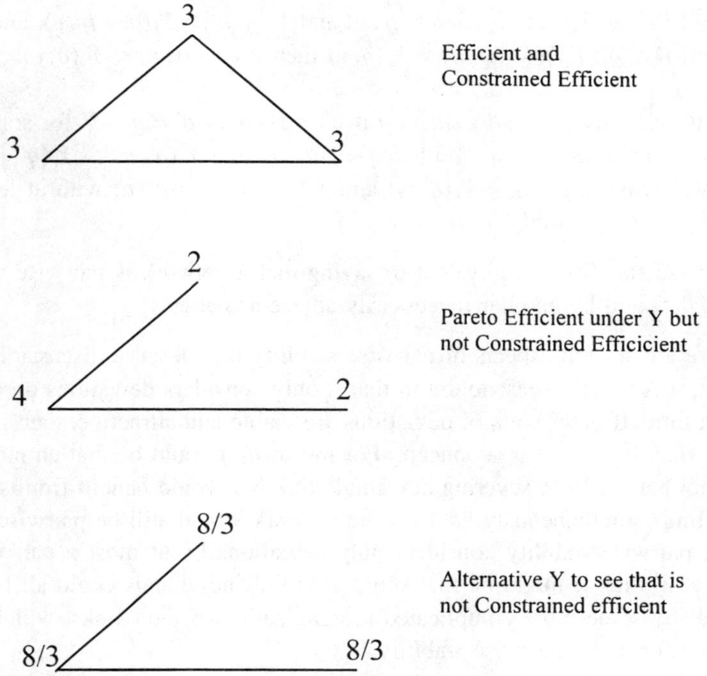

Fig. 3.

2 to each of the individuals with just one link and 4 to the individual with two links (and splits value equally among the two individuals in a link if there is just one link). The network $g = \{12, 23\}$ is Pareto efficient relative to v and Y, since any other network results in a lower payoff to at least one of the players (for instance, $Y_2(g, v) = 4$, while $Y_2(g^N, v) = 3$). The network g is not constrained efficient, since under the component balanced and anonymous rule $\overline{Y}$ such that $\overline{Y}_1(g, v) = \overline{Y}_2(g, v) = \overline{Y}_3(g, v) = 8/3$, all individuals prefer to be in the complete network g^N where they receive payoffs of 3. See Fig. 3.

4 Modeling Network Formation

A simple, tractable, and natural way to analyze the networks that one might expect to emerge in the long run is to examine a sort of equilibrium requirement that individuals not benefit from altering the structure of the network. A weak version of such a condition is the following pairwise stability notion defined by Jackson and Wolinsky (1996).

Pairwise Stability

A network g is *pairwise stable* with respect to allocation rule Y and value function v if

(i) for all $ij \in g$, $Y_i(g,v) \geq Y_i(g-ij,v)$ and $Y_j(g,v) \geq Y_j(g-ij,v)$, and
(ii) for all $ij \notin g$, if $Y_i(g+ij,v) > Y_i(g,v)$ then $Y_j(g+ij,v) < Y_j(g,v)$.

Let us say that g' is *adjacent* to g if $g' = g + ij$ or $g' = g - ij$ for some ij.
A network g' *defeats* g if either $g' = g - ij$ and $Y_i(g',v) > Y_i(g',v)$, or if $g' = g + ij$ with $Y_i(g',v) \geq Y_i(g',v)$ and $Y_i(g',v) \geq Y_i(g',v)$ with at least one inequality holding strictly.

Pairwise stability is equivalent to saying that a network is pairwise stable if it is not defeated by another (necessarily adjacent) network.

There are several aspects of pairwise stability that deserve discussion.
First, it is a very weak notion in that it only considers deviations on a single link at a time. If other sorts of deviations are viable and attractive, then pairwise stability may be too weak a concept. For instance, it could be that an individual would not benefit from severing any single link but would benefit from severing several links simultaneously, and yet the network would still be pairwise stable. Second, pairwise stability considers only deviations by at most a pair of individuals at a time. It might be that some group of individuals could all be made better off by some more complicated reorganization of their links, which is not accounted for under pairwise stability.
In both of these regards, pairwise stability might be thought of as a necessary but not sufficient requirement for a network to be stable over time. Nevertheless, we will see that pairwise stability already significantly narrows the class of networks to the point where efficiency and pairwise stability are already in tension at times.
There are alternative approaches to modeling network stability. One is to explicitly model a game by which links form and then to solve for an equilibrium of that game. Aumann and Myerson (1988) take such an approach in the context of communication games, where individuals sequentially propose links which are then accepted or rejected. Such an approach has the advantage that it allows one to use off-the-shelf game theoretic tools. However, such an approach also has the disadvantage that the game is necessarily ad hoc and fine details of the protocol (e.g., the ordering of who proposes links when, whether or not the game has a finite horizon, individuals are impatient, etc.) may matter. Pairwise stability can be thought of as a condition identifies networks that are the only ones that could emerge at the end of any well defined game where players where the process does not artificially end, but only ends when no player(s) wish to make further changes to the network.
Dutta and Mutuswami (1997) analyze the equilibria of a link formation game under various solution concepts and outline the relationship between pairwise stability and equilibria of that game. The game is one first discussed by Myerson (1991). Individuals simultaneously announce all the links they wish to be involved in. Links form if both individuals involved have announced that link. While such games have a multiplicity of unappealing Nash equilibria (e.g., nobody announces any links), using strong equilibrium and coalition-proof Nash

equilibrium, and variations on strong equilibrium where only pairs of individuals might deviate, lead to nicer classes of equilibria. The networks arising in variations of the strong equilibrium are in fact subsets of the pairwise stable networks.[22]

Finally, there is another aspect of network formation that deserves attention. The above definitions (including some of the game theoretic approaches) are both static and myopic. Individuals do not forecast how others might react to their actions. For instance, the adding or severing of one link might lead to the subsequent addition or severing of another link. Dynamic (but still myopic) network formation processes are studied by Watts (2001) and Jackson and Watts (1998), but a fully dynamic and forward looking analysis of network formation is still missing.[23]

Myopic considerations on the part of the individuals in a network are natural in large situations where individuals might be faced with the consideration of adding or severing a given link, but might have difficulty in forecasting the reactions to this. For instance, in deciding whether or not a firm wishes to connect its computer system to the internet, the firm might not forecast the impact of that decision on the future evolution of the internet. Likewise in forming a business contact or friendship, an individual might not forecast the impact of that new link on the future evolution of the network. Nevertheless, there are other situations, such as strategic alliances among airlines, where individuals might be very forward looking in forecasting how others will react to the decision. Such forward looking behavior has been analyzed in various contexts in the coalition formation literature (e.g., see Chwe (1994)), but is still an important issue for further consideration in the network formation literature.[24]

Existence of Pairwise Stable Networks

In some situations, there may not exist any pairwise stable network. It may be that each network is defeated by some adjacent network, and that these "improving paths" form cycles with no undefeated networks existing.[25]

[22] See Jackson and van den Nouweland (2000) for additional discussion of coalitional stability notions and the relationship to core based solutions.

[23] The approach of Aumann and Myerson (1988) is a sequential game and so forward thinking is incorporated to an extent. However, the finite termination of their game provides an artificial way by which one can put a limit on how far forward players have to look. This permits a solution of the game via backward induction, but does not seem to provide an adequate basis for a study of such forward thinking behavior. A more truly dynamic setting, where a network stays in place only if no player(s) wish to change it given their forecasts of what would happen subsequently, has not been analyzed.

[24] It is possible that with some forward looking aspects to behavior, situations are plausible where a network that is not pairwise stable emerges. For instance, individuals might not add a link that appears valuable to them given the current network, as that might in turn lead to the formation of other links and ultimately lower the payoffs of the original individuals. This is an important consideration that needs to be examined.

[25] Improving paths are defined by Jackson and Watts (1998), who provide some additional results on existence of pairwise stable networks.

An *improving path* is a sequence of networks $\{g_1, g_2, \ldots, g_K\}$ where each network g_k is defeated by the subsequent (adjacent) network g_{k+1}.

A network is pairwise stable if and only if it has no improving paths emanating from it. Given the finite number of networks, it then directly follows that if there does not exist any pairwise stable network, then there must exist at least one cycle, i.e., an improving path $\{g_1, g_2, \ldots, g_K\}$ where $g_1 = g_K$. The possibility of cycles and non-existence of a pairwise stable network is illustrated in the following example.

Example 7. *Exchange Networks – Non-existence of a Pairwise Stable Network* (Jackson and Watts (1998))

The society consists of $n \geq 4$ individuals who get value from trading goods with each other. In particular, there are two consumption goods and individuals all have the same utility function for the two goods which is Cobb-Douglas, $u(x, y) = xy$. Individuals have a random endowment, which is independently and identically distributed. An individual's endowment is either (1,0) or (0,1), each with probability 1/2.

Individuals can trade with any of the other individuals in the same component of the network. For instance, in a network $g = \{12, 23, 45\}$, individuals 1, 2 and 3 can trade with each other and individuals 4 and 5 can trade with each other, but there is no trade between 123 and 45. Trade flows without friction along any path and each connected component trades to a Walrasian equilibrium. This means, for instance, that the networks $\{12, 23\}$ and $\{12, 23, 13\}$ lead to the same expected trades, but lead to different costs of links.

The network $g = \{12\}$ leads to the following payoffs. There is a $\frac{1}{2}$ probability that one individual has an endowment of (1,0) and the other has an endowment of (0,1). They then trade to the Walrasian allocation of $(\frac{1}{2}, \frac{1}{2})$ each and so their utility is $\frac{1}{4}$ each. There is also a $\frac{1}{2}$ probability that the individuals have the same endowment and then there are no gains from trade and they each get a utility of 0. Expecting over these two situations leads to an expected utility of $\frac{1}{8}$. Thus, $Y_1(\{12\}) = Y_2(\{12\}) = \frac{1}{8} - c$, where c is the cost (in utils) of maintaining a link. One can do similar calculations for a network $\{12, 23\}$ and so forth.

Let the cost of a link $c = \frac{5}{96}$ (to each individual in the link).

Let us check that there does not exist a pairwise stable network. The utility of being alone is 0. Not accounting for the cost of links, the expected utility for an individual of being connected to one other is $\frac{1}{8}$. The expected utility for an individual of being connected (directly or indirectly) to two other individuals is $\frac{1}{6}$; and of being connected to three other individuals is $\frac{3}{16}$. It is easily checked that the expected utility of an individual is increasing and strictly concave in the number of other individuals that she is directly or indirectly connected to, ignoring the cost of links.

Now let us account for the cost of links and argue that there cannot exist any pairwise stable network. Any component in a pairwise stable network that connects k individuals must have exactly $k - 1$ links, as some additional link could be severed without changing the expected trades but saving the cost of the link. Also, any component in a pairwise stable network that involves three or more individuals cannot contain an individual who has just one link. This follows from the fact that an individual connected to some individual who is not connected to anyone else, loses at most $\frac{1}{6} - \frac{1}{8} = \frac{1}{24}$ in expected utility from trades by severing the link, but saves the cost of $\frac{5}{96}$ and so should sever this link. These two observations imply that a pairwise stable network must consist of pairs of connected individuals (as two completely unconnected individuals benefit from forming a link), with one unconnected individual if n is odd. However, such a network is not pairwise stable, since any two individuals in different pairs gain from forming a link (their utility goes from $\frac{1}{8} - \frac{5}{96}$ to $\frac{3}{16} - \frac{5}{96}$). Thus, there is no pairwise stable network. This is illustrated in Fig. 4.

(All payoffs are in 96-th's.)

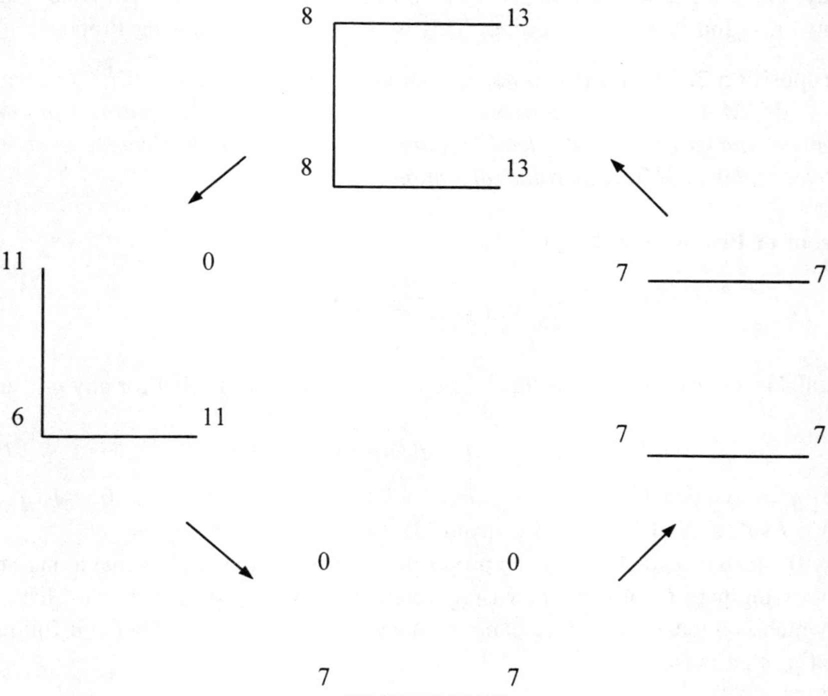

Fig. 4.

A cycle in this example is $\{12, 34\}$ is defeated by $\{12, 23, 34\}$ which is defeated by $\{12, 23\}$ which is defeated by $\{12\}$ which is defeated by $\{12, 34\}$.

Existence of Pairwise Stable Networks Under the Myerson Value

While the above example shows that pairwise stable networks may not exist in some settings for some allocation rules, there are interesting allocation rules for which pairwise stable networks always exist.

Existence of pairwise stable networks is straightforward for the egalitarian and component-wise egalitarian allocation rules. Under the egalitarian rule, any efficient network will be pairwise stable. Under the component-wise egalitarian rule, one can also always find a pairwise stable network. An algorithm is as follows:[26] find a component h that maximizes the payoff $Y_i^{ce}(h, v)$ over i and h. Next, do the same on the remaining population $N \setminus N(h)$, and so on. The collection of resulting components forms the network.[27]

What is less transparent, is that the Myerson value allocation rule also has very nice existence properties. Under the Myerson value allocation rule there always exists a pairwise stable network, all improving paths lead to pairwise stable networks, and there are no cycles. This is shown in the following Proposition.

Proposition 2. *There exists a pairwise stable network relative to Y^{MV} for every $v \in \mathscr{V}$. Moreover, all improving paths (relative to Y^{MV}) emanating from any network (under any $v \in \mathscr{V}$) lead to pairwise stable networks. Thus, there are no cycles under the Myerson value allocation rule.*

Proof of Proposition 2. Let

$$F(g) = \sum_{T \subset N} v(g|_T) \left(\frac{(|T| - 1)!(n - |T|)!}{n!} \right).$$

Straightforward calculations that are left to the reader verify that for any g, i and $ij \in g$ [28]

$$Y_i^{MV}(g, v) - Y_i^{MV}(g - ij, v) = F(g) - F(g - ij). \tag{3}$$

Let g^* maximize $F(\cdot)$. Thus $0 \geq F(g^* + ij) - F(g^*)$ and likewise $0 \geq F(g^* - ij) - F(g^*)$ for all ij. It follows from (3) that g^* is pairwise stable.

To see the second part of the proposition, note that (3) implies that along any improving path F must be increasing. Such an increasing path in F must lead to g which is a local maximizer (among adjacent networks) of F. By (3) it follows that g is pairwise stable.[29] □

---

[26] This is specified for component additive v's. For any other v, Y^e and Y^{ce} coincide.

[27] This follows the same argument as existence of core-stable coalition structures under the weak top coalition property in Banerjee, Konishi and Sönmez (2001). However, these networks are not necessarily stable in a stronger sense (against coalitional deviations). A characterization for when such strongly stable networks exist appears in Jackson and van den Nouweland (2001).

[28] It helps in these calculations to note that if $i \notin T$ then $g|_T = g - ij|_T$. Note that F is what is known as a potential function (see Monderer and Shapley (1996)). Based on some results in Monderer and Shapley (1996) (see also Quin (1996)), potential functions and the Shapley value have a special relationship; and it may be that there is a limited converse to Proposition 2.

[29] Jackson and Watts (1998, working paper version) show that for any Y and v there exist no cycles (and thus there exist pairwise stable networks and all improving paths lead to pairwise stable

5 The Compatibility of Efficiency and Stability

Let us now turn to the central question of the relationship between stability and efficiency of networks.

As mentioned briefly above, if one has complete control over the allocation rule and does not wish to respect component balance, then it is easy to guarantee that all efficient networks are pairwise stable: simply use the egalitarian allocation rule Y^e. While this is partly reassuring, we are also interested in knowing whether it is generally the case that some efficient network is pairwise stable without intervention, or with intervention that respects component balance. The following proposition shows that there is no component balanced and anonymous allocation rule for which it is always the case that some constrained efficient network is pairwise stable.

Proposition 3. *There does not exist any component balanced and anonymous allocation rule (or even a component balanced rule satisfying equal treatment of equals) such that for every v there exists a constrained efficient network that is pairwise stable.*

Proposition 3 strengthens Theorem 1 in Jackson and Wolinsky (1996) in two ways: first it holds under equal treatment of equals rather than anonymity, and second it applies to constrained efficiency rather than efficiency. Most importantly, the consideration of constrained efficiency is more natural that the consideration of the stronger efficiency notion, given that it applies to component balanced and anonymous allocation rules.

The proof of Proposition 3 shows that there is a particular v such that for every component balanced and anonymous allocation rule none of the constrained efficient networks are pairwise stable. It uses the same value function as Jackson and Wolinsky (1996) used to prove a similar proposition for efficient networks rather than constrained efficient networks. The main complication in the proof is showing that there is a unique constrained efficient architecture and that it coincides with the efficient architecture. As the structure of the value function is quite simple and natural, and the difficulty also holds for many variations on it, the proposition is disturbing. The proof appears in the appendix.

Proposition 3 is tight. If we drop component balance, then as mentioned above the egalitarian rule leads to $E(v) \subset PS(Y^e, v)$ for all v. If we drop anonymity (or equal treatment of equals), then a careful and clever construction of Y by Dutta and Mutuswami (1997) ensures that $E(v) \cap PS(Y, v) \neq \emptyset$ for a class of v. This is stated in the following proposition.

Let $\mathscr{V}^* = \{v \in \mathscr{V} \mid g \neq \emptyset \Rightarrow v(g) > 0\}$

networks) if and only if there exists a function $F : G \rightarrow \mathbb{R}$ such that g defeats g' if and only if $F(g) > F(g')$. Thus, the existence of the F satisfying (3) in this proof is actually a necessary condition for such nicely behaved improving paths.

Proposition 4. (Dutta and Mutuswami (1997)) *There exists a component balanced Y such that $E(v) \cap PS(Y,v) \neq \emptyset$ for all $v \in \mathscr{V}^*$. Moreover, Y is anonymous on some networks in $E(v) \cap PS(Y,v)$.*[30] [31]

This proposition shows that if one can design an allocation rule, and only wishes to satisfy anonymity on stable networks, then efficiency and stability are compatible.

While Proposition 4 shows that if we are willing to sacrifice anonymity, then we can reconcile stability with efficiency, there are also many situations where we need not go so far. That is, there are value functions for which there do exist component balanced and anonymous allocation rules for which some efficient networks are pairwise stable.

The Role of "Loose-Ends" in the Tension Between Stability and Efficiency

The following proposition identifies a very particular feature of the problem between efficiency and stability. It shows that if efficient networks are such that each individual has at least two links, then there is no tension. So, problems arise only in situations where efficient networks involve individuals who may be thought of as "loose ends."

A network g has *no loose ends* if for any $i \in N(g)$, $|\{j | ij \in g\}| \geq 2$.

Proposition 5. *There exists an anonymous and component balanced Y such that if v is anonymous and such that there exists $g^* \in E(v)$ with no loose ends, then $E(v) \cap PS(Y,v) \neq \emptyset$.*

The proof of Proposition 5 appears in the appendix. In a network with no loose ends individuals can alter the component structure by adding or severing links, but they cannot decrease the total number of individuals who are involved in the network by severing a link. This limitation on the ways in which individuals can change a network is enough to ensure the existence of a component balanced and anonymous allocation rule for which such an efficient network is stable, and is critical to the proof.

The proof of Proposition 5 turns out to be more complicated that one might guess. For instance, one might guess that the component wise egalitarian allocation rule Y^{ce} would satisfy the demands of the proposition.[32] However, this is not the case as the following example illustrates.

[30] The statement that Y is anonymous on some networks that are efficient and pairwise stable means that one needs to consider some other networks to verify the failure of anonymity.

[31] Dutta and Mutuswami actually work with a notion called strong stability, that is (almost) a stronger requirement than pairwise stability in that it allows for deviations by coalitions of individuals. They show that the strongly stable networks are a subset of the efficient ones. Strong stability is not quite a strengthening of pairwise stability, as it only considers one network to defeat another if there is a deviation by a coalitions that makes all of its members strictly better off; while pairwise stability allows one of the two individuals adding a link to be indifferent. However, one can check that the construction of Dutta and Mutuswami extends to pairwise stability as well.

[32] See the discussion of critical link monotonicity in Jackson and Wolinsky (1996) for a complete characterization of when Y^{ce} provides for efficient networks that are pairwise stable.

Example 8.

Let $n = 7$. Consider a component additive and anonymous v such that the value of a ring of three individuals is 6, the value of a ring of four individuals is 20, and the value of a network where a ring of three individuals with a single bridge to a ring of four individuals (e.g., $g^* = \{12, 23, 13, 14, 45, 56, 67, 47\}$) is 28. Let the value of other components be 0. The efficient network structure is g^*. Under the component wise egalitarian rule each individual gets a payoff of 4 under g^*, and yet if 4 severs the link 14, then 4 would get a payoff of 5 under any anonymous rule or one satisfying equal treatment of equals. Thus g^* would not be stable under the component-wise egalitarian rule. See Fig. 5.

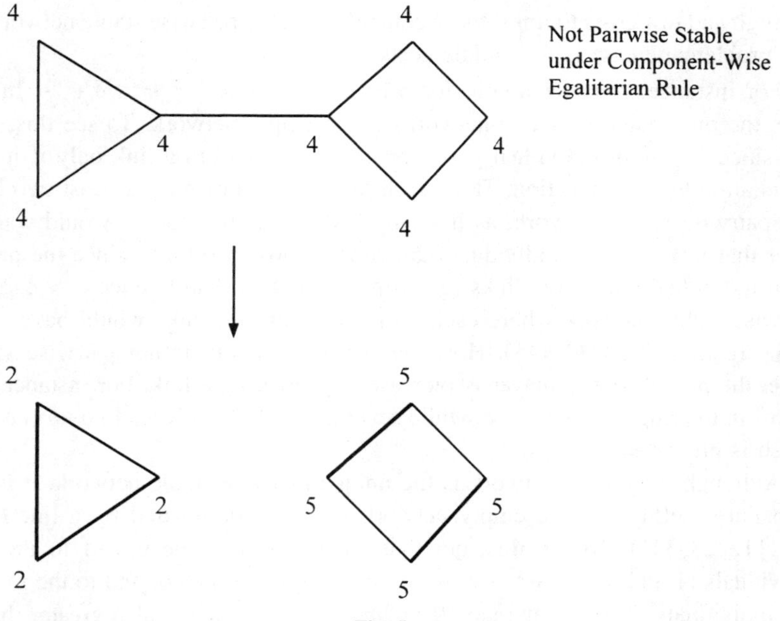

Fig. 5.

Thus, a Y that guarantees the pairwise stability of g^* will have to recognize that individual 4 can get a payoff of 5 by severing the link 14. This involves a carefully defined allocation rule, as provided in the appendix.

Taking the Allocation Rule as Given

As we have seen, efficiency and even constrained efficiency are only compatible with pairwise stability under certain allocation rules and for certain settings. Sometimes this involves quite careful design of the allocation rule, as under Propositions 4 and 5.

While there are situations where the allocation rule is an object of design, we are also interested in understanding when naturally arising allocation rules lead to pairwise stable networks that are (Pareto) efficient.

Let us examine some of some of the examples discussed previously to get a feeling for this.

Example 9. *Pareto Inefficiency in the Symmetric Connections Model.*

In the symmetric connections model (Example 1) efficient networks fall into three categories:

- empty networks when there are high costs to links,
- star networks ($n-1$ individuals all having 1 link to the n-th individual) when there are middle costs to links, and
- complete networks when there are low costs to links.

For high and low costs to links, these coincide with the pairwise stable networks.[33] The problematic case is for middle costs to links.

For instance, consider a situation where $n = 4$ and $\delta < c < \delta + \frac{\delta^2}{2}$. In this case, the only pairwise stable networks is the empty network. To see this, note that since $c > \delta$ an individual gets a positive payoff from a link only if it also offers an indirect connection. Thus, each individual must have at least two links in a pairwise stable network, as if i only had a link to j, then j would want to sever that link. Also an individual maintains at most two links, since the payoff to an individual with three links (given $n = 4$) is less than 0 since $c > \delta$. So, a pairwise stable network where each individual has two links would have to be a ring (e.g., $\{12, 23, 34, 14\}$). However, such a network is not pairwise stable since, the payoff to any player is increased by severing a link. For instance, 1's payoff in the ring is $2\delta + \delta^2 - 2c$, while severing the link 14 leads to $\delta + \delta^2 + \delta^3 - c$ which is higher since $c > \delta$.

Although the empty network is the unique pairwise stable network, it is not even Pareto efficient. The empty network is Pareto dominated by a line (e.g., $g = \{12, 23, 34\}$). To see this, not that under the line, the payoff to the end individuals (1 and 4) is $\delta + \delta^2 + \delta^3 - c$ which is greater than 0, and to the middle two individuals (2 and 3) the payoff is $2\delta + \delta^2 - 2c$ which is also greater than 0 since $c < \delta + \frac{\delta^2}{2}$.

Thus, there exist cost ranges under the symmetric connections model for which all pairwise stable networks are *Pareto* inefficient, and other cost ranges where all pairwise stable networks are efficient. There are also some cost ranges where some pairwise stable networks are efficient and some other pairwise stable networks are not even Pareto efficient.

Example 10. *Pareto Inefficiency in the Co-Author Model.*

Generally, the co-author model results in Pareto inefficient networks. To see this, consider a simple setting where $n = 4$. Here the only pairwise stable network is the complete network, as the reader can check with some straightforward

[33] The compatibility of pairwise stability and efficiency in the symmetric connections model is fully characterized in Jackson and Wolinsky (1996). The relationship with Pareto efficient networks is not noted.

calculations. The complete network leads to a payoff of 2.5 to each player. However, a network of two distinct linked pairs (e.g., $g = \{12, 34\}$) leads to payoffs of 3 for each individual. Thus, the unique pairwise stable network is Pareto inefficient.

Example 11. *Efficiency in the Corominas-Bosch Bargaining Networks*

While incentives to form networks do not always lead to efficiency in the connections model, the news is better in the bargaining model of Corominas-Bosch (Example 3). In that model the set of pairwise stable networks is often exactly the set of efficient networks, as it outlined in the following Proposition.

Proposition 6. *In the Corominas-Bosch model as outlined in Example 3, with costs to links* $1/2 > c_s > 0$ *and* $1/2 > c_b > 0$*, the pairwise stable networks are exactly the set of efficient networks.*[34] *The same is true if* $c_s > 1/2$ *and/or* $c_b > 1/2$ *and* $c_s + c_b \geq 1$*. If* $c_s > 1/2$ *and* $1 > c_s + c_b$*, or* $c_b > 1/2$ *and* $1 > c_s + c_b$*, then the only pairwise stable network is inefficient, but Pareto efficient.*

The proof of Proposition 6 appears in the appendix. The intuition for the result is fairly straightforward. Individuals get payoffs of either 0, 1/2 or 1 from the bargaining, ignoring the costs of links. An individual getting a payoff of 0 would never want to maintain any links, as they cost something but offer no payoff in bargaining. So, it is easy to show that all individuals who have links must get payoffs of 1/2. Then, one can show that if there are extra links in such a network (relative to the efficient network which is just linked pairs) that some particular links could be severed without changing the bargaining payoffs and thus saving link costs.

The optimistic conclusion in the bargaining networks is dependent on the simple set of potential payoffs to individuals. That is, either all linked individuals get payoffs of 1/2, or for every individual getting a payoff of 1 there is some linked individual getting a payoff of 0. The low payoffs to such individuals prohibit them from wanting to maintain such links. This would not be the case, if such individuals got some positive payoff. We see this next in the next example.

Example 12. *Pareto Inefficiency in Kranton and Minehart's Buyer-Seller Networks*

[34] Corominas-Bosch (1999) considers a different definition of pairwise stability, where a cost is incurred for creating a link, but none is saved for severing a link. Such a definition can clearly lead to over-connections, and thus a more pessimistic conclusion than that of Proposition 6 here. She also considers a game where links can be formed unilaterally and the cost of a link is incurred only by the individual adding the link. In such a setting, a buyer (say when there are more sellers than buyers) getting a payoff of 1/2 or less has an incentive to add a link to some seller who is earning a payoff of 0, which will then increase the buyer's payoff. As long as this costs the seller nothing, the seller is indifferent to the addition of the link. So again, Corominas-Bosch obtains an over-connection result. It seems that the more reasonable case is one that involves some cost for and consent of both individuals, which is the case treated in Proposition 6 here.

Despite the superficial similarities between the Corominas-Bosch and Kranton and Minehart models, the conclusions regarding efficiency are quite different. This difference stems from the fact that there is a possible heterogeneity in buyers' valuations in the Kranton and Minehart model, and so efficient networks are more complicated than in the simpler bargaining setting of Corominas-Bosch. It is generally the case that these more complicated networks are not pairwise stable.

Before showing that all pairwise stable networks may fail to be Pareto efficient, let us first show that they may fail to be efficient as this is a bit easier to see.

Consider Example 4, where there is one seller and up to n buyers.

The efficient network in this setting is one where $\frac{k}{k+1} - k(c_s + c_b)$ is maximized. This occurs where[35]

$$\frac{1}{k(k+1)} \geq c_s + c_b \geq \frac{1}{(k+1)(k+2)}.$$

Let us examine the pairwise stable networks. From (1) it follows that the seller gains from adding a new link to a network of with k links as long as

$$\frac{2}{(k+1)(k+2)} > c_s.$$

Also from (1) it follows that a buyer wishes to add a new link to a network of k links as long as

$$\frac{1}{k(k+1)} > c_b.$$

If we are in a situation where $c_s = 0$, then the incentives of the buyers lead to exactly the right social incentives: and the pairwise stable networks are exactly the efficient ones.[36] This result for $c_s = 0$ extends to situations with more than one seller and to general distributions over signals, and is a main result of Kranton and Minehart (1998).

However, let us also consider situations where $c_s > 0$, and for instance $c_b = c_s$. In this case, the incentives are not so well behaved.[37] For instance, if $c_s = 1/100 = c_b$, then any efficient network has six buyers linked to the seller ($k = 6$). However, buyers will be happy to add new links until $k = 10$, while sellers wish to add new links until $k = 13$. Thus, in this situation the pairwise stable networks would have 10 links, while networks with only 6 links are the efficient ones.

To see the intuition for the inefficiency in this example note that the increase in expected price to sellers from adding a link can be thought of as coming

[35] Or at n if such a $k > n$.

[36] Sellers always gain from adding links if $c_s = 0$ and so it is the buyers' incentives that limit the number of links added.

[37] See Kranton and Minehart (1998) for discussion of how a costly investment decision of the seller might lead to inefficiency. Although it is not the same as a cost per link, it has some similar consequences.

from two sources. One source is the expected increase in willingness to pay of the winning bidder due to an expectation that the winner will have a higher valuation as we see more draws from the same distribution. This increase is of social value, as it means that the good is going to someone who values it more. The other source of price increase to the seller from connecting to more buyers comes from the increased competition among the bidders in the auction. There is a greater number of bidders competing for a single object. This source of price increase is not of social value since it only increases the proportion of value which is transferred to the seller. Buyers' incentives are distorted relative to social efficiency since although they properly see the change in social value, they only bear part of the increase in total cost of adding a link.

While the pairwise stable networks in this example are not efficient (or even constrained efficient), they are Pareto efficient, and this is easily seen to be generally true when there is a single seller as then disconnected buyers get a payoff of 0. This is not true with more sellers as we now see.

Let us now show that it is possible for (non-trivial) pairwise stable networks in the Kranton-Minehart model to be Pareto inefficient. For this we need more than one seller.

Consider a population with two sellers and four buyers. Let individuals 1 and 2 be the sellers and 3,4,5,6, be the buyers. Let the cost of a link to a seller be $c_s = \frac{5}{60}$ and the cost of a link to a buyer be $c_b = \frac{1}{60}$.

Some straightforward (but tedious) calculations lead to the following payoffs to individuals in various networks:

$g^a = \{13\}$: $Y_1(g^a) = -\frac{5}{60}$ and $Y_1(g^a) = \frac{29}{60}$.

$g^b = \{13, 14\}$: $Y_1(g^b) = \frac{10}{60}$ and $Y_3 = Y_4(g^b) = \frac{9}{60}$.

$g^c = \{13, 14, 15\}$: $Y_1(g^c) = \frac{15}{60}$ and $Y_3 = Y_4 = Y_5(g^c) = \frac{4}{60}$.

$g^d = \{13, 14, 15, 16\}$: $Y_1(g^d) = \frac{16}{60}$ and $Y_3 = Y_4 = Y_5(g^d) = \frac{2}{60}$.

$g^e = \{13, 14, 25, 26\}$: $Y_1 = Y_2(g^e) = \frac{10}{60}$ and $Y_3 = Y_4 = Y_5 = Y_6(g^e) = \frac{9}{60}$.

$g^f = \{13, 14, 15, 25, 26\}$: $Y_1(g^f) = \frac{13}{60}$, $Y_2(g^f) = \frac{8}{60}$, and $Y_3 = Y_4(g^f) = \frac{6}{60}$, while $Y_5(g^f) = \frac{10}{60}$ and $Y_6(g^f) = \frac{11}{60}$.

$g^g = \{13, 14, 15, 24, 25, 26\}$: $Y_1 = Y_2(g^g) = \frac{9}{60}$ and $Y_3 = Y_4 = Y_5 = Y_6(g^g) = \frac{8}{60}$.

There are three types of pairwise stable networks here: the empty network, networks that look like g^d, and networks that look like g^g.[38] Both the empty network and g^g are not Pareto efficient, while g^d is. In particular, g^g is Pareto dominated by g^e. Also, g^d is not efficient nor is it constrained efficient.[39] In this example, one might hope that g^e would turn out to be pairwise stable, but as we see 1 and 5 then have an incentive to add a link; and then 2 and 4 which takes us to g^g. Thus, individuals have an incentive to over-connect as it increases their individual payoffs even when it is decreasing overall value.

---

38 The reader is left to check networks that are not listed here.

39 To see constrained inefficiency, consider an allocation rule that divides payoffs equally among buyers in a component and gives 0 to sellers. Under such a rule, g^e Pareto dominates g^d.

It is not clear whether there are examples where all pairwise stable networks are Pareto inefficient in this model, as there are generally pairwise stable networks like g^d where only one seller is active and gets his or her maximum payoff. But this is an open question, as with many buyers this may be Pareto dominated by networks where there are several active sellers. And as we see here, it is possible for active sellers to want to link to each others' buyers to an extent that is inefficient.

Pareto Inefficiency Under the Myerson Value

As we have seen in the above examples, efficiency and Pareto efficiency are properties that sometimes but not always satisfied by pairwise stable networks. To get a fuller picture of this, and to understand some sources of inefficiency, let us look at an allocation rule that will arise naturally in many applications. As equal bargaining power is a condition that may naturally arise in a variety of settings, the Myerson value allocation rule that is worthy of serious attention. Unfortunately, although it has nice properties with respect to the existence of pairwise stable networks, the pairwise stable networks are not always Pareto efficient networks.

The intuition behind the (Pareto) inefficiency under the Myerson value is that individuals can have an incentive to over-connect as it improves their bargaining position. This can lead to overall *Pareto* inefficiency. To see this in some detail, it is useful to separate costs and benefits arising from the network.

Let us write $v(g) = b(g) - c(g)$ where $b(\cdot)$ represents benefits and $c(\cdot)$ costs and both functions take on nonnegative values and have some natural properties.

$b(g)$ is *monotone* if

- $b(g) \geq b(g')$ if $g' \subset g$, and
- $b(\{ij\}) > 0$ for any ij.

$b(g)$ is *strictly monotone* if $b(g) > b(g')$ whenever $g' \subset g$.

Similar definitions apply to a cost function c.

Proposition 7. *For any monotone and anonymous benefit function b there exists a strictly monotone and anonymous cost function c such that all pairwise stable networks relative to Y^{MV} and $v = b - c$ are Pareto inefficient. In fact, the pairwise stable networks are over-connected in the sense that each pairwise stable network has some subnetwork that Pareto dominates it.*

Proposition 7 is a fairly negative result, saying that for any of a wide class of benefit functions there is some cost function for which individuals have incentives to over-connect the network, as they each try to improve their bargaining position and hence payoff.

Proposition 7 is actually proven using the following result, which applies to a narrower class of benefit functions but is more specific in terms of the cost functions.

Proposition 8. *Consider a monotone benefit function b for which there is some efficient network g^* relative to b ($g^* \in E(b)$) such that $g^* \neq g^N$. There exists $\overline{c} > 0$ such that for any cost function c such that $\overline{c} \geq c(g)$ for all $g \in G$, the pairwise stable networks relative to Y^{MV} and $v = b - c$ are all inefficient. Moreover, if b is anonymous and g^* is symmetric,*[40] *then each pairwise stable networks is Pareto dominated by some subnetwork.*

Proposition 8 says that for any monotone benefit function that has at least one efficient network under the benefit function that is not fully connected, if costs to links are low enough, then all pairwise stable networks will be over-connected relative to the efficient networks. Moreover, if the efficient network under the benefit function is symmetric does not involve too many connections, then all pairwise stable networks will be Pareto inefficient.

Proposition 8 is somewhat limited, since it requires that the benefit function have some network smaller than the complete network which is efficient. However, as there are many b's and c's that sum to the same v, this condition actually comes without much loss of generality, which is the idea behind the proof of Proposition 7. The proof of Propositions 7 and 8 appear in the appendix.

6 Discussion

The analysis and overview presented here shows that the relationship between the stability and efficiency of networks is context dependent. Results show that they are not always compatible, but are compatible for certain classes of value functions and allocation rules. Looking at some specific examples, we see a variety of different relationships even as one varies parameters within models.

The fact that there can be a variety of different relationships between stable and efficient networks depending on context, seems to be a somewhat negative finding for the hopes of developing a systematic theoretical understanding of the relationship between stability and efficiency that cuts across applications. However, there are several things to note in this regard. First, a result such as Proposition 5 is reassuring, since it shows that some systematic positive results can be found. Second, there is hope of tying incompatibility between individual incentives and efficiency to a couple of ideas that cut across applications. Let me outline this in more detail.

One reason why individual incentives might not lead to overall efficiency is one that economists are very familiar with: that of externalities. This comes out quite clearly in the failure exhibited in the symmetric connections model in Example 9. By maintaining a link an individual not only receives the benefits of that link (and its indirect connections) him or herself, but also provides indirect benefits to other individuals to whom he or she is linked. For instance, 2's decision of whether or not to maintain a link to 3 in a network $\{12, 23\}$ has

[40] A network g is symmetric if for every i and j there exists a permutation π such that $g = g^\pi$ and $\pi(j) = i$.

payoff consequences for individual 1. The absence of a proper incentive for 2 to evaluate 1's well being when deciding on whether to add or delete the link 23 is a classic externality problem. If the link 23 has a positive benefit for 1 (as in the connections model) it can lead to under-connection relative to what is efficient, and if the link 23 has a negative effect on 1 (as in the co- author model) it can lead to over-connection.

Power-Based Inefficiencies

There is also a second, quite different reason for inefficiency that is evident in some of the examples and allocation rules discussed here. It is what we might call a "power-based inefficiency". The idea is that in many applications, especially those related to bargaining or trade, an individual's payoff depends on where they sit in the network and not only what value the network generates. For instance, individual 2 in a network $\{12, 23\}$ is critical in allowing any value to accrue to the network, as deleting all of 2's links leaves an empty network. Under the Myerson value allocation rule, and many others, 2's payoff will be higher than that of 1 and 3; as individual 2 is rewarded well for the role that he or she plays. Consider the incentives of individuals 1 and 3 in such a situation. Adding the link 13 might lower the overall value of the network, but it would also put the individuals into equal roles in the network, thereby decreasing individual 2's importance in the network and resulting bargaining power. Thus, individual 1 and 3's bargaining positions can improve and their payoffs under the Myerson value can increase; even if the new network is less productive than the previous one. This leads 1 and 3 to over-connect the network relative to what is efficient. This is effectively the intuition behind the results in Propositions 7 and 8, which says that this is a problem which arises systematically under the Myerson value.

The inefficiency arising here comes not so much from an externality, as it does from individuals trying to position themselves well in the network to affect their relative power and resulting allocation of the payoffs. A similar effect is seen in Example 12, where sellers add links to new buyers not only for the potential increase in value of the object to the highest valued buyer, but also because it increases the competition among buyers and increases the proportion of the value that goes to the seller rather than staying in the buyers' hands.[41]

An interesting topic for further research is to see whether inefficiencies in network formation can always be traced to either externalities or power-based incentives, and whether there are features of settings which indicate when one, and which one, of these difficulties might be present.

[41] Such a source of inefficiency is not unique to network settings, but are also observed in, for example, search problems and bargaining problems more generally (e.g., see Stole and Zwiebel (1996) on intra-firm bargaining and hiring decisions). The point here is that this power-based source of inefficiency is one that will be particularly prevalent in network formation situations, and so it deserves particular attention in network analyses.

Some Other Issues for Further Study

There are other areas that deserve significant attention in further efforts to model the formation of networks.

First, as discussed near the definition of pairwise stability, it would be useful to develop a notion of network stability that incorporates farsighted and dynamic behavior. Judging from such efforts in the coalition formation literature, this is a formidable and potentially ad hoc task. Nevertheless, it is an important one if one wants to apply network models to things like strategic trade alliances.

Second, in the modeling here, allocation rules are taken as being separate from the network formation process. However, in many applications, one can see bargaining over allocation of value happening simultaneously with the formation of links. Intuitively, this should help in the attainment of efficiency. In fact, in some contexts it does, as shown by Currarini and Morelli (2000) and Mutuswami and Winter (2000). The contexts explored in those models use given (finite horizon) orderings over individual proposals of links, and so it would be interesting to see how robust such intuition is to the specification of bargaining protocol.

Third, game theory has developed many powerful tools to study evolutionary pressures on societies of players, as well as learning by players. Such tools can be very valuable in studying the dynamics of networks over time. A recent literature has grown around these issues, studying how various random perturbations to and evolutionary pressures on networks affects the long run emergence of different networks structures (e.g., Jackson and Watts (1998, 1999), Goyal and Vega-Redondo (1999), Skyrms and Pemantle (2000), and Droste, Gilles and Johnson (2000)). One sees from this preliminary work on the subject that network formation naturally lends itself to such modeling, and that such models can lead to predictions not only about network structure but also about the interaction that takes place between linked individuals. Still, there is much to be understood about individual choices, interaction, and network structure depend on various dynamic and stochastic effects.

Finally, experimental tools are becoming more powerful and well-refined, and can be brought to bear on network formation problems, and there is also a rich set of areas where network formation can be empirically estimated and some models tested. Experimental and empirical analyses of networks are well-founded in the sociology literature (e.g., see the review of experiments on exchange networks by Bienenstock and Bonacich (1993)), but is only beginning in the context of some of the recent network formation models developed in economics (e.g., see Corbae and Duffy (2000) and Charness and Corominas-Bosch (2000)). As these incentives-based network formation models have become richer and have many pointed predictions for wide sets of applications, there is a significant opportunity for experimental and empirical testing of various aspects of the models. For instance, the hypothesis presented above, that one should expect to see over-connection of networks due to the power-based inefficiencies under equal bargaining power and low costs to links, provides specific predictions that are testable and have implications for trade in decentralized markets.

In closing, let me say that the future for research on models of network formation is quite bright. The multitude of important issues that arise from a wide variety of applications provides a wide open landscape. At the same time the modeling proves to be quite tractable and interesting, and has the potential to provide new explanations, predictions, and insights regarding a host of social and economic settings and behaviors.

References

Arrow, K.J., Borzekowski, R. (2000) Limited Network Connections and the Distribution of Wages. mimeo: Stanford University.

Aumann, R., Myerson, R. (1988) Endogenous Formation of Links Between Players and Coalitions: An Application of the Shapley Value. In: A. Roth (ed.) *The Shapley Value*, Cambridge University Press, pp 175–191.

Bala, V., Goyal, S. (2000) A Strategic Analysis of Network Reliability. *Review of Economic Design* 5: 205–228.

Bala, V., Goyal, S. (2000a) Self-Organization in Communication Networks. *Econometrica* 68: 1181–1230.

Banerjee, S. (1999) Efficiency and Stability in Economic Networks. mimeo: Boston University.

Banerjee, S., Konishi, H., Sönmez, T. (2001) Core in a Simple Coalition Formation Game. *Social Choice and Welfare* 18: 135–154.

Bienenstock, E., Bonacich, P. (1993) Game Theory Models for Social Exchange Networks: Experimental Results. *Sociological Perspectives* 36: 117–136.

Bienenstock, E., Bonacich, P. (1997) Network Exchange as a Cooperative Game. *Rationality and Society* 9: 37-65.

Boorman, S. (1975) A Combinatorial Optimization Model for Transmission of Job Information through Contact Networks. *Bell Journal of Economics* 6: 216–249.

Bramoulle, Y. (2000) Congestion and Social Networks: an Evolutionary Analysis. mimeo: University of Maryland.

Burt, R. (1992) Structural Holes: The Social Structure of Competition, Harvard University Press.

Calvo-Armengol, A. (1999) Stable and Efficient Bargaining Networks. mimeo.

Calvo-Armengol, A. (2000) Job Contact Networks. mimeo.

Calvo-Armengol, A. (2001) Bargaining Power in Communication Networks. *Mathematical Social Sciences* 41: 69–88.

Charness, G., Corominas-Bosch, M. (2000) Bargaining on Networks: An Experiment. mimeo: Universitat Pompeu Fabra.

Chwe, M. S.-Y. (1994) Farsighted Coalitional Stability. *Journal of Economic Theory* 63: 299–325.

Corbae, D., Duffy, J. (2000) Experiments with Network Economies. mimeo: University of Pittsburgh.

Corominas-Bosch, M. (1999) On Two-Sided Network Markets, Ph.D. dissertation: Universitat Pompeu Fabra.

Currarini, S., Morelli, M. (2000) Network Formation with Sequential Demands. *Review of Economic Design* 5: 229–250.

Droste, E., Gilles, R., Johnson, C. (2000) Evolution of Conventions in Endogenous Social Networks. mimeo: Virginia Tech.

Dutta, B., and M.O. Jackson (2000) The Stability and Efficiency of Directed Communication Networks. *Review of Economic Design* 5: 251–272.

Dutta, B., and M.O. Jackson (2001) Introductory chapter. In: B. Dutta, M.O. Jackson (eds.) *Models of the Formation of Networks and Groups*, forthcoming from Springer-Verlag: Heidelberg.

Dutta, B., and S. Mutuswami (1997) Stable Networks. *Journal of Economic Theory* 76: 322–344.

Dutta, B., van den Nouweland, A., Tijs, S. (1998) Link Formation in Cooperative Situations *International Journal of Game Theory* 27: 245–256.

Ellison, G. (1993) Learning, Local Interaction, and Coordination. *Econometrica* 61: 1047–1071.

Ellison, G., Fudenberg, D. (1995) Word-of-Mouth Communication and Social Learning. *The Quarterly Journal of Economics* 110: 93–126.

Fafchamps, M., Lund, S. (2000) Risk-Sharing Networks in Rural Philippines. mimeo: Stanford University.

Goyal, S. (1993) Sustainable Communication Networks, Discussion Paper TI 93-250, Tinbergen Institute, Amsterdam-Rotterdam.

Goyal, S., Joshi, S. (2000) Networks of Collaboration in Oligopoly, Discussion Paper TI 2000-092/1, Tinbergen Institute, Amsterdam-Rotterdam.

Goyal, S., Vega-Redondo, F. (1999) Learning, Network Formation and Coordination. mimeo: Erasmus University.

Glaeser, E., Sacerdote, B., Scheinkman, J. (1996) Crime and Social Interactions. *Quarterly Journal of Economics* 111: 507–548.

Granovetter, M. (1973) The Strength of Weak Ties. *American Journal of Sociology* 78: 1360–1380.

Haller, H., Sarangi, S. (2001) Nash Networks with Heterogeneous Agents, mimeo: Virginia Tech and LSU.

Hendricks, K., Piccione, M., Tan, G. (1995) The Economics of Hubs: The Case of Monopoly, *Rev. Econ. Stud.* 62: 83–100.

Jackson, M.O., van den Nouweland, A. (2001) Efficient and stable networks and their relationship to the core, mimeo.

Jackson, M.O., Watts, A. (1998) The Evolution of Social and Economic Networks, forthcoming in *Journal of Economic Theory*.

Jackson, M.O., Watts, A. (1999) On the Formation of Interaction Networks in Social Coordination Games, forthcoming in *Games and Economic Behavior*.

Jackson, M.O., Wolinsky, A. (1996)A Strategic Model of Social and Economic Networks. *Journal of Economic Theory* 71: 44–74.

Johnson, C. and R.P. Gilles (2000) Spatial Social Networks. *Review of Economic Design* 5: 273–300.

Katz, M., Shapiro, C. (1994) Systems Competition and Networks Effects. *Journal of Economic Perspectives* 8: 93–115.

Kirman, A. (1997) The Economy as an Evolving Network *Journal of Evolutionary Economics* 7: 339–353.

Kirman, A., Oddou, C., Weber, S. (1986) Stochastic Communication and Coalition Formation. *Econometrica* 54: 129–138.

Kranton, R., Minehart, D. (2001) A Theory of Buyer-Seller Networks, *American Economic Review* 91: 485–524.

Kranton, R., Minehart, D. (1996) Link Patterns in Buyer-Seller Networks: Incentives and Allocations in Graphs. mimeo: University of Maryland and Boston University.

Kranton, R., Minehart, D. (2000) Competition for Goods in Buyer-Seller Networks. *Review of Economic Design* 5: 301–332.

Liebowitz, S., Margolis, S. (1994) Network Externality: An Uncommon Tragedy. *Journal of Economic Perspectives* 8: 133–150.

Monderer, D., Shapley, L. (1996) Potential Games. *Games and Economic Behavior* 14: 124–143.

Montgomery, J. (1991) Social Networks and Labor Market Outcomes. *The American Economic Review* 81: 1408–1418.

Mutuswami, S., Winter, E. (2000) Subscription Mechanisms for Network Formation. mimeo: CORE and Hebrew University in Jerusalem.

Myerson, R (1977) Graphs and Cooperation in Games. *Math. Operations Research* 2: 225–229.

Myerson, R. (1991) *Game Theory: Analysis of Conflict*. Harvard University Press: Cambridge, MA.

Qin, C-Z. (1996) Endogenous Formation of Cooperation Structures. *Journal of Economic Theory* 69: 218–226.

Roth, A., Sotomayor, M. (1989) *Two Sided Matching*, Econometric Society Monographs No. 18: Cambridge University Press.

Skyrms, B., Pemantle, R. (2000) A Dynamic Model of Social Network Formation. *Proceedings of the National Academy of Sciences* 97: 9340–9346.

Slikker, M. (2000) *Decision Making and Cooperation Structures* CentER Dissertation Series: Tilburg.

Slikker, M., R.P. Gilles, H. Norde, and S. Tijs (2001) Directed Networks, Allocation Properties and Hierarchy Formation, mimeo.

Slikker, M., van den Nouweland, A. (2000) Network Formation Models with Costs for Establishing Links. *Review of Economic Design* 5: 333–362.

Slikker, M., van den Nouweland, A. (2001) *Social and Economic Networks in Cooperative Game Theory*. Forthcoming from Kluwer publishers.

Slikker, M., van den Nouweland, A. (2001b) A One-Stage Model of Link Formation and Payoff Division. *Games and Economic Behavior* 34: 153–175.

Starr, R.M., Stinchcombe, M.B. (1992) Efficient Transportation Routing and Natural Monopoly in the Airline Industry: An Economic Analysis of Hub-Spoke and Related Systems. UCSD dp 92-25.

Starr, R.M., Stinchcombe, M.B. (1999) Exchange in a Network of Trading Posts. In: G. Chichilnisky (ed.), *Markets, Information and Uncertainty*, Cambridge University Press.

Stole, L., Zweibel, J. (1996) Intra-Firm Bargaining under Non-Binding Constraints. *Review of Economic Studies* 63: 375–410.

Tesfatsion, L. (1997) A Trade Network Game with Endogenous Partner Selection. In: H. Amman et al. (eds.), *Computational Approaches* to Economic Problems, Kluwer Academic Publishers, 249–269.

Tesfatsion, L. (1998) Gale-Shapley matching in an Evolutionary Trade Network Game. Iowa State University Economic Report no. 43.

Topa, G. (2001) Social Interactions, Local Spillovers and Unemployment. *Review of Economic Studies* 68: 261–296.

Wang, P., Wen, Q. (1998) Network Bargaining. mimeo: Penn State University.

Wasserman, S., Faust, K. (1994) *Social Network Analysis: Methods and Applications*. Cambridge University Press.

Watts, A. (2001) A Dynamic Model of Network Formation. *Games and Economic Behavior* 34: 331–341.

Watts, D.J. (1999) *Small Worlds: The Dynamics of Networks between Order and Randomness*. Princeton University Press.

Weisbuch, G., Kirman, A., Herreiner, D. (1995) Market Organization. mimeo, École Normal Superieure.

Young, H.P. (1998) *Individual Strategy and Social Structure*. Princeton University Press, Princeton.

Appendix

Proof of Proposition 3. The proof uses the same value function as Jackson and Wolinsky (1996), and is also easily extended to more individuals. The main complication is showing that the constrained efficient and efficient networks coincide. Let $n = 3$ and the value of the complete network be 12, the value of a single link 12, and the value of a network with two links 13.

Let us show that the set of constrained efficient networks is exactly the set of networks with two links. First consider the complete network. Under any component balanced Y satisfying equal treatment of equals (and thus anonymity), each individual must get a payoff of 4. Consider the component balanced and anonymous Y which gives each individual in a two link network $13/3$. Then $g = \{12, 23\}$ offers each individual a higher payoff than g^N, and so the complete network is not constrained efficient. The empty network is similarly ruled out as being constrained efficient. Next consider the network $g' = \{12\}$ (similar arguments hold for any permutation of it). Under any component balanced and Y satisfying equal treatment of equals, $Y_1(g', v) = Y_2(g', v) = 6$. Consider $g'' = \{13, 23\}$ and a component balanced and anonymous Y such that $Y_1(g'', v) = Y_2(g'', v) = 6.25$ and $Y_3(g'', v) = .5$. All three individuals are better off under g''

than g' and so g' is not constrained efficient. The only remaining networks are those with two links, which are clearly efficient and thus constrained efficient.

To complete the proof, we need to show that any component balanced Y satisfying equal treatment of equals results in none of the two link networks being pairwise stable.

As noted above, under any component balanced Y satisfying equal treatment of equals, each individual in the complete network gets a payoff of 4, and the two individuals with connections in the single link network each get a payoff of 6. So consider the network $g = \{12, 23\}$ (or any permutation of it) and let us argue that it cannot be pairwise stable. In order for individual 2 not to want to sever a link, 2's payoff must be at least 6. In order for individuals 1 and 3 not to both wish to form a link (given equal treatment of equals) their payoffs must be at least 4. Thus, in order to have g be pairwise stable it must be that $Y_1(g, v) + Y_2(g, v) + Y_3(g, v) \geq 14$, which is not feasible. □

Proof of Proposition 5. Let $N^*(g) = |C(g)| + n - |N(g)|$. Thus, $N^*(g)$ counts the components of g, and also counts individuals with no connections. So if we let a component* be either a component or isolated individual, then N^* counts component*'s. For instance, under this counting the empty network has one more component* than the network with a single link.

Let

$$B(g) = \{i \,|\, \exists j \text{ s.t. } |N^*(g - ij)| > |N^*(g)|\}.$$

Thus $B(g)$ is the set of individuals who form bridges under g, i.e., those individuals who by severing a link can alter the component structure of g. Let[42]

$$SB(g) = \{i \,|\, \exists j \text{ s.t. } |N^*(g - ij)| > |N^*(g)| \text{ and } i \in N(h_i), h_i \in C(g - ij), h_i \text{ is symmetric}\}.$$

$SB(g)$ identifies the individuals who form bridges and who by severing the bridge end up in a symmetric component.

Claim 1. If g is connected ($|C(g)| = 1$) and has no loose ends, then $i \in SB(g)$ implies that i has at most one bridge in g. Also, for any such g, $|N(g)|/3 \geq |SB(g)|$, and if $\{i,j\} \subset SB(g)$ and $ij \in g$, then $\{i,j\} = B(g)$.

Proof of claim: Since there are no loose ends under g, each $i \in N(g)$ has at least two links. This implies that if $i \in SB(g)$ severs a link and ends up in a symmetric component h of $g - ij$, that h will have at least three individuals since each must have at least two links. Also $N(h) \cap SB(g) = \{i\}$. To see this note that if not, then there exists some $k \neq i$, $k \in N(h)$, such that k has a bridge under h. However, given the symmetry of h and the fact that each individual has at least two links, there are at least two distinct paths connecting any two individuals in the component, which rules out any bridges. Note this implies that i has at most one bridge. As we have shown that for each $i \in SB(g)$ there are at least two other individuals in $N(g^*) \setminus SB(g)$ and so $|N(g)|/3 \geq |SB(g)|$. If $\{i,j\} \subset SB(g)$

[42] Recall that a network g is symmetric if for every i and j there exists a permutation pi such that $g = g^\pi$ and $\pi(j) = i$.

and $ij \in g$, then given the symmetry of the component from severing a bridge, it must be that ij is the bridge for both i and j and that severing this results in two symmetric components with not bridges. This completes the claim.

Pick g^* to be efficient under v and have no loose ends. Also, choose g^* so that if $h^* \in C(g^*)$ then $v(h^*) > 0$. (Simply replace any $h^* \in C(g^*)$ such that $0 \geq v(h^*)$ with an empty component, which preserves efficiency.)

Consider any i that is non-isolated under g^* and the component $h_i^* \in C(g^*)$ with $i \in N(h_i^*)$. Define $\widehat{Y}(h_i^*, v)$ as follows.

$$\widehat{Y}_i(h_i^*, v) = \begin{cases} \max[Y^{ce}(g^*, v), Y_i^{ce}(h_i, v)] & \text{if } i \in SB(h), \text{ where } h_i \text{ is the symmetric component when } i \text{ severs his bridge} \\ \frac{v(h_i^*) - \sum_{k \in SB(h)} \widehat{Y}_k(h_i^*, v)}{|N(h_i^*) \setminus SB(h_i^*)|} & \text{if } i \in N(h_i^*) \setminus SB(h_i^*). \end{cases}$$

Let $\widehat{Y}(g^*, v)$ be the component balanced allocation rule defined on g^* from $\widehat{Y}$ defined above.

Claim 2. $\widehat{Y}_i(g^*, v) > 0$ for all $i \in N(g^*)$.

This is clear for $i \in SB(h_i^*)$ since i gets at least $Y_i^{ce}(h_i^*, v) > 0$. Consider $i \in N(h^*) \setminus SB(h_i^*)$. From the definition of $\widehat{Y}$, we need only show that $v(h^*) > \sum_{k \in SB(h^*)} \widehat{Y}_k(h^*, v)$. Given that by Claim 1 we know $|N(h^*)|/3 \geq |SB(h^*)|$, it is sufficient to show that $\frac{2v(h^*)}{|N(h^*)|} \geq \widehat{Y}_k(h^*, v)$ for any $k \in SB(h^*)$. Let h_k be the symmetric component obtained when k severs his bridge. By efficiency of g^* and anonymity of v

$$v(h^*) \geq v(h_k) \left(\frac{|N(h^*)|}{|N(h_k)|} \right)^-$$

where $(\cdot)^-$ rounds down.

$$\frac{v(h^*)}{|N(h_k)| \left(\frac{|N(h^*)|}{|N(h_k)|} \right)^-} \geq \frac{v(h_k)}{|N(h_k)|}.$$

Also note that $|N(h_k)| \left(\frac{|N(h^*)|}{|N(h_k)|} \right)^- \geq \frac{1}{2}$. Thus,

$$\frac{2v(h^*)}{|N(h^*)|} \geq \frac{v(h^*)}{|N(h_k)| \left(\frac{|N(h^*)|}{|N(h_k)|} \right)^-} \geq \frac{v(h_k)}{|N(h_k)|}.$$

So, from the definition of $\widehat{Y}$, we know that for any $k \in SB(h^*)$ that $\frac{2v(h^*)}{|N(h^*)|} \geq \widehat{Y}_k(h^*, v)$. As argued above, this completes the proof of the claim.

Now let us define $\widehat{Y}$ on other networks to satisfy the Proposition.

For a component of a network h let the symmetry groups be coarsest partition of $N(h)$, such that if i and j are in the same symmetry group, then there exists a permutation π with $\pi(i) = j$ and $h^\pi = h$. Thus, individuals in the same symmetry group are those who perform the same role in a network architecture and must be given the same allocation under an anonymous allocation rule when faced with an anonymous v.

For g adjacent to g^*, so that $g = g^* + ij$ or $g = g^* - ij$ for some ij, set $\widehat{Y}$ as follows.

Consider $h \in C(g)$

Case 1. There exists $k \in N(h)$ such that k is not in the symmetry group of either i nor j under g: split $v(h)$ equally among the members of k's symmetry group within h, and 0 to other members of $N(h)$.

Case 2. Otherwise, set $\widehat{Y}(h, v) = Y^{ce}(h, v)$.

For anonymous permutations of g^* and its adjacent networks define $\widehat{Y}$ according to the corresponding permutations of $\widehat{Y}$ defined above. For any other g let $\widehat{Y} = Y^{ce}$.

Let us verify that g^* is pairwise stable under $\widehat{Y}$.

Consider any $ij \in g^*$ and $g = g^* - ij$. Consider $h_i \in C(g)$ such that $i \in N(h_i)$. We show that i (and hence also j since the labels are arbitrary) cannot be better off.

If h_i falls under Case 1 above, then i gets 0 which by Claim 2 cannot be improving.

Next consider case where h_i has a single symmetry group. If $N(h_i) \cap SB(g^*) = \emptyset$, then ij could not have been a bride and so $N(h_i)$ was the same group of individuals i was connected to under g^* ($N(h_i) = N(h_i^*)$). Thus i got $Y_i^{ce}(g^*, v)$ under g^* and now gets $Y_i^{ce}(g, v)$, and so by efficiency this cannot be improving since i is still connected to the same group of individuals. If $N(h_i) \cap SB(g^*) \neq \emptyset$, then it must be that $i \in SB(g^*)$ and ij was i's bridge. In this case it follows from the definition of $\widehat{Y}_i(g^*, v)$ that the deviation could not be improving.

The remaining case is where $N(h_i) \subset N_i \cup N_j$, where N_i and N_j are the symmetry groups of i and j under g, and $N_i \cap N_j = \emptyset$. If i and j are both in $N(h_i)$ it must be that $N(h_i) = N(h_i^*)$, and that $N(h_i) \cap SB(g^*) = \emptyset$. [To see this suppose the contrary. ij could not be a bridge since i and j are both in $N(h_i)$. Thus, there is some $k \notin \{i, j\}$ with $k \in SB(g^*)$. But then there is no path from i to j that passes through k. Thus i and j are in the same component when k severs a bridge, which is either the component of k - which cannot be since then k must be in a different symmetry group from i and j under g - or in the other component. But then $k \in SB(g)$. This implies that either $i \in SB(g)$ or $j \in SB(g)$ but not both. Take $i \in SB(g)$. By severing i's bridge under g, i's component must be symmetric and include j (or else j also has a bridge under g and there must be more than two symmetry groups which would be a contradiction). There is some $l \neq j$ connected to i who is not i's bridge. But l and j cannot be in the same symmetry group under g since l is connected to some $i \in SB(g)$ and j cannot be (by claim 1) as $ij \notin g$. Also, l is not in i's symmetry group (again the proof of claim 1), and so his is a contraction.] Thus i got $Y_i^{ce}(g^*, v)$ under g^* and now gets $Y_i^{ce}(g, v)$, and so by efficiency this cannot be improving since i is still connected to the same group of individuals. If i and j are in different components under g, then it must be that they are in identical architectures given that $N(h_i) \subset N_i \cup N_j$. In this case ij was a bridge and since

h_i (and h_j) are not symmetric and $N(h_i) \subset N_i \cup N_j$, it follows the component of g^* containing i and j had no members of $SB(g^*)$. Thus $\widehat{Y}_i(g^*, v) = Y_i^{ce}(g^*, v)$ and also $\widehat{Y}_i(g, v) = Y_i^{ce}(g, v)$. Since the two components that are obtained when ij is severed are identical, by efficiency it follows that the payoffs to i (and j) are at least as high under g^* as under g.

Next, consider any $ij \in g^*$ and $g = g^* + ij$. Consider $h_i \in C(g)$ such that $i \in N(h_i)$. We show that if i is better off, then j must be worse off.

If h_i falls under Case 1 above, then i gets 0 which by Claim 2 makes i no better off.

Next consider case where h_i has a single symmetry group. Then since ij was added, and each individual had two links to begin with, it follows that $N(h_i) \cap SB(g^*) = \emptyset$. Moreover, it must be that $N(h_i) = N(h_i^*)$, where h_i^* is i's component under g^*. This implies that i got $Y_i^{ce}(g^*, v)$ under g^* and now gets $Y_i^{ce}(g, v)$. By efficiency, this cannot be improving for i.

The remaining case is where h_i is not symmetric and $N(h_i) \subset N_i \cup N_j$, where N_i and N_j are the symmetry groups of i and j under g, and $N_i \cap N_j = \emptyset$. As argued below, $N(h_i) \cap SB(g^*) = \emptyset$. Also, it follows again that $N(h_i) = N(h_i^*)$, and so the argument from the case above applies again. So to complete the proof we need only show that $N(h_i) \cap SB(g^*) = \emptyset$. First, note that ij cannot be a bridge as by the arguments of claim 1 there must be some $l \notin B(g)$, which would then put l is a different symmetry group than either i or j which would be a contradiction of this case. Consider the case where $B(g) = B(g^*)$. Then it must be that either $i \in SB(g^*)$ or $j \in B(g^*)$, but not both (given only two symmetry groups under g). Take $i \in SB(g^*)$. Then by severing i's bridge, the resulting component (given the addition of ij under g) is not symmetric. But this means there is some l in that component not in j's symmetry class, and also not in $B(g)$ and so l is in a third symmetry class which is a contradiction. Thus $B(g) \neq B(g^*)$. This means that ij is a link that connects two components that were only connected via some other link kl under g^*. Given there are only two symmetry classes N_i and N_j under h_i, then it must be that every individual is involved in such a duplicate bridge and that the duplicate ij was not present in g^*, which contradicts the fact that some individual in $N(h_i)$ is in $SB(g^*)$. □

Proof of Proposition 6. Under (i) from Example 3, it follows that any buyer (or seller) who gets a payoff of 0 from the bargaining would gain by severing any link, as the payoff from the bargaining would still be at least 0, but at a lower cost. Thus, in any pairwise stable network g all individuals who have any links must get payoffs of 1/2. Thus, from (iii) from Example 3, it follows that there is some number $K \geq 0$ such that there are exactly K buyers collectively linked to exactly K sellers and that we can find some subgraph g' with exactly K links linking all buyers to all sellers. Let us show that it must be that $g = g'$. Consider any buyer or seller in $N(g)$. Suppose that buyer (seller) has two or more links. Consider a link for that buyer (seller) in $g \setminus g'$. If that buyer (seller) severs that link, the resulting network will still be such that any subgroup of k buyers in the component can be matched with at least k distinct sellers and vice versa, since

g' is still a subset of the resulting network. Thus, under (iii) that buyer (seller) would still get a payoff of 1/2 from the trading under the new network, but would save a cost c_b (or c_s) from severing the link, and so g cannot be pairwise stable.

Thus, we have shown that all pairwise stable networks consist of $K \geq 0$ links connecting exactly K sellers to K buyers, and where all individuals who have a link get a payoff of 1/2.

To complete the proof, note that if there is any pair of buyer and seller who each have no link and each cost is less than 1/2, then both would benefit from adding a link, and so that cannot be pairwise stable. Without loss of generality assume that the number of buyers is at least the number of sellers. We have shown that any pairwise stable network is such that each seller is connected to exactly one buyer, and each seller to a different buyer. It is easily checked (by similar arguments) that any such network is pairwise stable. Since this is exactly the set of efficient networks for these cost parameters, the first claim in the Proposition follows.

The remaining two claims in the proposition follow from noting that in the case where $c_s > 1/2$ or $c_b > 1/2$, then K must be 0. Thus, the empty network is the only pairwise stable network in those cases. It is always Pareto efficent in these cases since someone must get a payoff less than 0 in any other network in this case. It is only efficient if $c_s + c_b \geq 1$. □

Proof of Proposition 8. The linearity of the Shapley value operator, and hence the Myerson value allocation rule,[43] implies that $Y_i(v, g) = Y_i(b, g) - Y_i(c, g)$. It follows directly from (2) that for monotone b and c, that $Y_i(b, g) \geq 0$ and likewise $Y_i(c, g) \geq 0$. Since $\sum_i Y_i(b, g) = b(g)$, and each $Y_i(b, g)$ is nonnegative it also follows that $b(g) \geq Y_i(b, g) \geq 0$ and likewise that $c(g) \geq Y_i(c, g) \geq 0$.

Let us show that for any monotone b and small enough $\overline{c} \geq c(\cdot)$, that the *unique* pairwise stable network is the complete network ($PS(Y^{MV}, v = b - c) = \{g^N\}$). We first show that for any network $g \in G$, if $ij \notin g$, then

$$Y_i(g + ij, b) \geq Y_i(g, b) + \frac{2b(\{ij\})}{n(n-1)(n-2)} \tag{4}$$

From (2) it follows that

$$Y_i^{MV}(g, b) - Y_i(g - ij, b) = \sum_{S \subset N \setminus \{i\}: j \in S} (b(g + ij|_{S \cup i}) - b(g|_{S \cup i})) \frac{\#S!(n - \#S - 1)!}{n!}.$$

Since b is monotone, it follows that $b(g + ij|_{S \cup i}) - b(g|_{S \cup i}) \geq 0$ for every S. Thus,

$$Y_i^{MV}(g, b) - Y_i(g - ij, b) \geq (b(g + ij|_{\{i,j\}}) - b(g|_{\{i,j\}})) \frac{\#2!(n-3)!}{n!}.$$

Since $b(g + ij|_{S \cup i}) - b(g|_{S \cup i}) = b(\{ij\}) > 0$, (4) follows directly.

Let $\overline{c} < \min_{ij} \frac{2b(\{ij\})}{n(n-1)(n-2)}$. (Note that for a monotone b, $b(\{ij\}) > 0$ for all ij.) Then from (4)

[43] This linearity is also easily checked directly from (2).

$$Y_i(g+ij,v) - Y_i(g,v) \geq \frac{2b(\{ij\})}{n(n-1)(n-2)} - (Y_i(g+ij,c) - Y_i(g,c)).$$

Note that since $\overline{c} \geq c(g) \geq Y_i(c,g) \geq 0$ for all g', it follows that $\overline{c} \geq Y_i(g+ij,c) - Y_i(g,c)$. Hence, from our choice of $\overline{c}$ it follows that $Y_i(g+ij,v) - Y_i(g,v)$ for all g and $ij \notin g$. This directly implies that the only pairwise stable network is the complete network.

Given that $g^* \neq g^N$ is efficient under b and c is strictly monotone, then it follows that the complete network is not efficient under v. This establishes the first claim of the proposition.

If b is such that $g^* \subset g \subset g^N$ for some symmetric $g \neq g^N$, then given that b is monotone it follows that g is also efficient for b. Also, the symmetry of g and anonymity of Y^{MV} implies that $Y_i(g,b) = Y_j(g,b)$ for all i and j. Since this is also true of g^N, it follows that $Y_i(g,b) \geq Y_i(g^N,b)$ for all i. For a strictly monotone c, this implies that $Y_i(g,b-c) > Y_i(g^N,b-c)$ for all i. Thus, g^N is Pareto dominated by g. Since g^N is the unique pairwise stable network, this implies the claim that $PS(Y^{MV},v) \cap PE(Y^{MV},v) = \emptyset$. □

Proof of Proposition 7. Consider b that is anonymous and monotone. Consider a symmetric g such that $C(g) = g$ and $N(g) = N$ and $g \neq g^N$. Let $b'(g') = \min[b(g'), b(g)]$. Note that b' is monotone and that g is efficient for b'. Find a strictly monotone c' according to Proposition 8, for which the unique pairwise stable network under $b' - c'$ is the complete network while the Pareto efficient networks are incomplete. Let $c = c' + b - b'$. It follows that c is strictly monotone. Also, $v = b - c = b' - c'$ and so the unique pairwise stable network under $b' - c'$ is the complete network while the Pareto efficient networks are incomplete. □

A Noncooperative Model of Network Formation

Venkatesh Bala[1], **Sanjeev Goyal**[2,*]

[1] Dept. of Economics, McGill University, 855 Sherbrooke St. W., Montreal H3A 2T7, Canada; e-mail: vbala2001@yahoo.com; http:rr www.arts.mcgill.ca
[2] Econometric Institute, Erasmus University, 3000 DR Rotterdam, The Netherlands
e-mail: goyal@few.eur.nl; http://www.few.eur.nl/few/people/goyal

Abstract. We present an approach to network formation based on the notion that social networks are formed by individual decisions that trade off the costs of forming and maintaining links against the potential rewards from doing so. We suppose that a link with another agent allows access, in part and in due course, to the benefits available to the latter via his own links. Thus individual links generate externalities whose value depends on the level of decay/delay associated with indirect links. A distinctive aspect of our approach is that the costs of link formation are incurred only by the person who initiates the link. This allows us to formulate the network formation process as a noncooperative game.

We first provide a characterization of the architecture of equilibrium networks. We then study the dynamics of network formation. We find that individual efforts to access benefits offered by others lead, rapidly, to the emergence of an equilibrium social network, under a variety of circumstances. The limiting networks have simple architectures, e.g., the wheel, the star, or generalizations of these networks. In many cases, such networks are also socially efficient.

Key Words: Coordination, learning dynamics, networks, noncooperative games.

1 Introduction

The Importance of Social and Economic Networks has been extensively documented in empirical work. In recent years, theoretical models have high-lighted their role in explaining phenomena such as stock market volatility, collective

---

* A substantial portion of this research was conducted when the first author was visiting Columbia University and New York University, while the second author was visiting Yale University. The authors thank these institutions for their generous hospitality.

We are indebted to the [Econometrica] editor and three anonymous referees for detailed comments on earlier versions of the paper. We thank Arun Agrawal, Sandeep Baliga, Alberto Bisin, Francis Bloch, Patrick Bolton, Eric van Damme, Prajit Dutta, David Easley, Yossi Greenberg, Matt Jackson, Maarten Janssen, Ganga Krishnamurthy, Thomas Marschak, Andy McLennan, Dilip Mookherjee, Yaw Nyarko, Hans Peters, Ben Polak, Roy Radner, Ashvin Rajan, Ariel Rubinstein, Pauline Rutsaert, and Rajeev Sarin for helpful comments. Financial support from SSHRC and Tinbergen Institute is acknowledged. Previous versions of this paper, dating from October 1996, were circulated under the title, "Self-Organization in Communication Networks."

action, the career profiles of managers, and the diffusion of new products, technologies and conventions.[1] These findings motivate an examination of the process of network formation.

We consider a setting in which each individual is a source of benefits that others can tap via the formation of costly pairwise links. Our focus is on benefits that are *nonrival*.[2] We suppose that a link with another agent allows access, in part and in due course, to the benefits available to the latter via his own links. Thus individual links generate externalities whose value depends on the level of decay/delay associated with indirect links. A distinctive aspect of our approach is that the costs of link formation are incurred only by the person who initiates the link. This allows us to model the network formation process as a noncooperative game, where an agent's strategy is a specification of the set of agents with whom he forms links. The links formed by agents define a social network.[3]

We study both one-way and two-way flow of benefits. In the former case, the link that agent i forms with agent j yields benefits solely to agent i, while in the latter case, the benefits accrue to both agents. In the benchmark model, the benefit flow across persons is assumed to be *frictionless*: if an agent i is linked with some other agent j via a sequence of intermediaries, $\{j_1, \ldots, j_s\}$, then the benefit that i derives from j is insensitive to the number of intermediaries. Apart from this, we allow for a general class of individual payoff functions: the payoff is strictly increasing in the number of other people accessed directly or indirectly and strictly decreasing in the number of links formed.

Our first result is that *Nash networks are either connected or empty*.[4] Connectedness is, however, a permissive requirement: for example, with one-way flows a society with 6 agents can have upwards of 20,000 Nash networks representing more than 30 different architectures.[5] This multiplicity of Nash equilibria motivates an examination of a stronger equilibrium concept. If an agent has multiple best responses to the equilibrium strategies of the others, then this may make the network less stable as the agent may be tempted to switch to a payoff-equivalent strategy. This leads us to study the nature of networks that can be supported in a strict Nash equilibrium.

---

[1] For empirical work see Burt (1992), Coleman (1966), Frenzen and Davis (1990), Granovetter (1974), and Rogers and Kincaid (1981). The theoretical work includes Allen (1982), Anderlini and Ianni (1996), Baker and Iyer (1992), Bala and Goyal (1998), Chwe (1998), Ellison (1993), Ellison and Fudenberg (1993), Goyal and Janssen (1997), and Kirman (1997).

[2] Examples include information sharing concerning brands/products among consumers, the opportunities generated by having trade networks, as well as the important advantages arising out of social favors.

[3] The game can be interpreted as saying that agents incur an initial fixed cost of forging links with others – where the cost could be in terms of time, effort, and money. Once in place, the network yields a flow of benefits to its participants.

[4] A network is connected if there is a path between every pair of agents. In recent work on social learning and local interaction, connectedness of the society is a standard assumption; see, e.g., Anderlini and Ianni (1996), Bala and Goyal (1998), Ellison (1993), Ellison and Fudenberg (1993), Goyal and Janssen (1997). Our results may be seen as providing a foundation for this assumption.

[5] Two networks have the same architecture if one network can be obtained from the other by permuting the strategies of agents in the other network.

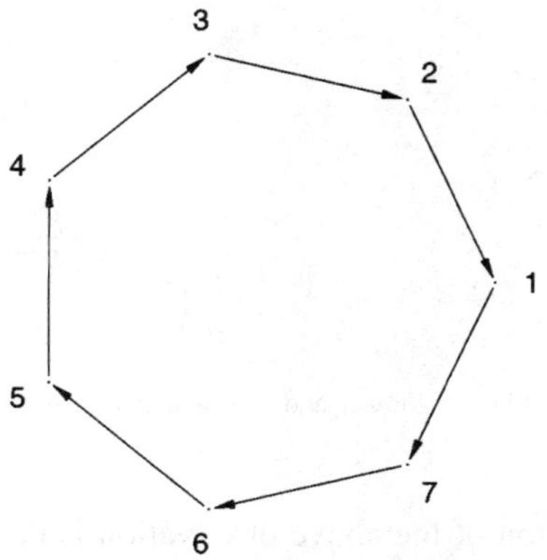

Fig. 1a. Wheel network

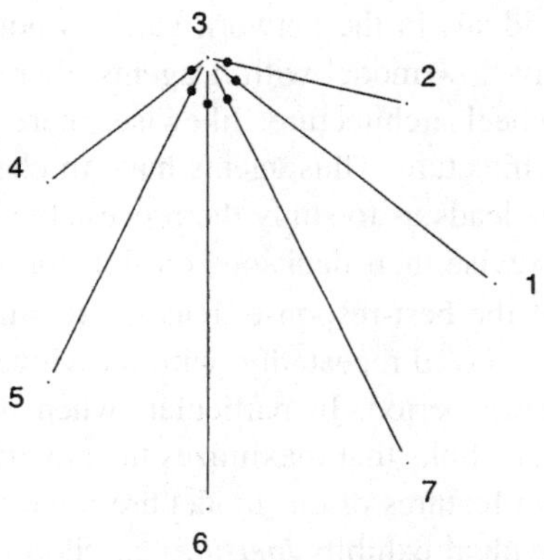

Fig. 1b. Center-sponsored star

We find that the refinement of strictness is very effective in our setting: *in the one-way flow model, the only strict Nash architectures are the wheel and the empty network.* Figure 1a depicts a wheel, which is a network where each agent forms exactly one link, represented by an arrow pointing to the agent. (The arrow also indicates the direction of benefit flow). The empty network is one where there are no links. In the *two-way flow model, the only strict Nash architectures are the center-sponsored star and the empty network.* Figure 1b depicts a center-sponsored star, where one agent forms all the links (agent 3 in the figure, as represented by the filled circles on each link adjacent to this agent).

These results exploit the observation that in a network, if two agents i and j have a link with the same agent k, then one of them (say) i will be indifferent between forming a link with k or instead forming a link with j. We know that Nash networks are either connected or empty. This argument implies that in the one-way flow model a nonempty strict Nash network has exactly n links. Since the wheel is the unique such network, the result follows. In the case of the two-way model, if agent i has a link with j, then no other agent can have a link with j. As a Nash network is connected, this implies that i must be the center of a

Fig. 1c. Flower and linked star networks

star. A further implication of the above observation is that every link in this star must be made or "sponsored" by the center.

While these findings restrict the set of networks sharply, the coordination problem faced by individuals in the network game is not entirely resolved. For example, in the one-way flow model with n agents, there are $(n-1)!$ networks corresponding to the wheel architecture; likewise, there are n networks corresponding to the star architecture. Thus agents have to choose from among these different equilibria. This leads us to study the process by which individuals learn about the network and revise their decisions on link formation, over time.

We use a version of the best-response dynamic to study this issue. The network formation game is played repeatedly, with individuals making investments in link formation in every period. In particular, when making his decision an individual chooses a set of links that maximizes his payoffs given the network of the previous period. Two features of our model are important: one, there is some probability that an individual exhibits *inertia*, i.e., chooses the same strategy as in the previous period. This ensures that agents do not perpetually miscoordinate. Two, if more than one strategy is optimal for some individual, then he *randomizes* across the optimal strategies. This requirement implies, in particular, that a non-strict Nash network can never be a steady state of the dynamics. The rules on individual behavior define a Markov chain on the state space of all networks; moreover, the set of absorbing states of the Markov chain coincides with the set of strict Nash networks of the one-shot game.[6]

Our results establish that the dynamic process converges to a limit network. In the one-way flow model, *for any number of agents and starting from any initial network, the dynamic process converges to a wheel or to the empty network, with probability 1*. The proof exploits the idea that well-connected people generate positive externalities. Fix a network g and suppose that there is an agent i who accesses all people in g, directly or indirectly. Consider an agent j who is not critical for agent i, i.e., agent i is able to access everyone even if agent j deletes all his links. Allow agent j to move; he can form a single link with agent i and access all the different individuals accessed by agent i. Thus if forming

[6] Our rules do not preclude the possibility that the Markov chain cycles permanently without converging to a strict Nash network. In fact, it is easy to construct examples of two-player games with a unique strict Nash equilibrium, where the above dynamic cycles.

links is at all profitable for agent j, then one best-response strategy is to form a single link with agent i. This strategy in turn makes agent j well-connected. We now consider some person k who is not critical for j and apply the same idea. Repeated application of this argument leads to a network in which everyone accesses everyone else via a single link, i.e., a wheel network. We observe that in a large set of cases, in addition to being a limit point of the dynamics, the *wheel is also the unique efficient architecture.*

In the two-way flow model, *for any number of agents and starting from any initial network, the dynamic process converges to a center-sponsored star or to the empty network, with probability 1.* With two-way flows the extent of the externalities are even greater than in the one-way case since, in principle, a person can access others without incurring any costs himself. We start with an agent i who has the maximum number of direct links. We then show that individual agents who are not directly linked with this agent i will, with positive probability, eventually either form a link with i or vice-versa. Thus, in due course, agent i will become the center of a star.[7] In the event that the star is not already center-sponsored, we show that a certain amount of miscoordination among 'spoke' agents leads to such a star. We also find that *a star is an efficient network for a class of payoff functions.*

The value of the results on the dynamics would be compromised if convergence occurred very slowly. In our setting, there are $2^{n(n-1)}$ networks with n agents. With $n = 8$ agents for example, this amounts to approximately 7×10^{16} networks, which implies that a slow rate of convergence is a real possibility. Our simulations, however, suggest that the *speed of convergence to a limiting network is quite rapid.*

The above results are obtained for a benchmark model with no *frictions*. The introduction of decay/delay complicates the model greatly and we are obliged to work with a linear specification of the payoffs. We suppose that each person potentially offers benefits V and that the cost of forming a link is c.We introduce decay in terms of a parameter $\delta \in [0, 1]$. We suppose that if the shortest path from agent j to agent i in a network involves q links, then the value of agent j's benefits to i is given by $\delta^q V$. The model without friction corresponds to $\delta = 1$.

We first show that *in the presence of decay, strict Nash networks are connected.* We are, however, unable to provide a characterization of strict Nash and efficient networks, analogous to the case without decay. The main difficulty lies in specifying the agents' best response correspondence. Loosely speaking, in the absence of decay a best response consists of forming links with agents who are connected with the largest number of other individuals. With decay, however, the distances between agents also becomes relevant, so that the entire structure

[7] It would seem that the center-sponsored star is an attractor because it reduces distance between different agents. However, in the absence of frictions, the distance between agents is not payoff relevant. On the other hand, among the various connected networks that can arise in the dynamics, this network is the only one where a single agent forms all the links, with everyone else behaving as a free rider. This property of the center-sponsored star is crucial.

of the network has to be considered. We focus on low levels of decay, where some properties of best responses can be exploited to obtain partial results.

In the one-way flow case, we identify *a class of networks with a flower architecture that is strict Nash* (see left-hand side of Fig. 1c). Flower networks trade-off the higher costs of more links (as compared to a wheel) against the benefits of shorter distance between different agents that is made possible by a "central agent." The wheel and the star[8] are special cases of this architecture. In the case of two-way flows, we find that *networks with a single star and linked stars are strict Nash* (see right-hand side of Fig. 1c).[9] We also provide a characterization of efficient networks and find that the *star is the unique efficient network for a wide range of parameters.* Simulations of the dynamics for both one-way and two-way models show that convergence to a limit (strict Nash) network is nearly universal and usually occurs very rapidly.

The arguments we develop can be summarized as follows: in settings where potential benefits are widely dispersed, individual efforts to access these benefits lead fairly quickly to the emergence of an equilibrium social network. The limiting networks have simple architectures, e.g., the wheel, the star, or generalizations of these networks. Moreover, in many instances these networks are efficient.

Our paper is a contribution to the theory of network formation. There is a large literature in economics, as well as in computer science, operations research, and sociology on the subject of networks; see, e.g., Burt (1992), Marshak and Radner (1972), Wellman and Berkowitz (1988). Much of this work is concerned with the efficiency aspects of different network structures and takes a planner's viewpoint.[10] By contrast, we consider network formation from the perspective of individual incentives. More specifically, the current paper makes two contributions.

The *first* contribution is our model of link formation. In the work of Boorman (1975), Jackson and Wolinsky (1996), among others, a link between two people requires that both people make some investments and the notion of stable networks therefore rests on pairwise incentive compatibility. We refer to this as a model with *two-sided link formation.* By contrast, in the present paper, link-formation is one-sided and noncooperative: an individual agent can form links with others by incurring some costs. This difference in modelling methodology is substantive since it allows the notion of Nash equilibrium and related refinements to be used in the study of network formation.[11]

[8] Star networks can also be defined with one-way flows and should not be confused with the star networks that arise in the two-way flows model.

[9] The latter structure resembles some empirically observed networks, e.g., the communication network in village communities (Rogers and Kincaid (1981, p. 175)).

[10] For recent work in this tradition, see Bolton and Dewatripont (1994) and Radner (1993). Hendricks, Piccione, and Tan (1995) use a similar approach to characterize the optimal flight network for a monopolist.

[11] The model of one-sided and noncooperative link formation was introduced and some preliminary results on the static model were presented in Goyal (1993).

The difference in formulation also alters the results in important ways. For instance, Jackson and Wolinsky (1996) show that with two-sided link formation the star is efficient but is not stable for a wide range of parameters. By contrast, in our model with noncooperative link formation, we find that the star is the unique efficient network and is also a strict Nash network for a range of values (Propositions 5.3–5.5). To see why this happens, suppose that $V < c$. With two-sided link formation, the central agent in a star will be better off by deleting his link with a spoke agent. In our framework, however, a link can be formed by a 'spoke' agent on his own. If there are enough persons in the society, this will be worthwhile for the 'spoke' agent and a star is sustainable as a Nash equilibrium.

The *second* contribution is the introduction of learning dynamics in the study of network formation.[12] Existing work has examined the relationship between efficient networks and strategically stable networks, in static settings. We believe that there are several reasons why the dynamics are important. One reason is that a dynamic model allows us to study the process by which individual agents learn about the network and adjust their links in response to their learning.[13] Relatedly, dynamics may help select among different equilibria of the static game: the results in this paper illustrate this potential very well.

In recent years, considerable work has been done on the theory of learning in games. One strand of this work studies the myopic best response dynamic; see e.g., Gilboa and Matsui (1991), Hurkens (1995), and Sanchirico (1996), among others. Gilboa and Matsui study the local stability of strategy profiles. Their approach allows for mixing across best responses, but does not allow for transitions from one strategy profile to another based on one player choosing a best response, while all others exhibit inertia. Instead, they require changes in social behavior to be continuous.[14] This difference with our formulation is significant. They show that every strict Nash equilibrium is a socially stable strategy, but that the converse is not true. This is because in some games a Nash

---

The literature on network games is related to the research in coalition formation in game-theoretic models. This literature is surveyed in Myerson 1991 and van den Nouweland (1993). Jackson and Wolinsky (1996) present a detailed discussion of the relationship between the two research programs. Dutta and Mutuswamy (1997) and Kranton and Minehart (1998) are some other recent papers on network formation. An alternative approach is presented in a recent paper by Mailath, Samuelson, and Shaked (1996), which explores endogenous structures in the context of agents who play a game after being matched. They show that partitions of society into groups with different payoffs can be evolutionary stable.

[12] Bala (1996) initially proposed the use of dynamics to select across Nash equilibria in a network context and obtained some preliminary results.

[13] Two earlier papers have studied network evolution, but in quite different contexts from the model here. Roth and Vande Vate (1990) study dynamics in a two-sided matching model. Linhart, Lubachevsky, Radner, and Meurer (1994) study the evolution of the subscriber bases of telephone companies in response to network externalities created by their pricing policies.

[14] Specifically, they propose that a strategy profile s is accessible from another strategy profile s' if there is a continuous smooth path leading from s' to s that satisfies the following property: at each strategy profile along the path, the direction of movement is consistent with each of the different players choosing one of their best responses to the current strategy profile. A set of strategy profiles S is 'stable' if no strategy profile $s' \notin S$ is accessible from any strategy profile $s \in S$, and each strategy profile in S is accessible from every other strategy profile in S.

equilibrium in mixed strategies is socially stable. By contrast, under our dynamic process, the set of strict Nash networks is equivalent to the set of absorbing networks.

Hurkens (1995) and Sanchirico (1996) study variants of best response learning in general games. They show that if the dynamic process satisfies certain properties, which include randomization across best responses, then it 'converges' to a minimal curb set, i.e., a set that is closed under the best response operation, in the long run. These results imply that weak Nash equilibria are not limit points of the dynamic process. However, in general games, a minimal curb set often consists of more than one strategy profile and there are usually several such sets. The games we analyze are quite large and the main issue here is the nature of minimal curb sets. Our results characterize these sets as well as show convergence of the dynamics.[15]

The rest of the paper is organized as follows. Section 2 presents the model. Section 3 analyzes the case of one-way flows, while Section 4 considers the case of two-way flows. Section 5 studies network formation in the presence of decay. Section 6 concludes.

2 The Model

Let $N = \{1, \ldots, n\}$ be a set of agents and let i and j be typical members of this set. To avoid trivialities, we shall assume throughout that $n \geq 3$. For concreteness in what follows, we shall use the example of gains from information sharing as the source of benefits. Each agent is assumed to possess some information of value to himself and to other agents. He can augment this information by communicating with other people; this communication takes resources, time, and effort and is made possible via the setting up of *pair-wise* links.

A *strategy* of agent $i \in N$ is a (row) vector $g_i = (g_{i,1}, \ldots, g_{i,i-1}, g_{i,i+1}, \ldots, g_{i,n})$ where $g_{i,j} \in \{0, 1\}$ for each $j \in N \backslash \{i\}$. We say agent i has a *link* with j if $g_{i,j} = 1$. A link between agent i and j can allow for either one-way (asymmetric) or two-way (symmetric) flow of information. With one-way communication, the link $g_{i,j} = 1$ enables agent i to access j's information, but not vice-versa.[16] With two-way communication, $g_{i,j} = 1$ allows both i and j to access each other's information.[17] The set of all strategies of agent i is denoted by G_i. Throughout the paper we restrict our attention to pure strategies. Since agent i has the option of forming or not forming a link with each of the remaining $(n - 1)$ agents, the number of strategies of agent i is clearly $|G_i| = 2^{n-1}$. The set $G = G_1 \times \cdots \times G_n$ is the space of pure strategies of all the agents. We now consider the game played by the agents under the two alternative assumptions concerning information flow.

---

[15] For a survey of recent research on learning in games, see Marimon (1997).

[16] For example, i could access j's website, or read a paper written by j.

[17] Thus, i could make a telephone call to j, after which there is information flow in both directions.

2.1 One-way Flow

In the one-way flow model, we can depict a strategy profile $g = (g_1, \ldots, g_n)$ in G as a *directed network*. The link $g_{i,j} = 1$ is represented by an *edge* starting at j with the arrowhead pointing at i. Figure 2a provides an example with $n = 3$ agents. Here agent 1 has formed links with agents 2 and 3, agent 3 has a link with agent 1 while agent 2 does not link up with any other agent. Note that there is a one-to-one correspondence between the set of all directed networks with n vertices and the set G.

Define $N^d(i; g) = \{k \in N | g_{i,k} = 1\}$ as the set of agents with whom i maintains a link. We say there is a *path* from j to i in g either if $g_{i,j} = 1$ or there exist distinct agents $j_1, \ldots, j_m$ different from i and j such that $g_{i,j_1} = g_{j_1,j_2} = \cdots = g_{j_m,j} = 1$. For example, in Fig. 2a there is a path from agent 2 to agent 3. The notation "$j \xrightarrow{g} i$" indicates that there exists a path from j to i in g. Furthermore, we define $N(i; g) = \{k \in N | \xrightarrow{g} i\} \cup \{i\}$. This is the set of all agents whose information i accesses either through a link or through a sequence of links. We shall typically refer to $N(i; g)$ as the set of agents who are observed by i. We use the convention that $i \in N(i; g)$, i.e. agent i observes himself. Let $\mu_j^d : G \to \{0, \ldots, n-1\}$ and $\mu_i : G \to \{1, \ldots, n\}$ be defined as $\mu_i^d(g) \equiv |N^d(i; g)|$ and $\mu_i(g) \equiv |N(i; g)|$ for $g \in G$. Here, $\mu_i^d(g)$ is the number of agents with whom i has formed links while $\mu_i(g)$ is the number of agents observed by agent i.

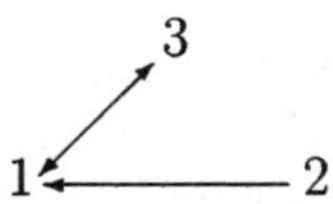

Fig. 2a.

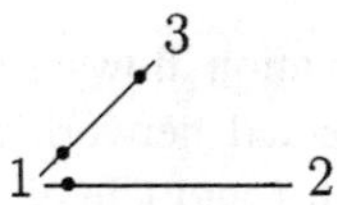

Fig. 2b.

To complete the definition of a normal-form game of network formation, we specify a class of payoff functions. Denote the set of nonnegative integers by Z_+. Let $\Phi : Z_+^2 \to R$ be such that $\Phi(x, y)$ is strictly increasing in x and strictly decreasing in y. Define each agent's payoff function $\Pi_i : G \to R$ as

$$\Pi_i(g) = \Phi(\mu_i(g), \mu_i^d(g)) . \tag{2.1}$$

Given the properties we have assumed for the function Φ, $\mu_i(g)$ can be interpreted as providing the "benefit" that agent i receives from his links, while $\mu_i^d(g)$ measures the "cost" associated with maintaining them.

The payoff function in (2.1) implicitly assumes that the value of information does not depend upon the number of individuals through which it has passed, i.e., that there is no information decay or delay in transmission. We explore the consequences of relaxing this assumption in Section 5.

A special case of (2.1) is when payoffs are linear. To define this, we specify two parameters $V > 0$ and $c > 0$, where V is regarded as the *value* of each agent's information (to himself and to others), while c is his *cost* of link formation. Without loss of generality, V can be normalized to 1. We now define $\Phi(x, y) = x - yc$, i.e.

$$\Pi_i(g) = \mu_i(g) - \mu_i^d(g)c \ . \tag{2.2}$$

In other words, agent i's payoff is the number of agents he observes less the total cost of link formation. We identify three parameter ranges of importance. If $c \in (0, 1)$, then agent i will be willing to form a link with agent j for the sake of j's information alone. When $c \in (1, n-1)$, agent i will require j to observe some additional agents to induce him to form a link with j. Finally, if $c > n-1$, then the cost of link formation exceeds the total benefit of information available from the rest of society. Here, it is a dominant strategy for i not to form a link with any agent.

2.2 Two-way Flow

In the two-way flow model, we depict the strategy profile $g = (g_1, \ldots, g_n)$ as a *nondirected network*. The link $g_{i,j} = 1$ is represented by an *edge* between i and j: a filled circle lying on the edge near agent i indicates that it is this agent who has initiated the link. Figure 2b below depicts the example of Fig. 2a for the two-way model. As before, agent 1 has formed links with agents 2 and 3, agent 3 has formed a link with agent 1 while agent 2 does not link up with any other agent.[18] Every strategy-tuple $g \in G$ has a unique representation in the manner shown in the figure.

To describe information flows formally, it is useful to define the *closure* of g: this is a nondirected network denoted $\bar{g} = cl(g)$, and defined by $\bar{g}_{i,j} = \max\{g_{i,j}, g_{j,i}\}$, for each i and j in N.[19] We say there is a *tw-path* (for two-way) in g between i and j if either $\bar{g}_{i,j} = 1$ or there exist agents $j_1, \ldots, j_m$ distinct from each other and i and j such that $\bar{g}_{i,j_1} = \cdots = \bar{g}_{j_m,j} = 1$. We write $i \overset{\bar{g}}{\leftrightarrow} j$ to indicate a tw-path between i and j in g. Let $N^d(i; g)$ and $\mu_j^d(g)$ be defined as in Sect. 2.1. The set $N(i; \bar{g}) = \{k | i \overset{\bar{g}}{\leftrightarrow} k\} \cup \{i\}$ consists of agents that i observes in g under two-way communication, while $\mu_j(\bar{g}) \equiv |N(i; \bar{g})|$ is its cardinality. The payoff accruing to agent i in the network g is defined as

$$\bar{\Pi}_i(g) = \Phi(\mu_j(\bar{g}), \mu_j^d(g)) \ . \tag{2.3}$$

where $\Phi(\cdot, \cdot)$ is as in Section 2.1. The case of linear payoffs is $\Phi(x, y) = x - yc$ as before. We obtain, analogously to (2.2):

$$\bar{\Pi}_i(g) = \mu_j(\bar{g}) - \mu_j^d(g)c \ . \tag{2.4}$$

The parameter ranges $c \in (0, 1)$, $c \in (1, n-1)$, and $c > n-1$ have the same interpretation as in Section 2.1.

[18] Since agents choose strategies independently of each other, two agents may simultaneously initiate a two-way link, as seen in the figure.

[19] Note that $\bar{g}_{i,j} = \bar{g}_{j,i}$ so that the order of the agents is irrelevant.

2.3 Nash and Efficient Networks

Given a network $g \in G$, let g_{-i} denote the network obtained when all of agent i's links are removed. The network g can be written as $g = g_i \oplus g_{-i}$ where the '$\oplus$' indicates that g is formed as the union of the links in g_i and g_{-i}. Under one-way communication, the strategy g_i is said to be a best-response of agent i to g_{-i} if

$$\Pi_i(g_i \oplus g_{-i}) \geq \Pi_i(g_i' \oplus g_{-i}), \quad \text{for all } g_i' \in G_i \ . \tag{2.5}$$

The set of all of agent i's best responses to g_{-i} is denoted $BR_i(g_{-i})$. Furthermore, a network $g = (g_1, \ldots, g_n)$ is said to be a *Nash network* if $g_i \in BR_i(g_{-i})$ for each i, i.e. agents are playing a Nash equilibrium. A *strict* Nash network is one where each agent gets a strictly higher payoff with his current strategy than he would with any other strategy. For two-way communication, the definitions are the same, except that $\overline{\Pi}_i$ replaces Π_i everywhere. The best-response mapping is likewise denoted by $\overline{BR}_i(\cdot)$.

We shall define our welfare measure in terms of the sum of payoffs of all agents. Formally, let $W : G \to R$ be defined as $W(g) = \sum_{i=1}^{n} \Pi_i(g)$ for $g \in G$. A network g is *efficient* if $W(g) \geq W(g')$ for all $g' \in G$. The corresponding welfare function for two-way communication is denoted $\overline{W}$. For the linear payoffs specified in (2.2) and (2.4), an efficient network is one that maximizes the total value of information made available to the agents, less the aggregate cost of communication.

Two networks $g \in G$ and $g' \in G$ are equivalent if g' is obtained as a permutation of the strategies of agents in g. For example, if g is the network in Fig. 2a, and g' is the network where agents 1 and 2 are interchanged, then g and g' are equivalent. The equivalence relation partitions G into classes: each class is referred to as an *architecture*.[20]

2.4 The Dynamic Process

We describe a simple process that is a modified version of naive best response dynamics. The network formation game is assumed to be repeated in each time period $t = 1, 2, \ldots$. In each period $t \geq 2$, each agent observes the network of the previous period.[21] With some fixed probability $r_i \in (0, 1)$, agent i is assumed to i exhibit 'inertia', i.e. he maintains the strategy chosen in the previous period.

[20] For example, consider the one-way flow model. There are n possible 'star' networks, all of which come under the equivalence class of the star architecture. Likewise, the wheel architecture is the equivalence class of $(n-1)!$ networks consisting of all permutations of n agents in a circle.

[21] As compared to models where, say, agents are randomly drawn from large populations to play a two-player game, the informational requirements for agents to compute a best response here are much higher. This is because the links formed by a single agent can be crucial in determining a best response. Some of our results on the dynamics can be obtained under somewhat weaker requirements. For instance, in the one-way flow model, the results carry over if, in a network g, an agent i knows only the sets $N(k; (g_{-i})$, and not the structure of links of every other agent k in the society. Further analysis under alternative informational assumptions is available in a working paper version, which is available from the authors upon request.

Furthermore, if the agent does not exhibit inertia, which happens with probability $p_i = 1 - r_i$, he chooses a myopic pure strategy best response to the strategy of all other agents in the previous period. If there is more than one best response, each of them is assumed to be chosen with positive probability. The last assumption introduces a certain degree of 'mixing' in the dynamic process and in particular rules out the possibility that a weak Nash equilibrium is an absorbing state.[22]

Formally, for a given set A, let $\Delta(A)$ denote the set of probability distributions on A. We suppose that for each agent i there exists a number $p_i \in (0, 1)$ and a function $\phi_i : G \rightarrow \Delta(G_i)$ where ϕ_i satisfies, for all $g = g_i \oplus g_{-i} \in G$:

$$\phi_i(g) \in \text{Interior } \Delta(BR_i(g_{-i})) \,. \tag{2.6}$$

For $\hat{g}_i$ in the support of $\phi_i(g)$, the notation $\phi_i(g)(\hat{g}_i)$ denotes the probability assigned to $\hat{g}_i$ by the probability measure $\phi_i(g)$. If the network at time $t \geq 1$ is $g^t = g_i^t \oplus g_{-i}^t$, the strategy of agent i at time $t + 1$ is assumed to be given by

$$g_i^{t+1} = \begin{cases} \hat{g}_i \in \text{ support } \phi_i(g), & \text{with probability } p_i \times \phi_i(g)(\hat{g}_i), \\ g_i^t, & \text{with probability } 1 - p_i \,. \end{cases} \tag{2.7}$$

Equation 2.7 states that with probability $p_i \in (0, 1)$, agent i chooses a naive i best response to the strategies of the other agents. It is important to note that under this specification, an agent may switch his strategy (to another best-response strategy) even if he is currently playing a best-response to the existing strategy profile. The function ϕ_i defines how agent i randomizes between best responses if more than one exists. Furthermore, with probability $1 - p_i$ agent i exhibits 'inertia', i.e. maintains his previous strategy.

We assume that the choice of inertia as well as the randomization over best responses by different agents is independent across agents. Thus our decision rules induce a transition matrix T mapping the state space G to the set of all probability distributions $\Delta(G)$ on G. Let $\{X_i\}$ be the stationary Markov chain starting from the initial network $g \in G$ with the above transition matrix. The process $\{X_t\}$ describes the dynamics of network evolution given our assumptions on agent behavior.

The dynamic process in the two-way model is the same except that we use the best-response mapping $\overline{BR_i}(\cdot)$ instead of $BR_i(\cdot)$.

---

[22] We can interpret the dynamics as saying that the links of the one-shot game, while durable, must be renewed at the end of each period by fresh investments in social relationships. An alternative interpretation is in terms of a fixed-size overlapping-generations population. At regular intervals, some of the individuals exit and are replaced by an equal number of new people. In this context, p_i is the probability that an agent is replaced by a new agent. Upon entry an agent looks around and informs himself about the connections among the set of agents. He then chooses a set of people and forms links with them, with a view to maximizing his payoffs. In every period that he is around, he renews these links via regular investments in personal relations. This models the link formation behavior of students in a school, managers entering a new organization, or families in a social setting.

3 The One-way Flow Model

In this section, we analyze the nature of network formation when information flow is one-way. Our results provide a characterization of strict Nash and efficient networks and also show that the dynamic process converges to a limit network, which is a strict Nash network, in all cases.

3.1 Static Properties

Given a network g, a set $C \subset N$ is called a *component* of g if for every distinct pair of agents i and j in C we have $j \overset{g}{\rightarrow} i$ (equivalently, $j \in N(i;g)$) and there is no strict superset C' of C for which this is true. A component C is said to be *minimal* if C is no longer a component upon replacement of a link $g_{i,j} = 1$ between two agents i and j in C by $g_{i,j} = 0$, ceteris paribus. A network g is said to be *connected* if it has a unique component. If the unique component is minimal, g is called *minimally connected.* A network that is not connected is referred to as disconnected. A network is said to be *empty* if $N(i;g) = \{i\}$ and it is called *complete* if $N^d(i;g) = N\backslash\{i\}$ for all $i \in N$. We denote the empty and the complete network by g^e and g^c, respectively. A *wheel* network is one where the agents are arranged as $\{i, \ldots, i_n\}$ with $g_{i_2,i_1} = \cdots = g_{i_1,i_{n-1}} = g_{i_1,i_n} = 1$ and there are no other links. The wheel network is denoted g^w. A *star* network has a central agent i such that $g_{i,j} = g_{j,i} = 1$ for all $j \in N\backslash\{i\}$ and no other links.

The (geodesic) distance from agent j to agent i in g is the number of links in the shortest path from j to i, and is denoted $d(i,j;g)$. We set $d(i,j;g) = \infty$ if there is no path from j to i in g. These definitions are taken from Bollobas (1978).

Our first result highlights a general property of Nash networks when agents are symmetrically positioned vis-a-vis information and the costs of access: in equilibrium, either there is no social communication or every agent has access to all the information in the society.

Proposition 3.1. *Let the payoffs be given by (2.1). A Nash network is either empty or minimally connected.*

The proof is given in Appendix A; the intuition is as follows. Consider a nonempty Nash network, and suppose that agent i is the agent who observes the largest number of agents in this network. Suppose i does not observe everyone. Then there is some agent j who is not observed by i and who does not observe i (for otherwise j would observe more agents than i). Since i gets values from his links, and payoffs are symmetric, j must also have some links. Let j deviate from his Nash strategy by forming a link with i alone. By doing so, j will observe strictly more agents than i does, since he has the additional benefit of observing i. Since j was observing no more agents than i in his original strategy, j increases his payoff by his deviation. The contradiction implies that i must observe every agent in the society. We then show that every other agent will have an incentive to either link with i or to observe him through a sequence of links, so that the network is connected. If the network is not minimally connected, then some agent could delete a link and still observe all agents, which would contradict Nash.

Figures 3a and 3b depict examples of Nash networks in the linear payoffs case specified by (2.2) with $c \in (0, 1)$. The number of Nash networks increases quite rapidly with n; for example, we compute that there are 5, 58, 1069, and in excess of 20,000 Nash networks as n takes on values of 3, 4, 5, and 6, respectively.

A Nash network in which some agent has multiple best responses is likely to be unstable since this agent can decide to switch to another payoff-equivalent

Fig. 3a. The star and the wheel (one-way model)

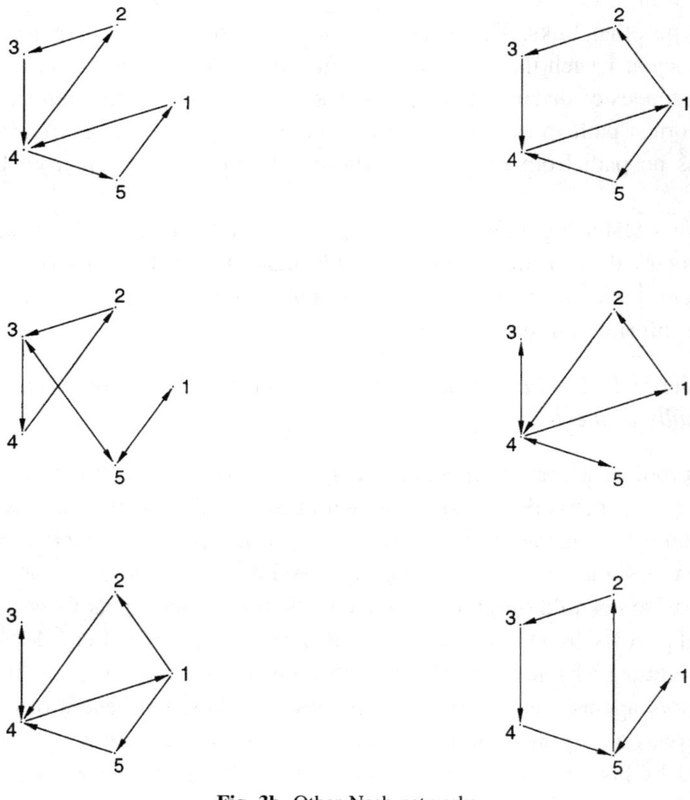

Fig. 3b. Other Nash networks

strategy. This motivates an examination of strict Nash networks. It turns out there are only two possible architectures for such networks.

Proposition 3.2. *Let the payoffs be given by (2.1). A strict Nash network is either the wheel or the empty network. (a) If* $\Phi\{\hat{x}+1,\hat{x}\} > \Phi(1,0)$ *for some* $\hat{x} \in \{1,\dots,n-1\}$*, then the wheel is the unique strict Nash. (b) If* $\Phi(x+1,x) < \Phi(1,0)$ *for all* $x \in (1,\dots,n-1)$ *and* $\Phi(n,1) > \Phi(1,0)$*, then the empty network and the wheel are both strict Nash. (c) If* $\Phi(x+1,x) < \Phi(1,0)$ *holds for all* $x \in \{1,\dots,n-1\}$ *and* $\Phi(n,1) < \Phi(1,0)$*, then the empty network is the unique strict Nash.*

Proof. Let $g \in G$ be strict Nash, and assume it is not the empty network. We show that for each agent k there is one and only one agent i such that $g_{i,k} = 1$. Since g is Nash, it is minimally connected by Proposition 3.1. Hence there is an agent i who has a link with k. Suppose there exists another agent j such that $g_{j,k} = 1$. As g is minimal we have $g_{i,j} = 0$, for otherwise i could delete the link with k and g would still be connected. Let $\hat{g}_i$ be the strategy where i deletes his link with k and forms one with j instead, ceteris paribus. Define $\hat{g} = \hat{g}_i \oplus g_{-i}$, where $\hat{g} \neq g$. Then $\mu_i^d(\hat{g}) = \mu_j^d(g)$. Furthermore, since $k \in N^d(j;g) = N^d(j;\hat{g})$, clearly $\mu_j(\hat{g}) \geq \mu_i(g)$ as well. Hence i will do at least as well with the strategy $\hat{g}_i$ as with his earlier strategy g_j, which violates the hypothesis that g_i is the unique best response to g_{-i}. As each agent has exactly one other agent who has a link with him, g has exactly n links. It is straightforward to show that the only connected network with n links is the wheel. Parts a–c now follow by direct verification. *Q.E.D.*

For the linear payoff case $\Pi_i(g) = \mu_j(g) - \mu_j^d(g)c$ of (2.2), Proposition 3.2(a) reduces to saying that the wheel is the unique strict Nash when $c \in (0,1]$. Proposition 3.2(b) implies that the wheel and the empty network are strict Nash in the region $c \in (1, n-1)$, while Proposition 3.2(c) implies that the empty network is the unique strict Nash when $c > n-1$. The final result in this subsection characterizes efficient networks.

Proposition 3.3. *Let the payoffs be given by (2.1). (a) If* $\Phi(n,1) > \Phi(1,0)$*, then the wheel is the unique efficient architecture, while (b) if* $\Phi(n,1) < \Phi(1,0)$*, then the empty network is the unique efficient architecture.*

Proof. Consider part (a) first. Let Γ be the set of values $(\mu_i(g), \mu_i^d(g))$ as g ranges over G. If $\mu_i^d(g) = 0$, then $\mu_i(g) = 1$, while if $\mu_i^d(g) \in \{1,\dots,n-1\}$, then $\mu_i(g) \in \{\mu_i^d(g)+1, n\}$. Thus, $\Gamma \subset \{1,\dots,n\} \times \{1,\dots,n-1\} \cup \{(1,0)\}$. Given $(x,y) \in \Gamma \backslash \{(1,0)\}$, we have $\Phi(n,1) \geq \Phi(n,y) \geq \Phi(x,y)$ since Φ is decreasing in its second argument and increasing in its first. For the wheel network g^w, note that $\mu_i(g^w) = n$ and $\mu_i^d(g^w) = 1$. Next consider a network $g \neq g^w$: for each $i \in N$, if $\mu_i^d(g) \geq 1$, then $\mu_i(g) \leq n$, while if $\mu_i^d(g) = 0$, then $\mu_i(g) = 1$. In either case,

$$\Pi_i(g^w) = \Phi(n,1) \geq \Phi(\mu_i(g), \mu_i^d(g)) = \Pi_i(g)\,, \tag{3.1}$$

where we have used the assumption that $\Phi(n,1) > \Phi(1,0)$. It follows that $W(g^w) = \sum_{i \in N} \Phi(n,1) \geq \sum_{i \in N} \Phi(\mu_i(g), \mu_i^d(g)) = W(g)$ as well. Thus g^w is an efficient architecture. To show uniqueness, note that our assumptions on Φ imply that equation (3.1) holds with strict inequality if $\mu_i^d(g) \neq 1$ or if $\mu_i(g) < n$. Let $g \neq g^w$ be given: if $\mu_i^d(g) \neq 1$ for even one i, then the inequality (3.1) is strict, and $W(g^w)) > W(g)$. On the other hand, suppose $\mu_i^d(g) = 1$ for all $i \in N$. As the wheel is the only connected network with n agents, and $g \neq g^w$, there must be an agent j such that $\mu_j(g) < n$. Thus, (3.1) is again a strict inequality for agent j and $W(g^w) > W(g)$, proving uniqueness.

In part (b), let g be different from the empty network g^e. Then there exists some agent j such that $\mu_j^d(g) \geq 1$. For this agent $\Pi_j(g^e) = \Phi(1,0) > \Phi(n,1) \geq \Phi(\mu_j(g), \mu_j^d(g)) = \Pi_j(g)$ while for all other agents i, $\Pi_i(g^e) = \Phi(1,0) \geq \Phi(\mu_i(g), \mu_i^d(g)) = \Pi_i(g)$. The result follows by summation. *Q.E.D.*

3.2 Dynamics

To get a first impression of the dynamics, we simulate a sample trajectory with $n = 5$ agents, for a total of twelve periods (Fig. 4).[23] As can be seen, the choices of agents evolve rapidly and settle down by period 11: the limit network is a wheel.

The above simulation raises an interesting question: under what conditions – on the structure of payoffs, the size of the society, and the initial network – does the dynamic process converge? Convergence of the process, if and when it occurs, is quite appealing from an economic perspective since it implies that agents who are myopically pursuing self-interested goals, without any assistance from a central coordinator, are nevertheless able to *evolve* a stable pattern of communication links over time. The following result shows that convergence occurs irrespective of the size of the society or the initial network.

Theorem 3.1. *Let the payoff functions be given by equation (2.1) and let g be the initial network. (a) If there is some* $\hat{x} \in \{1, \ldots, n-1\}$ *such that* $\Phi(\hat{x}+1, \hat{x}) \geq \Phi(1,0)$*, then the dynamic process converges to the wheel network, with probability 1. (b) If instead,* $\Phi(x+1, x) < \Phi(1,0)$ *for all* $x \in \{1, \ldots, n-1\}$ *and* $\Phi(n,1) > \Phi(1,0)$*, then the process converges to either the wheel or the empty network, with probability 1. (c) Finally, if* $\Phi(x+1,x) < \Phi(1,0)$ *for all* $x \in \{1, \ldots, n-1\}$ *and* $\Phi(n,1) < \Phi(1,0)$*, then the process converges to the empty network, with probability 1.*

Proof. The proof relies on showing that given an arbitrary network g there is a positive probability of transiting to a strict Nash network in finite time, when agents follow the rules of the process. As strict Nash networks are absorbing

[23] We suppose that payoffs have the linear specification (2.2) and that $c \in (0, 1)$. The initial network (labelled $t = 1$) has been drawn at random from the set of all directed networks with 5 agents. In period $t \geq 2$, the choices of agents who exhibit inertia have been drawn with solid lines, while the links of those who have actively chosen a best response are drawn with dashed lines.

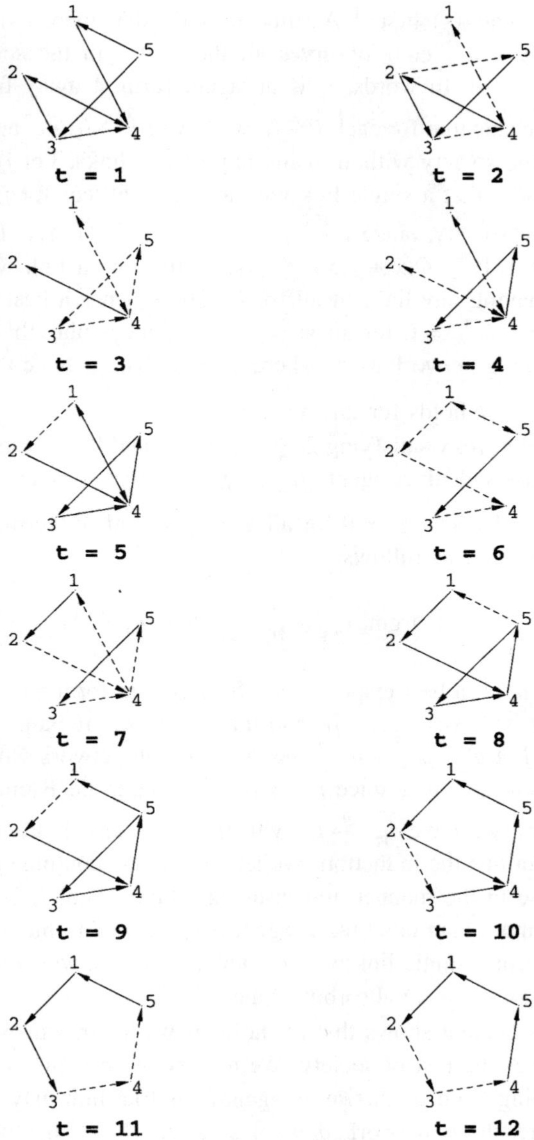

Fig. 4. Sample path (one-way model)

states, the result will then follow from the standard theory of Markov chains. By (2.7) there is a positive probability that all but one agent will exhibit inertia in a given period. Hence the proof will follow if we can specify a sequence of networks where at each stage of the sequence only one (suitably chosen) agent selects a best response. In what follows, unless specified otherwise, when we allow an agent to choose a best response, we implicitly assume that all other agents exhibit inertia.

We consider part (a) first.[24] Assume initially that there exists an agent j_i for whom $\mu_{j_1}(g) = n$, i.e. j_1 observes all the agents in the society. Let $j_2 \in \text{argmax}_{m \in N} d(j_1, m; g)$. In words, j_2 is an agent furthest away from j_1 in g. In particular, this means that for each $i \in N$ we have $i \xrightarrow{g-j_2} j_1$, i.e. agent j_1 observes every agent in the society without using any of j_2's links. Let j_2 now choose a best response. Note that a single link with agent j_1 suffices for j_2 to observe all the agents in the society, since $i \xrightarrow{g-j_2} j_1$ for all $i \in N \backslash \{j_1, j_2\}$. Furthermore, as $\Phi(n, 1) \geq \Phi(\hat{x} + 1, 1) \geq \Phi(\hat{x} + 1, \hat{x}) \geq \Phi(1, 0)$, forming a link with j_1 (weakly) dominates not having any links at all for j_2. Thus, j_2 has a best response $\hat{g}_{j_2}$ of the form $\hat{g}_{j_2,j_1} = 1, \hat{g}_{j_2,m} = 0$ for all $m \neq j_1$. Let agent j_2 play this best response. Denote the resulting network as g^1 where $g^1 = \hat{g}_{j_2} \oplus g_{-j_2}$. Note that the property $i \xrightarrow{g^1} j_1$ for all $i \in N$ holds for this network.

More generally, fix s satisfying $2 \leq s \leq n - 1$, and let g^{s-1} be the following network: there are s distinct agents $j_1, \ldots, j_s$ such that for each $q \in \{2, \ldots, s\}$ we have $g^{s-1}_{j_q, j_{q-1}} = 1$ and $g^{s-1}_{q,m} = 0$ for all $m \neq j_{q-1}$, and furthermore, $i \xrightarrow{g^{s-1}} j_1$ for all $i \in N$. Choose j_{s+1} as follows:

$$j_{s+1} \in \text{argmax}_{m \in N \backslash \{j_1, \ldots, j_s\}} d(j_1, m; g^{s-1}) \,. \tag{3.2}$$

Note that given g^{s-1}, a best response $\hat{g}_{j_{s+1}}$ for j_{s+1} is to form a link with j_s alone. By doing so, he observes $j_s, \ldots, j_1$, and through j_1, the remaining agents in the society as well. Let $g^s = \hat{g}_{j_{s+1}} \oplus g^{s-1}_{-j_{s+1}}$ be the resulting network when j_{s+1} chooses this strategy. Note also that since j_{s+1}'s link formation decision is irrelevant to j_1 observing him, we have $j_{s+1} \xrightarrow{g^s} j_1$, with the same also holding for $j_s, \ldots, j_2$. Thus we can continue the induction. We let the process continue until j_n chooses his best response in the manner above: at this stage, agent j_1 is the only agent with (possibly) more than one link. If agent j_1 is allowed to move again, his best response is to form a single link with j_n, which creates a wheel network g^w. By Proposition 3.2(a), g^w is an absorbing state.

The above argument shows that (a) holds if we assume there is some agent in g who observes the rest of society. We now show that this is without loss of generality. Starting from g, choose an agent i' and let him play a best response $\hat{g}_{i'}$. Label the resulting network $\hat{g}_{i'} \oplus g_{-i'}$ as g'. Note that we can suppose $\mu^d_{i'}(g') \geq 1$. This is because zero links yield a payoff no larger than forming $\hat{x}$ links and observing $\hat{x}+1$ (or more) agents. If $\mu_{i'}(g') = n$ we are done. Otherwise, if $\mu_{i'}(g') < n$, choose $i'' \notin N(i'; g')$ and let him play a best response $\hat{g}_{i'}$. Define $g'' = \hat{g}_{i''} \oplus g'_{-i''}$. As before, we can suppose without loss of generality that $\hat{g}_{i''}$ involves at least one link. We claim that $\mu_{i''}(g'') \geq \mu_{i'}(g')+1$. Indeed, by forming a link with i', agent i'' can observe i' and all the other agents that i' observes, and thereby guarantee himself a payoff of $\Phi(\mu_{i'}(g')+1, 1)$. The claim now follows because $\Phi(\mu_{i'}(g') + 1, 1) > \Phi(x, y)$ for any (x, y) pair satisfying $x \leq \mu_{i'}(g')$ and

[24] We thank an anonymous referee for suggesting the following arguments, which greatly simplify our original proof.

$y \geq 1$. Repeating this argument if necessary, we eventually arrive at a network where some agent observes the entire society, as required.

We now turn to parts (b) and (c). If $\Phi(n,1) < \Phi(1,0)$, it is a dominant strategy for each agent not to form any links. Statement (c) follows easily from this observation. We consider (b) next. Note from Proposition 3.2(b) that the wheel is strict Nash for this payoff regime. Suppose there exists an agent $i \in N$ such that $\mu_i(g) = n$. Then the argument employed in part (a) ensures convergence to the wheel with positive probability. If, instead, $\mu_i(g) < n$, let $\hat{x} \geq 2$ be the largest number such that $\Phi(\hat{x},1) \leq \Phi(1,0)$. Note that $\hat{x} \leq n-1$ since $\Phi(n,1) > \Phi(1,0)$. Suppose there exists $i \in N$ such that $\mu_i(g) \in \{\hat{x}, \ldots, n-1\}$. Then the argument used in the last part of the proof of part (a) can be applied to eventually yield an agent who observes every agent in the society. The last possibility is that for all agents i in g we have $\mu_i(g) < \hat{x}$. Choose an agent i' and consider the network g' formed after he chooses his best response. Suppose $\mu^d_{i'}(g') \geq 1$ and $\mu_{i'}(g') < \hat{x}$. Then $\Pi_{i'}(g') = \Phi(\mu_{i'}(g'), \mu^d_{i'}(g')) < \Phi(\hat{x},1) \leq \Phi(1,0)$ and forming no links does strictly better. Hence, if i' has a best response involving the formation of at least one link, he must observe at least $\hat{x}$ agents (including himself) in the resulting network. Thus we let each agent play in turn – either they will all choose to form no links, in which case the process is absorbed into the empty network, or some agent eventually observes at least $\hat{x}$ agents. In the latter event, we can employ the earlier arguments to show convergence with positive probability to a wheel.

Q.E.D.

In the case of linear payoffs $\Pi_i(g) = \mu_i(g) - \mu^d_i(g)c$, Theorem 3.1 says that when costs are low ($0 < c \leq 1$) the dynamics converge to the wheel, when costs are in the intermediate range ($1 < c < n-1$), the dynamics converge to either the wheel or the empty network, while if costs are high ($c > n-1$), then the system collapses into the empty network.

Under the hypotheses of Theorem 3.1(b), it is easy to demonstrate *path dependence*, i.e. a positive probability of converging to either the wheel or the empty network from an initial network. Consider a network where agent 1 has $n-1$ links and no other agent has any links. If agent 1 moves first, then $\Phi(x+1,x) < \Phi(1,0)$ for all $x \in \{1, \ldots, n-1\}$ implies that his unique best response is not to form any links, and the process collapses to the empty network. On the other hand, if the remaining agents play one after another in the manner specified by the proof of the above theorem, then convergence to the wheel occurs.

Recall from Proposition 3.3 that when $\Phi(n,1) > \Phi(1,0)$, the unique efficient network is the wheel, while if $\Phi(n,1) < \Phi(1,0)$ the empty network is uniquely efficient. Suppose the condition $\Phi(\hat{x}+1,\hat{x}) \geq \Phi(1,0)$ specified in Theorem 3.1(a) holds. Then as $\Phi(n,1) \geq \Phi(\hat{x}+1,1) \geq \Phi(\hat{x}+1,\hat{x})$ with at least one of these inequalities being strict, we get $\Phi(n,1) > \Phi(1,0)$. Thus we have the following corollary.

Corollary 3.1. *Suppose the hypothesis of Theorem 3.1(a) or Theorem 3.1(c) holds. Then starting from any initial network, the dynamic process converges to the unique efficient architecture with probability 1.*

Efficiency is not guaranteed in Theorem 3.1(b): while the wheel is uniquely efficient, the dynamics may converge to the empty network instead. However, as the proof of the theorem illustrates, there are many initial networks from which convergence to the efficient architecture occurs with positive probability.

Rates of Convergence. We take payoffs according to the linear model (2.2), i.e. $\Pi_i(g) = \mu_i(g) - \mu_i^d(g)c$. We focus upon two cases: $c \in (0,1)$ and $c \in (1,2)$. In the former case, Theorem 3.1(a) shows that convergence to the wheel always occurs, while in the latter case, Theorem 3.1(b) indicates that either the wheel or the empty network can be the limit.

In the simulations we assume that $p_i = p$ for all agents. Furthermore, let $\hat{\phi}$ be such that it assigns equal probability to all best responses of an agent given a network g. We assume that all agents have the same function $\hat{\phi}$. The initial network is chosen by the method of equiprobable links: a number $k \in \{0,\ldots,n(n-1)\}$ is first picked at random, and then the initial network is chosen randomly from the set of all networks having a total of k links.[25] We simulate the dynamic process starting from the initial network until it converges to a limit. Our simulations are with $n = 3$ to $n = 8$ agents, for $p = 0.2$, 0.5, and 0.8. For each (n,p) pair, we run the process for 500 simulations and report the average convergence time. Table 1 summarizes the results when $c \in (0,1)$ and $c \in (1,2)$. The standard errors are in parentheses.

Table 1. Rates of convergence in one-way flow model

	$c \in (0,1)$			$c \in (1,2)$		
n	$p = 0.2$	$p = 0.5$	$p = 0.8$	$p = 0.2$	$p = 0.5$	$p = 0.8$
3	15.29(0.53)	7.05 (0.19)	6.19 (0.19)	8.58 (0.35)	4.50 (0.17)	5.51 (0.24)
4	23.23(0.68)	12.71(0.37)	13.14 (0.42)	11.52(0.38)	5.98 (0.18)	6.77 (0.22)
5	28.92(0.89)	17.82(0.54)	28.99 (1.07)	15.19(0.40)	9.16 (0.27)	14.04 (0.59)
6	38.08(1.02)	26.73(0.91)	55.93 (2.30)	19.93(0.57)	12.68(0.41)	28.81 (1.16)
7	45.90(1.30)	35.45(1.19)	119.57(5.13)	25.46(0.71)	18.51(0.57)	57.23 (2.29)
8	57.37(1.77)	54.02(2.01)	245.70(10.01)	27.74(0.70)	26.24(0.89)	121.99(5.62)

Table 1 suggests that the rates of convergence are very rapid. In a society with 8 agents we find that with $p = 0.5$, the process converges to a strict Nash in less than 55 periods on average.[26] Secondly, we find that in virtually all the cases (except for $n = 3$) the average convergence time is higher if $p = 0.8$ or $p = 0.2$ compared to $p = 0.5$. The intuition for this finding is that when p is small, there is a very high probability that the state of the system does not change very much from one period to the next, which raises the convergence time. When p is very large, there is a high probability that "most" agents move

[25] An alternative approach specifies that each network in G is equally likely to be chosen as the initial one. Simulation results with this approach are similar to the findings reported here.

[26] The precise significance of these numbers depends on the duration of the periods and more generally on the particular application under consideration.

simultaneously. This raises the likelihood of miscoordination, which slows the process. The convergence time is thus lowest for intermediate values of p where these two effects are balanced. Thirdly, we find that the average convergence time increases relatively slowly as n increases. So, for instance, as we increase the size of the society from three agents to eight agents, the number of networks increases from 64 to more than 10^{16} networks. Yet the average convergence time (for $p = 0.5$) only increases from around 8 periods to around 54 periods. Finally, we note that the average times are even lower when the communication cost is higher, as seen when $c \in (1, 2)$. This is not simply a reflection of the possibility of absorption into the empty network when $c > 1$: for example, with $n = 8$ this occurred in no more than 3% of all simulations. Instead, it seems to be due to the fact that the set of best responses decreases with higher costs of communication.

4 Two-way Flow Model

In this section, we study network formation when the flow of information is two-way. Our results provide a characterization of strict Nash networks and efficient networks. We also show that the dynamic process converges to a limit network that is a strict Nash network, for a broad class of payoff functions.

4.1 Static Properties

Let the network g be given. A set $C \subset N$ is called a *tw-component* of g if for all i and j in C there is a tw-path between them, and there does not exist a tw-path between an agent in C and one in $N \backslash C$. A tw-component C is called *minimal* if (a) there does not exist a *tw-cycle* within C, i.e. $q \geq 3$ agents $\{j_1, \ldots, j_q\} \subset C$ such that $\overline{g}_{j_1,j_2} = \cdots = \overline{g}_{j_q,j_1} = 1$, and (b) $g_{i,j} = 1$ implies $g_{j,i} = 0$ for any pair of agents i, j in C. The network g is called *tw-connected* if it has a unique tw-component C. If the unique tw-component C is minimal, we say that g is *minimally tw-connected*. This implies that there is a unique tw-path between any two agents in N. The *tw-distance* between two agents i and j in g is the length of the shortest tw-path between them, and is denoted by $d(i, j; \overline{g})$. We begin with a preliminary result on the structure of Nash networks.

Proposition 4.1. *Let the payoffs be given by (2.3). A Nash network is either empty or minimally tw-connected.*

We make some remarks in relation to the above result. First, by the definition of payoffs, while one agent bears the cost of a link, both agents obtain the benefits associated with it. This asymmetry in payoffs is relevant for defining the architecture of the network. As an illustration, we note that there are now three types of 'star' networks, depending upon which agents bear the costs of the links in the network. For a society with $n = 5$ agents, Figs. 5a–c illustrate these types. Figure 5a shows a *center-sponsored* star, Fig. 5b a *periphery-sponsored* star, and Fig. 5c depicts a *mixed-type* star.

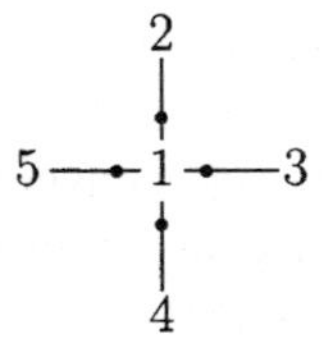

Fig. 5a. Center-sponsored

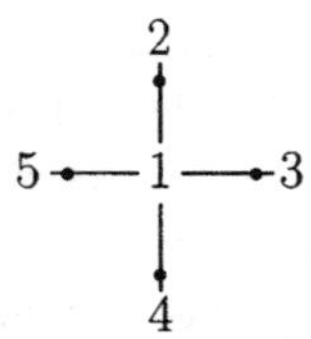

Fig. 5b. Periphery-sponsored

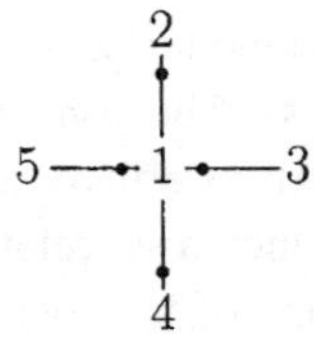

Fig. 5c. Mixed-type

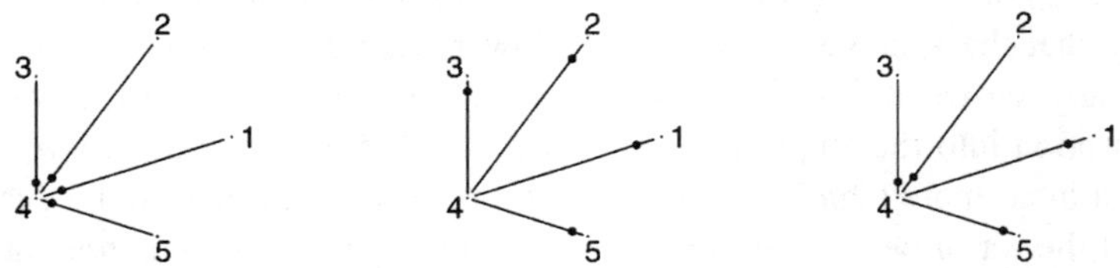

Fig. 6a. Star networks (two-way model)

Second, there can be a large number of Nash equilibria. For example, consider the linear specification (2.4) with $c \in (0,1)$. With $n = 3, 4, 5$, and 6 agents there are 12, 128, 2000, and 44 352 Nash networks, respectively. Figures 6a and 6b present some examples of Nash networks.

We now show that the set of strict Nash equilibria is significantly more restrictive.

Proposition 4.2. *Let the payoffs be given by (2.3). A strict Nash network is either a center-sponsored star or the empty network. (a) A center-sponsored star is strict Nash if and only if* $\Phi(n, n-1) > \Phi(x+1, x)$ *for all* $x \in \{0, \ldots, n-2\}$. *(b) The empty network is strict Nash if and only if* $\Phi(1,0) > \Phi(x+1, x)$ *for all* $x \in \{1, \ldots, n-1\}$.

Proof. Suppose g is strict Nash and is not the empty network. Let $\overline{g} = cl(g)$. Let i and j be agents such that $g_{i,j} = 1$. We claim that $\overline{g}_{j,j'} = 0$ for any $j' \notin \{i, j\}$. If this were not true, then i can delete his link with j and form one with j' instead, and receive the same payoff, which would contradict the assumption that g is strict Nash. Thus any agent with whom i is directly linked cannot have any other links. As g is minimally tw-connected by Proposition 4.1, i must be the center of a star and $g_{j,i} = 0$. If $j' \neq j$ is such that $g_{j',i} = 1$, then j' can switch to j and get the same payoff, again contradicting the supposition that g is strict Nash. Hence, the star must be center-sponsored.

Under the hypothesis in (a) it is clear that a center-sponsored star is strict Nash, while the empty network is not Nash. On the other hand, let g be a center-sponsored star with i as center, and suppose there is some $x \in \{0, \ldots, n-2\}$ such that $\Phi(x+1, x) \geq \Phi(n, n-1)$. Then i can delete all but x links and do at least as well, so that g cannot be strict Nash. Similar arguments apply under the hypotheses in (b). *Q.E.D.*

For the linear specification (2.4), Proposition 4.2 implies that when $c \in (0,1)$ the unique strict Nash network is the center-sponsored star, and when $c > 1$ the unique strict Nash network is the empty network.

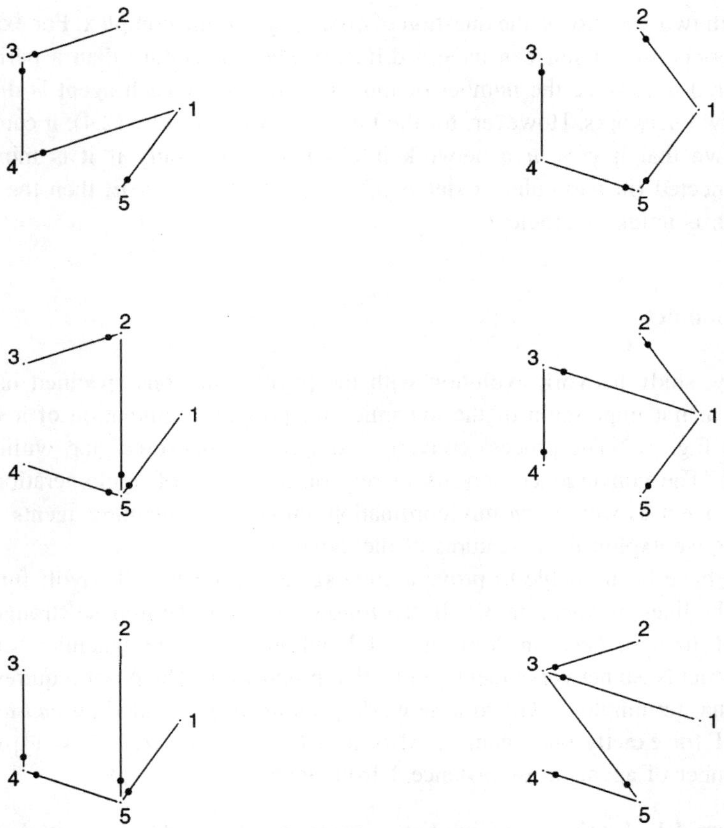

Fig. 6b. Other Nash networks

We now turn to the issue of efficiency. In general, an efficient network need not be either tw-connected or empty.[27] We provide the following partial characterization of efficient networks.

Proposition 4.3. *Let the payoffs be given by (2.3). All tw-components of an efficient network are minimal. If* $\Phi(x+1, y+1) \geq \Phi(x, y)$, *for all* $y \in \{0, \ldots, n-2\}$ *and* $x \in \{y+1, \ldots, n-1\}$, *then an efficient network is tw-connected.*

As the intuition provided below is simple, a formal proof is omitted. Minimality is a direct consequence of the absence of frictions. In the second part, tw-connectedness follows from the hypothesis that an additional link to an unobserved agent is weakly preferred by individual agents; since information flow is two-way, such a link generates positive externalities in addition and therefore increases social welfare.

[27] For example, consider a society with 3 agents. Let $\Phi(1,0) = 6.4$, $\Phi(2,0) = 7$, $\Phi(3,0) = 7.1$, $\Phi(2,1) = 6$, $\Phi(3,1) = 6.1$, $\Phi(3,2) = 0$. Then the network $g_{1,2} = 1$, and $g_{i,j} = 0$ for all other pairs of agents (and its permutations) constitutes the unique efficient architecture.

With two-way flows, the question of efficiency is quite complex. For example, a center-sponsored star can have a different level of welfare than a periphery-sponsored one, since the number of links maintained by each agent is different in the two networks. However, for the linear payoffs given by (2.4), it can easily be shown that if $c \leq n$ a network is efficient if and only if it is minimally tw-connected (in particular, a star is efficient), while if $c > n$, then the empty network is uniquely efficient.

4.2 Dynamics

We now study network evolution with the payoff functions specified in (2.3). To get a first impression of the dynamics we present a simulation of a sample path in Fig. 7.[28] The process converges to a center-sponsored star, within nine periods. The convergence appears to rely on a process of agglomeration on a central agent as well as on miscoordination among the remaining agents. In our analysis we exploit these features of the dynamic.

We have been unable to prove a convergence result for all payoff functions along the lines of Theorem 3.1. In the following result, we impose stronger versions of the hypotheses in Proposition 4.2 and prove that the dynamics converge to the strict Nash networks identified by that proposition. The proof requires some additional terminology. Given a network g, an agent j is called an *end-agent* if $\overline{g}_{j,k} = 1$ for exactly one agent k. Also, let $\sigma(i;\overline{g}) = |\{k|d(i,k;\overline{g}) = 1\}|$ denote the number of agents at tw-distance 1 from agent i.

Theorem 4.1. *Let the payoff functions be given by (2.3) and fix any initial network g. (a) If $\Phi(x+1, y+1) > \Phi(x,y)$ for all $y \in \{0,1,\ldots,n-2\}$ and $x \in \{y+1,\ldots,n-1\}$, then the dynamic process converges to the center-sponsored star, with probability 1. (b) If $\Phi(x+1, y+1) < \Phi(x,y)$ for all $y \in \{0,1,\ldots,n-2\}$ and $x \in \{y+1,\ldots,n-1\}$, then the dynamic process converges to the empty network, with probability 1.*

Proof. As with Theorem 3.1, the broad strategy behind the proof is to show that there is a positive probability of transition to a strict Nash network in finite time. We consider part (a) first. Note that the hypothesis on payoffs implies that $\Phi(n,n-1) > \max_{\{0 \leq x \leq n-2\}} \Phi(x+1,x)$, which, by Proposition 4.2(a), implies that the center-sponsored star is the unique strict Nash network. Starting from g, we allow every agent to move in sequence, one at a time. Lemma 4.1 in Appendix B shows that after all agents have moved, the resulting network is either minimally tw-connected or is the empty network. Suppose first that the network is empty. Then we allow a single agent to play. As $\Phi(n,n-1) > \max_{\{0 \leq x \leq n-2\}} \Phi(x+1,x)$, the agent's unique best response is to form links with all the others. This results

[28] Here, the payoffs are given by the linear model (2.4) with $c \in (0,1)$. The initial network (labelled $t = 1$) has been drawn at random from the set of all directed networks with 5 agents. In period $t \geq 2$, the choices of agents who exhibit inertia have been drawn with solid lines, while those whose choices are best responses have been drawn using dashed lines.

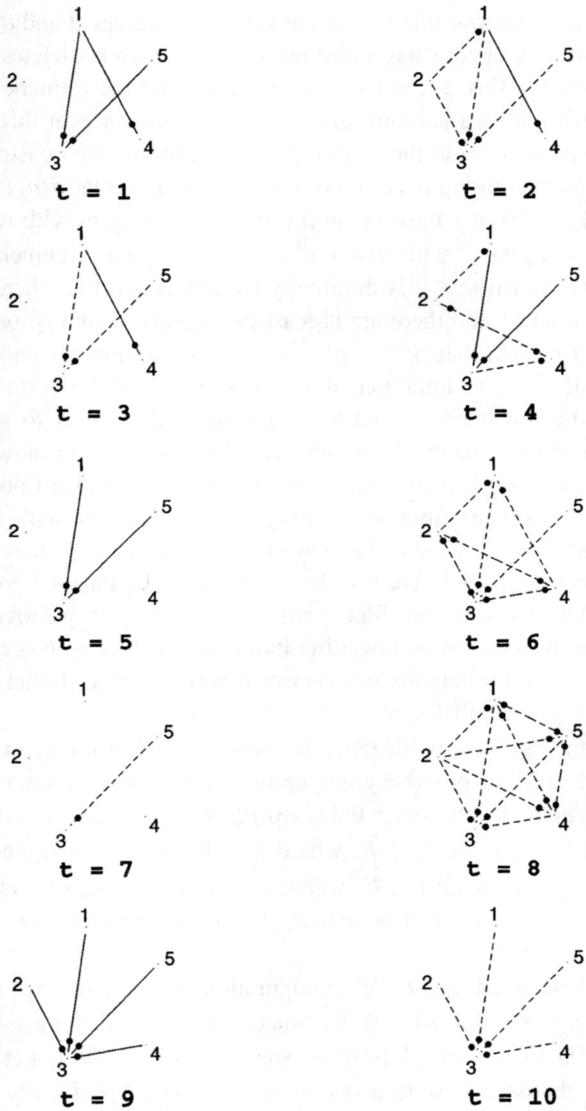

Fig. 7. Sample path (two-way model)

in a center-sponsored star, and (a) will follow. There is thus no loss of generality in supposing that the initial network g itself is minimally tw-connected.

Let agent $n \in \text{argmax}_{i \in N}\, \sigma(i; \overline{g})$. Since g is tw-connected, $\sigma(n; \overline{g}) \geq 2$. Furthermore, as g is also minimal, there is a unique tw-path between agent n and every other agent. Thus if $i \neq n$ then either $\overline{g}_{n,i} = 1$ or there exist $\{i_1, \ldots, i_q\}$ such that $\overline{g}_{n,i_1} = \cdots = \overline{g}_{i_q,i} = 1$. We shall say that i is *outward-pointing* with respect to n, if $g_{i,n} = 1$ in the former case and $g_{i,i_q} = 1$ in the latter case. Likewise, i is *inward-pointing* with respect to n if $g_{n,i} = 1$ in the former case and $g_{i_q,i} = 1$

in the latter case. Suppose that i is an outward-pointing agent and $d(i,n;\overline{g}) \geq 2$. It can be shown that agent i has a best response in which he deletes the link g_{i,i_q} and instead forms a link $g_{i,n} = 1$ (see Lemma 4.2 and the Remark in Appendix B). Let all such outward-pointing agents move in sequence in this manner and form a link with n. Denote the resulting network by g^1. By construction of g^1, for every outward-pointing agent i vis-a-vis n, it is true that $d(n,i;\overline{g}^1 = 1$; thus if $d(n,j;g^{-1}) \geq 2$, then j must be an inward-pointing agent with respect to n.

Consider an agent j, with $d(n,j;g^1) \geq 3$. Using the argument of Lemma 4.1, it is easily shown that $g^1$is minimally tw-connected; thus there is a unique path between n and j and there are at least two agents, j_1 and j_2, on the tw-path between j and n such that $g^1_{j_1,j_2} = g^1_{j_2,j} = 1$. From Lemma 4.2 and the Remark in Appendix B, we can infer that it is a best response for j_1 to maintain all links except the link with j_2, and to *switch* the link with j_2 to a link with j instead. Let g' denote the resulting network. Note that j_2 is an outward-pointing agent vis-a-vis n in g'. The arguments above concerning outward-pointing agents apply and it is a best response for agent j_2 to delete his link with j and instead form a link $g_{j_2,n} = 1$. Denote the new network by g^2. We have thus shown that $\sigma(n;\overline{g}^2) = \sigma(n;\overline{g}^1) + 1$. We use the argument of Lemma 4.1 to deduce that g^2 is minimally tw-connected. Since $\sigma(n;\cdot)$ increases with positive probability as long as the furthest away inward-pointing agent is at distance $q \geq 3$, we eventually arrive at a minimally tw-connected network g^3 such that $\sigma(n;\overline{g}^3) \geq 2$ and $d(n,j;\overline{g}^3) \leq 2$ for all $j \in N$.

As all agents are at a tw-distance no larger than 2 from agent n, it can be seen that there are four possible configurations for an agent i linked with agent n. (a) $g^3_{i,n} = 1$ and i has no other links. (b) $g^3_{n,i} = 1$ and i has no other links. (c) $g^3_{i,n} = 1$ and $g^3_{i,j} = 1$ for all $j \in E$, where E is the set of end-agents of i.[29] (d) $g^3_{n,i} = 1$ and $g^3_{i,j} = 1$ for all $j \in E$, where E is again the set of end-agents of i. We also note that case (d) can be reduced to case (c) by applying the switching argument presented above.

Suppose there is an agent i in configuration (c) with end-agents E so that $g^3_{i,n} = 1$ and $g^3_{i,j} = 1$ for all $j \in E$. Since $\sigma(n;\overline{g}^3) \geq 2$ there is at least one other agent k at tw-distance 1 from n. Suppose that $g^3_{k,n} = 1$. Let agent i and agent k both choose a best response simultaneously. Specifically, it is a best response for i to maintain his links with the agents in E and switch his link from agent n to agent k. Likewise, it is a best response for k to switch his link from n to i. In the resulting network n no longer has a tw-path with either agent: thus k and i *miscoordinate*. We now allow agent n to choose a best response. It is easily checked (using Lemma 4.2 and the Remark in Appendix B), that it is a best response for him to form a link with some agent $j \in E$, ceteris paribus. Now, if i and k again move simultaneously, i can delete his links $g_{i,j} = g_{i,k} = 1$ and only form links with the agents in $E \backslash \{j\}$ in addition to forming a link with n. Likewise, k will not form any links (in particular, he will delete his link with i). Finally, if n moves again, he will form a link with

[29] These are the set of end-agents in g^3 whose unique link is with agent i.

k, ceteris paribus. Label the resulting network g^4. Since n now also has a link with j, in addition to links with i and k we get $\sigma(n;\overline{g}^4) = \sigma(n;\overline{g}^3)+1$. The other combinations of cases (a), (b), and (c) can be analyzed with a combination of switching and miscoordination arguments to eventually reach a minimally tw-connected network g^* where $\sigma(n;\overline{g}^*) = n-1$. If g^* is a center-sponsored star, we are done. Otherwise, miscoordination arguments can again be used to show transition to a center-sponsored star.

Part (b) of the result is proved using similar arguments; a sketch is presented in Appendix B. *Q.E.D.*

For linear payoffs (2.4), Theorem 4.1(a) implies convergence to the center-sponsored star when $c \in (0,1)$, while Theorem 4.1(b) implies convergence to the empty network for $c > 1$. In particular, since a star is efficient for $c \leq n$ and the empty network is efficient for $c > n$, the limit network if efficient when $c < 1$ or $c > n$.

Rates of Convergence. We study the rates of convergence for the linear specification in (2.4), i.e. $\Pi_i(g) = \mu_i(\overline{g}) - \mu_i^d(g)c$. We shall suppose $c \in (0,1)$. Our simulations are carried out under the same assumptions as in the one-way model, with 500 simulations for each n and for four different values of p. Table 2 summarizes the findings.

We see that when $p = 0.5$, average convergence times are extremely high, but come down dramatically as p increases. When $n = 8$ for example, it takes more than 1600 periods to converge when $p = 0.5$, but when $p = 0.95$, it requires only slightly more than 10 periods on average to reach the center-sponsored star. The intuition can be seen by initially supposing that $p = 1$. If we start with the empty network, all agents will simultaneously choose to form links with the rest of society. Thus, the complete network forms in the next period. Since this gives rise to a perfect opportunity for free riding, each agent will form no links in the subsequent period. Thus, the dynamics will oscillate between the empty and the complete network. When p is close to 1, a similar phenomenon occurs (as seen in Fig. 7, where $p = 0.75$) except there is now a chance that all but one agent happen to move, leaving that agent as the center of a center-sponsored star. On the other hand, when p is small, few agents move simultaneously. This makes rapid oscillations unlikely, and greatly reduces the speed of convergence.

Table 2. Rates of convergence in two-way flow model

n	$p = 0.5$	$p = 0.65$	$p = 0.8$	$p = 0.95$
3	191.12(16.89)	47.78(4.22)	17.34(1.43)	18.19(0.71)
4	318.23(22.93)	71.34(4.93)	17.55(1.02)	14.83(0.53)
5	613.28(36.08)	70.08(4.49)	16.27(0.83)	13.23(0.46)
6	753.88(43.94)	89.84(5.07)	17.90(0.88)	11.89(0.37)
7	1010.64(54.86)	123.44(6.78)	22.11(1.02)	10.28(0.35)
8	1625.63(87.52)	174.62(9.40)	27.87(1.24)	10.34(0.34)

5 Decay

In the analysis above, we exploit the assumption that information obtained through indirect links has the same value as that obtained through direct links. This assumption is strong; in general, there will be delays as well as lowering of quality, as information is transmitted through a series of agents. In this section, we study the effects of relaxing the no-decay assumption. Since this is a difficult and voluminous topic, we shall assume a specific functional form for the payoffs, and also largely restrict our study to "small" societies.

5.1 One-way Flow Model with Decay

We consider a modification of the linear payoff structure given by (2.2), i.e. where the value of information is $V \equiv 1$ and its cost is $c > 0$. We measure the level of decay by a parameter $\delta \in (0, 1]$. Given a network g, it is assumed that if an agent i has a link with another agent j, i.e. $g_{i,j} = 1$, then agent i receives information of value δ from j. More generally if the shortest path in the network from j to i has $q \geq 1$ links, then the value of agent j's information to i is δ^q. The cost of link formation is still taken to be c per link. The payoff to an agent i in the network g is then given by

$$\Pi_i(g) = 1 + \sum_{j \in N(i;g) \setminus \{i\}} \delta^{d(i,j;g)} - \mu_i^d(g)c \,, \tag{5.1}$$

where $d(i,j;g)$ is the geodesic distance from j to i. The linear model of (2.2) corresponds to $\delta = 1$. Henceforth, we shall always assume $\delta < 1$ unless specified otherwise.

Nash Networks. The trade-off between the costs of link formation and the benefits of having short information channels to overcome transmission losses is central to an understanding of the architecture of networks in this setting. If $c < \delta - \delta^2$, the incremental payoff from replacing an indirect link by a direct one exceeds the cost of link formation; hence it is a dominant strategy for an agent to form links with everyone, and the complete network g^c is the unique (strict) Nash equilibrium. Suppose next that $\delta - \delta^2 < c < \delta$. Since $c < \delta$, an agent has an incentive to directly or indirectly access everyone. Furthermore, $c > \delta - \delta^2$ implies the following: if there is some agent who has links with every other agent, then the rest of society will form a single link with him. Hence a star is always a (strict) Nash equilibrium. Third, it follows from continuity and the fact that the wheel is strict Nash when $\delta = 1$ that it is also strict Nash for δ close to 1. Finally it is obvious that if $c > \delta$, then the empty network is strict Nash. The following result summarizes the above observations and also derives a general property of strict Nash networks.

Proposition 5.1. *Let the payoffs be given by (5.1). Then a strict Nash network is either connected or empty. Furthermore, (a) the complete network is strict Nash*

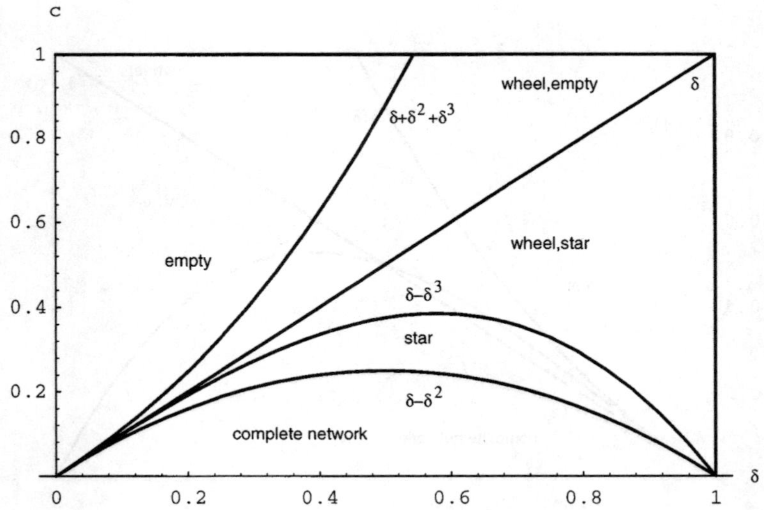

Fig. 8a. Strict Nash networks (one-way model, ($n = 4$)

if and only if $0 < c < \delta - \delta^2$, *(b) the star network is strict Nash if and only if* $\delta - \delta^2 < c < \delta$, *(c) if* $c \in (0, n-1)$, *then there exists* $\delta(c) \in (0,1)$ *such that the wheel is strict Nash for all* $\delta \in (\delta(c), 1)$, *(d) the empty network is strict Nash if and only if* $c > \delta$.

Appendix C provides a proof for the statement concerning connectedness, while parts (a)–(d) can be directly verified.[30] Figure 8a provides a characterization of strict Nash equilibria, for a society with $n = 4$ agents.[31]

Ideally, we would like to have a characterization of strict Nash for all n. This appears to be a difficult problem and we have been unable to obtain such results. Instead, we focus on the case where information decay is "small" and identify an important and fairly general class of networks that are strict Nash. To motivate this class, consider the networks depicted in Figures 9a–c. Assume that $c \in (0, 1)$ and consider the network in Fig. 9a. Here, agent 5 has formed three links, while all others have only one. Thus, agent 5's position is similar to that of a "central coordinator" in a star network. When $\delta = 1$, agent 1 (say) does not receive any additional benefit from a link with agent 5 as compared to a link with agent 2 or 3 or 4 instead. Hence this network cannot be strict Nash. However, when δ falls below one, agent 1 strictly benefits from the link with agent 5 as compared to a link with any other agent, since agent 5 is at a shorter distance from the rest of the society. Similar arguments apply for agent 2 and agent 4 to have a link with

[30] In the presence of decay, a nonempty Nash network is not necessarily connected. Suppose $n = 6$. Let $\delta + \delta^2 < 1$ and $\delta + \delta^2 - \delta^3 < c < \delta + \delta^2$. Then it can be verified that the network given by the links, $g_{1,2} = g_{2,4} = g_{4,3} = g_{3,2} = g_{5,2} = g_{6,5} = g_{2,6} = 1$ is Nash. It is clearly nonempty and it is disconnected since agent 1 is not observed by anyone.

[31] To show that the networks depicted in the different parameter regions are strict Nash is straightforward. Incentive considerations in each region (e.g. that the star is not strict Nash when $c > \delta$) rule out other architectures.

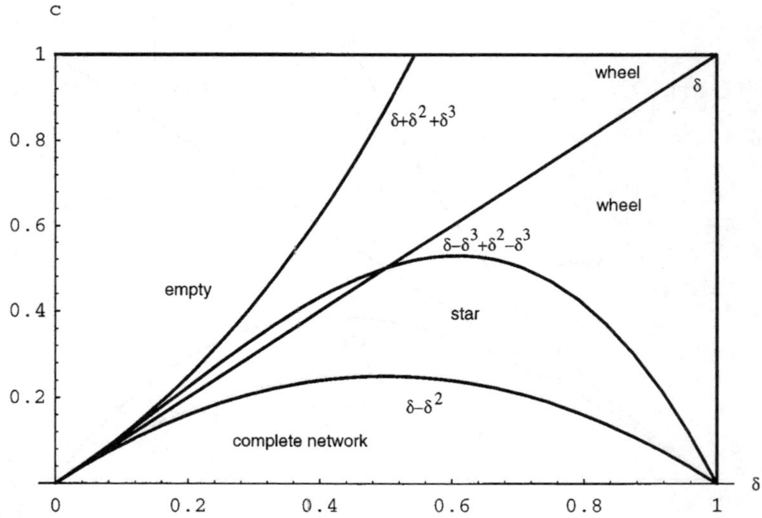

Fig. 8b. Efficient networks one-way model, (n = 4)

agent 5. Thus, decay creates a role for "central" agents who enable closer access to other agents. At the same time, the logic underlying the wheel network – of observing the rest of the society with a single link – still operates. For example, under low decay, agent 3's unique best response will be to form a single link with agent 2. The above arguments suggest that the network of Fig. 9a can be supported as strict Nash for low levels of decay. Analogous arguments apply for the network in Fig. 9b. More generally, the trade-off between cost and decay leads to strict Nash networks where a central agent reduces distances between agents, while the presence of small wheels enables agents to economize on the number of links.

Formally, a *flower* network g partitions the set of agents N into a central agent (say agent n) and a collection $\mathscr{P} = \{\mathscr{P}_1, \ldots, \mathscr{P}_q\}$ where each $P \in \mathscr{P}$ is nonempty. A set $P \in \mathscr{P}$ of agents is referred to as a *petal*. Let $u = |P|$ be the cardinality of petal P, and denote the agents in P as $\{j_1, \ldots, j_u\}$. A flower network is then defined by setting $g_{j_1,n} = g_{j_2,j_1} = \cdots = g_{j_u,j_{u-1}} = g_{n,j_u} = 1$ for each petal $P \in \mathscr{P}$ and $g_{i,j} = 0$ otherwise. A petal P is said to be a *spoke* if $|P| = 1$. A flower network is said to be of *level* $s \geq 1$ if every petal of the network has at least s agents and there exists a petal with exactly s agents. Note that a star is a

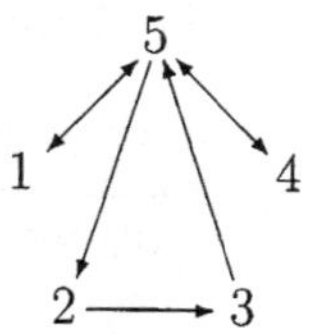

Fig. 9a.

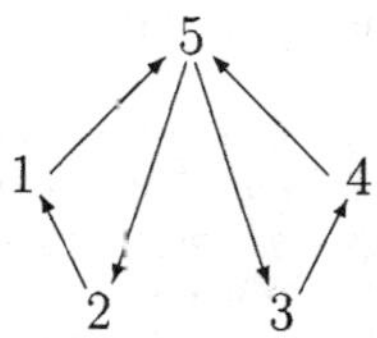

Fig. 9b.

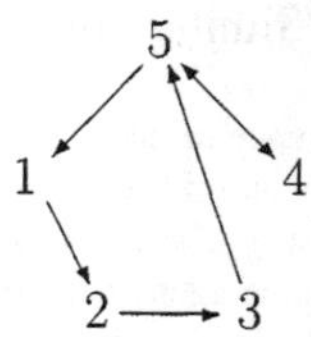

Fig. 9c.

flower network of level 1 with $n - 1$ spokes, while a wheel is a flower network of level $n - 1$ with a single petal.

We are interested in finding conditions under which flower networks can be supported as strict Nash. However, we first exclude a certain type of flower network from our analysis. Figure 9c provides an example. Here agent 5 is the central agent, and there are exactly two petals. Moreover, one petal is a spoke, so that it is a flower network of level 1. Note that agent 4 will be indifferent between forming a link with any of the remaining agents, since their position is completely symmetric. Thus, this network can never be strict Nash. In what follows, a flower network g with exactly two petals, of which at least one is a spoke, will be referred to as the "exceptional case."

Proposition 5.2. *Suppose that the payoffs are given by (5.1). Let $c \in (s-1, s)$ for some $s \in \{1, 2, \ldots, n-1\}$ and let g be a flower network (other than the exceptional case) of level s or higher. Then there exists a $\delta(c, g) < 1$ such that, for all $\delta \in (\delta(c, g), 1)$, g is a strict Nash network. Furthermore, no flower network of a level lower than s is Nash for any $\delta \in (0, 1]$.*

The proof is given in Appendix C. When $s > 1$ the above proposition rules out any networks with spokes as being strict Nash. In particular, the star cannot be supported when $c > 1$.

Finally, we note the impact of the size of the society on the architecture of strict Nash networks. As n increases, distances in the wheel network become larger, creating greater scope for central agents to reduce distances. This suggests that intermediate flower networks should become more prominent as the society becomes larger. Our simulation results are in accord with this intuition.

Efficient Networks. The welfare function is taken to be $W(g) \sum_{i=1}^{n} \Pi_i(g)$, where Π_i is specified by equation (5.1). Figure 8b characterizes the set of efficient networks when $n = 4$.[32] The trade-off between costs and decay mentioned above also determines the structure of efficient networks. If the costs are sufficiently low, efficiency dictates that every agent should be linked with every other agent. For values of δ close to one, and/or if the costs of link formation are high, the wheel is still efficient. For intermediate values of cost and decay, the star strikes a balance between these forces.

A comparison between Figures 8a and 8b reveals that there are regions where strict Nash and efficient networks coincide (when $c < \delta - \delta^2$ or $c > \delta + \delta^2 + \delta^3$). The figures suggest, however, that the overall relationship is quite complicated.

Dynamics. We present simulations for low values of decay, i.e., δ close to 1, for a range of societies from $n = 3$ to $n = 8$.[33] This helps to provide a robustness check

---

[32] The assertions in the figure are obtained by comparing the welfare levels of all possible network architectures to obtain the relevant parameter ranges. We used the list of architectures given in Harary (1972).

[33] For $n = 4$ it is possible to prove convergence to strict Nash in all parameter regions identified in Fig. 8a. The proof is provided in an earlier working paper version. For general n, it is not difficult to show that, from every initial network, the dynamic process converges almost surely to the complete network when $c < \delta - \delta^2$ and to the empty network when $c > \delta + (n-2)\delta^2$.

for the convergence result of Theorem 3.1 and also gives some indication about the relative frequencies with which different strict Nash networks emerge. For each n, we consider a 25×25 grid of (δ, c) values in the region $[0.9, 1) \times (0, 1)$, but discard points where $c \leq \delta - \delta^2$ or $c \geq \delta$. For the remaining 583 grid values, we simulate the process for a maximum of 20,000 periods, starting from a random initial network. We also set $p = 0.5$ for all the agents.

Figure 10 depicts some of the limit networks that emerge. In many cases, these are the wheel, the star, or other flower networks. However, some variants of flower networks (left-hand side network for $n = 6$ and right-hand side network for $n = 7$) also arise. Thus, in the $n = 7$ case, agent 2 has an additional link with agent 6 in order to access the rest of the society at a closer distance. Since $c = 0.32$ is relatively small, this is worthwhile for the agent. Likewise, in the $n = 6$ example, two petals are "fused," i.e. they share the link from agent 6 to agent 3. Other architectures can also be limits when c is small, as in the left-hand side network for $n = 8$.[34]

Table 3 (below) provides an overall summary of the simulation results. Column 2 reports the average time and standard error, *conditional upon convergence to a limit network in 20,000 periods*. Columns 3–6 show the relative likelihood of different strict Nash networks being the limit, while the last column shows the likelihood of a limit cycle.[35] With the exception of $n = 4$, the average convergence times are all relatively small. Moreover, the chances of eventual convergence to a limit network are fairly high. The wheel and the star become less likely, while other flower networks as well as nonflower networks become more important as n increases. This corresponds to the intuition presented in the discussion on flower networks. We also see that when $n = 8$, 56.6% of the limit networks are not flower networks. In this category, 45.7% are variants of flower networks (e.g. with fused petals, or with an extra link between the central agent and the final agent in a petal) while the remaining 10.9% are networks of the type seen in the left-hand side network for $n = 8$. Thus, flower networks or their variants occur very frequently as limit points of the dynamics.

Table 3. Dynamics in one-way flow model with decay

n	Avg. Time (Std. Err.)	Flower Networks: Wheel	Star	Other	Other Networks	Limit Cycles
3	6.5(0.2)	100.0%	0.0%	0.0%	0.0%	0.0%
4	234.2(61.7)	71.9%	27.8%	0.0%	0.0%	0.3%
5	28.1(6.2)	20.6%	11.5%	58.7%	4.6%	4.6%
6	26.4(3.6)	3.6%	6.3%	58.8%	27.1%	4.1%
7	94.3(14.7)	0.9%	4.1%	56.1%	28.0%	11.0%
8	66.5(8.5)	0.7%	3.8%	37.2%	56.6%	1.7%

34 Due to space constraints, we do not investigate such networks in this paper.

35 We assume that the process has entered a limit cycle if convergence to a limit network does not occur within the specified number of periods.

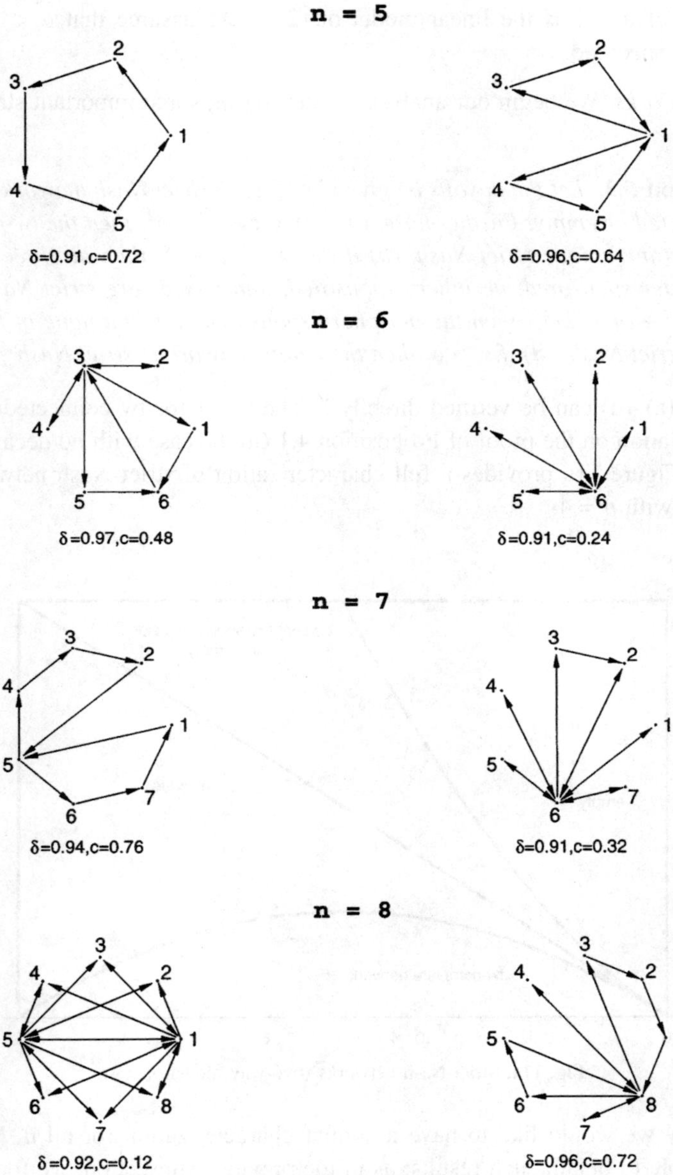

Fig. 10. Limit networks (one-way model)

5.2 Two-way Flow Model with Decay

This section studies the analogue of (5.1) with two-way flow of information. The payoffs to an agent i from a network g are given by

$$\overline{\Pi}_i(g) = 1 + \sum_{j \in N(i;\overline{g}) \setminus \{i\}} \delta^{d(i,j,\overline{g})} - \mu_i^d(g)c \ . \tag{5.2}$$

The case of $\delta = 1$ is the linear model of (2.4). We assume that $\delta < 1$ unless otherwise specified.

Nash Networks. We begin our analysis by describing some important strict Nash networks.

Proposition 5.3. *Let the payoffs be given by (5.2). A strict Nash network is either tw-connected or empty. Furthermore, (a) if $0 < c < \delta - \delta^2$, then the tw-complete network is the unique strict Nash, (b) if $\delta - \delta^2 < c < \delta$, then all three types of stars (center-sponsored, periphery-sponsored, and mixed) are strict Nash, (c) if $\delta < c < \delta + (n-2)\delta^2$, then the periphery-sponsored star, but none of the other stars, is strict Nash, (d) if $c > \delta$, then the empty network is strict Nash.*

Parts (a)–(d) can be verified directly.[36] The proof for tw-connectedness is a slight variation on the proof of Proposition 4.1 (in the case with no decay) and is omitted. Figure 11a provides a full characterization of strict Nash networks for a society with $n = 4$.

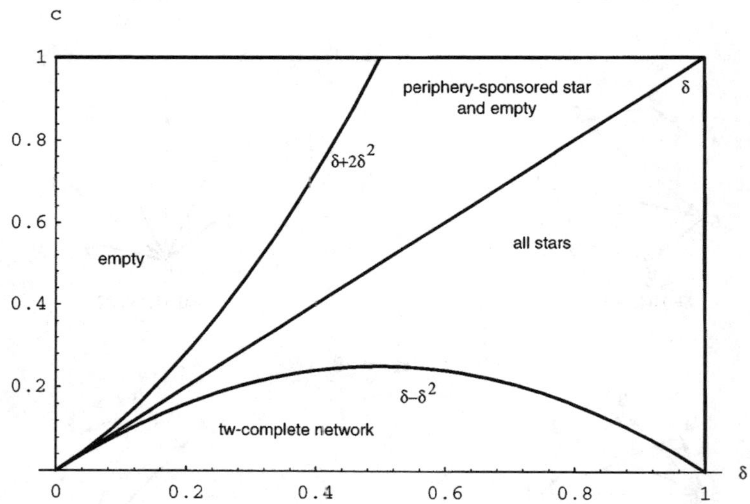

Fig. 11a. Strict Nash networks (two-way model, $n = 4$)

Ideally we would like to have a similar characterization for all n. We have been unable to obtain such results; as in the previous subsection, we focus upon low levels of decay. When $c \in (0, 1)$ we can identify an important class of networks, which we label as *linked stars*. Figures 12a–c provide examples of such networks.

Linked stars are described as follows: Fix two agents (say agent 1 and n) and partition the remaining agents into nonempty sets S_1 and S_2, where $|S_1| \geq 1$ and $|S_2| \geq 2$. Consider a network g such that $g_{i,j} = 1$ implies $g_{j,i} = 0$. Further suppose that $g_{i,n} = 1$. Lastly, suppose one of the three mutually exclusive conditions (a),

[36] A tw-complete network g is one where, for all i and j in N, we have $d(i,j;\overline{g}) = 1$ and $g_{i,j} = 1$ implies $g_{j,i} = 0$.

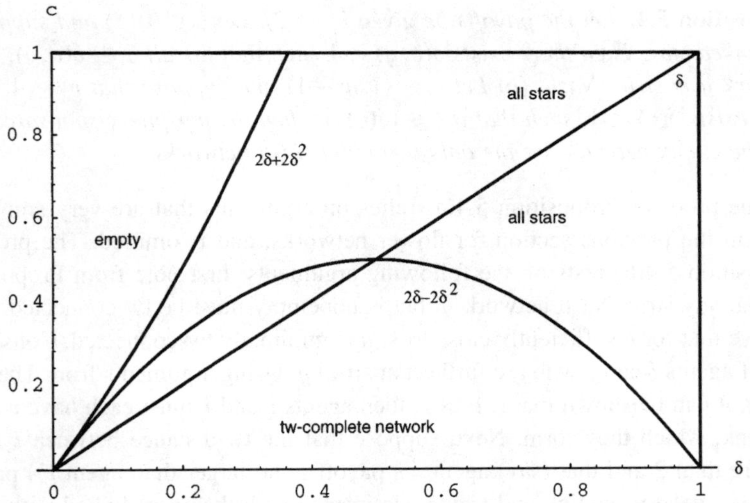

Fig. 11b. Efficient networks (two-way model, $n = 4$)

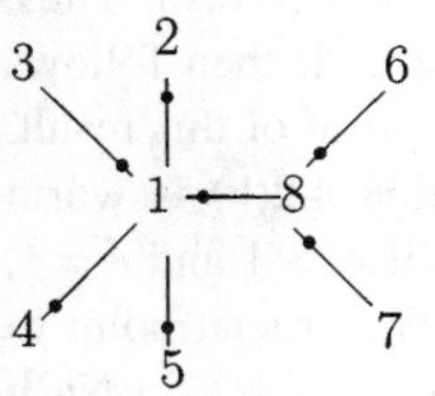

Fig. 12a. $|S_1| > |S_2| + 1$

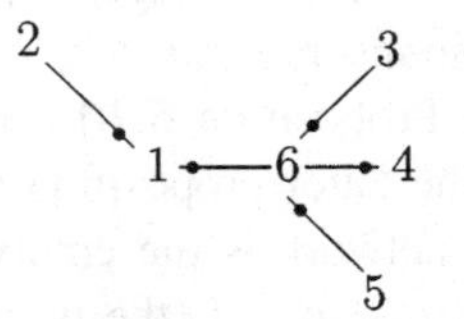

Fig. 12b. $|S_1| < |S_2| - 1$

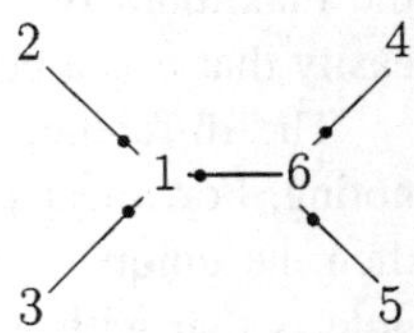

Fig. 12c. $|S_1| = |S_2|$

(b), or (c) holds: (a) If $|S_1| > |S_2| + 1$, then $\max\{g_{1,i}, g_{i,1}\} = 1$ for all $i \in S_1$ and $g_{n,j} = 1$ for all $j \in S_2$. (b) If $|S_1| < |S_2| - 1$, then $\max\{g_{n,j}, g_{j,n}\} = 1$ for all $j \in S_2$ and $g_{1,i} = 1$ for all $i \in S_1$. (c) If $||S_1| - |S_2|| \leq 1$, then $g_{1,i} = 1$ for all $i \in S_1$ and $g_{n,j} = 1$ for all $j \in S_2$.

The agents 1 and n constitute the "central" agents of the linked star. If δ is sufficiently close to 1, a spoke agent will not wish to form any links (if the central agent has formed one with him) and otherwise will form at most one link. Conditions (a) and (b) ensure that the spoke agents of a central agent will not wish to switch to the other central agent.[37]

If $c > 1$ and decay is small, it turns out that there are at most two strict Nash networks. One of them is, of course, the empty network. The other network is the periphery-sponsored star. These observations are summarized in the next result.

[37] Thus, note that in Fig. 12a, if $g_{7,8} = 1$ rather than $g_{8,7} = 1$, then agent 7 would strictly prefer forming a link with agent 1 instead, since agent 1 has more links than agent 8. Likewise, in Fig. 12b, each link with an agent in S_1 must be formed by agent 1 for otherwise the corresponding 'spoke' agent will gain by moving his link to agent n instead. The logic for condition (c) can likewise be seen in Fig. 12c. We also see why $|S_2| \geq 2$. In Figure 12c, if agent 5 were not present, then agent 1 would be indifferent between a link with agent 6 and one with agent 4. Lastly, we observe that since $|S_1| \geq 1$ and $|S_2| \geq 2$, the smallest n for which a linked star exists is $n = 5$.

Proposition 5.4. *Let the payoffs be given by (5.2). Let $c \in (0,1)$ and suppose g is a linked star. Then there exists $\delta(c,g) < 1$ such that for all $\delta \in (\delta(c,g),1)$ the network g is strict Nash. (b) Let $c \in (1, n-1)$ and suppose that $n \geq 4$. Then there exists $\delta(c) < 1$, such that if $\delta \in (\delta(c),1)$ then the periphery-sponsored star and the empty network are the only two strict Nash networks.*

The proof of Proposition 5.4(a) relies on arguments that are very similar to those in the previous section for flower networks, and is omitted. The proof of Proposition 5.4(b) rests on the following arguments: first note from Proposition 5.3 that any strict Nash network g that is nonempty must be tw-connected. Next observe that for δ sufficiently close to 1, g is minimally tw-connected. Consider a pair of agents i and j who are furthest apart in g. Using arguments from Theorem 4.1(b), it can be shown that if $c > 1$, then agents i and j must each have exactly one link, which they form. Next, suppose that the tw-distance between i and j is more than 2 and that (say) agent i's payoff is no larger than agent j's payoff. Then if i deletes his link and forms one instead with the agent linked with j, his tw-distance to all agents apart from j (and himself) is the same as j, and he is also closer to j. Then i strictly increases his payoff, contradicting Nash. Thus, the maximum tw-distance between two agents in g must be 2. It then follows easily that g is a periphery-sponsored star. We omit a formal proof of this result.

The difference between Proposition 5.4(b) and Proposition 4.2(b) is worth noting. For linear payoffs, the latter proposition implies that if $c > 1$ and $\delta = 1$, then the unique strict Nash network is the empty network. The crucial point to note is that with $\delta = 1$, and $c < n - 1$, the periphery-sponsored star is a Nash but not a strict Nash network, since a 'spoke' agent is indifferent between a link with the central agent and another 'spoke' agent. This indifference breaks down in favor of the central agent when $\delta < 1$, which enables the periphery-sponsored star to be strict Nash (in addition to the empty network).

Efficient Networks. We conclude our analysis of the static model with a characterization of efficient networks.

Proposition 5.5. *Let the payoffs be given by (5.2). The unique efficient network is (a) the complete network if $0 < c < 2(\delta - \delta^2)$, (b) the star if $2(\delta - \delta^2) < c < 2\delta + (n-2)\delta^2$, and (c) the empty network if $c > 2\delta + (n-2)\delta^2$.*

The proof draws on arguments presented in Proposition 1 of Jackson and Wolinsky (1996) and is given in Appendix C. The nature of networks – complete, stars, empty – is the same, but the range of values for which these networks are efficient is different. This contrast arises out of the differences in the way we model network formation: Jackson and Wolinsky assume two-sided link formation, unlike our framework. Figure 11b displays the set of efficient networks for $n = 4$ in different parameter regions.

Dynamics. We now turn to simulations to study the convergence properties of the dynamics. As in the one-way case, for each n we consider a 25×25 grid of (δ, c) values in the region $[0.9, 1) \times (0, 1)$, with points satisfying $c \leq \delta - \delta^2$ or

Table 4. Dynamics in two-way flow model with decay

n	Avg. Time (Std. Err.)	Stars: Center	Stars: Mixed	Stars: Periphery	Linked Stars	Other Networks	Limit Cycles
3	166.5(14.2)	100.0%	0.0%	0.0%	0.0%	0.0%	0.0%
4	5.2(0.2)	37.6%	56.9%	5.5%	0.0%	0.0%	0.0%
5	8.9(0.4)	34.0%	53.7%	3.6%	8.7%	0.0%	0.0%
6	8.8(0.3)	26.8%	42.9%	4.3%	26.1%	0.0%	0.0%
7	10.2(0.4)	20.4%	43.4%	3.9%	24.8%	3.8%	3.6%
8	12.3(0.4)	16.6%	34.6%	6.0%	34.5%	7.4%	0.9%

$c \geq \delta$ being discarded. As earlier, there are a total of 583 grid values for each n. We also fix $p = 0.5$ as in the one-way model.[38]

Figure 13 depicts some of the limit networks. In most cases, they are stars of different kinds or linked stars. However, as the right-hand side network for $n = 7$ shows, other networks can also be limits. To see this, note that the maximum geodesic distance between two agents in a linked star is 3, whereas agents 5 and 7 are four links apart in this network. We also note that limit cycles can occur.[39]

Table 4 provides an overall summary of the simulations. For $n \leq 6$, convergence to a limit network occurred in 100% of the simulations, while for $n = 7$ and $n = 8$ there is a positive probability of being absorbed in a limit cycle. Column 2 reports the average convergence time and the standard error, *conditional upon convergence to a limit network*. Columns 3–8 show the frequency with which different networks are the limits of the process. Among stars, mixed-type ones are the most likely. Linked stars become increasingly important as n rises, while other kinds of networks (such as the right-hand-side network when $n = 7$) may also emerge. Limit cycles are more common when $n = 7$ than when $n = 8$. In contrast to Table 2 concerning the two-way model without decay, convergence occurs very rapidly even though $p = 0.5$. A likely reason is that under decay an agent has a strict rather than a weak incentive to link to a well-connected agent: his choice increases the benefit for other agents to do so as well, leading to quick convergence. Absorption into a limit network is also much more rapid as compared to Table 3 for the one-way model, for perhaps the same reason.

---

[38] For $n = 4$ convergence to strict Nash can be proved for all parameter regions identified in Fig. 11a. For general n, it is not difficult to show that, starting from any initial network, the dynamic process is absorbed almost surely into the tw-complete network when $c < \delta - \delta^2$ and into the empty network when $c > \delta + (n-2)\delta^2$.

[39] To see how this can happen, consider the left-hand side network for $n = 7$ in Fig. 13, which is strict Nash. However, if it is agent 3 rather than agent 5 who forms the link between them in the figure, we see that agent 3 can obtain the same payoff by switching this link to agent 1 instead, while all other agents have a unique best response. Thus, the dynamics will oscillate between two Nash networks.

For $n \leq 6$ it is not difficult to show that given $c \in (0, 1)$, the dynamics will always converge to a star or a linked star for all δ sufficiently close to 1. Thus, $n = 7$ is the smallest value for which a limit cycle occurs.

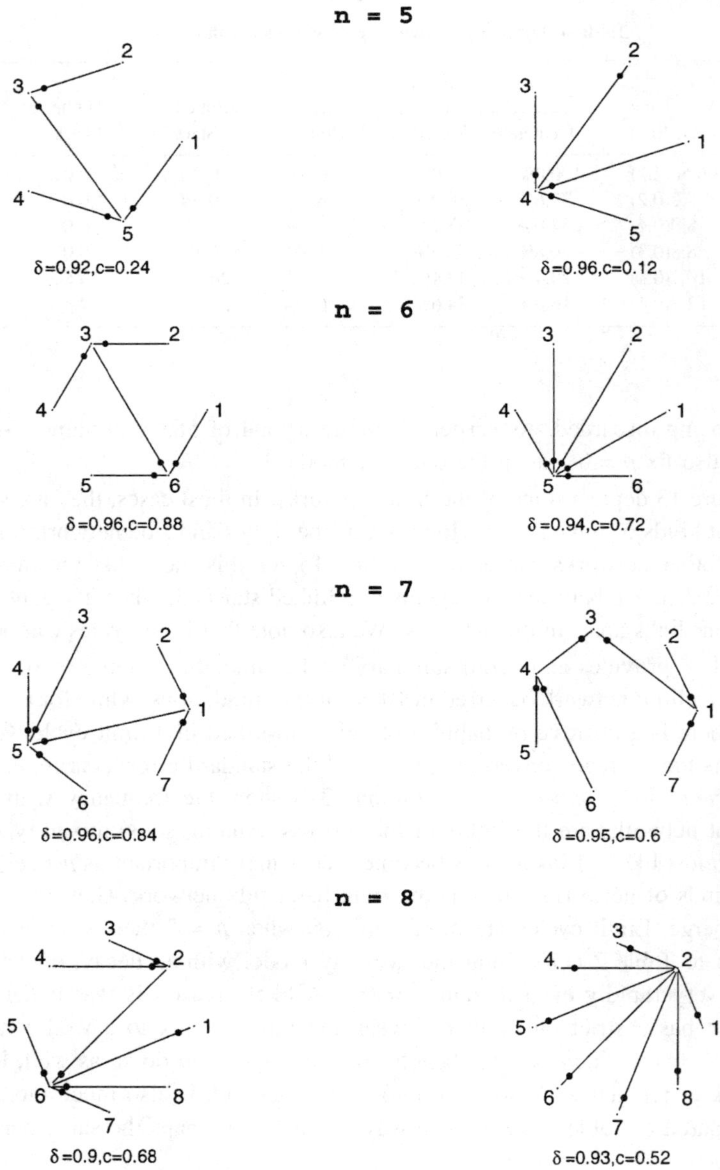

Fig. 13. Limit networks (two-way model)

6 Conclusion

In this paper, we develop a noncooperative model of network formation where we consider both one-way and two-way flow of benefits. In the absence of decay, the requirement of strict Nash sharply delimits the case of networks to the empty network and the one other architecture: in the one-way case, this is a wheel network, where every agent bears an equal share of the cost, while in the two-way case it is a center-sponsored star, where as the name suggests, a single agent bears the full cost. Moreover, in both models, a simple dynamic process converges to a strict Nash network under fairly general conditions, while simulations indicate that convergence is relatively rapid. For low levels of decay, the set of strict Nash equilibria expands both in the one-way and two-way models. Many of the new strict equilibria are natural extensions of the wheel and the center-sponsored star, and also appear frequently as limits of simulated sample paths of the dynamic process. Notwithstanding the parallels between the results for the one-way and two-way models, prominent differences also exist, notably concerning the kinds of architectures that are supported in equilibrium.

Our results motivate an investigation into different aspects of network formation. In this paper, we have assumed that agents have no "budget" constraints, and can form any number of links. We have also supposed that contacting a well-connected person costs the same as contacting a relatively idle person. Moreover, in revising their strategies, it is assumed that individuals have full information on the existing social network of links. Finally, an important assumption is that the benefits being shared are nonrival. The implications of relaxing these assumptions should be explored in future work.

Appendix A

Proof of Proposition 3.1. Let g be a Nash network. Suppose first that $\Phi(n, 1) < \Phi(1, 0)$. Let $i \in N$. Note that $\mu_i(g) \leq n$. Thus $\mu_i^d(g) \geq 1$ implies $\Pi_i(g) = \Phi(\mu_i(g), \mu_i^d(g)) \leq \Phi(n, \mu_i^d(g)) \leq \Phi(n, 1) < \Phi(1, 0)$, which is impossible since g is Nash. Hence it is a dominant strategy for each agent to have no links, and g is the empty network. Consider the case $\Phi(n, 1) = \Phi(1, 0)$. An argument analogous to the one above shows that $\mu_i^d(g) \in \{0, 1\}$ for each $i \in N$. Furthermore $\mu_j^d(g) = 1$ can hold if $\mu_i(g) = n$. It is now simple to establish that if g is nonempty, it must be the wheel network, which is connected.[40]

Henceforth assume that $\Phi(n, 1) > \Phi(1, 0)$. Assume that g is not the empty network. Choose $i \in \text{argmax}_{i' \in N}\, \mu_{i'}(g)$. Since g is nonempty, $x_i \equiv \mu_i(g) \geq 2$ and $y_i \equiv \mu_i^d(g) \geq 1$. Furthermore, since g is Nash, $\Pi_i(g) = \Phi(x_i, y_i) \geq \Phi(1, 0)$. We claim that i observes everyone, i.e. $x_i = n$. Suppose instead that $x_i < n$. Then there exists $j \notin N(i; g)$. Clearly, $i \notin N(j; g)$ either, for otherwise $N(i; g)$ would be a strict subset of $N(j; g)$ and $\mu_j(g) > x_i = \mu_i(g)$, contradicting the definition of i. If $y_j \equiv \mu_j^d(g) = 0$ let j deviate and form a link with i, ceteris paribus. His payoff will be $\Phi(x_i + 1, 1) > \Phi(x_i, 1) \geq \Phi(x_i, y_i) \geq \Phi(1, 0)$, so that he can do strictly better. Hence $y_j \geq 1$. By definition of i, we have $x_j \equiv \mu_j(g) \leq x_i$. Let j delete all his links and form a single link with i instead. His new payoff will be $\Phi(x_i + 1, 1) > \Phi(x_i, 1) \geq \Phi(x_j, 1) \geq \Phi(x_j, y_j)$, i.e. he does strictly better. The contradiction implies that $x_i = n$ as required, i.e. there is a path from every agent in the society to agent i.

Let i be as above. An agent j is called *critical* (to i) if $\mu_i(g_{-j}) < \mu_i(g)$; if instead $\mu_i(g_{-j}) = \mu_i(g)$, agent j is called *noncritical*. Let E be the set of noncritical agents. If $j \in \text{argmax}_{i' \in N}\, d(i, i'; g)$, clearly j is noncritical, so that E is nonempty. We show that $j \in E$ implies $\mu_j(g) = n$. Suppose this were not true. If $y_j \equiv \mu_j^d(g) = 0$, then j can deviate and form a link with i. His new payoff will be $\Phi(n, 1) > \Phi(1, 0)$. Thus $y_j \geq 1$. If $x_j \equiv \mu_j(g) < n$, let j delete his links and form a single link with i. Since he is noncritical, his new payoff will be $\Phi(n, 1) > \Phi(x_j, 1) \geq \Phi(x_j, y_j)$, i.e. he will again do better. It follows that $\mu_j(g) = n$ as required.

We claim that for every agent $j_1 \notin E \cup \{i\}$, there exists $j \in E$ such that $j \in N(j_1; g)$. Since j_1 is critical, there exists $j_2 \in N(j_1; g)$ such that every path from j_2 to i in g involves agent j_1.

[40] This assertion requires the assumption that $n \geq 3$. If $n = 2$ and $\Phi(2, 1) = \Phi(1, 0)$, then the disconnected network $g_{1,2} = 1$, $g_{2,1} = 0$ is a Nash network.

Hence $d(i,j_2;g) > d(i,j_1;g)$. If $j_2 \in E$ we are done; otherwise, by the same argument, there exists $j_3 \in N(j_2;g)$ such that $d(i,j_3;g) > d(i,j_2;g)$. Since i observes every agent and N is finite, repeating the above process no more than $n-2$ times will yield an agent $j \in E$ such that $j \in N(j_1;g)$. Since we have shown $\mu_j(g) = n$, we have $\mu_{j_1}(g) = n$ as well. Hence g is connected. If g were not minimally connected, then some agent could delete a link and still observe every agent in the society, thereby increasing his payoff, in which case g is not Nash. The result follows. *Q.E.D.*

Appendix B

Proof of Proposition 4.1. Let g be a nonempty Nash network and suppose it is not tw-connected. Since g is nonempty there exists a tw-component C such that $|C| \equiv x \geq 2$. Choose $i \in C$ satisfying $\mu_i^d(g) \geq 1$. Then we have $\Phi(x,1) \geq \Phi(x,\mu_i^d(g)) = \Phi(\mu_i(\overline{g}),\mu_i^d(g)) = \overline{\Pi}_i(g)$. Note that g_{-i} can be regarded as the network where i forms no links. Since g is Nash, $\overline{\Pi}_i(g) \geq \overline{\Pi}_i(g_{-i}) = \Phi(\mu_i(g_{-i}),0) \geq \Phi(1,0)$. Thus, $\Phi(x,1) \geq \Phi(1,0)$. As g is not tw-connected, there exists $j \in N$ such that $j \notin C$. If j is a singleton tw-component then the payoff to agent j from a link with i is $\Phi(x+1,1) > \Phi(x,1) \geq \Phi(1,0)$, which violates the hypothesis that agent j is choosing a best response. Suppose instead that j lies in a tw-component D where $|D| \equiv w \geq 2$. By definition there is at least one agent in D who forms links; assume without loss of generality that j is this agent. As with agent i we have $\Phi(w,1) \geq \overline{\Pi}_j(g)$.

Suppose without loss of generality that $w \leq x = |C|$. Suppose agent j deletes all his links and instead forms a single link with agent $i \in C$. Then his payoff is at least $\Phi(x+1,1) > \Phi(w,1) \geq \overline{\Pi}_j(g)$. This violates the hypothesis that agent j is playing a best response. The contradiction implies g is tw-connected. If g is not minimally tw-connected, there exists an agent who can delete a link and still have a tw-path with every other agent, so that g is not Nash. The result follows. *Q.E.D.*

Lemma 4.1. *Let the payoffs be given by (2.3). Starting from any initial network g, the dynamic process (2.7) moves with positive probability either to a minimally tw-connected network or to the empty network, in finite time.*

Proof. We first show that the process transits with positive probability to a network all of whose components are tw-minimal. Starting with agent 1, let each agent choose a best response one after the other and let g' denote the network after all agents have moved. Let C be a tw-component of g'. Suppose there is a tw-cycle in C, i.e. there are $q \geq 3$ agents $\{j_1,\ldots,j_q\} \subset C$ such that $\overline{g}'_{j_1,j_2} = \cdots = \overline{g}'_{j_q,j_1} = 1$. Let $S \subset \{j_1,\ldots,j_q\}$ consist of those agents who have formed at least one link within the tw-cycle. Note that S is nonempty. Let j_s be the agent who has played most recently amongst those in S, and assume without loss of generality that $g'_{j_s,j_{s-1}} = 1$. Let g'' be the network prior to agent j_s's move. By definition of j_s we have

$$\overline{g}''_{j_{s+1},j_{s+2}} = \cdots = \overline{g}''_{j_q,j_1} = \cdots = \overline{g}''_{j_{s-2},j_{s-1}} = 1\,. \tag{B.1}$$

Consider agent j_s's best response to $\overline{g}''_{-j_s}$. There are two possibilities: either $\overline{g}''_{j_{s+1},j_s} = 1$, or $\overline{g}''_{j_{s+1},j_s} = 0$. In the former case, by virtue of (B.1), agent j_s can get the same information as before without forming the link with j_{s-1}. In the latter case, j_s forms links with both j_{s-1} and j_{s+1} as part of his best response. However, (B.1) again implies that he is strictly better off by forming a link with only one of them. This contradiction shows that C cannot have a tw-cycle. A similar argument shows that $g'_{i,j} = 1$ for two agents i and j in C implies $g'_{j,i} = 0$. Since C is an arbitrary tw-component of g', every such tw-component must be minimal.

Let C_1 be the largest tw-component in g'. If $|C_1| = n$ or $|C_1| = 1$ we are done. Suppose instead that $|C_1| = x$ where $1 < x < n$. Denote the agents in $N \backslash C_1$ as S. There are now two cases (1) and (2).

(1) The unique best response of every agent in S is not to form any links: let all agents in S move simultaneously, with all the agents in C_1 exhibiting inertia. Call the resulting network g^1. Clearly, g^1 has one nonsingleton tw-component C_1 and $|S|$ singleton tw-components. Let $j \in S$. (1a) Suppose j's unique best response is not to form any links. Then $\Phi(x+u,u) \leq \Phi(1,0)$ for all $u \in \{1,\ldots,|S|\}$ since he has the option of forming links with any subset of the remaining $|S|$ tw-components. If

$i \in C_1$ has formed any links, the highest payoff from $u \geq 1$ links is $\Phi(x+u,u) \leq \Phi(1,0)$ so that to delete all links is a best response. If all the agents in C_1 who have links are allowed to move simultaneously, the empty network results. (1b) Suppose instead that all of j's best responses involve forming one or more links. Since C_1 is the unique nonsingleton tw-component, any best response $\hat{g}_j$ must involve forming a link with C_1. Define $g^2 = \hat{g}_j \oplus g^1_{-j}$. Using above arguments it is easily seen that all tw-components of g^2 are minimal. Let C_2 be the largest tw-component in g^2. Clearly, $C_1 \subset C_2$ with the inclusion being strict. Now proceed likewise with the other singleton tw-components to arrive at a minimal tw-connected network.

(2) There exists an agent j in S all of whose best responses to g' involve forming one or more links: as is (1b), if we let j choose a best response, we obtain a new network g'' where the largest component C_2 satisfies $C_1 \subset C_2$ with the inclusion being strict. Moreover, it can be seen that all tw-components of g'' are minimal. We repeat (1) or (2) with g'' in place of g' and so on until either the empty network or a minimal tw-connected network is obtained. *Q.E.D.*

Lemma 4.2. *Let g be a minimally tw-connected network. Suppose $\mu_i^d(g) = u \geq 0$. If agent i deletes $s \leq u$ links, then the resulting network has $s+1$ minimal tw-components, $C_1, \ldots, C_{s+1}$, with $i \in C_{s+1}$.*

Proof. Let g' be the network after i deletes s links, say, with agents $\{j_1, \ldots, j_s\}$. Since g is minimally tw-connected there is a unique tw-path between every pair of agents i and j in g. In particular, if i deletes s links, then each of the s agents $j_1, j_2, j_3, \ldots, j_s$, have no tw-path linking them with agent i as well as no tw-path linking them with each other either. Thus each of the s agents and agent i must lie in a distinct tw-component, implying that there are at least $s+1$ tw-components in the network g'.

We now show that there cannot be more than $s+1$ tw-components. Suppose not. Let $j_1, j_2, \ldots, j_s$ and i belong to the first $s+1$ tw-components and consider an agent k who belongs to the $s+2$th tw-component. Since g is minimally tw-connected there is a unique tw-path between i and k in g; the lack of any such tw-path in g' implies that the unique tw-path between i and k must involve a now deleted link g_{i,j_q} for some $q = 1, 2, \ldots, s$. Thus in g there must be a tw-path between k and j_q, which does not involve agent i. Since only agent i moves in the transition from g to g', there is also a tw-path between k and j_q in g'. This contradicts the hypothesis that k lies in the $s+2$th tw-component. The minimality of each tw-component in g' follows directly from the hypothesis that g' is obtained by deleting links from a minimally tw-connected network. *Q.E.D.*

Lemma 4.2 implies that the following strategy is a best response.

Remark. Suppose $\Phi(x+1, y+1) > \Phi(x,y)$ for all $y \in \{0, \ldots, n-2\}$ and $x \in \{y+1, \ldots, n-1\}$. Let g and $C_1, \ldots, C_{s+1}$ be as in Lemma 4.2 above. Define g_i^* as $g_{i,k}^* = 1$ for one and only one k in each of $C_1, \ldots, C_s$ and $g_{i,k}^* = g_{i,k}$ for all $k \in N \backslash (C_1 \cup \cdots C_2 \cup \{i\})$. Then g_i^* is a best response to g_{-i}.

Proof of Theorem 4.1(b) (Sketch). The hypothesis on the payoffs implies that $\Phi(1,0) > \max_{(1 \leq x \leq n-1)} \Phi(x+1, x)$. Proposition 4.2(b) then implies that the empty network is the unique strict Nash network, and hence is an absorbing state for the dynamic process. Next note that if $\Phi(1,0) \geq \Phi(n,1)$, then it is a weakly dominant strategy for a player to form no links. In this case convergence to the empty network is immediate. We focus on the case where $\Phi(n,1) > \Phi(1,0)$.

Fix an initial nonempty network g. From Lemma 4.1 we can assume without loss of generality that g is a minimal tw-connected network. Let $n = \text{argmax}_{i \in N}\, \sigma(i; \overline{g})$ and let $P(n;g)$ and $I(n;g)$ be the set of outward-pointing agents and the set of inward-pointing agents vis-a-vis n, respectively. In addition, define $E(n;g)$ as the end-agents in the network g and let $P^e(n;g) = E(n;g) \cap P(n;g)$ be the set of outward-pointing end-agents. Since $\Phi(n,1) > \Phi(1,0)$, we can apply the argument for outward-pointing agents in part (a) of Theorem 4.1 to have every agent $j \in P^e(n;g)$ form a link $g_{j,n} = 1$. Let g' be the network that results after every $j \in P^e(n;g)$ has moved, and formed a link with n. Define $P^e(n;g')$ analogously, and proceed as before, with every $j \in P^e(m;g')$. Repeated application of this argument leads us eventually to either the periphery-sponsored star or a network in which all end-agents more than one link away from agent n are inward-pointing with respect to n. In the former case a simple variant of the miscoordination argument establishes convergence to the empty network. In the latter case, label the network as g^1 and proceed as follows.

Note that the hypothesis on payoffs implies that if agent i has a link with an end-agent, i's best response must involve deleting that link. Let j be the agent furthest away from n in g^1. Since g^1 is minimally tw-connected, there is a unique path between j and n. Then either $g^1_{n,j} = 1$ or there is an agent $j_q \neq n$ on the path between n and j, such that $g_{j_q,j} = 1$. In the former case, g^1 must be a star: if n chooses a best response, he will delete all his links, after which a miscoordination argument ensures that the empty network results. In the latter case, let j_q choose a best response and let g^2 denote the resulting network. Clearly j_2 will delete his link with j, in which case j will become a singleton component. Moreover, if j_2 forms any link at all, we can assume without loss of generality that he will form it with n. Let S_2 and S_1 be the set of agents in singleton components in g^2 and g^1, respectively. We have $S_1 \subset S_2$ where the inclusion is strict. Repeated application of the above arguments leads us to a network in which either an agent is a singleton component or is part of a star. If every agent falls in the former category, then we are at the empty network while in the latter case we let agent n move and delete all his links. Then a variant of the miscoordination argument (applied to the periphery-sponsored star) leads to the empty network. *Q.E.D.*

Appendix C

Proof of Proposition 5.1. (Sketch). If $c < \delta$, then it is immediate that a Nash network is connected. In the proof we focus on the case $c \geq \delta$. The proof is by contradiction. Consider a strict Nash network g that is nonempty but disconnected. Then there exists a pair of agents i_1 and i_2 such that $g_{i_1,i_2} = 1$. Moreover, since $c \geq \delta$ and g is strict Nash, there is an agent $i_3 \neq i_1$ such that $g_{i_2,i_3} = 1$. The same property must hold for i_3; continuing in this way, since N is finite, there must exist a cycle of agents, i.e. a collection $\{i_1, \ldots, i_q\}$ of three or more agents such that $g_{i_1,i_2} = \cdots = g_{i_q,i_1} = 1$. Denote the component containing this cycle as C. Since g is not connected there exists at least one other component D. We say there is a path from C to D if there exists $i \in C$ and $j \in D$ such that $i \overset{g}{\to} j$. There are two cases: (1) there is no path from C to D or vice-versa, and (2) either $C \overset{g}{\to} D$ or $D \overset{g}{\to} C$.

In case (1), let $i \in C$ and $j \in D$. Since g is strict Nash we get

$$\Pi_i(g_i \oplus g_{-i}) > \Pi_i(g'_i \oplus g_{-i}), \quad \text{for all } g'_i \in G_i, \text{ where } g'_i \neq g_i \ , \tag{C.1}$$

$$\Pi_j(g_j \oplus g_{-j}) > \Pi_j(g'_j \oplus g_{-j}), \quad \text{for all } g'_j \in G_j, \text{ where } g'_j \neq g_j \ , \tag{C.2}$$

Consider a strategy g^*_i such that $g^*_{i,k} = g_{j,k}$ for all $k \notin \{i,j\}$ and $g^*_{i,j} = 0$. The strategy g^*_i thus "imitates" that of agent j. By hypothesis, $j \notin N(i;g)$ and $i \notin N(j;g)$. This implies that the strategy of agent i has no bearing on the payoff of agent j or vice-versa. Hence, i's payoff from g^*_i satisfies

$$\Pi_i(g^*_i \oplus g_{-i}) \geq \Pi_j(g_j \oplus g_{-j}) \ . \tag{C.3}$$

Likewise, the payoff to agent j from the corresponding strategy g^*_j that imitates i satisfies

$$\Pi_j(g^*_j \oplus g_{-j}) \geq \Pi_i(g_i \oplus g_{-i}) \ . \tag{C.4}$$

We know that C is not a singleton. This immediately implies that the strategies g_i and g^*_i must be different. Putting together equations (C.2)–(C.4) with g^*_i in place of g'_i and g^*_j in place of g'_j yields

$$\Pi_i(g_i \oplus g_{-i}) > \Pi_i(g^*_i \oplus g_{-i}) \geq \Pi_j(g_j \oplus g_{-j}) > \Pi_j(g^*_j \oplus g_{-j}) \geq \Pi_i(g_i \oplus g_{-i}) \ . \tag{C.5}$$

The contradiction completes the argument for case (1). In case (2) we choose an $i' \in N(i;g)$ who is furthest away from $j \in D$ and apply a similar argument to that in case (1) to arrive at a contradiction. The details are omitted. The rest of the proposition follows by direct verification. *Q.E.D.*

Proof of Proposition 5.2. Consider the case of $s = 1$ and $c \in (0,1)$ first. Let g be a flower network with central agent n. Let $M = \max_{i,j \in N} d(i,j;g)$. Note that $2 \leq M \leq n-1$ by the definition of a flower network. Choose $\delta(c,g) \in (c,1)$ such that for all $\delta \in [\delta(c,g), 1]$ we have $(n-2)(\delta - \delta^M) < c$. Henceforth fix $\delta \in [\delta(c,g), 1]$. Suppose $P = \{j_1, \ldots, j_u\}$ is a petal of g. Since $c < \delta$ and no other agent has a link with j_u, agent n will form a link with him in his best response. If n formed any more links than those in g, an upper bound on the additional payoff he can obtain is $(n-2)(\delta - \delta^M) - c < 0$;

thus, n is playing a best response in g. The same argument ensures that agents $j_2, \ldots, j_u$ are also playing their best response. It remains to show the same for j_1. If there is only a single petal (i.e. g is a wheel) symmetry yields the result. Suppose there are two or more petals. For j_1 to observe all the other agents in the society, it is necessary and sufficient that he forms a link with either agent n or some agent $j' \in P'$, where $P' \neq P$ is another petal. Given such a link, the additional payoff from more links is negative, by the same argument used with agent n. If he forms a link with j' rather than n, agent j_1 will get the same total payoff from the set of agents $P' \in \{n\}$ since the sub-network of these agents is a wheel. However, the link with j' means that to access other petals (including the remaining agents in P, if any) agent j_1 must first go through all the agents in the path from n to j', whereas with n he can avoid these extra links. Hence, if there are at least three petals, forming a link with j' will make j strictly worse compared to forming it with n, so that g is a strict Nash network as required. If g contains only two petals P and P', both of level 2 or higher, j_1's petal will contain at least one more agent, and the argument above applies. Finally, if there are two petals P and P' and g is of level 1, then g is the exceptional case, and it is not a strict Nash. Thus, unless g is the exceptional case, it is a strict Nash for all $\delta \in [\delta(c, g), 1)$.

Next, consider $c \in (s-1, s)$ for some $s \in \{1, \ldots, n-1)\}$. If g is a flower network of level less than s, there is some petal $P = \{j_1, \ldots, j_{s'}\}$ with $s' \leq s-1$. Clearly the central agent n can increase his payoff by deleting his link with $j_{s'}$, ceteris paribus. Hence, a flower network of level smaller than s cannot be Nash.

Let g now be a flower network of level s or more. Let $M = \max_{i,j \in N} d(i, j; g)$. Choose $\delta(c, g)$ to ensure that for all $\delta \in [\delta(c, g), 1]$ both (1) $(n-2)(\delta - \delta^M) < \delta$ and (2) $\sum_{q=1}^{s} \delta^q - c > 0$ are satisfied. Let $P = \{j_1, \ldots, j_u\}$ be a petal with $u \geq s$. The requirement (2) ensures that agent n will wish to form a link with j_u. The requirement (1) plays the same role as in $s = 1$ above to ensure that n will not form more than one link per petal. If g has only one petal (i.e. it is a wheel) we are done. Otherwise, analogous arguments show that $\{j_2, \ldots, j_p\}$ are playing their best responses in g. Finally, for j_1, note that $u \geq 2$ implies that each petal is not a spoke. In this event, the argument used in part (a) shows that j_1 will be strictly worse off by forming a link with an agent other than agent n. The result (1) follows. *Q.E.D.*

Proof of Proposition 5.5. Consider a network g, and suppose that there is a pair of agents i and j, such that $g_{i,j} \neq 1$. If agent i forms a link $g_{i,j} = 1$, then the additional payoffs to i and j will be at least $2(\delta - \delta^2)$. If $c < 2(\delta - \delta^2)$, then this is clearly welfare enhancing. Hence, the unique efficient network is the complete network.

Fix a network g and consider a tw-component C_1, with $|C_1| = m$. If $m = 2$ then the nature of a component in an efficient network is obvious. Suppose $m \geq 3$ and let $k \geq m-1$ be the number of links in $|C_1|$. The social welfare of this component is bounded above by $m + k(2\delta - c) + [m(m-1) - 2k]\delta^2$. If the component is a star, then the social welfare is $(m-1)[2\delta - c + (m-2)\delta^2] + m$. Under the hypothesis that $2(\delta - \delta^2) < c$, the former can never exceed the latter and is equal to the latter only if $k = m-1$. It can be checked that the star is the only network with m agents and $m-1$ links, in which every pair of agents is at a distance of at most 2. Hence the network g must have at least one pair of agents i and j at a distance of 3. Since the number of direct links is the same while all indirect links are of length 2 in a star, this shows that the star welfare dominates every other network with $m-1$ links. Hence the component must be a star.

Clearly, a tw-component in an efficient network must have nonnegative social welfare. It can be calculated that the social welfare from a network with two distinct components of m and m' agents, respectively, is strictly less than the social welfare from a network where these distinct stars are merged to form a star with $m + m'$ agents. It now follows that a single star maximizes the social welfare in the class of all nonempty networks. An empty network yields the social welfare n. Simple calculations reveal that the star welfare dominates the empty network if and only if $2\delta + (n-2)\delta^2 > c$. This completes the proof.

Q.E.D.

References

Allen, B. (1982) A Stochastic Interactive Model for the Diffusion of Innovations. *Journal of Mathematical Sociology* 8: 265–281.

Anderlini, L., Iannni, A. (1996) Path-dependence and Learning from Neighbors. *Games and Economic Behaviour* 13: 141–178.
Baker, W., Iyer, A. (1992) Information Networks and Market Behaviour. *Journal of Mathematical Sociology* 16: 305–332.
Bala, V. (1996) Dynamics of Network Formation. Unpublished notes, McGill University.
Bala, V., Goyal, S. (1998) Learning from Neighbours. *Review of Economic Studies* 65: 595–621.
Bollobas, B. (1978) An Introduction to Graph Theory. Berlin: Springer Verlag.
Bolton, P., Dewatripont, M. (1994) The Firm as a Communication Network. *Quarterly Journal of Economics* 109: 809–839.
Boorman, S. (1975) A Combinatorial Optimization Model for Transmission of Job Information through Contact Networks. *Bell Journal of Economics* 6: 216–249.
Burt, R. (1992) Structural Holes: The Social Structure of Competition. Cambridge, MA: Harvard University Press.
Chwe, M. (1998) Structure and Strategy in Collective Action. mimeo, University of Chicago.
Coleman, J. (1966) Medical Innovation: A Diffusion Study, Second Edition. New York: Bobbs-Merrill.
Dutta, B., Mutuswami, S. (1997) Stable Networks. *Journal of Economic Theory* 76: 322–344.
Ellison, G. (1993) Learning, Local Interaction and Coordination. *Econometrica* 61: 1047–1072.
Ellison, G., Fudenberg, D. (1993) Rules of Thumb for Social Learning. *Journal of Political Economy* 101: 612–644.
Frenzen, J.K., Davis, H.L. (1990) Purchasing Behavior in Embedded Markets. *Journal of Consumer Research* 17: 1–12.
Gilboa, I., Matsui, A. (1991) Social Stability and Equilibrium. *Econometrica* 59: 859–869.
Goyal, S. (1993) Sustainable Communication Networks. Tinbergen Institute Discussion Paper 93-250.
Goyal, S., Janssen, M. (1997) Non-Exclusive Conventions and Social Coordination. *Journal of Economic Theory* 77: 34–57.
Granovetter, M. (1974) Getting a Job: A Study of Contacts and Careers. Cambridge, MA: Harvard University Press.
Harary, F. (1972) Network Theory. Reading, Massachusetts: Addison-Wesley Publishing Company.
Hendricks, K., Piccione, M., Tan, G. (1995) The Economics of Hubs: The Case of Monopoly. *Review of Economic Studies* 62: 83–101.
Hurkens, S. (1995) Learning by Forgetful Players. *Games and Economic Behavior* 11: 304–329.
Jackson, M., Wolinsky, A. (1996) A Strategic Model of Economic and Social Networks. *Journal of Economic Theory* 71: 44–74.
Kirman, A. (1997) The Economy as an Evolving Network. *Journal of Evolutionary Economics* 7: 339–353.
Kranton, R.E., Minehart, D.F. (1998) A Theory of Buyer-Seller Networks. mimeo, Boston University.
Linhart, P.B., Lubachevsky, B., Radner, R., Meurer, M.J. (1994) Friends and Family and Related Pricing Strategies. mimeo., AT&T Bell Laboratories.
Mailath, G., Samuelson, L., Shaked, A. (1996) Evolution and Endogenous Interactions. mimeo., Social Systems Research Institute, University of Wisconsin.
Marimon, R. (1997) Learning from Learning in Economics. In: D. Kreps and K. Wallis (eds.) *Advances in Economics and Econometrics: Theory and Applications, Seventh World Congress*, Cambridge: Cambridge University Press.
Marshak, T., Radner, R. (1972) The Economic Theory of Teams. New Haven: Yale University Press.
Myerson, R. (1991) Game Theory: Analysis of Conflict. Cambridge, MA: Harvard University Press.
Nouweland, A. van den (1993) Games and Networks in Economic Situations. Unpublished Ph.D Dissertation, Tilburg University.
Radner, R. (1993) The Organization of Decentralized Information Processing. *Econometrica* 61: 1109–1147.
Rogers, E., Kincaid, L.D. (1981) Communication Networks: Toward a New Paradigm for Research. New York: Free Press.
Roth, A., Vande Vate, J.H. (1990) Random Paths to Stability in Two-Sided Matching. *Econometrica* 58: 1475–1480.
Sanchirico, C.W. (1996) A Probabilistic Method of Learning in Games. *Econometrica* 64: 1375–1395.
Wellman, B., Berkowitz, S. (1988) Social Structure: A Network Approach. Cambridge: Cambridge University Press.

The Stability and Efficiency of Directed Communication Networks

Bhaskar Dutta[1], **Matthew O. Jackson**[2]

[1] Indian Statistical Institute, 7 SJS Sansanwal Marg, New Delhi 110016, India
(e-mail: dutta@isid.ac.in)
[2] Division of the Humanities and Social Sciences, Caltech, Pasadena, CA 91125, USA
(e-mail: jacksonm@hss.caltech.edu)

Abstract. This paper analyzes the formation of directed networks where self-interested individuals choose with whom they communicate. The focus of the paper is on whether the incentives of individuals to add or sever links will lead them to form networks that are efficient from a societal viewpoint. It is shown that for some contexts, to reconcile efficiency with individual incentives, benefits must either be redistributed in ways depending on "outsiders" who do not contribute to the productive value of the network, or in ways that violate equity (i.e., anonymity). It is also shown that there are interesting contexts for which it is possible to ensure that efficient networks are individually stable via (re)distributions that are balanced across components of the network, anonymous, and independent of the connections of non-contributing outsiders.

JEL Classification: A14, D20, J00

Key Words: Networks, stability, efficiency, incentives

1 Introduction

Much of the communication that is important in economic and social contexts does not take place via centralized institutions, but rather through networks of decentralized bilateral relationships. Examples that have been studied range from the production and transmission of gossip and jokes, to information about job opportunities, securities, consumer products, and even information regarding the returns to crime. As these networks arise in a decentralized manner, it is important

Matthew Jackson gratefully acknowledges financial support under NSF grant SBR 9507912. We thank Anna Bogomolnaia for providing the proof of a useful lemma. This paper supercedes a previous paper of the same title by Jackson.

to understand how they form and to what degree the resulting communication is efficient.

This paper analyzes the formation of such directed networks when self-interested individuals choose with whom they communicate. The focus of the paper is on whether the incentives of individuals will lead them to form networks that are efficient from a societal viewpoint. Most importantly, are there ways of allocating (or redistributing) the benefits from a network among individuals in order to ensure that efficient networks are stable in the face of individual incentives to add or sever links?

To be more precise, networks are modeled as directed graphs among a finite set of individual players. Each network generates some total productive value or utility. We allow for situations where the productive value or utility may depend on the network structure in general ways, allowing for indirect communication and externalities.

The productive value or utility is allocated to the players. The allocation may simply be the value that players themselves realize from the network relationships. It may instead represent some redistribution of that value, which might take place via side contracts, bargaining, or outside intervention by a government or some other player. We consider three main constraints on the allocation of productive value or utility. First, the allocation must be anonymous so that the allocation depends only on a player's position in a network and how his or her position in the network affects overall productive value, but the allocation may not depend on a player's label or name. Second, the allocation must respect component balance: in situations where there are no externalities in the network, the network's value should be (re)distributed inside the components (separate sub-networks) that generate the value. Third, if an outsider unilaterally connects to a network, but is not connected to by any individual in that network, then that outsider obtains at most her marginal contribution to the network. We will refer to this property as *outsider independence*.

The formation of networks is analyzed via a notion of individual stability based on a simple game of network formation in such a context: each player simultaneously selects a list of the other players with whom she wishes to be linked. Individual stability then corresponds to a (pure strategy) Nash equilibrium of this game.

We show that there is an open set of value functions for which *no* allocation rule satisfies anonymity, component balance, and outsider independence, and still has at least one efficient (value maximizing) network being individually stable. However, this result is not true if the outsider independence condition is removed. We show that there exists an allocation rule which is anonymous, component balanced and guarantees that some efficient network is individually stable. This shows a contrast with the results for non-directed networks. We go on to show that for certain classes of value functions an anonymous allocation rule satisfying component balance and outsider independence can be constructed such that an efficient network is individually stable. Finally, we show that when value accumulates from connected communication, then the value function is in this class and so there is an allocation rule that satisfies anonymity, component balance, and outsider independence, and still ensures that at least one (in fact all) efficient networks are individually stable.

Relationship to the Literature

There are three papers that are most closely related to the analysis conducted here: Jackson and Wolinsky (1996), Dutta and Mutuswami (1997), Bala and Goyal (2000).[1]

The relationship between efficiency and stability was analyzed by Jackson and Wolinsky (1996) in the context of non-directed networks. They noted a tension between efficiency and stability of networks under anonymity and component balance, and also identified some conditions under which the tension disappeared or could be overcome via an appropriate method of redistribution.

There are two main reasons for revisiting these questions in the context of directed networks. The most obvious reason is that the set of applications for the directed and non-directed models is quite different. While a trading relationship, marriage, or employment relationship necessarily requires the consent of two individuals, an individual can mail (or email) a paper to another individual without the second individual's consent. The other reason for revisiting these questions is that incentive properties turn out to be different in the context of directed networks. Thus, the theory from non-directed networks cannot simply be cut and pasted to cover directed networks. There turn out to be some substantive similarities between the contexts, but also some significant differences. In particular, the notion of an outsider to a network is unique to the directed network setting. The differences between the directed and non-directed settings are made evident through the theorems and propositions, below.

Dutta and Mutuswami (1997) showed that if one weakens anonymity to only hold on stable networks, then it is possible to carefully construct a component balanced allocation rule for which an efficient network is pairwise stable. Here the extent to which anonymity can be weakened in the directed network setting is explored. It is shown that when there is a tension between efficiency and stability, then anonymity must be weakened to hold only on stable networks. Moreover, only some (and not all) permutations of a given network can be supported even when all permutations are efficient. So, certain efficient networks can be supported as being individually stable by weakening anonymity, but not efficient network architectures.

This paper is also related to a recent paper by Bala and Goyal (2000), who also examine the formation of directed communication networks. The papers are, however, quite complementary. Bala and Goyal focus on the formation of networks in the context of two specific models (the directed connections and

[1] Papers by Watts (1997), Jackson and Watts (2002), and Currarini and Morelli (2000) are not directly related, but also analyze network formation in very similar contexts and explore efficiency of emerging networks.

hybrid connections models discussed below) without the possibility of reallocating of any of the productive value.[2] Here, the focus is instead on whether there exist equitable and (component) balanced methods of allocating (or possibly reallocating) resources to provide efficient incentives in the context of a broad set of directed network models. Results at the end of this paper relate back to the directed connections and hybrid connections models studied by Bala and Goyal, and show that the individual stability of efficient networks in those models can be ensured (only) if reallocation of the productive value of the network is possible.

2 Definitions and Examples

Players

$\{1, \ldots, N\}$ is a finite set of players. The network relations among these players are formally represented by graphs whose *nodes* are identified with the players.

Networks

We model directed networks as digraphs.

A *directed network* is an $N \times N$ matrix g where each entry is in $\{0, 1\}$. The interpretation of $g_{ij} = 1$ is that i is *linked* to j, and the interpretation of $g_{ij} = 0$ is that i is *not linked* to j. Note that $g_{ij} = 1$ does not necessarily imply that $g_{ji} = 1$. It can be that i is linked to j, but that j is not linked to i. Adopt the convention that $g_{ii} = 0$ for each i, and let G denote the set of all such directed networks. Let g_i denote the vector $(g_{i1}, \ldots, g_{iN})$.

For $g \in G$ let $N(g) = \{i \mid \exists j \text{ s.t. } g_{ij} = 1 \text{ or } g_{ji} = 1\}$. So $N(g)$ are the active players in the network g, in that either they are linked to someone or someone is linked to them.

For any given g and ij let $g+ij$ denote the network obtained by setting $g_{ij} = 1$ and keeping other entries of g unchanged. Similarly, let $g-ij$ denote the directed network obtained by setting $g_{ij} = 0$ and keeping other entries of g unchanged.

Paths

A *directed path in g connecting* i_1 to i_n is a set of distinct nodes $\{i_1, i_2, \ldots, i_n\} \subset N(g)$ such that $g_{i_k i_{k+1}} = 1$ for each k, $1 \leq k \leq n - 1$.

A *non-directed path in g connecting* i_1 to i_n is a set of distinct nodes $\{i_1, i_2, \ldots, i_n\} \subset N(g)$ such that either $g_{i_k i_{k+1}} = 1$ or $g_{i_{k+1} i_k} = 1$ for each k, $1 \leq k \leq n - 1$.[3]

Components

A network g' is a *sub-network* of g if for any i and j $g'_{ij} = 1$ implies $g_{ij} = 1$.

[2] Also, much of Bala and Goyal's analysis is focussed on a dynamic model of formation that selects strict Nash equilibria in the link formation game in certain contexts where there also exist Nash equilibria that are not strict.

[3] Non-directed paths are sometimes referred to as semipaths in the literature.

A non-empty sub-network of g, g', is a *component of* g if for all $i \in N(g')$ and $j \in N(g')$, $i \neq j$, there exists a non-directed path in g' connecting i and j, and for any $i \in N(g')$ and $j \in N(g)$ if there is a non-directed path in g between i and j, then $j \in N(g')$. The set of components of a network g is denoted $C(g)$.

A network g is *completely connected* (or the complete network) if $g_{ij} = 1$ for all ij.

A network g is *connected* if for each distinct i and j in N there is a non-directed path between i and j in g.

A network g' is a *copy* of g if there exists a permutation π of N such that $g' = g^{\pi}$.

Specific Network Structures

A network g is a *star* if there is i such that $g_{kl} = 1$ only if $i \in \{k, l\}$. That is, a star is a network in which all connections involve a *central* node i.

A network g is a *k-person wheel* if there is a sequence of players $\{i_1, \ldots, i_k\}$ such that $g_{i_k i_1} = g_{i_j i_{j+1}} = 1$ for all $j = 1, \ldots, k-1$, and $g_{ij} = 0$ otherwise.

Value Functions

A *value function* $v : G \to I\!R$, assigns a value $v(g)$ to each network g. The set of all value functions is denoted V.

In some applications the value of a network is an aggregate of individual utilities or productions, so that $v(g) = \sum_i u_i(g)$ for some profile of $u_i : G \to I\!R$.

The concepts above are illustrated in the context of the following examples.

Example 1. The Directed Connections Model.[4] The value function $v^d(\cdot)$ is the sum of utility functions ($u_i(\cdot)$'s) that describe the benefit (net of link costs) that players obtain from direct and indirect communication with others. Each player has some information that has a value 1 to other players.[5] The factor $\delta \in [0, 1]$ captures decay of information as it is transmitted. If a player i has $g_{ij} = 1$, then i obtains δ in value from communication with j. There are different interpretations of this communication: sending or receiving. Player i could be getting value from receiving information that i has accessed from j (e.g., contacting j's web site), or it could be that i is getting value from sending j information (e.g., mailing research papers or advertising). In either case, it is i who incurs the cost of communication and is benefiting from the interaction. If the shortest directed path between i and j contains 2 links (e.g., $g_{ik} = 1$ and $g_{kj} = 1$), then i gets a value of δ^2 from the indirect communication with j. Similarly, if the shortest directed path between i and j contains m links, then i gets a value of δ^m from

---

[4] This model is considered by Bala and Goyal (2000), and is also related to a model considered by Goyal (1993). The name reflects the relationship to the non-directed "connections model" discussed in Jackson and Wolinsky (1996).

[5] Bala and Goyal consider a value V. Without loss of generality this can be normalized to 1 since it is the ratio of this V to the cost c that matters in determining properties of networks, such as identifying the efficient network or considering the incentives of players to form links.

the indirect communication with j. If there is no directed path from i to j, then i gets no value from communication with j.

Note that information only flows one way on each link. Thus, j gets no value from the link $g_{ij} = 1$. This also means that i gets no value from j if there is exists a non-directed path between i and j, but no directed path from i to j.

Player i incurs a cost $c > 0$ of maintaining each direct link. Player i can benefit from indirect communication without incurring any cost beyond i's direct links.

Let $N(i,g)$ denote the set of players j for which there is a directed path from i to j. For i and any $j \in N(i,g)$, let $d(ij,g)$ denote the number of links in the minimum-length directed path from i to j. Let $n_d(i,g) = \#\{j \mid g_{ij} = 1\}$ represent the number of direct links that i maintains. The function u_i can be expressed as[6]

$$u_i(g) = \sum_{j \in N(i,g)} \delta^{d(ij,g)} - n_d(i,g)c.$$

Then $v^d(g) = \sum_i u_i(g)$.

Example 2. The Hybrid Connections Model. Consider a variation on the directed connections model where players still form directed links, but where information flows both ways along any link. This model is studied in Bala and Goyal (1999), who mention telephone calls as an example of such communication. One player initiates the link and incurs the cost, but both share the communication benefits (or losses). Another example that would fit into this hybrid model would be physical connections on a computer network like the internet. A player (who may be an individual, a university, company, or some other collection of users) incurs the cost for connecting to a network, and then others already interconnected can communicate with the player.

Let $\overline{N}(i,g)$ denote the set of players j for which there is a non-directed path between i and j. For i and any $j \in \overline{N}(i,g)$, let $\overline{d}(ij,g)$ denote the number of links in the minimum-length non-directed path from i to j. The function u_i can be expressed as

$$u_i(g) = \sum_{j \in \overline{N}(i,g)} \delta^{\overline{d}(ij,g)} - n_d(i,g)c,$$

and $v^h(g) = \sum_i u_i(g)$.

Strong Efficiency

A network $g \subset g^N$ is *strongly efficient* if $v(g) \geq v(g')$ for all $g' \subset g^N$.

[6] Player i gets no value from his or her own information. This is simply a normalization so that the value of the empty network is 0.

The term strong efficiency indicates maximal total value, rather than a Paretian notion.[7] Of course, these are equivalent if value is transferable across players. In situations where Y represents a redistribution, and not a primitive utility, then implicitly value is transferable and strong efficiency is an appropriate notion.

Allocation Functions

An *allocation rule* $Y : G \times V \to I\!R^N$ describes how the value associated with each network is distributed to the individual players.

$Y_i(g, v)$ is the payoff to player i from graph g under the value function v.

In the directed connections model (without any redistribution) $Y_i(g, v) = u_i(g)$, so that players obtain exactly the utility of their communication. The definition of an allocation rule, however, also allows for situations where $Y_i(g, v) \neq u_i(g)$, so that transfers or some redistribution is considered.

Anonymity of a Value Function

A value function v is *anonymous* if $v(g^\pi) = v(g)$ for all g and π.

Anonymity of a value function states that the value of a network depends only on the pattern of links in the network, and not on the labels of the players who are in given positions in the network.

Anonymity of an Allocation Function

For any value function v and permutation of players π, let the value function v^π be defined by $v^\pi(g^\pi) = v(g)$ for each g.

An allocation rule Y is *anonymous* relative to a network g and value function v if, for any permutation π, $Y_{\pi(i)}(g^\pi, v^\pi) = Y_i(g, v)$. Y is anonymous, if it is anonymous relative to each network g and value function v.

Anonymity of an allocation rule states that if all that has changed is the names of the agents (and not anything concerning their relative positions or production values in some network), then the allocations they receive should not change. In other words, the anonymity of Y requires that the information used to decide on allocations be obtained from the value function v and the particular network g, and not from the label of a player.

Note that anonymity of an allocation rule implies that individuals who are in symmetric positions in a network are assigned the same allocation, *if* the underlying v is anonymous, but not necessarily otherwise.[8] For instance if g is such that $g_{12} = g_{21} = 1$ and $g_{ij} = 0$ for all other ij, then provided v is anonymous[9] it follows that $Y_1(g, v) = Y_2(g, v)$.

---

[7] The term strong efficiency is used by Jackson and Wolinsky (1996), Dutta and Mutuswami (1997), and Jackson and Watts (2002). This is referred to as efficiency by Bala and Goyal (2000). We stick with the term strong efficiency to distinguish the notion from Pareto efficiency.

[8] This is the only implication of anonymity that is needed to establish the negative results in what follows.

[9] More explicitly, for this network the conclusion follows if $v^{\pi(12)} = v$, where $\pi(12)$ is the permutation such that 1 and 2 are switched and all other players are mapped to themselves.

Balance and component balance

An allocation rule Y is *balanced* if $\sum_i Y_i(g, v) = v(g)$ for all value functions v and networks g.

A stronger notion of balance, component balance, requires Y to allocate resources generated by any component to that component.

A value function v is *component additive* if $v(g) = \sum_{h \in C(g)} v(h)$ for each network g.[10]

An allocation rule Y is *component balanced* if $\sum_{i \in N(h)} Y_i(g, v) = v(h)$ for every $g \in G$ and $h \in C(g)$ and component additive $v \in V$.

Component balance requires that the value generated by a given component be redistributed only among the players in that component. It is important that the definition of component balance *only* applies when v is component additive. Thus, it is only required to hold when there are no externalities across components.

Outsiders

A stronger version of component balance turns out to be important in the context of directed networks. The following definition of outsider is important in that definition and outsider independence.

A player i is an *outsider* of a network g if

(i) $g_{ij} = 1$ for some $j \in N(g)$,
(ii) $g_{ki} = 0$ for all $k \in N(g)$, and
(iii) for every $j \neq i, j \in N(g)$, there exists $k \neq i$ with $k \in N(g)$ such that $g_{kj} = 1$.

Thus, an outsider is a player who has linked to some other players in a network, but to whom no other player has linked. Furthermore, a player is considered an outsider only when all other players in the network have someone (other than the outsider) linked to them, so the outsider is not important in connecting anyone else to the network. This last condition avoids the trivial case of calling player 1 an outsider in the network g where $g_{12} = 1$ and $g_{ij} = 0$ for all other ij. It also implies that there is at most one outsider to a network.

Directed Component Balance

Let $g - i$ denote the network obtained from network g by deleting each of player i's links, but not the links from any player $j \neq i$ to player i. That is, $(g - i)_{ij} = 0$ for all j, and $(g - i)_k = g_k$ whenever $k \neq i$.

The allocation rule Y satisfies *directed component balance* if it is component balanced, and for any component additive value function v, network g, and outsider i to g, if $v(g) = v(g - i)$, then $Y(g) = Y(g - i)$.

[10] This definition implicitly requires that the value of disconnected players is 0. This is not necessary. One can redefine components to allow a disconnected player to be a component. One has also to extend the definition of v so that it assigns values to such components.

The situation covered by directed component balance but not by component balance is one where a single player i is initially completely unconnected under $g - i$, then connects to some other players resulting in g, but does not change the value of the network. The directed component balance condition requires that the allocation rule not change due to the addition of such an outsider. This directed version of component balance is in the same spirit as component balance. The reasoning is that a player who unilaterally links up to a component whose members are already interconnected, and who does not change the productive value of the network *in any way*, effectively should not be considered to be part of that component for the purposes of allocating productive value.

Network Formation and Individual Stability

Let $D_i(g) = \{g' | g'_j = g_j \forall j \neq i\}$. These are the networks that i can reach from g by a unilateral change in strategy.

A network g is *individually stable* relative to Y and v if $Y_i(g', v) \leq Y_i(g, v)$ for all $g' \in D_i(g)$. [11]

The idea of individual stability is quite straightforward. A network is individually stable if no player would benefit from changing his or her directed links. The set of individually stable networks corresponds to the networks that are pure strategy Nash equilibrium outcomes of a link formation game where each player simultaneously writes down the list of players who he or she will link to, and those links are then formed. [12]

3 Individual Stability and Strong Efficiency

Theorem 1. *If $N \geq 3$, then there is no Y which satisfies anonymity and directed component balance and is such that for each v at least one strongly efficient graph is individually stable.*

Proof. Let $N = 3$ and consider any Y which satisfies anonymity and directed component balance. The theorem is verified by showing that there exists a v such that no strongly efficient graph is individually stable.

Let g be such that $g_{12} = g_{23} = g_{31} = 1$ and all other $g_{ij} = 0$, and g' be such that $g'_{13} = g'_{32} = g'_{21} = 1$ and all other $g'_{ij} = 0$. Thus, g and g' are the 3-person wheels.

Let v be such that $v(g) = v(g') = 1 + \epsilon$ and $v(g'') = 1$ for any other graph g''. Therefore, the strongly efficient networks are the wheels, g and g'.

Consider g'' such that $g''_{12} = g''_{21} = 1$ and all other $g''_{ij} = 0$.

[11] This notion is called 'sustainability' by Bala and Goyal (2000). The term stability is used to be consistent with a series of definitions from Jackson and Wolinsky (1996) and Dutta and Mutuswami (1997) for similar concepts with non-directed graphs.

[12] This link formation process is a variation of the game defined by Myerson (1991, page 448). Similar games are used to model link formation by Qin (1996), Dutta, et al. (1998), Dutta and Mutuswami (1996), and Bala and Goyal (2000).

It follows from anonymity and component balance that $Y_1(v, g'') = Y_2(v, g'') = 1/2$.

It follows from directed component balance that $Y_1(v, g'' + 31) = Y_2(v, g'' + 31) = 1/2$.

It follows from anonymity and balance that $Y_1(g, v) = Y_2(g, v) = Y_3(g, v) = \frac{1+\epsilon}{3}$.

Consider the strategy profile leading to g in the link formation game. If $\epsilon < 1/6$, then this strategy profile is not a Nash equilibrium, since player 2 will benefit by deviating and adding 21 and deleting 23. (Notice that $g'' + 31$ is obtained from g by adding 21 and deleting 23.) A similar argument shows that the strategy profile leading to g' in the link formation game does not form a Nash equilibrium. The case of $N > 3$ is easily handled by extending the above v so that components with more than three players have no value. □

The proof of Theorem 1 necessarily follows a different line of reasoning from the proof of the analogous theorem for the non-directed case in Jackson and Wolinsky (1996). This reflects the difference between individual stability in the directed setting and pairwise stability in the non-directed setting that naturally arises due to the possibility of unilateral link formation in the directed network context. In the proof here, the problematic efficient network is an anonymous one and the contradiction is reached via a comparison to the network g'' which makes use of directed component balance. In the non-directed case, the proof examines a situation where the efficient network is not anonymous, and reaches a contradiction via comparisons to anonymous super- and sub-networks. The difference between the directed and non-directed settings is further explored below.

For the case of non-directed networks, one of the main points of Dutta and Mutuswami's (1997) analysis is that one can weaken anonymity to require that it only hold on stable networks and thereby overcome the incompatibility between efficiency and stability noted by Jackson and Wolinsky (1996). This is based on an argument that one is normatively less concerned with what occurs on unstable networks (out of equilibrium), provided one expects to see stable networks form. So Dutta and Mutuswami use non-anonymous rewards and punishments out of equilibrium to support an anonymous stable allocation. It can be shown, however, that in the non-directed case there is no Y that is component balanced and for which a strongly efficient network is pairwise stable,[13] as are all anonymous permutations of that network when v is anonymous. (This follows from Theorem 1' and its proof in the appendix of Jackson and Wolinsky (1996).) The implication of this is that in order to have at least one strongly efficient network be pairwise stable and satisfy component balance, it can be that only one of the strongly efficient networks is pairwise stable even though anonymous permuta-

[13] In the context of non-directed networks it takes the consent of two individuals to form a link. Pairwise stability requires that no individual benefit from severing one link, and no two individuals benefit (one weakly and one strictly) from adding a link. A precise definition is given in Jackson and Wolinsky (1996).

tions of it are also strongly efficient. Thus, pairwise stability may apply just to a specific efficient network with players in a fixed relationship (and not to a network structure). For example, in certain contexts one can construct a component balanced allocation rule for which a star with player 1 at the center is strongly efficient and pairwise stable, but one cannot at the same time ensure that a star with player 2 at the center is also pairwise stable even though it generates exactly the same total productive value as the star with player 1 at the center, and thus is also strongly efficient. [14] This may not be objectionable, as long as one can at least ensure an anonymous set of payoffs to players, as Dutta and Mutuswami do. But the fact that only specific efficient networks can be supported, and not a given efficient network structure, gives a very precise idea of the extent to which anonymity must be weakened in order to reconcile efficiency and stability in the face of component balance. This is stated in the context of directed networks as follows.

Theorem 2. *If $N \geq 3$, then there is no Y that satisfies anonymity relative to individually stable networks, directed component balance, has an anonymous set of individually stable networks when v is anonymous,*[15] *and is such that for each v at least one strongly efficient network is individually stable.*

Proof. Let $N = 3$ and consider any Y which satisfies anonymity on individually stable networks, directed component balance, has an anonymous set of stable networks when v is anonymous. The theorem is proven by showing that there exists a v such that no strongly efficient network is individually stable.

Consider g, g', g'', and v from the proof of Theorem 1. Suppose the contrary, so that either g or g' is individually stable. Since v is anonymous and g and g' are anonymous permutations of each other, it follows that both g and g' are individually stable.

Thus, anonymity on individually stable networks and balance imply that $Y_1(g, v) = Y_2(g, v) = Y_3(g, v) = \frac{1+\epsilon}{3}$ and likewise that $Y_1(v, g') = Y_2(v, g') = Y_3(v, g') = \frac{1+\epsilon}{3}$.

Also, it follows from directed component balance that $Y(v, g''+31) = Y(v, g'')$ and that $Y(v, g'' + 32) = Y(v, g'')$.

Case 1: $Y_1(v, g'') \geq 1/2$. Consider the strategy profile leading to g' in the link formation game. If $\epsilon < 1/6$, then this strategy profile is not a Nash equilibrium, since player 1 will benefit by deviating and adding 12 and deleting 13 (which results in $g'' + 32$). This is a contradiction.

Case 2: $Y_2(v, g'') > 1/2$. Consider the strategy profile leading to g in the link formation game. If $\epsilon \leq 1/6$, then this strategy profile is not a Nash equilibrium, since player 2 will benefit by deviating and adding 21 and deleting 23 (which results in $g'' + 31$). This is a contradiction.

By component balance, these two cases are exhaustive. □

---

[14] Again, see the proof of Theorem 1' in the appendix of Jackson and Wolinsky (1996).

[15] g^π is individually stable whenever g, for any permutation π.

4 Outsiders

We consider next, a condition that states one cannot shift too much value to an outsider: no more than their marginal contribution to the network. A reason for exploring the role of outsiders in detail is that the value function used in the proof of Theorems 1 and 2 is special. In particular, several networks all have the same value even though their architectures are different. Moreover, that fact is important to the application of directed component balance in the proof of Theorems 1 and 2. This reliance on specific value functions is really only due to the weak way in which outsiders are addressed in directed component balance. If directed component balance is replaced by the following outsider independence condition which is more explicit about the treatment of outsiders, then the results of Theorems 1 and 2 hold for open sets of value functions.

Outsider Independence

An allocation rule Y satisfies *outsider independence* if for all $g \in G$ and $v \in V$ and $i \in N(g)$ who is an outsider of g such that $v(g) \geq v(g-i)$, then $Y_j(g, v) \geq Y_j(g-i, v)$ for each $j \neq i$.

Outsider independence states that an outsider obtains at most her marginal contribution to the value of a network. The idea is that if a set of players has formed a network, and cannot prevent an outsider from linking to it, then the players should not suffer because of the outsider's actions. Such a condition is clearly satisfied in the directed connections model, and in any setting where the outsider's actions have no externalities.

Outsider independence is only required to hold in situations where the outsider's presence does not decrease the value of the network. Normatively, one might argue for it more generally.

Theorem 3. *If $N \geq 3$, there is an open subset*[16] *of the anonymous value functions for which any Y that satisfies anonymity on individually stable networks, component balance, and outsider independence, and has an anonymous set of individually stable networks when v is anonymous, cannot have any strongly efficient network be individually stable.*

The proof of Theorem 3 is a straightforward extension of the proofs of Theorems 1 and 2 and therefore is omitted.

It is easily seen that Theorems 1, 2, and 3 are tight in that dropping anonymity invalidates the results. For example, let $\overline{Y}$ be the equal split of value within components rule as defined below. Define Y by picking a strongly efficient g^*, and let $Y(g^*, v) = \overline{Y}(g^*, v)$. For any g such that $g_j = g_j^*$ for all $j \neq i$ for some i, set $Y_i(g, v) \leq \overline{Y}_i(g^*, v)$, set $Y_j(g, v) = \overline{Y}_j(g, v)$ for $j \notin N(h_i)$ where h_i is the

[16] Given that the set of networks G is a finite set, a value function can be represented as a finite vector. Here, open is relative to the subspace of anonymous value functions.

component of g containing i, and let $Y_j(g,v) = \frac{v(h_i)-Y_i(g,v)}{\#N(h_i)-1}$ for $j \in N(h_i), j \neq i$. For any other g set $Y(g,v) = \overline{Y}(g,v)$.

Next, we show that weakening directed component balance or ignoring outsider independence invalidates Theorems 1, 2, and 3. If value can be allocated to outsiders without regards to their contribution to the value of a network, then it is possible to sustain efficient networks as being individually stable.

Theorem 4. *There exists an allocation rule Y that is anonymous, component balanced and such that for each v there is some strongly efficient network that is individually stable.*

Theorem 4 shows that there are important differences between the directed and non-directed network contexts. In the directed case it is always possible for any player unilaterally to become part of a network. If the allocation rule can shift value to outsiders, even when they contribute nothing to the value of a network, then one can overcome the difficulties imposed by component balance.

The proof of Theorem 4 is constructive and appears in the appendix. Here, we provide some intuition underlying the constructive proof.

Let Y be an allocation rule that we are designing to support a given strongly efficient network g^* as individually stable. So, it must be that for all i, $Y_i(g^*,v) \geq \max_{g \in D_i(g)} Y_i(g,v)$. At the same time we need to make sure that Y is anonymous and component balanced. To get a feeling for the impact of those restrictions, consider the following example.

Example 3. There are 5 players. The value function v is anonymous. A strongly efficient network g^* is such that $g^*_{12} = g^*_{23} = g^*_{34} = g^*_{45} = 1$ and $g^*_{ij} = 0$ for other ij. So, g^* is a directed line. Suppose that $v(g^*) = 5$ and that $v(g) = 5$ if g is a copy of g^*.

Let us consider the restrictions on Y imposed by anonymity, component balance, and guaranteeing that g^* is individually stable.

First, player 5 can deviate from g^* by adding the link 51, to result in the network $g^* + 51$. Let us denote that network as g^5. So, g^5 is a wheel. Since a wheel is symmetric, it must be that $Y_5(g^5,v) = v(g^5)/5$. Then, to ensure that g^* is individually stable, we need to have $Y_5(g^*,v) \geq v(g^5)/5$.

Next, player 4 can deviate from g^* by deleting link 45 and adding link 41. The resulting network, denoted g^4 is a four person wheel. Here, to ensure that g^* is individually stable and Y is anonymous and component balanced, we need to have $Y_4(g^*,v) \geq v(g^4)/4$.

Also, player 3 can deviate from g^* by deleting link 34 and adding link 31. The resulting network, denoted g^3 is a three person wheel among 1,2, and 3, together with the extra link 45. Here, to ensure that g^* is individually stable and Y is anonymous and component balanced, we need to have $Y_3(g^*,v) \geq v(h^3)/3$, where h^3 is the three person wheel among 1, 2, and 3.

There is a similar requirement for player 2. These requirements are different for different players, and so an allocation rule that simply equally splits value does not work. The proof involves showing that these requirements can all be satisfied simultaneously, and that the type of requirements arising in this example are those arising more generally and can always be handled.

5 Efficiency and the Connections Models

The above results indicate that in order to find an allocation rule that reconciles individual stability and strong efficiency in general, in some cases one needs to allocate some value to non-productive outsiders. However, there are still interesting settings where strong efficiency and individual stability can be reconciled, while preserving anonymity, directed component balance, and outsider independence. We explore some such settings here.

Given a value function v and a set $K \subset N$, let $g_v^*(K)$ be a selection of a strongly efficient network restricted to the set of players K (so $N(g^*(K)) \subset K$). If there is more than one such strongly efficient network among the players K, then select one which minimizes the number of players in $N(g)$.

A value function v has *non-decreasing returns to scale* if for any $K' \subset K \subset N$

$$\frac{v(g_v^*(K))}{\#N(g_v^*(K))} \geq \frac{v(g_v^*(K'))}{\#N(g_v^*(K'))}.$$

If a value function has non-decreasing returns to scale, then *per-capita* value of the efficient network is non-decreasing in the number of individuals available. This does not imply anything about the structure of the efficient network, except that larger groups can be at least as productive per capita in an efficient configuration as smaller groups. As we shall see shortly, it is satisfied by some natural value functions.

Theorem 5. *If a component additive value function v has nondecreasing returns to scale, then there exists an allocation rule Y satisfying anonymity, directed component balance and outsider independence for which at least one strongly efficient networks is individually stable relative to v.*

The proof of Theorem 5 is given in the appendix.

The proof of Theorem 5 relies on the following allocation rule $\widehat{Y}$, which is a variant on a component-wise egalitarian rule $\overline{Y}$. Such a rule is attractive because of its strong equity properties. To be specific, define $\overline{Y}$ as follows. Consider any g and a component additive v. If i is in a component h of g (which is by definition necessarily non-empty), then $\overline{Y}_i(g, v) = \frac{v(h)}{\#N(h)}$, and if i is not in any component then $\overline{Y}_i(g, v) = 0$. For any v that is not component additive, let $Y_i(g, v) = \frac{v(g)}{N}$ for all i. $\overline{Y}$ is a component-wise egalitarian rule, and is component balanced and anonymous. It divides the value generated by a given component equally among all the members of that component, provided v is component additive (and divides resources equally among all players otherwise). It is shown in the appendix that under non-decreasing returns to scale, all strongly efficient networks are individually stable relative to $\overline{Y}$.

Unfortunately, $\overline{Y}$ does not always satisfy outsider independence. For instance, in the directed connections model it fails outsider independence for ranges of values of δ and c.[17] However, a modification of $\overline{Y}$ results in the allocation rule $\widehat{Y}$ that satisfies anonymity, directed component balance, outsider independence, and for which all strongly efficient networks are individually stable for v's that have non-decreasing returns to scale. The modified allocation rule $\widehat{Y}$ is defined as follows. For any v and strongly efficient network g^*, set $\widehat{Y}(g^*, v) = \overline{Y}(g^*, v)$. For any other g: if g has an outsider i then set $\widehat{Y}_j(g, v) = \max[\overline{Y}_j(g - i, v), \overline{Y}_j(g, v)]$ for $j \neq i$ and $\widehat{Y}_i(g, v) = v(g) - \sum_{j \neq i} \widehat{Y}_j(g, v)$; and otherwise set $\widehat{Y}(g, v) = \overline{Y}(g, v)$. As there is at most one outsider to a network, $\widehat{Y}$ is well-defined.

Both the directed connections and hybrid connections models have non-decreasing returns to scale:

Proposition 1. *The directed and hybrid connections models (v^d and v^h) have non-decreasing returns to scale. Thus, all strongly efficient networks are individually stable in the connections models, relative to the anonymous, directed component balanced and outsider independent allocation rule $\widehat{Y}$.*

The re-allocation of value under $\widehat{Y}_i$ compared with u_i^d and u_i^h is important to Proposition 1. Without any re-allocation of value, in both the directed and hybrid connections models the set of individually stable and strongly efficient networks do not intersect for some ranges of parameter values. For instance, Bala and Goyal (1999) show in the context of the directed connections model that if $N = 4$ and $\delta < c < \delta + \delta^2 - 2\delta^3$, then stars and "diamonds"[18] are the strongly efficient network structures, but are not individually stable. Similarly, in the context of the hybrid connections model if $N = 4$ and $\delta + 2\delta^2 < c < 2\delta + 2\delta^2$ then a star[19] is the strongly efficient network structure but is not individually stable. As Proposition 1 shows, reallocation of value under $\widehat{Y}$ overcomes this problem.

Let us make a couple of additional remarks about the result above. First, anonymity of $\widehat{Y}$ implies that the set of individually stable networks will be an anonymous set, so that all anonymous permutations of a given individually stable network are also individually stable. Second, in situations where $c > \delta$ (in any of the connections model) the empty network is individually stable relative to $\widehat{Y}$, even though it is not strongly efficient. The difficulty is that a single link generates negative value and so no player will want to add a link (or set of links) given that none exist. It is not clear whether an anonymous, component balanced, and outsider independent Y exists for which the set of individually stable networks exactly coincides with the set of strongly efficient networks (when $c > \delta$) in these

[17] For example, let $N = 4$, $\delta < 1/4$ and c be close to 0 in the directed connections model. Consider the network where $g_{12} = g_{13} = g_{21} = g_{31} = 1$. Adding the link 41 results in $\overline{Y}_1(g{+}41, v^d) < \overline{Y}_1(g, v^d)$ even though 4 is an outsider to g.

[18] For instance a star with 1 at the center has $g_{12} = g_{13} = g_{14} = g_{21} = g_{31} = g_{41} = 1$, while a diamond has $g_{12} = g_{13} = g_{21} = g_{23} = g_{32} = g_{41} = 1$.

[19] Here, given the two-way communication on a directed link, $g_{31} = g_{21} = g_{41}$ would constitute a star, as would $g_{13} = g_{12} = g_{14}$, etc.

connections models. Such a Y would necessarily involve careful subsidization of links, in some cases violating individual rationality constraints.

Appendix

For each i, let $H^i(g) = \{h^i | h^i \in C(g^i), i \in N(h^i), g^i \in D_i(g)\}$.

Let $n_d^{-1}(i,g) = \#\{j \; |g_{ji} = 1\}$ represent the number of individuals who maintain a link with i.

We begin by stating some Lemmas that will be useful in some of the proofs that follow.

We are most grateful to Anna Bogomolnaia who provided the proof of Lemma 1.

Lemma 1. *Let $\{a_1, \ldots, a_n\}$ be any sequence of nonnegative numbers such that $\sum_{k \in S} a_k \leq a_n$ for any $S \subset \{1, \ldots, n\}$ such that $\sum_{k \in S} k \leq n$. Then,*

$$\sum_{i=1}^{n} \frac{a_i}{i} \leq a_n. \tag{1}$$

Proof. : We construct a set of n inequalities whose sum will be the left hand side of (1). We label the i-th inequality in this set as (i').

First, for each i, let (r_i, j_i) be the unique pair such that: $n = r_i i + j_i$, r_i is some integer, and $0 \leq j_i < i$.

For each $i > \frac{n}{2}$, write inequality (i') as

$$\frac{a_i}{i} + \frac{a_{n-i}}{i} \leq \frac{a_n}{i} \tag{2}$$

(Here, we adopt the convention that $a_0 = 0$.)

Now, consider any $i \leq \frac{n}{2}$, and suppose all inequalities from (n') to $(i+1')$ have been defined. Let H_i be the sum of the coefficients of a_i in inequalities (n') to $(i+1')$. Let us now show that $h_i = \frac{1}{i} - H_i \geq 0$.

Claim. For each $i \leq \frac{n}{2}, h_i \geq 0$.

Proof of Claim. First, we prove that

$$\#\{q \text{ is an integer } | qj + i = n \text{ for } j \text{ being an integer}, i < q\} \leq \frac{n-i}{i} \tag{3}$$

Let $P = n - i$, and note that for j being an integer, $\#\{q | q > i, P = jq\} = \#\{\frac{P}{j} | \frac{P}{j} > i, P = jq\} = \#\{\frac{P}{j} | \frac{P}{i} > j, P = jq\} = \#\{j | \frac{P}{i} > j, P = jq\} = \#\{j : \frac{P}{i} > j\} \leq \frac{P}{i}$.

So, each i appears in at most $\frac{(n-i)}{i}$ inequalities. Choose $q > i$ such that $qr_q + i = n$. Then, from (3), the coefficient of a_i in (q') is $\frac{h_q}{r_q}$. Note that since $H_q = \frac{1}{q} - h_q \geq 0$, we must have $\frac{1}{q} \geq h_q$. Hence, $\frac{h_q}{r_q} \leq \frac{1}{qr_q} = \frac{1}{n-i}$. Using (4), we get $H_i \leq (\frac{n-i}{i})(\frac{1}{n-i}) = \frac{1}{i}$.

This completes the proof of the claim. □

By (1) it follows that that $r_i a_i + a_{j_i} \leq a_n$. Thus, write (i') as

$$h_i a_i + \frac{h_i}{r_i} \le \frac{h_i}{r_i} a_n \tag{4}$$

Note that by construction, the sum of the coefficients of a_i in inequalities (n') to (i') equals $H_i + h_i = \frac{1}{i}$, and that a_i does not figure in any inequality (k') for $k < i$. So, we have proved that the sum of the left hand side of the set of inequalities (i') equals the left hand side of (1).

To complete the proof of the lemma, we show that the sum of the right hand side of the inequalities (i') is an expression that must be less than or equal to a_n. The right hand side of the sum of the inequalities (i') is of the form Ca_n, where C is independent of the values $\{a_1, \ldots, a_n\}$. Let $a_i = \frac{i}{n}$ for all i. Then the inequalities (i') hold with equality. But, this establishes that $C = 1$ and completes the proof of the lemma. □

For any g, let $D(g) = \cup_i D_i(g)$.

Let

$$X(g, g') = \{i | \exists g'' \in D_i(g) \text{ s.t. } g'' \text{ is a copy of } g'\}.$$

So, $X(g, g')$ is the set of players who via a unilateral deviation can change g into a copy of g'.

Say that $S \subset N$ is a *dead end under* $g \in G$ if for any i and j in S, $i \neq j$, there exists a directed path from i to j, and for each $k \notin S$ $g_{ik} = 0$ for each $i \in S$.

For any g and $i \in N(g)$, either there is a directed path from i to a dead end S under g, or i is a member of a dead end of g. (Note that a completely disconnected player forms a dead end.)

Observation. Suppose that $\{S_1, \ldots, S_\ell\}$ are the dead ends of $g \in G$. Consider i and g' such that $g' \in D_i(g)$. If $i \notin S_k$ for any k, then S_k is still a dead end in g'. If $i \in S_k$ for some k, and i has a link to some $j \notin S_k$ under g', then $\{S_1, \ldots, S_\ell\} \setminus \{S_k\}$ are the dead ends of g'.

To see the second statement, note that there exists a path from every $l \in S_k$, $l \neq i$ to i, and so under g' all of the players in S_k have a directed path to j. If j is in a dead end, then the statement follows. Otherwise, there is a directed path from j to a dead end, and the statement follows.

Lemma 2. *Consider a player l, $g' \in G$, $g \in D_l(g')$ and corresponding $h^l \in C(g)$ such that $N(h^l) \subset X(g, g')$. If $C(g) \neq C(g')$, then there exists a directed path from any $i \in N(h^l)$ to any $j \in N(h^l)$.*

Proof of Lemma 2. Let $Z = N \setminus N(h)$. Consider $i \in N(h)$ and suppose that $g' \in D_i(g)$. Let $S_1, \ldots, S_\ell$ be the dead ends of g. If i is in a dead end S_k under g, then since $C(g) \neq C(g')$, i must be linked to some player in Z under g'. (Note that since i is in a dead end, there is a directed path from every player in S_k to i, and so i can only change the component structure of g by adding a link to a player outside of S_k.) From the observation above, it then follows that $\{S_1, \ldots, S_\ell\} \setminus \{S_k\}$ are the dead ends of g'.

Suppose the contrary of the lemma. This implies that there is a dead end of g, $S_k \subset N(h)$, and $\{i,j\} \subset N(h)$ such that $i \notin S_k$ and $j \in S_k$. From the Observation it follows that if $g^i \in D_i(g)$ is a copy of g', then g' has at least ℓ dead ends. However, if $g^j \in D_j(g)$ is a copy of g', then from the arguments above it follows that g^j has at most $\ell - 1$ dead ends. This implies that g^i and g^j could not both be copies of g'. This is a contradiction of the fact that $N(h) \subset X(g, g')$. □

Lemma 3. *Suppose g' is connected. Choose any $i,j \in N(g')$ with $i \neq j$, and take $g^i \in D_i(g')$, $g^j \in D_j(g')$, and corresponding $h^i \in C(g^i)$ and $h^j \in C(g^j)$. If $N(h^i)$ and $N(h^j)$ are intersecting but neither is a subset of the other, then $N(h^i) \not\subset X(g^i, g')$ and $N(h^j) \not\subset X(g^j, g')$.*

Proof of Lemma 3. Suppose to the contrary of the Lemma that, say, $N(h^i) \subset X(g^i, g')$.

Consider the case where $j \notin N(h^i)$. By Lemma 2, for any $k \in N(h^i)$ with $k \neq i$, there is a directed path from k to i in h^i. Since $g'_l = h^i_l = h^j_l$ for all $l \neq i,j$, this must be a directed path in h^j as well. Hence, $i \in N(h^j)$. By this reasoning, there is a directed path from every $l \in N(h^i) \setminus \{i\}$ to i in h^i, and hence in h^j. So, $N(h^i)$ is then a subset of $N(h^j)$, which contradicts the supposition that $N(h^i)$ and $N(h^j)$ are intersecting but neither is a subset of the other.

So, consider the case where $j \in N(h^i)$. We first show that $i \in N(h^j)$. Since $N(h^j)$ is not a subset of $N(h^i)$, there exists $k \in N(h^j)$ with $k \notin N(h^i)$. Since $k \notin N(h^i)$, the only paths (possibly non-directed) connecting j and k in g' must pass through i. Thus, under g' there is a path connecting i to k that does not include j. So, since $k \in N(h^j)$, it follows that $i \in N(h^j)$. Next, for any $l \in N(h^i) \setminus \{i\}$, by Lemma 2 there is a directed path from l to i in h^i. If this path passes through j, then there is a directed path from l to j in g' (not passing through i) and so $l \in N(h^j)$. If this path does not involve j, then it is also a path in h^j. Thus, $l \in N(h^j)$ for every $l \in N(h^i) \setminus \{i\}$. Since $i \in N(h^j)$, we have contradicted the fact that $N(h^i)$ is not a subset of $N(h^j)$ and so our supposition was incorrect. □

Lemma 4. *Consider i, g and g', with $g \in D_i(g')$, and $h^i \in C(g)$ such that $i \in N(h^i)$.*[20] *If $N(h^i) \subset X(g, g')$, then $N(h^i) \subset N(h')$ for some $h' \in C(g')$.*

Proof of Lemma 4. Suppose the contrary, so that there exists $j \in N(h^i)$ with $j \notin N(h')$, where $i \in N(h')$ and $h' \in C(g')$. Note, this implies that $C(g) \neq C(g')$. Either j is a dead end under g, or there is a path leading from j to a dead end under g. So, there exists a dead end S in h^i with $i \notin S$. This contradicts lemma 2. □

Proof of Theorem 4. If $v \in V$ is not component additive, then the allocation rule defined by $Y_i(g, v) = v(g)/N$ for each player i and $g \in G$ satisfies the desired properties. So, let us consider the case where v is component additive.

[20] Adopt the convention that a disconnected player is considered their own component.

Fix a v and pick some network g^* that is strongly efficient. Define Y^* relative to v as follows.[21]

Consider $g \in D(g^*)$. For any i, let $h^i \in C(g)$ be such that $i \in N(h^i)$.

If $i \in X(g, g^*)$, let $Y_i^*(g, v) = \overline{Y}_i(g, v)$ if $N(h^i) \subset X(g, g^*)$ and $Y_i^*(g, v) = 0$ otherwise.

If $i \notin X(g, g^*)$, let $Y_i^*(g, v) = \frac{v(h^i)}{\#\{j | j \in N(h^i), j \notin X(g, g^*)\}}$.

Let $r_i = \max_{g \in D_i(g^*)} Y_i^*(g, v)$.

Claim. $\sum_{j \in N(h^*)} r_j \leq v(h^*)$ for each $h^* \in C(g^*)$.

We return to prove the claim below.

Set $Y_i^*(g^*, v) = r_i + \frac{v(h^{*i}) - \sum_{j \in N(h^{*i})} r_j}{\#N(h^{*i})}$. Set Y^* on g^π which is a copy of $g \in D(g^*) \cup \{g^*\}$ according to anonymity, whenever $v(g^\pi) = v(g)$. For all remaining g, set $Y^*(g, v) = \overline{Y}(g, v)$.

By the definition of Y^*, $Y_i^*(g^*, v) \geq \max_{g \in D_i(g^*)} Y_i^*(g, v)$ for all $i \in N(g^*)$. Hence, g^* is individually stable. Also Y^* is component balanced and anonymous.

To complete the proof, we need only verify the claim.

Proof of Claim. By the definition of r_i it follows that $r_i > 0$ only if $N(h^i) \subset X(g, g^*)$. By Lemma 4, this implies that $N(h^i) \subset N(h^*)$ for some $h^* \in C(g^*)$.

For each $h^* \in C(g^*)$, let $I(h^*) = \{i \in N(h^*) | r_i > 0\}$. For each $i \in I(h^*)$, let $\bar{h}^i$ be such that $r_i = \frac{v(\bar{h}^i)}{\#N(\bar{h}^i)}$. Then, the argument in the previous paragraph establishes that each $\bar{h}^i$ is such that $N(\bar{h}^i) \subset N(h^*)$. Hence, applying lemma 3 to h^*, the set $\{\bar{h}^i \mid i \in I(h^*)\}$ can be partitioned into $\{H_1, \ldots, H_k\}$ such that
(i) Each $\bar{h}^i$ in H_1 is disjoint from every other $\bar{h}^j, j \neq i$.
(ii) For all $k = 2, \ldots K$, $H_k = \{h_1, \ldots, h_\ell\}$ is such that $N(\bar{h}_1) \subset N(\bar{h}_2) \subset \cdots N(\bar{h}_\ell)$, and elements in H_k are disjoint from elements in $H_{k'}$ if $k \neq k'$.

Define $\Delta = v(g) - \sum_{\bar{h}^i \in H^1} v(\bar{h}^i)$.

Since g^* is strongly efficient, $v(h^*) \geq v(h)$ for all h such that $N(h) \subset N(h^*)$. Now, one can use Lemma 1 and the fact that v is component additive, to deduce that there are numbers $\{\Delta_2, \ldots, \Delta_K\}$ such that

(i) $\sum_{k=2}^{K} \Delta_k \leq \Delta$.

(ii) For each $k = 2, \ldots, K$, $\Delta_k \geq \sum_{\bar{h}^j \in H_k} \frac{v(\bar{h}^j)}{\#N(\bar{h}^j)}$.

These inequalities prove the claim. □

Proof of Theorem 5.

[21] To ensure anonymity, work with equivalence classes of v with v^π for each π defined via the anonymity property.

$\widehat{Y}$ satisfies anonymity by definition. Since an outsider is necessarily unique, $\widehat{Y}$ satisfies directed component balance, and outsider independence relative to any $g \neq g^*$. These conditions relative to g^*, follow from the claim below.

Fix a component additive v that has non-decreasing returns to scale. We now show that $g_v^*(N)$ is individually stable relative to $\widehat{Y}$.

The following claim is useful.

Claim. Consider any component additive v that has non-decreasing returns to scale. If g is a component of $g_v^*(N)$, then for any g'

$$\frac{v(g)}{\#N(g)} \geq \frac{v(g')}{\#N(g')}.$$

Proof of Claim. Note that

$$\frac{v(g_v^*(N))}{\#N(g_v^*(N))} = \frac{\sum_{h \in C(g_v^*(N))} v(h)}{\sum_{h \in C(g_v^*(N))} \#N(h)}.$$

By non-decreasing returns,

$$\frac{\sum_{h \in C(g_v^*(N))} v(h)}{\sum_{h \in C(g_v^*(N))} \#N(h)} \geq \frac{v(h')}{\#N(h')}$$

for each $h' \in C(g_v^*(N))$. Thus,

$$\sum_{h \in C(g_v^*(N)), h \neq h'} v(h) \geq \frac{\sum_{h \in C(g_v^*(N)), h \neq h'} \#N(h)}{\#N(h')} v(h'). \tag{5}$$

Also, by non-decreasing returns,

$$\frac{\sum_{h \in C(g_v^*(N))} v(h)}{\sum_{h \in C(g)} \#N(h)} \geq \frac{\sum_{h \in C(g_v^*(N)), h \neq h'} v(h)}{\sum_{h \in C(g_v^*(N)), h \neq h'} \#N(h)},$$

for each $h' \in C(g_v^*(N))$. Thus,

$$v(h') \geq \frac{\#N(h')}{\sum_{h \in C(g_v^*(N)), h \neq h'} \#N(h)} \sum_{h \in C(g_v^*(N)), h \neq h'} v(h), \tag{6}$$

for each $h' \in C(g_v^*(N))$. Inequalities (5) and (6) then imply that

$$\frac{v(h')}{\#N(h')} = \frac{v(g_v^*(N))}{\#N(g_v^*(N))}.$$

for every $h' \in C(g_v^*(N))$. The desired conclusion then follows from non-decreasing returns. □

Consider $g^*(N)$ and some deviation by a player i, resulting in the network $g_{-i}^*(N), \overline{g}_i$. It then follows from the claim that $\overline{Y}_i(g^*(N)) \geq \overline{Y}_i(g_{-i}^*(N), \overline{g}_i)$ and $\overline{Y}_i(g^*(N)) \geq \overline{Y}_i((g_{-i}^*(N), \overline{g}_i) - j)$ for any j. Thus, if i is not an outsider at $g_{-i}^*(N), \overline{g}_i$, then from the definition of $\widehat{Y}$ it follows that $\widehat{Y}_i(g^*(N)) \geq$

$\widehat{Y}_i(g^*_{-i}(N), \overline{g}_i)$. If i is an outsider at $g^*_{-i}(N), \overline{g}_i$, then from the definition of $\widehat{Y}$, $\overline{Y}_i(g^*_{-i}(N), \overline{g}_i) \geq \widehat{Y}_i(g^*_{-i}(N), \overline{g}_i)$. So, $\widehat{Y}_i(g^*(N)) = \overline{Y}_i(g^*(N)) \geq \overline{Y}_i(g^*_{-i}(N), \overline{g}_i) \geq \widehat{Y}_i(g^*_{-i}(N), \overline{g}_i)$. Thus, $g^*_v(N)$ is individually stable. □

Proof of Proposition 1. The following claim is stronger than the stated property.

Claim. *Fix δ and c. If $g^*(K)$ is any strongly efficient network with a number*[22] *K players relative to the directed connections model, and g is any network with $K \geq \#N(g) > 0$, then $\frac{v^d(g^*(K))}{K} \geq \frac{v^d(g)}{\#N(g)}$. The same is true of the hybrid connections model, substituting v^h for v^d.*

Proof of the Claim. It is clear that $\frac{v^d(g^*(K))}{K} \geq 0$ ($\frac{v^h(g^*(K))}{K} \geq 0$), since the empty network is always feasible. The claim is established by showing that for each $K > 2$, $\frac{v^d(g^*(K))}{K} \geq \frac{v^d(g^*(K-1))}{K-1}$, (and $\frac{v^h(g^*(K))}{K} \geq \frac{v^h(g^*(K-1))}{K-1}$), where $g^*(K)$ denotes any selection of a strongly efficient network with K players. This implies the claim.

First, consider the directed connections model. Consider K players, with players $1, \ldots, K-1$ arranged as in $g^*(K-1)$. If $g^*(K-1)$ is empty, then the claim is clear. So suppose that $g^*(K-1)$ is not empty and consider $i \in N(g^*(K-1))$ such that $u_i(g^*(K-1)) \geq u_j(g^*(K-1))$ for all $j \in N(g^*(K-1))$, where u_i is as defined in Example 1. Thus, $u_i(g^*(K-1)) \geq \frac{v^d(g^*(K-1))}{K-1}$ Consider the network g, where $g_j = g^*_j(K-1)$ for all $j < K$, and where $g_K = g^*_i(K-1)$. It follows that $u_j(g) = u_j(g^*(K-1))$ for all $j < K$, and that $u_K(g) = u_i(g^*(K-1)) \geq \frac{v^d(g^*(K-1))}{K-1}$. Since $v^d(g) = \sum_k u_k(g)$, it follows that $v^d(g) \geq v^d(g^*(K-1)) + \frac{v^d(g^*(K-1))}{K-1}$. This implies that $v^d(g) \geq v^d(g^*(K-1)) + \frac{v^d(g^*(K-1))}{K-1}$. So $v^d(g) \geq \frac{Kv^d(g^*(K-1))}{K-1}$, and thus $\frac{v^d(g)}{K} \geq \frac{v^d(g^*(K-1))}{K-1}$.

Next, consider the hybrid connections model. Again, suppose that $K > 2$.

If $2\delta + (K-3)\delta^2 \leq c$, then a strongly efficient network for $K-1$ players, $g^*(K-1)$ is an empty network, (or when $2\delta + (K-3)\delta^2 = c$ then it is possible that $g^*(K-1)$ is nonempty, but still $v^h(g^*(K-1)) = 0$).[23] The result follows directly.

If $c \leq \delta - \delta^2$ then the efficient networks are those that have either $g^*_{ij} = 1$ or $g^*_{ji} = 1$ (but not both) for each ij (or when $c = \delta - \delta^2$ has a value equivalent to such a network). Then $v^h(g^*(K-1)) = (K-1)(K-2)(\delta - \frac{c}{2})$ and $v^h(g^*(K)) = (K)(K-1)(\delta - \frac{c}{2})$. This establishes the claim, since it implies that $\frac{v^h(g^*(K))}{K} = (K-1)(\delta - \frac{c}{2}) \geq \frac{v^h(g^*(K-1))}{K-1} = (K-2)(\delta - \frac{c}{2})$, and $c < 2\delta$ (or else $c = \delta = 0$ in which case $v^h(g) = 0$ for all g).

If $\delta - \delta^2 < c < 2\delta + (K-3)\delta^2$, a star is the strongly efficient network structure for $K-1$ players. Here, $v^h(g^*(K-1)) = (K-2)(2\delta + (K-3)\delta^2 - c)$. The value of

[22] As the connections models are anonymous we need only consider the number of players and not their identities.

[23] See Jackson and Wolinsky (1996) Proposition 1 for a proof of the characterization of efficient networks in the connections model. This translates into the hybrid connections model as noted by Bala and Goyal (1999) Proposition 5.2.

$g^*(K)$ is at least the value of a star, so that $v^h(g^*(K)) \geq (K-1)(2\delta+(K-2)\delta^2-c)$, which establishes the claim. □

References

1. Bala, V., Goyal, S. (2000) A Noncooperative Model of Network Formation. *Econometrica* 68: 1181–1229 originally circulated as Self-organization in communication networks.
2. Currarini, S., Morelli, M. (2000) Network formation with sequential demands. *Review of Economic Design* 3: 229–249
3. Dutta, B., Mutuswami, S. (1997) Stable networks. *Journal of Economic Theory* 76: 322–344
4. Dutta, B., van den Nouweland, A. Tijs, S. (1998) Link formation in cooperative situations. *International Journal of Game Theory* 27: 245–256
5. Goyal, S. (1993) Sustainable communication networks. Discussion Paper TI 93-250, Tinbergen Institute, Amsterdam-Rotterdam.
6. Jackson, M., Wolinsky, A. (1996) A strategic model of social and economic networks. *Journal of Economic Theory* 71: 44–74
7. Jackson, M., Watts, A. (2002) *The evolution of social and economic networks. Journal of Economic Theory* (forthcoming)
8. Myerson, R. (1991) *Game theory: analysis of conflict.* Harvard University Press, Cambridge, MA
9. Qin, C-Z. (1996) Endogenous formation of cooperation structures. *Journal of Economic Theory* 69: 218-226
10. Watts, A. (1997) *A dynamic model of network formation.* mimeo, Vanderbilt University

Endogenous Formation of Links Between Players and of Coalitions: An Application of the Shapley Value

Robert J. Aumann[1], **Roger B. Myerson**[2]

[1] Research by Robert J. Aumann supported by the National Science Foundation at the Institute for Mathematical Studies in the Social Sciences (Economics), Stanford University, under Grant Number IST 85-21838.
[2] Research by Roger B. Myerson supported by the National Science Foundation under grant number SES 86-05619.

1 Introduction

Consider the coalitional game v on the player set (1,2,3) defined by

$$v(S) = \begin{cases} 0 & \text{if } |S| = 1, \\ 60 & \text{if } |S| = 2, \\ 72 & \text{if } |S| = 3, \end{cases} \tag{1}$$

were $|S|$ denotes the number of players in S. Most cooperative solution concepts "predict" (or assume) that the all-player coalition $\{1, 2, 3\}$ will form and divide the payoff 72 in some appropriate way. Now suppose that P_1 (player 1) and P_2 happen to meet each other in the absence of P_3. There is little doubt that they would quickly seize the opportunity to form the coalition $\{1, 2\}$ and collect a payoff of 30 each. This would happen in spite of its inefficiency. The reason is that if P_1 and P_2 were to invite P_3 to join the negotiations, then the three players would find themselves in effectively symmetric roles, and the expected outcome would be $\{24, 24, 24\}$. P_1 and P_2 would not want to risk offering, say, 4 to P_3 (and dividing the remaining 68 among themselves), because they would realize that once P_3 is invited to participate in the negotiations, the situation turns "wide open" – anything can happen.

All this holds if P_1 and P_2 "happen" to meet. But even if they do not meet by chance, it seems fairly clear that the players in this game would seek to form pairs for the purpose of negotiation, and not negotiate the all-player framework.

The preceding example is due to Michael Maschler (see Aumann and Dreze 1974, p. 235, from which much of this discussion is cited). Maschler's example is particularly transparent because of its symmetry. Even in unsymmetric cases, though, it is clear that the framework of negotiations plays an important role in the outcome, so individual players and groups of players will seek frameworks that are advantageous to them. The phenomenon of seeking an advantageous framework for negotiating is also well known in the real world at many levels – from decision making within an organization, such as a corporation or university, to international negotiations. It is not for nothing that governments think hard and often long-about "recognizing" or not recognizing other governments; that the question of whether, when, and under what conditions to negotiate with terrorists is one of the utmost substantive importance; and that at this writing the government of Israel is tottering over the question not of *whether* to negotiate with its neighbors, but of the *framework* for such negotiations (broad-base international conference or direct negotiations).

Maschler's example has a natural economic interpretation in terms of S-shaped production functions. The first player alone can do nothing because of setup costs. Two players can produce 60 units of finished product. With the third player, decreasing returns set in, and all three together can produce only 72. The foregoing analysis indicates that the form of industrial organization in this kind of situation may be expected to be inefficient.

The simplest model for the concept "framework of negotiations" is that of a *coalition structure*, defined as a partition of the player set into disjoint coalitions. Once the coalition structure has been determined, negotiations take place only within each of the coalitions that constitute the structure; each such coalition B divides among its members the total amount $v(B)$ that it can obtain for itself. Exogenously given coalition structures were perhaps first studied in the context of the bargaining set (Aumann and Maschler 1964), and subsequently in many contexts; a general treatment may be found in Aumann and Dreze (1974). Endogenous coalition formation is implicit already in the von Neumann–Morgenstern (1944) theory of stable sets; much of the interpretive discussion in their book and in subsequent treatments of stable sets centers around which coalitions will "form". However, coalition structures do not have a formal, explicit role in the von Neumann–Morgenstern theory. Recent treatments that consider endogenous coalition structures explicitly within the context of a formal theory include Hart and Kurz (1983), Kurz (1988), and others.

Coalition structures, however, are not rich enough adequately to capture the subtleties of negotiation frameworks. For example, diplomatic relations between countries or governments need not be transitive and, therefore, can not be adequately represented by a partition; thus both, Syria and Israel have diplomatic relations with the United States but not with each other. For another example, in salary negotiations within an academic department, the chairman plays a special role; members of the department cannot usually negotiate directly with each other, though certainly their salaries are not unrelated.

To model this richer kind of framework, Myerson (1977) introduced the notion of a *cooperation structure* (or *cooperation graph*) in a coalitional game. This graph is simply defined as one whose vertices are the players. Various interpretations are possible; the one we use here is that a link between two players (an edge of the graph) exists if it is possible for these two players to carry on meaningful direct negotiations with each other. In particular, ordinary coalition structures $(B_1, B_2, \ldots, B_k)$ (with disjoint B_j) may be modeled within this framework by defining two players to be linked if and only if they belong to the same B_j. (For generalizations of this cooperation structure concept, see Myerson 1980.)

Shapley's 1953 definition of the value of a coalitional game v may be interpreted as evaluating the players' prospects when there is full and free communication among all of them – when the cooperation structure is "full," when any two players are linked. When this is not so, the prospects of the players may change dramatically. For an extreme example, a player i who is totally isolated – is linked to no other player – can expect to get nothing beyond his own worth $v(\{i\})$; in general, the more links a player has with other players, the better one may expect his prospects to be. To capture this intuition, Myerson (1977) defined an extension of the Shapley value of a coalitional game v to the case of an arbitrary cooperation structure g. In particular, if g is the complete graph on the all-player set N (any two players are directly linked), then Myerson's value coincides with Shapley's. Moreover, if the cooperation graph g corresponds to the coalition structure $(B_1, B_2, \ldots, B_k)$ in the sense indicated here, then the Myerson value of a member i of B_j is the Shapley value of i as a player of the game $v|B_j$ (v restricted to B_j).

This chapter suggests a model for the endogenous formation of cooperation structures. Given a coalitional game v, what links may be expected to form between the players? Our approach differs from that of previous writers on endogenous coalition formation in two respects: First, we work with cooperation graphs rather than coalition structures, using the Myerson value to evaluate the pros and cons of a given cooperation structure for any particular player. Second, we do not use the usual myopic, here-and-now kind of equilibrium condition. When a player considers forming a link with another one, he does not simply ask himself whether he may expect to be better off with this link than without it, given the previously existing structure. Rather, he looks ahead and asks himself, "Suppose we form this new link, will other players be motivated to form further new links that were not worthwhile for them before? Where will it all lead? Is the *end result* good or bad for me?"

In Sect. 2 we review the Myerson value and illustrate the "lookahead" reasoning by returning to the three-person game that opened the chapter. The formal definitions are set forth in Sect. 3, and the following sections are devoted to examples and counterexamples. The final section contains a general discussion of various aspects of this model, particularly of its range of application.

No new theorems are proved. Our purpose is to study the conceptual implications of the Shapley value and Myerson's extension of it to cooperation structures in examples that are chosen to reflect various applied contexts.

2 Looking Ahead with the Myerson Value

We start by reviewing the Myerson value. Let v be a coalitional game with N as player set, and g a graph whose vertices are the players. For each player i the *value* $\phi_i^g = \phi_i^g(v)$ is determined by the following axioms.

Axiom 1. If a graph g is obtained from another graph h by adding a single link, namely the one between players i and j, then i and j gain (or lose) equally by the change; that is,

$$\phi_i^g - \phi_i^h = \phi_j^g - \phi_j^h .$$

Axiom 2. If S is a connected component of g, then the sum of the values of the players in S is the worth of S; that is,

$$\sum_{i \varepsilon S} \phi_i^g(v) = v(S)$$

(Recall that a *connected component* of a graph is a maximal set of vertices of which any two may be joined by a chain of linked vertices.)

That this axiom system indeed determines a unique value was demonstrated by Myerson (1977). Moreover, he showed that if v is superadditive, then two players who form a new link never lose by it: The two sides of the equation in Axiom 1 are nonnegative. He also established[1] the following practical method for calculating the value: Given v and g, define a coalitional game v^g by

$$v^g(S) := \sum v(S_j^g) , \tag{2}$$

where the sum ranges over the connected component S_j^g of the graph $g|S$ (g restricted to S). Then

$$\phi_i^g(v) = \phi_i(v^g) , \tag{3}$$

where ϕ_i denotes the ordinary Shapley value for player i.

We illustrate with the game v defined by (1). If P_1 and P_2 happen to meet in the absence of P_3, then the graph g may be represented by

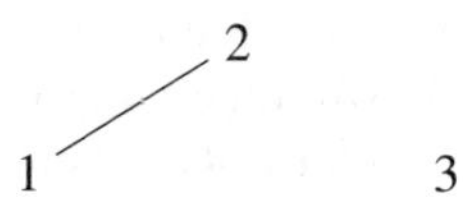

(4)

with only P_1 and P_2 connected. Then $\phi^g(v) = (30, 30, 0)$; we have already seen that in this situation it is not worthwhile for P_1 and P_2 to bring P_3 into the negotiations, because that would make things entirely symmetric, so P_1 and P_2 would get only 24 each, rather than 30. But P_2, say, might consider offering to form a link with P_3. The immediate result would be the graph

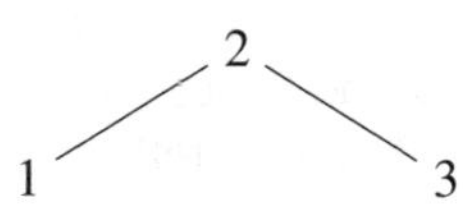

(5)

This graph is not at all symmetric; the central position of P_2 – all communication must pass through him – gives him a decided advantage. This advantage is reflected nicely in the corresponding value, (14,44,14). Thus P_2 stands to gain

[1] These statements are proved in the appendix, and they imply the assertions about the Myerson value that we made in the introduction.

from forming this link, so it would seem that he should go ahead and do so. But now in this new situation, it would be advantageous for P_1 and P_3 to form a link; this would result in the complete graph

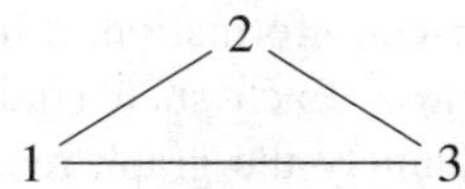

(6)

which is again symmetric and so corresponds to a payoff (24,24,24). Therefore, whereas it originally seemed worth while for P_2 to forge a new link, on closer examination it turns out to lead to a net loss of 6 (he goes from 30 to 24). Thus the original graph, with only P_1 and P_2 linked, would appear to be in some sense "stable" after all.

Can this reasoning be formalized and put into a more general context? It is true that if P_2 offers to link up with P_3, then P_1 also will, but wouldn't P_1 do this anyway? To make sense of the argument, must one assume that P_1 and P_2 explicitly agree not to bring P_3 in? If so, under what conditions would such an agreement come about?

It turns out that no such agreement is necessary to justify the argument. As we shall see in the next section, the argument makes good sense in a framework that is totally noncooperative (as far as link formation is concerned; once the links are formed, enforceable agreements may be negotiated).

3 The Formal Model

Given a coalitional game v with n players, construct an auxiliary *linking game* as follows: At the beginning of play there are no links between any players. The game consists of pairs of players being offered to form links, the offers being made one after the other according to some definite rule; the rule is common knowledge and will be called the *rule of order*. To form a link, *both* potential partners must agree; once formed, a link cannot be destroyed, and, at any time, the entire history of offers, acceptances, and rejections is known to all players (the game is of perfect information). The only other requirements for the rule of order are that it lead to a finite game, and that after the last link has been formed, each of the $n(n-1)/2$ pairs must be given a final opportunity to form an additional link (as in the bidding stage of bridge). At this point some cooperation graph g has been determined; the payoff to each player i is then defined as $\phi_i^g(v)$.

Most of the analysis in the sequel would not be affected by permitting the rule of order to have random elements as long as perfect information is maintained. It does, however, complicate the analysis, and we prefer to exclude chance moves at this stage.

Note that it does not matter in which order the two players in a pair decide whether to agree to a link; in equilibrium, either order (with perfect information) leads to the same outcome as simultaneous choice.

In practice, the initiative for an offer may come from one of the players rather than from some outside agency. Thus the rule of order might give the initiative to some particular player and have it pass from one player to another in some specified way.

Because the game is of perfect information, it has subgame perfect equilibria (Selten 1965) in pure strategies.[2] Each such equilibrium is associated with a unique cooperation graph g, namely the graph reached at the end of play. Any such g (for any choice of the order on pairs) is called a *natural structure* for v (or a *natural outcome* of the linking game).

Rather than starting from an initial position with no links, one may start from an exogenously given graph g. If all subgame perfect equilibria of the resulting game (for any choice of order) dictate that no additional links form, then g is called *stable*.

4 An Illustration

We illustrate with the game defined by (1). To find the subgame perfect equilibria, we use "backwards induction". Suppose we are already at a stage in which there are two links. Then, as we saw in Sect. 2, it is worthwhile for the two players who have not yet linked up to do so; therefore we may assume that they will. Thus one may assume that an inevitable consequence of going to two links is a graph with three links. Suppose now there is only one link in the graph, say that between P_1 and P_2 [as in (4)]. P_2 might consider offering to link up with P_3 [as in (5)], but we have just seen that this necessarily leads to the full graph [as in (6)]. Because P_2 gets less in (6) than in (4), he will not do so.

Suppose, finally, that we are in the initial position, with no links at all. At this point the way in which the pairs are ordered becomes important; [3] suppose it is 12, 23, 13. Continuing with our backwards induction, suppose the first two pairs have refused. If the pair 13 also refuses, the result will be 0 for all; if, on the other hand, they accept, it will be (30,0,30). Therefore they will certainly accept. Going back one step further, suppose that the pair 12 – the first pair in the order – has refused, and the pair 23 now has an opportunity to form a link. P_2 will certainly wish to do so, as otherwise he will be left in the cold. For P_3, though, there is no difference, because in either case he will get 30; therefore there is a subgame perfect equilibrium at which P_3 turns down this offer. Finally, going back to the first stage, similar considerations lead to the conclusion that the linking game has three natural outcomes, each consisting of a single link between two of the three players.

This argument, especially its first part, is very much in the spirit of the informal story in Sect. 2. The point is that the formal definition clarifies what

[2] Readers unfamiliar with German and the definition of subgame perfection will find the latter repeated, in English, in Selten (1975), though this reference is devoted mainly to the somewhat different concept of "trembling hand" perfection (even in games of perfect information, trembling hand perfect equilibria single out only *some* of the subgame perfect equilibria).

[3] For the analysis, not the conclusion.

lies behind the informal story and shows how this kind of argument may be used in a general situation.

5 Some Weighted Majority Games

Weighted majority games are somewhat more involved than the one considered in the previous section, and we will go into less detail. We start with a fairly typical example. Let v be the five-person weighted majority game $[4; 3, 1, 1, 1, 1]$ (4 votes are needed to win; one player has three votes, the other four have one vote each). Let us say that the coalition S has *formed* if g is the complete graph on the members of S (two players are linked if both are members of S). We start by tabulating the values for the complete graphs on various kinds of coalitions, using an obvious notation.

$$\begin{array}{ll} \{1,1,1,1,\} & \{0,\frac{1}{4},\frac{1}{4},\frac{1}{4},\frac{1}{4}\} \\ \{3,1\} & \{\frac{1}{2},\frac{1}{2},0,0,0\} \\ \{3,1,1\} & \{\frac{2}{3},\frac{1}{6},\frac{1}{6},0,0\} \\ \{3,1,1,1\} & \{\frac{3}{4},\frac{1}{12},\frac{1}{12},\frac{1}{12},0\} \\ \{3,1,1,1,1\} & \{\frac{3}{5},\frac{1}{10},\frac{1}{10},\frac{1}{10},\frac{1}{10}\} \end{array}$$

Intuitively, one may think of a parliament with one large party and four small ones. To form a government, the large party needs only one of the small ones. But it would be foolish actually to strive for such a narrow government, because then it (the large party) would be relatively weak *within* the government, the small party could topple the government at will; it would have veto power within the government. The more small parties join the government, the less the large party depends on each particular one, and so the greater the power of the large party. This continues up to the point where there are so many small parties in the government that the large party itself loses its veto power; at that point the large party's value goes *down*. Thus with only one small party, the large party's value is $\frac{1}{2}$; it goes up to $\frac{2}{3}$ with two small parties and to $\frac{3}{4}$ with three, but then drops to $\frac{3}{5}$ with four small parties, because at that point the large party itself loses its veto power within the government. Note, too, that up to a point, the fewer small parties there are in the government, the better for those that are, because there are fewer partners to share in the booty.

We proceed now to an analysis by the method of Sect. 3. It may be verified that any natural outcome of this game is necessarily the complete graph on some set of players; if a player is linked to another one indirectly, through a "chain" of other linked players, then he must also be linked to him directly. In the analysis, therefore, we may restrict attention to "complete coalitions" – coalitions within which all links have formed.

As before, we use backwards induction. Suppose a coalition of type $\{3, 1, 1, 1\}$ has formed. If any of the "small" players in the coalition links up with the single small player who is not yet in, then, as noted earlier, the all-player coalition will form. This is worthwhile both for the small player who was previously "out" and for the one who was previously "in" (the latter's payoff goes up from $\frac{1}{12}$ to $\frac{1}{10}$. Therefore such a link will indeed form, and we conclude that a coalition of type $\{3, 1, 1, 1\}$ is unstable, in that it leads to $\{3, 1, 1, 1, 1\}$.

Next, suppose that a coalition of type $\{3, 1, 1\}$ has formed. If any player in the coalition forms a link with one of the small players outside it, then this will lead to a coalition of the form $\{3, 1, 1, 1\}$, and, as we have just seen, this in turn will lead to the full coalition. This means that the large player will end up with $\frac{3}{5}$ (rather than the $\frac{2}{3}$ he gets in the framework of $\{3, 1, 1\}$) and the small players with $\frac{1}{10}$ (rather than the $\frac{1}{6}$ they get in the framework of $\{3, 1, 1\}$). Therefore none of the players in the coalition will agree to form any link with any player outside it, and we conclude that a coalition of type $\{3, 1, 1\}$ is stable.

Suppose next that a coalition of type $\{3, 1\}$ has formed. Then the large player does have an incentive to form a link with a small player outside it. For this will lead to a coalition of type $\{3, 1, 1\}$, which, as we have seen, is stable. Thus the large player can raise his payoff from the $\frac{1}{2}$ he gets in the framework of $\{3, 1\}$ to the $\frac{2}{3}$ he gets in the framework of $\{3, 1, 1\}$. This is certainly worth while for him, and therefore $\{3, 1\}$ is unstable.

Finally, suppose no links at all have as yet been formed. If the small players all turn down all offers of linking up with the large player but do link up with each other, then the result is the coalition $\{1, 1, 1, 1\}$, and each one will end up with $\frac{1}{4}$. If, on the other hand, one of them links up with the large player, then the immediate consequence is a coalition of type $\{3, 1\}$; this in turn leads to a coalition of type $\{3, 1, 1\}$, which is stable. Thus for a small player to link up with the large player in evitably leads to a payoff of $\frac{1}{6}$ for him, which is less than the $\frac{1}{4}$ he could get in the framework of $\{1, 1, 1, 1\}$. Therefore considerations of subgame perfected equilibrium lead to the conclusion that starting from the initial position (no links), all small players reject all overtures from the large player, and the final result is that the coalition $\{(1, 1, 1, 1\}$ forms.

This conclusion is typical for weighted majority games with one "large" player and several "small" players of equal weight. Indeed, we have the following general result.

Theorem A. *In a superadditive weighted majority game of the form* $[q; w, 1, \ldots, 1]$ *with* $q > w > 1$ *and without veto players, a cooperation structure is natural if and only if it is the complete graph on a minimal winning coalition consisting of "small" players only.*

The proof, which will not be given here, consists of a tedious examination of cases. There may be a more direct proof, but we have not found it.

The situation is different if there are two large players and many small ones, as in $[4; 2, 2, 1, 1, 1]$ or $[6; 3, 3, 1, 1, 1, 1, 1]$. In these cases, either the two large players get together or one large player forms a coalition with *all* the small ones (not minimal winning!). We do not have a general result that covers all games of this type.

Our final example is the game $[5; 3, 2, 2, 1, 1]$. It appears that there are two types of natural coalition structure: one associated with coalitions of type $\{2, 2, 1, 1\}$, and one with coalitions of type $\{3, 2, 1, 1\}$. Note that neither one is minimal winning.

In all these games some coalition forms; that is, the natural graphs all are "internally complete". As we will see in the next section, that is not the case in general. For simple games, however, and in particular for weighted majority games, we do not know of any counter example.

6 A Natural Structure That is not Internally Complete

Define v as the following sum of three voting games:

$$v := [2; 1, 1, 1, 0] + [3; 1, 1, 1, 0] + [5; 3, 1, 1, 2].$$

That is, v is the sum of a three-person majority game in which P_4 is a dummy, a three-person unanimity game in which P_4 is again a dummy, and a four-person voting game in which the minimal winning coalitions are $\{1, 2, 3\}$ and $\{1, 4\}$. The *sum* of these games is defined as any sum of functions, so the worth $v(S)$ of a coalition S is the number of component games in which S wins. For example, $v(\{2, 3\}) = 1$ and $v(\{1, 2, 4\}) = 2$.

The unique natural structure for this game is

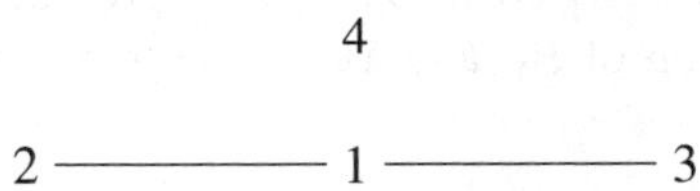

That is, P_1 links up with P_2 and P_3, but P_2 and P_3 do not link up with each other, and no player links up with P_4. The Myerson value of this game for this cooperation structure is $\left(\frac{5}{3}, \frac{5}{6}, \frac{5}{6}, 0\right)$.

The Shapley value of this game, which is also the Myerson value for the complete graph on all the players, is $\left(\frac{5}{4}, \frac{3}{4}, \frac{3}{4}, \frac{1}{4}\right)$. Notice that P_1, P_2, and P_3 all do strictly worse with the Shapley value than with the Myerson value for the natural structure described earlier. It can be verified that for any other graph either the value equals the Shapley value or there is at least one pair of players who are not linked and would do strictly better with the Shapley value. This implies inductively that if any pair of players forms a link that is not in the natural structure, then additional links will continue to form until every player is left with his Shapley value. To avoid this outcome, P_1, P_2, and P_3 will refuse to form any links beyond the two already shown.

For example, consider what happens if P_2 and P_3 add a link so that the graph becomes

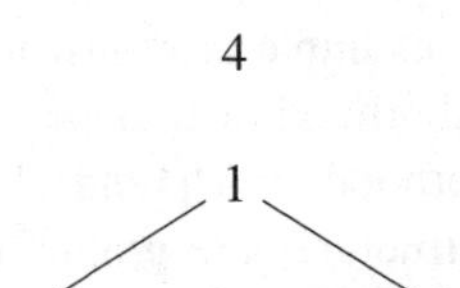

The value for this graph is $(1, 1, 1, 0)$, which is better than the Shapley value for P_2 and P_3, but worse than the Shapley value for P_1. To rebuild his claim to a higher payoff than P_2 and P_3, P_1 then has an incentive to form a link with P_4.

Intuitively, P_1 needs both P_2 and P_3 in order to collect the payoff from the unanimity game $[3; 1, 1, 1, 0]$. They, in turn, would like to keep P_4 out because he is comparatively strong in the weighted voting game $[5; 3, 1, 1, 2]$, whose Shapley value is $\left(\frac{7}{12}, \frac{1}{12}, \frac{1}{12}, \frac{3}{12}\right)$. With P_4 out, all three remaining players are on the same footing, because all three are then needed to form a winning coalition. Therefore P_1 and P_2 may each expect to get $\frac{1}{3}$ from this game, which is $\frac{1}{4}$ more than the $\frac{1}{12}$ they were getting with P_4 in. On the other hand, excluding P_4 *lowers* P_1's value by $\frac{1}{4}$ from $\frac{7}{12}$ to $\frac{1}{3}$, and P_1 will therefore want P_4 in.

This is where the three-person majority game $[2; 1, 1, 1, 0]$ enters the picture. If P_2 and P_3 refrain from linking up with each other, then P_1's centrality makes him much stronger in *this* game, and his Myerson value in it is then $\frac{2}{3}$ (rather than $\frac{1}{3}$ the Shapley value). This gain of $\frac{1}{3}$ more than makes up for the loss of $\frac{1}{4}$ suffered by P_1 in the game $[5; 3, 1, 1, 2]$, so he is willing to keep P_4 out. On the other hand, P_2 and P_3 also gain thereby, because the $\frac{1}{4}$ each gains in $[5; 3, 1, 1, 2]$ more than makes up for the $\frac{1}{6}$ each loses in the three-person majority game. Thus P_2 and P_3 are motivated to refrain from forming a link with each other, and all are motivated to refrain from forming links with P_4.

In brief, P_2 and P_3 gain by keeping P_4 isolated; but they must give P_1 the central position in the $\{1, 2, 3\}$ coalition so as to provide an incentive for him to go along with the isolation of P_4, and a credible threat if he doesn't.

7 Natural Sructures That Depend on the Rule of Order

The natural outcome of the link-forming game may well depend on the rule of order. For example, let u be the majority game $[3; 1, 1, 1, 1]$, let $w := [2; 1, 1, 0, 0]$, and let $w' := [2; 0, 0, 1, 1]$. Let $v := 24u + w + w'$. If the first offer is made to $\{1, 2\}$, then either $\{1, 2, 3\}$ or $\{1, 2, 4\}$ will form; if it is made to $\{3, 4\}$, then either $\{1, 3, 4\}$ or $\{2, 3, 4\}$ will form.

The underlying idea here is much like in the game defined by (1). The first two players to link up are willing to admit one more player in order to enjoy the proceeds of the four-person majority game u; but the resulting coalition is not willing to admit the fourth player, who would take a large share of those proceeds and himself contribute comparatively little. The difference between this game and (1) is that here each player in the first pair to get an *opportunity* to link up is positively motivated to seize that opportunity, which was not the case in (1).

The nonuniqueness in this example is robust to small changes in the game. That is, there is an open neighborhood of four-person games around v such that, for all games in this neighborhood, if P_1 and P_2 get the first opportunity to form a link then the natural structures are graphs in which P_1 , P_2, and P_3 are connected to each other but not to P_4; but if P_3 and P_4 get the first opportunity to form a link, then the natural structures are graphs in which P_2, P_3, and P_4 are connected to each other but not to P_1. (Here we use the topology that comes from identifying the set of n-person coalitional games with euclidean space of dimension $2^n - 1$.)

Each example in this chapter is also robust in the phenomenon that it is designed to illustrate. That is, for all games in a small open neighborhood of the example in Sect. 4, the natural outcomes will fail to be Pareto optimal; and for all games in a small open neighborhood of the example in Sect. 6, the natural outcomes will not be complete graphs on any coalition.

8 Discussion

The theory presented here makes no pretense to being applicable in all circumstances. The situations covered are those in which there is a preliminary period that is devoted to link formation only, during which, for one reason or another, one cannot enter into binding agreements of any kind (such as those relating to subsequent division of the payoff, or even conditional link-forming, or nonforming, deals of the kind "I won't link up with Adams if you don't link up with Brown"). After this preliminary period one carries out negotiations, but then new links can no longer be added.

An example is the formation of a coalition government in a parliamentary democracy in which no single party has a majority (Italy, Germany, Israel, France during the Fifth Republic, even England at times). The point is that a government, once formed, can only be altered at the cost of a considerable upheaval, such as new elections. On the other hand, one cannot really negotiate in a meaningful way on substantive issues before the formation of the government, because one does not know what issues will come up in the future. Perhaps one does know something about some of the issues, but even then one cannot make *binding* deals about them. Such deals, when attempted, are indeed often eventually circumvented or even broken outright; they are to a large extent window dressing, meant to mollify the voter.

An important assumption is that of perfect information. There is nothing to stop us from changing the definition by removing this assumption – something we might well wish to try – but the analysis of the examples would be quite different. Consider, for example, the game $[4; 3, 1, 1, 1, 1]$ treated at the beginning of Sect. 5. Suppose that the rule of order initially gives the initiative to the large player. That is, he may offer links to each of the small players in any order he wants; links are made public once they are forged, but rejected offers do not become known. This is a fairly reasonable description of what may happen in the negotiations formulation of governments in parliamentary democracies of the kind described here. In this situation the small players lose the advantage that was conferred on them by perfect information; formation of a coalition of type $\{3, 1, 1\}$ becomes a natural outcome. Intuitively, a small player will refuse an offer from the large player only if he feels reasonably sure that all the small players will refuse. Such a feeling is justified if it is common knowledge that all the others have already refused, and from there one may work one's way backward by induction. But the induction is broken if refused offers do not become known; and then the small players may become suspicious of each other – quite likely rightfully, as under imperfect information, mutual suspicion becomes an equilibrium outcome. We hasten to add that mutual trust – all small players refusing others from the large one – remains in equilibrium; but unlike in the case of perfect information, where everything is open and above board, it is no longer the *only* equilibrium. In short, secrecy breeds mistrust – justifiable mistrust.

Which model is the "right" one (i.e., perfect or imperfect information) is moot. Needless to say, the perfect information model is not being suggested as a universal model for all negotiations. But one may feel that the secrecy in the imperfect information model is a kind of irrelevant noise that muddies the waters and detracts from our ability properly to analyze power relationships. On the other hand, one may feel that the backwards induction in the perfect information model is an artificiality that overshadows and dominates the analysis, much as in the finitely repeated Prisoner's Dilemma, and again obscures the "true" power relationships. Moreover, the outcome predicted by the perfect information model in the game $[4; 3, 1, 1, 1, 1]$ (formation of the coalition of all small players) *is* somewhat strange and anti-intuitive. On the contrary, one would have thought that the large player has a better chance than each individual small player to get in to the ruling coalition; one might expect *him* to "form the government," so to speak.

In brief, there is no single "right" model. Each model has something going for it and something going against it. You pay your money, and you take your choice.

We end with an anecdote. This chapter is based on a correspondence that took place between the authors during the first half of 1977. That spring, there were elections in Israel, and they brought the right to power for the first time since the foundation of the state almost thirty years earlier. After the election, one of us used the perfect information model proposed here to try to predict which government would form. He was disappointed when the government that actually did form after about a month of negotiations did not conform to the prediction of the model, in that it failed to contain Professor Yigael Yadin's new "Democratic Party for Change". Imagine his delight when Yadin did after all join the government about four months later!

Appendix

We state and prove here the main result of Myerson (1977).

For any graph g, any set of players S, and any two players j and k in S, we say that j and k are connected in S by g if and only if there is a path in g that goes from j to k and stays within S. That is, j and k are connected in S

by g if there exists some sequence of players $i_1, i_2, \ldots, i_M$ such that $i_1, = j, i_M = k, (i_1, i_2, \ldots, i_M) \subseteq S$ and every pair (i_n, i_{n+1}) corresponds to a link in g. Let S/g denote the partition of S into the sets of players that are connected in S by g. That is,

$$S/g = \{\{k|j \text{ and } k \text{ are connected in } S \text{ by} g\}|j \in S\}\,.$$

With this notation, the definition of v^g from (2) becomes

$$v^g(S) = \sum_{T \in S/g} v(T) \tag{A1}$$

for any coalition S. Then the main result of Myerson (1977) is as follows

Theorem. *Given a coalitional game v, Axims 1 and 2 (as stated in Sect. 2) are satisfied for all graphs if and only if, for every graph g and every player i,*

$$\phi_i^g(v) = \phi_i(v^g). \tag{A2}$$

where ϕ_i denotes the ordinary Shapley value for player i. Furthermore, if i is superadditive and if g is a graph obtained from another graph h by adding a single link between players i and j, then $\phi_i(v^g) - \phi_i(v^h) \geq 0$, so the differences in Axiom 1 are nonnegative.

Proof. For any given graph g, Axiom 1 gives us as many equations as there are links in g, and Axiom 2 gives us as many equations as there are connected components of g. When g contains cycles, some of these equations may be redundant, but it is not hard to show that these two axioms give us at least as many independent linear equations in the values ϕ_i^g as there are players in the game. Thus, arguing by induction on the number of links in the graph (starting with the graph that has no links), one can show that there can be at most one value satisfying Axioms 1 and 2 for all graphs.

The usual formula for the Shapley (1953) value implies that

$$\phi_i(v^g) - \phi_j(v^g) = \sum_{S \subseteq N \setminus \{i,j\}} \frac{|S|!(|N| - |S| - 2)!}{(|N| - 1)!}(v^g(S \cup \{i\}) - v^g(S \cup \{j\})).$$

Notice that a coalition's worth in v^g depends only on the links in g that are between two players both of whom are in the coalition Thus, when S does not contain i or j, the worths $v^g(S \cup \{i\})$ and $v^g(S \cup \{j\})$ would not be changed if we added or deleted a link in g between players i and j. Therefore, $\phi_i(v^g) - \phi_j(v^g)$ would be unchanged if we added or deleted a linking between players i and j. Thus (A2) implies Axiom 1.

Given any coalition S and graph g, let the games u^s and w^s be defined by $u^s(T) = v^g(T \cap S)$ and $w^s(T) = v^g(T \setminus S)$ for any $T \subseteq N$. Notice that S is a carrier of u^s, and all players in S are dummies in w^s. Furthermore, if S is a connected component of g, then $v^g = u^S + w^s$. Thus, if S is a connected component of g, then

$$\sum_{i \in S} \phi_i(v^g) = \sum_{i \in S} \phi_i(u^s) = u^s(S) = v^g(S),$$

and so (A2) implies Axiom 2.

Now suppose that the graph g is obtained from the graph h by adding a single link between players i and j. If v is superadditive and $i \in S$, then $v^g(S) \geq v^h(S)$, because S/g is either the same as S/h or a coarser partition than S/h. On the other hand, if $i \notin S$, then $v^g(S) = v^h(S)$. Thus, by the montonicity of the Shapley value, $\phi_i(v^g) \geq \phi_i(v^h)$ if v is superadditive. *Q.E.D.*

References

Aumann, R.J., Dreze, J.H. (1974) Cooperative Games with Coalition Structures. *International Journal of Game Theory* 3: 217–237.

Aumann, R.J., Maschler, M. (1964) The Bargaining Set for Cooperative Games. In: Dresher, Shapley, Tucker (eds.) pp. 443–476.

Dresher, M., Shapley, L.S. Tucker, A.W. (1964) (eds.) *Advances in Game Theory*. Annals of Mathematics Studies No. 52, Princeton: Princeton University Press.

Hart, S., Kurz, M. (1983) Endogenous Formation of Coalitions. *Econometrica* 51: 1047–1064.

Kuhn, H.W., Tucker, A.W. (1953) (eds.) Contributors to the Theory of Games, Vol. 11. Annals of Mathematics Studies No. 28, Princeton: Princeton University Press.

Kurz, M. (1988) Coalitional Value. In A. Roth (ed.) *The Shapley Value*, Cambridge University Press, 155–173.

Myerson, R.B. (1977) Graphs and Cooperation in Games. *Mathematics of Operations Research* 2: 225–229.

Myerson, R.B. (1980) Conference Structures and Fair Allocation Rules. *International Journal of Game Theory* 9: 169–182.

von Neumann, J., Morgenstern, O. (1944) Theory of Games and Economic Behavior. Princeton: Princeton University Press.

Selten, R.C. (1965) Spieltheoretische Behandlung eines Oligopolmodells mit Nachfragetraegheit. *Zeitschrift fuer die gesamte Staatswissenschaft* 121: 301–324, 667–689.

Selten, R.C. (1975) Reexamination of the Perfectness Concept for Equilibrium Points in Extensive Games. *International Journal of Game Theory* 4: 22–55.

Shapley, L.S. (1953) A Value for n-Person Games. In: Kuhn, Tucker (eds.) pp. 307–317.

Link Formation in Cooperative Situations

Bhaskar Dutta[1], **Anne van den Nouweland**[2], **Stef Tijs**[3]

[1] Indian Statistical Institute, 7SJS Sansanwal Marg, New Delhi-110016, India
(e-mail dutta@isid.ernet.in)
[2] Department of Economics, 435 PLC, 1285 University of Oregon, Eugene, OR 97403-1285, USA
(e-mail Annev@oregon.uoregon.edu)
[3] Department of Econometrics and Center for Economic Research, Tilburg University, PO Box 90153, 5000 LE Tilburg, The Netherlands
(e-mail Tijs@KUB.NL)

Abstract. In this paper we study the endogenous formation of cooperation structures or communication graphs between players in a superadditive TU game. For each cooperation structure that is formed, the payoffs to the players are determined by an exogenously given solution. We model the process of cooperation structure formation as a game in strategic form. It is shown that several equilibrium refinements predict the formation of the complete cooperation structure or some structure which is payoff-equivalent to the complete structure. These results are obtained for a large class of solutions for cooperative games with cooperation structures.

Key Words: Link formation, TU game, exogenous solution

1 Introduction

The main goal of this paper is to analyse the pattern of cooperation between players in a cooperative game. A full-blown analysis would require a simultaneous determination of the coalition structure as well as the payoffs associated with each coalition structure. However, this is an extremely complicated task. Following Hart and Kurz (1983), we address ourselves to the simpler task of analysing the equilibrium pattern of cooperation between players, assuming an *exogeneously* given rule or solution which specifies the distribution of payoffs corresponding to each pattern of cooperation.

In contrast to Hart and Kurz (1983), who dealt with coalition structures, we focus attention on Myerson's (1977) *cooperation structures*[1], rather than coalition

The authors are grateful to the anonymous referee and Associate Editor for helpful suggestions and comments.

[1] See van den Nouweland (1993) for a survey of recent research on games with cooperation structures.

structures. A cooperation structure is a graph whose vertices are identified with the players. A link between two players means that these players can carry on meaningful direct negotiations with each other. Notice that a coalition structure is a special kind of cooperation structure where two members i and j are linked if and only if they are in the same coalition.[2] Following Aumann and Myerson (1988), we model situations in which the eventual distribution of payoffs is determined in *two* distinct stages. The first period is devoted to link formation only. During this period, the players cannot enter into binding agreements of any kind, either on the nature of the link formation, or on the subsequent division of payoffs. In the second period, no new links can be formed, but players negotiate over the division of the payoff, *given* the cooperation structure which has formed in the first stage.

We assume that agents' decisions on whether or not to form a link with other agents can be represented as a game in strategic form.[3] In the link formation game, each player announces a set of players with whom he or she wants to form a link. A link is formed between i and j if both players want the link. Given the announcements of the n players, this specification gives the cooperation structure. Suppose there is a rule or solution which determines a distribution of payoffs for each cooperation structure. This, then, also gives the payoff function of the strategic form game. Since this is a well-defined strategic form game, we can use any noncooperative equilibrium concept to analyse the game.

Suppose now that the rule which determines payoffs for each cooperation structure has the property that no agent wants to unilaterally *break* a link with any player. Since no player wants to break a link, and it needs the consent of *two* players to form an additional link, any cooperation structure can be sustained as a Nash equilibrium. We, therefore, use refinements of the Nash equilibrium concept. In particular, we employ undominated Nash equilibrium, coalition-proof Nash equilibrium, and strong Nash equilibrium. Our principal conclusion is that for a wide class of solutions, the first two equilibrium refinements lead to the formation of the full cooperation structure or cooperation structures which are *payoff-equivalent* to this structure. This is also true when the equilibrium concept is that of strong Nash equilibrium, *provided* a strong Nash equilibrium exists. However, we show that there are games in which reasonable solution concepts fail to guarantee the existence of a strong Nash equilibrium.

The plan of this paper is as follows. In Sect. 2 we provide some basic definitions, including those of cooperation structures and solutions for games with cooperation structures. Also, we introduce the properties Component Efficiency, Weak Link Symmetry, and Improvement Property, which we believe to be 'reasonable' properties on such solutions, and we derive some implications of these properties. Section 3 contains the model of link formation studied in this paper

---

[2] Aumann and Myerson (1988) give examples of negotiation situations which can be modelled by cooperation structures, but not by coalition structures.

[3] This game was originally introduced by Myerson (1991) (p. 448). See also Hart and Kurz (1983), who discuss a similar strategic-form game in the context of the endogenous formation of coalition structures.

and the definitions of the equilibrium concepts used to analyze the model. Endogenous cooperation structures corresponding to undominated Nash equilibrium and coalition-proof Nash equilibrium are determined in Sect. 4. We conclude in Sect. 5.

2 Cooperation Structures and Solutions

Let (N, v) be a *TU* coalitional game, where $N = \{1, 2, \ldots, n\}$ denotes the finite player set and v is a real-valued function on the family 2^N of all subsets of N with $v(\emptyset) = 0$. *Throughout this paper, we will assume that v is superadditive*[4].

A *cooperation structure* is a graph $g = (N, L)$ where N is the set of vertices, and L is the *edge set*. An edge will also be called a *link*, and denoted by l, l', etc. For any $S \subseteq N$, we say that players $i, j \in S$ are *connected* in S if there exists a path from i to j that uses only vertices in S. The relation 'connected in N' is an equivalence relation on N. The equivalence classes of this relation are the *connected components* of the graph g.

We follow Aumann and Myerson (1988) in interpreting a link between two players as meaning that these players can carry on meaningful direct negotiations with each other. The negotiation to form links takes place in a preliminary period when "for one reason or another, one cannot enter into binding agreements of any kind (such as those relating to subsequent divisions of the payoff ...)".[5]

A *solution* is a mapping γ which assigns an element in $\mathbb{R}^n$ to each *TU* game (N, v) and cooperation structure $g = (N, L)$. Since there will be no ambiguity about the underlying game (N, v), we will simply write $\gamma(L)$, $\gamma(L')$, etc., instead of writing $\gamma(N, v, L)$, $\gamma(N, v, L')$, etc.

A solution can for example be generated for any graph g by applying the usual or familiar cooperative solution concepts to the 'graph-restricted game' (N, v^g). This game is defined as follows. Let $S \backslash g$ denote the partition of S into subsets of players that are connected in S by g. That is,

$$S \backslash g = \{\{i \mid j \text{ and } i \text{ are connected in } S \text{ by } g\} \mid j \in S\} \tag{1}$$

Now, define $v^g : 2^N \to \mathbb{R}$ by

$$v^g(S) = \sum_{T \in S \backslash g} v(T) \tag{2}$$

For instance, for any $g = (N, L)$, the Shapley value of the associated game (N, v^g) is a solution for (N, v, L), and has come to be called the *Myerson value*.[6] Similarly, weighted Myerson values of (N, v, L) are the weighted Shapley values[7] of (N, v^g).

[4] v is superadditive if for all $S, T \in 2^N$ with $S \cap T = \emptyset$, $v(S) + v(T) \leq v(S \cup T)$.

[5] Aumann and Myerson (1988), page 187. See also Myerson (1977).

[6] Myerson (1977) contains a characterization of the Myerson value. See also Jackson and Wolinsky (1994).

[7] See Kalai and Samet (1988).

A class of solutions which will play a prominent role in this paper is the class satisfying the following 'reasonable' properties on a solution γ below.

Component efficiency (CE): For all cooperation structures (N, L) and all $S \in 2^N$, if S is a connected component of (N, L), then $\sum_{i \in S} \gamma_i(L) = v(S)$.

Weak link symmetry (WLS): For all $i, j \in N$, and all cooperation structures (N, L), if $\gamma_i(L \cup \{i,j\}) > \gamma_i(L)$, then $\gamma_j(L \cup \{i,j\}) > \gamma_j(L)$.

Improvement property (IP): For all $i, j \in N$ and all cooperation structures (N, L), if for some $k \in N \backslash \{i,j\}$, $\gamma_k(L \cup \{i,j\}) > \gamma_k(L)$, then $\gamma_i(L \cup \{i,j\}) > \gamma_i(L)$ or $\gamma_j(L \cup \{i,j\}) > \gamma_j(L)$.

These properties all have very simple interpretations. Component efficiency, which was originally used by Myerson (1977), states that the players in a connected component S split the value $v(S)$ amongst themselves. The second property is a very weak form of symmetry. It says that if a new link between players i and j makes i *strictly* better off, then it must also strictly improve the payoff of player j. Finally, the improvement property states that if a new link between players i and j strictly improves the payoff of any other player k, then the payoff of *either* i *or* j must also strictly improve.

The class of *weighted Myerson values* satisfies all the properties listed above. There are also others. For instance, if (N, v) is a *convex* game, then the *egalitarian solution* of Dutta and Ray (1989) corresponding to the associated game (N, v^g) also satisfies these properties.

The three properties together imply an interesting fourth property. This is the content of the next lemma.

Lemma 1. *Let γ be any solution satisfying CE, WLS and IP. Then, for all $i, j \in N$, and all cooperation structures (N, L),*

$$\gamma_i(L \cup \{i,j\}) \geq \gamma_i(L). \tag{3}$$

Proof. Suppose for some $i, j \in N$ and (N, L), $\gamma_i(L) > \gamma_i(L \cup \{i,j\})$. Then, by WLS, we must also have $\gamma_j(L) \geq \gamma_j(L \cup \{i,j\})$. But then, since v is superadditive, and γ satisfies CE, there must exist $k \notin \{i,j\}$ such that $\gamma_k(L) < \gamma_k(L \cup \{i,j\})$. This shows that γ violates IP since $\gamma_i(L) > \gamma_i(L \cup \{i,j\})$ and $\gamma_j(L) \geq \gamma_j(L \cup \{i,j\})$. □

Remark 1. We will denote the property incorporated in equation (3) by *Link Monotonicity*. Note that Link Monotonicity is an appealing property in its own right. It says that a player i should not be worse-off as a result of forming a new link with some player j.[8]

Remark 2. It is easy to construct examples to show that the Component Efficiency, Weak Link Symmetry, and Improvement properties are independent.

[8] Note that the game is superadditive.

Another consequence of these three properties is derived in the next lemma. We show that if the formation of a link $\{i,j\}$ affects the payoff of some other player k, then it must also affect the payoffs of both players that formed the new link. This property will be used later on in the paper.

Lemma 2. *Let γ satisfy CE, WLS and IP. Then, for all $i,j \in N$, and all cooperation structures (N,L), if for some $k \in N \setminus \{i,j\}$, $\gamma_k(L \cup \{i,j\}) \neq \gamma_k(L)$, then $\gamma_i(L \cup \{i,j\}) > \gamma_i(L)$ and $\gamma_j(L \cup \{i,j\}) > \gamma_j(L)$.*

Proof. Suppose for some $i,j \in N$ and $k \in N \setminus \{i,j\}$, $\gamma_k(L \cup \{i,j\}) \neq \gamma_k(L)$. If $\gamma_k(L \cup \{i,j\}) > \gamma_k(L)$, then from WLS and IP we must have $\gamma_i(L \cup \{i,j\}) > \gamma_i(L)$ and $\gamma_j(L \cup \{i,j\}) > \gamma_j(L)$.

Suppose $\gamma_k(L \cup \{i,j\}) < \gamma_k(L)$. From WLS, either $\gamma_i(L \cup \{i,j\}) > \gamma_i(L)$ and $\gamma_j(L \cup \{i,j\}) > \gamma_j(L)$, or $\gamma_i(L \cup \{i,j\}) \leq \gamma_i(L)$ and $\gamma_j(L \cup \{i,j\}) \leq \gamma_j(L)$. But, in the latter case, CE and superadditivity imply that there exists a $l \notin \{i,j\}$ such that $\gamma_l(L \cup \{i,j\}) > \gamma_l(L)$. This would violate IP, so it must hold that $\gamma_i(L \cup \{i,j\}) > \gamma_i(L)$ and $\gamma_j(L \cup \{i,j\}) > \gamma_j(L)$. This establishes the lemma. □

While the three properties are all appealing and are satisfied by a large class of solutions, there are other solutions outside this class that seem to be appealing. One such solution is defined below.

For any i and L, let $L_i = \{\{i,j\} \mid j \in N, \{i,j\} \in L\}$, the set of links that are adjacent to i, and $l_i = |L_i|$. Let $S_i(L)$ denote the connected component of L containing i. Then, the *Proportional Links Solution*, denoted γ^P, is given by

$$\gamma_i^P(L) = \begin{cases} \frac{l_i}{\sum_{j \in S_i(L)} l_j} v(S_i(L)) & \text{if } |S_i(L)| \geq 2 \\ v(\{i\}) & \text{if } S_i(L) = \{i\} \end{cases} \tag{4}$$

for all L and all $i \in N$. The solution γ^P captures the notion that the more links a player has with other players, the better are his *relative* prospects in the subsequent negotiations over the division of the payoff. Notice that this makes sense only when the players are equally 'powerful' in the game (N,v). Otherwise, a *big* player may get more than *small* players even if he has fewer links. We leave it to the reader to check that γ^P satisfies CE and IP, but not WLS.

3 Modelling Negotiation Processes

As we have remarked before, we model the process of link formation as a game in strategic form.[9] The specific strategic form game that we will construct was first defined by Myerson (1991), and has subsequently been used by Qin (1996). This model is described below.

Let γ be a solution. Then, the *linking game* $\Gamma(\gamma)$ associated with γ is given by the $(n+2)$-tuple $(N; S_1, \ldots, S_n; f^\gamma)$ where for each $i \in N$, S_i is player i's strategy

[9] In contrast, Aumann and Myerson (1988) use an extensive form approach. See Dutta et al. (1995) for a discussion of the two approaches.

set with $S_i = 2^{N\setminus\{i\}}$, and the payoff function is the mapping $f^\gamma : \Pi_{i\in N} S_i \to \mathbb{R}^n$ given by

$$f_i^\gamma(s) = \gamma_i(L(s)) \tag{5}$$

for all $s \in \Pi_{i\in N} S_i$, with

$$L(s) = \{\{i,j\} \mid j \in s_i,\ i \in s_j\} \tag{6}$$

The interpretation of (5) and (6) is straightforward. A typical strategy of player i in $\Gamma(\gamma)$ consists of the set of players with whom i wants to form a link. Then (6) states that a link between i and j is formed if and only if they both want to form this link. Thus, each strategy vector s gives rise to a unique cooperation structure $L(s)$. Finally, the payoff to player i associated with s is simply $\gamma_i(L(s))$[10], the payoff that γ associates with the cooperation structure $L(s)$.

We will let $\bar{s} = (\bar{s}_1, \ldots, \bar{s}_n)$ denote the strategy vector such that $\bar{s}_i = N\setminus\{i\}$ for all $i \in N$, while $\bar{L} = \{\{i,j\} \mid i \in N,\ j \in N\} = L(\bar{s})$ denotes the complete edge set on N. A cooperation structure L is *essentially complete for* γ if $\gamma(L) = \gamma(\bar{L})$. Hence, if L is essentially complete for γ, but $L \neq \bar{L}$, then the links which are not formed in L are inessential in the sense that their absence does not change *the payoff vector* from that corresponding to $\bar{L}$. Notice that the property of "essentially complete" is specific to the solution γ - a cooperation structure L may be essentially complete for γ, but not for γ'.

We now define some equilibrium concepts for any $\Gamma(\gamma)$ that will be used in section 4 below.

The first equilibrium concept that we consider is the undominated Nash equilibrium. For any $i \in N, s_i$ *dominates* s_i' iff for all $s_{-i} \in S_{-i}$, $f_i^\gamma(s_i, s_{-i}) \geq f_i^\gamma(s_i', s_{-i})$ with the inequality being strict for *some* s_{-i}. Let $S_i^u(\gamma)$ be the set of undominated strategies for i in $\Gamma(\gamma)$, and $S^u(\gamma) = \Pi_{i\in N} S_i^u(\gamma)$. A strategy tuple s is an *undominated Nash equilibrium* of $\Gamma(\gamma)$ if s is a Nash equilibrium and, moreover, $s \in S^u(\gamma)$.

The second equilibrium concept that will be discussed is the Coalition-Proof Nash Equilibrium. In order to define the concept of Coalition-Proof Nash Equilibrium of $\Gamma(\gamma)$, we need some more notation. For any $T \subset N$ and $s_T^* \in S_T := \Pi_{i\in T} S_i$, let $\Gamma(\gamma, s_{N\setminus T}^*)$ denote the game induced on subgroup T by the actions $s_{N\setminus T}^*$. So,

$$\Gamma(\gamma, s_{N\setminus T}^*) = \langle T, \{S_i\}_{i\in T}, \tilde{f}^\gamma \rangle$$

where for all $j \in T, \tilde{f}_j^\gamma : \Pi_{i\in T} S_i \to \mathbb{R}$ is given by $\tilde{f}_j^\gamma((s_i)_{i\in T}) = f_j^\gamma((s_i)_{i\in T}, s_{N\setminus T}^*)$ for all $(s_i)_{i\in T} \in S_T$.

The Coalition-Proof Nash Equilibrium is defined inductively as follows:
In a single player game, $s^* \in S$ is a *Coalition-Proof Nash Equilibrium* (CPNE) of $\Gamma(\gamma)$ iff s_i^* maximizes $f_i^\gamma(s)$ over S. Now, let $\Gamma(\gamma)$ be a game with n players, where $n > 1$, and assume that Coalition-Proof Nash Equilibria have been defined

[10] We again remind the reader that we have suppressed the underlying *TU* game (N, v) in order to simplify the notation.

for games with less than n players. Then, a strategy tuple $s^* \in S_N := \Pi_{i\in N} S_i$ is called *self-enforcing* if for all $T \subsetneqq N$, s_T^* is a CPNE in the game $\Gamma(\gamma, s_{N\setminus T}^*)$. A strategy tuple $s^* \in S_N$ is a CPNE of $\Gamma(\gamma)$ if it is self-enforcing and, moreover, there does not exist another self-enforcing strategy vector $s \in S_N$ such that $f_i^\gamma(s) > f_i^\gamma(s^*)$ for all $i \in N$.

Let CPNE (γ) denote the set of CPNE of $\Gamma(\gamma)$.[11] Notice that the notion of CPNE incorporates a kind of 'farsighted' thought process on the part of players since a coalition when contemplating a deviation takes into consideration the possibility of further deviations by subcoalitions.[12]

The third equilibrium concept that we consider is that of strong Nash equilibrium. A strategy tuple s is a *Strong Nash Equilibrium* (SNE) of $\Gamma(\gamma)$ if there is no coalition $T \subseteq N$ and strategies $s'_T \in S_T$ such that

$$f_i^\gamma((s'_T, s_{N\setminus T}) > f_i^\gamma(s) \text{ for all } i \in T.$$

We denote the set of SNE of $\Gamma(\gamma)$ by SNE (γ).

4 Equilibrium Cooperation Structures

In this section, we characterize the sets of equilibrium cooperation structures under the equilibrium concepts defined in the previous section.

We consider refinements of Nash equilibrium because Nash equilibrium itself does not enable us to distinguish between different cooperation structures. If a solution satisfies the properties listed in section 2, then no player wants to unilaterally break a link because of link monotonicity. Further, it needs the consent of *two* players to form a link. Because of these two facts, any cooperation structure can be sustained in a Nash equilibrium.

Proposition 1. *Let γ be a solution that satisfies CE, WLS, and IP. Then any cooperation structure can be sustained in a Nash equilibrium.*

Proof. Let $g = (N, L)$ be a cooperation structure. Define for each player $i \in N$ the strategy $s_i = \{j \in N\setminus\{i\} \mid \{i,j\} \in L\}$. That is, each player announces that he wants to form links with exactly those players to which he is directly connected in g. It is easily seen that $s = (s_i)_{i\in N}$ is a Nash equilibrium of $\Gamma(\gamma)$, because for all $i, j \in N$ it holds that $j \in s_i$ if and only if $i \in s_j$. Further, $L(s) = L$. □

Our principal objective is to show that the equilibrium concepts of undominated Nash equilibrium and coalition-proof Nash equilibrium both lead to essentially complete cooperation structures for solutions satisfying the properties that are listed in Sect. 2.

---

[11] See Bernheim, Peleg and Whinston (1987) for discussion of Coalition-Proof Nash Equilibrium.

[12] We mention this because Aumann and Myerson (1988) state that they do not use the 'usual, myopic, here-and-now kind of equilibrium condition', but a 'look ahead' one. Of course, farsightedness can be modelled in many different ways.

Theorem 1. *Let γ be a solution that satisfies CE, WLS and IP. Then, $\bar{s}$ is an undominated Nash equilibrium of $\Gamma(\gamma)$. Moreover, if s is an undominated Nash equilibrium of $\Gamma(\gamma)$, then $L(s)$ is essentially complete for γ.*

Proof. First, we show that $\bar{s}_i$ is undominated for all $i \in N$.

So, choose $i \in N, s_i \in S_i$ and $s_{-i} \in S_{-i}$ arbitrarily. Let $L = L(\bar{s}_i, s_{-i})$ and $L' = L(s_i, s_{-i})$. Note that, since $s_i \subseteq \bar{s}_i$, $L' \subseteq L$. Also, if $l \in L \backslash L'$, then $i \in l$. So, from repeated application of link monotonicity (see lemma 1),

$$f_i^{\gamma}(\bar{s}_i, s_{-i}) = \gamma_i(L) \geq \gamma_i(L') = f_i^{\gamma}(s_i, s_{-i}) \tag{7}$$

Since s_i and s_{-i} were chosen arbitrarily, this shows that $\bar{s}_i \in S_i^u(\gamma)$. Further, putting $s_{-i} = \bar{s}_{-i}$ in (7), we also get that $\bar{s}$ is a Nash equilibrium of $\Gamma(\gamma)$. So, we may conclude that $\bar{s} \in S^u(\gamma)$.

Now, we show that $L(s)$ is essentially complete for an undominated Nash equilibrium s. Choose $s \neq \bar{s}$ arbitrarily. Without loss of generality, let $\{i \in N \mid s_i \neq \bar{s}_i\} = \{1, 2, ..., K\}$. Construct a sequence $\{s^0, s^1, \ldots, s^K\}$ of strategy tuples as follows.

(i) $s^0 = s$

(ii) $s_k^k = \bar{s}_k$ for all $k = 1, 2, ..., K$.

(iii) $s_j^k = s_j^{k-1}$ for all $k = 1, 2, ..., K$, and all $j \neq k$.

Clearly, $s^K = \bar{s}$. Consider any s^{k-1} and s^k. By construction, $s_j^{k-1} = s_j^k$ for all $j \neq k$, while $s_k^k = \bar{s}_k$ and $s_k^{k-1} = s_k$. So, using link monotonicity, we have

$$f_k^{\gamma}(s^k) = \gamma_k(L(s^k)) \geq \gamma_k(L(s^{k-1})) = f_k^{\gamma}(s^{k-1}) \tag{8}$$

Suppose (8) holds with strict inequality. Then, we have demonstrated the existence of strategies s_{-k} such that

$$f_k^{\gamma}(\bar{s}_k, s_{-k}) > f_k^{\gamma}(s_k, s_{-k}) \tag{9}$$

But, (7) and (9) together show that $\bar{s}_k$ *dominates* s_k. So, if $s \in S^u(\gamma)$, then (8) must hold with equality. Then it follows from lemma 2 that the payoffs to all players remain unchanged when going from s^{k-1} to s^k, so

$$\gamma(L(s^k)) = \gamma(L(s^{k-1})) \tag{10}$$

Since this argument can be repeated for $k = 1, 2, ..., K$, we get $\gamma(L(s^0)) = \gamma(L(s^1)) = \cdots = \gamma(L(\bar{s}))$. Hence, if $s \in S^u(\gamma)$, then $L(s)$ is essentially complete. □

The following example shows that link monotonicity alone does not guarantee the validity of the statements in theorem 1. It is easily seen from the proof of the theorem that $\bar{s}$ is an undominated Nash equilibrium of $\Gamma(\gamma)$ if γ is link monotonic, so the first part of the theorem only requires link monotonicity of γ. However, the second part of the theorem might be violated even if γ is link monotonic.

Example 1. Consider the TU game v on the player set $\{1, 2, 3\}$ defined by

$$v(S) = \begin{cases} 5 \text{ if } S = N \\ 1 \text{ if } |S| = 2 \\ 0 \text{ otherwise} \end{cases}$$

and the component efficient solution γ defined for this game by $\gamma(\{1,2\}) = \gamma(\{2,3\}) = (0,1,0)$, $\gamma(\{1,3\}) = (0,0,1)$, $\gamma(\{1,2\},\{1,3\}) = (2,2,1)$, $\gamma(\{1,2\},\{2,3\}) = (1,4,0)$, and $\gamma(\{1,3\},\{2,3\}) = \gamma(\bar{L}) = (1,3,1)$. It is not hard to see that γ satisfies IP and link monotonicity but fails to satisfy WLS. Further, strategy $s_3 = \{1\}$ is an undominated strategy for player 3, and strategies $s_1 = \{2,3\}$ and $s_2 = \{1,3\}$ are undominated strategies for players 1 and 2, respectively. Hence, $s = (s_1, s_2, s_3)$ is an undominated Nash equilibrium of the game $\Gamma(\gamma)$. Note that $L(s)$ is not essentially complete for γ.

In the following theorem we consider Coalition-Proof Nash Equilibria.

Theorem 2. *Let γ be a solution satisfying CE, WLS and IP. Then $\bar{s} \in$ CPNE (γ). Moreover, if $s \in$ CPNE (γ), then $L(s)$ is essentially complete for γ.*

Proof. In fact, we will prove a slightly generalized version of the theorem and show that for each coalition $T \subseteq N$ and all $s_{N\setminus T} \in S_{N\setminus T}$ it holds that $\bar{s}_T \in$ CPNE $(\gamma, s_{N\setminus T})$ and that for all $s_T^* \in$ CPNE $(\gamma, s_{N\setminus T})$ it holds that $f^\gamma(s_T^*, s_{N\setminus T}) = f^\gamma(\bar{s}_T, s_{N\setminus T})$. We will follow the definition of Coalition-Proof Nash Equilibrium and proceed by induction on the number of elements of T. Throughout the following, we will assume $s_{N\setminus T} \in S_{N\setminus T}$ to be arbitrary.

Let $T = \{i\}$. Then by repeated application of Link Monotonicity we know that $f_i^\gamma(\bar{s}_i, s_{N\setminus\{i\}}) \geq f_i^\gamma(s_i, s_{N\setminus\{i\}})$ for all $s_i \in S_i$. From this it readily follows that $\bar{s}_i \in$ CPNE $(\gamma, s_{N\setminus\{i\}})$. Now, suppose $s_i^* \in$ CPNE $(\gamma, s_{N\setminus\{i\}})$. Then, since $f_i^\gamma(s_i^*, s_{N\setminus\{i\}}) \leq f_i^\gamma(\bar{s}_i, s_{N\setminus\{i\}})$, it follows that $f_i^\gamma(s_i^*, s_{N\setminus\{i\}}) = f_i^\gamma(\bar{s}_i, s_{N\setminus\{i\}})$ must hold. Now we use lemma 2 and see that $f^\gamma(s_i^*, s_{N\setminus\{i\}}) = f^\gamma(\bar{s}_i, s_{N\setminus\{i\}})$.

Now, let $|T| > 1$ and assume that we already proved that for all R with $|R| < |T|$ and all $s_{N\setminus R} \in S_{N\setminus R}$ it holds that $\bar{s}_R \in$ CPNE $(\gamma, s_{N\setminus R})$ and that for all $s_R^* \in$ CPNE $(\gamma, s_{N\setminus R})$ it holds that $f^\gamma(s_R^*, s_{N\setminus R}) = f^\gamma(\bar{s}_R, s_{N\setminus R})$. Then it readily follows from the first part of the induction hypothesis that $\bar{s}_R \in$ CPNE $(\gamma, \bar{s}_{T\setminus R}, s_{N\setminus T})$ for all $R \subsetneq T$. This shows that $\bar{s}_T$ is self-enforcing.

Suppose $s_T^* \in S_T$ is also self-enforcing, i.e. $s_R^* \in$ CPNE $(\gamma, s_{T\setminus R}^*, s_{N\setminus T})$ for all $R \subsetneq T$. We will start by showing that $f_i^\gamma(\bar{s}_T, s_{N\setminus T}) \geq f_i^\gamma(s_T^*, s_{N\setminus T})$ for all $i \in T$, which proves that $\bar{s}_T \in$ CPNE $(\gamma, s_{N\setminus T})$. So, let $i \in T$ be fixed for the moment. Then repeated application of Link Monotonicity implies that $f_i^\gamma(\bar{s}_T, s_{N\setminus T}) \geq f_i^\gamma(s_i^*, \bar{s}_{T\setminus i}, s_{N\setminus T})$. Further, since $s_{T\setminus\{i\}}^* \in$ CPNE $(\gamma, s_i^*, s_{N\setminus T})$, it follows from the second part of the induction hypothesis that $f^\gamma(s_i^*, \bar{s}_{T\setminus\{i\}}, s_{N\setminus T}) = f^\gamma(s_T^*, s_{N\setminus T})$. Combining the two last (in)equalities we find that $f_i^\gamma(\bar{s}_T, s_{N\setminus T}) \geq f_i^\gamma(s_T^*, s_{N\setminus T})$.

Note that we will have completed the proof of the theorem if we show that, in addition to $f_i^\gamma(\bar{s}_T, s_{N\setminus T}) \geq f_i^\gamma(s_T^*, s_{N\setminus T})$ for all $i \in T$, it holds that either $f_i^\gamma(\bar{s}_T, s_{N\setminus T}) > f_i^\gamma(s_T^*, s_{N\setminus T})$ for all $i \in T$ (and, consequently, $s_T^* \notin$ CPNE $(\gamma, s_{N\setminus T})$) or $f_i^\gamma(\bar{s}_T, s_{N\setminus T}) = f_i^\gamma(s_T^*, s_{N\setminus T})$ for all $i \in T$ (and $s_T^* \in$ CPNE $(\gamma, s_{N\setminus T})$). So, suppose $i \in T$ is such that $f_i^\gamma(\bar{s}_T, s_{N\setminus T}) > f_i^\gamma(s_T^*, s_{N\setminus T})$. Because s_T^* is self-enforcing, we know that $s_{T\setminus\{j\}}^* \in$ CPNE $(\gamma, s_j^*, s_{N\setminus T})$ for each $j \in T$, and it follows from the induction hypothesis that $f^{\gamma}(s_T^*, s_{N\setminus T}) = f^{\gamma}(s_j^*, \bar{s}_{T\setminus j}, s_{N\setminus T})$ for each $j \in T$. Let $j \in T\setminus\{i\}$ be fixed. Then we have just shown that $f_i^{\gamma}(\bar{s}_T, s_{N\setminus T}) > f_i^{\gamma}(s_T^*, s_{N\setminus T}) = f_i^{\gamma}(s_j^*, \bar{s}_{T\setminus j}, s_{N\setminus T})$. We know by repeated application of Link Monotonicity that $f_j^{\gamma}(\bar{s}_T, s_{N\setminus T}) \geq f_j^{\gamma}(s_j^*, \bar{s}_{T\setminus j}, s_{N\setminus T})$. However, if this should hold with equality, $f_j^{\gamma}(\bar{s}_T, s_{N\setminus T}) = f_j^{\gamma}(s_j^*, \bar{s}_{T\setminus j}, s_{N\setminus T})$, then repeated application of lemma 2 would imply that $f^{\gamma}(\bar{s}_T, s_{N\setminus T}) = f^{\gamma}(s_j^*, \bar{s}_{T\setminus j}, s_{N\setminus T})$, which contradicts that $f_i^{\gamma}(\bar{s}_T, s_{N\setminus T}) > f_i^{\gamma}(s_j^*, \bar{s}_{T\setminus j}, s_{N\setminus T})$. Hence, we may conclude that $f_j^{\gamma}(\bar{s}_T, s_{N\setminus T}) > f_j^{\gamma}(s_j^*, \bar{s}_{T\setminus j}, s_{N\setminus T})$. Since $f_j^{\gamma}(s_j^*, \bar{s}_{T\setminus j}, s_{N\setminus T}) = f_j^{\gamma}(s_T^*, s_{N\setminus T})$, we now know that $f_j^{\gamma}(\bar{s}_T, s_{N\setminus T}) > f_j^{\gamma}(s_T^*, s_{N\setminus T})$.

This shows that either $f_i^{\gamma}(\bar{s}_T, s_{N\setminus T}) > f_i^{\gamma}(s_T^*, s_{N\setminus T})$ for all $i \in T$ or $f_i^{\gamma}(\bar{s}_T, s_{N\setminus T}) = f_i^{\gamma}(s_T^*, s_{N\setminus T})$ for all $i \in T$. □

Remark 3. We have an example of a solution satisfying CE, WLS and IP, for which CPNE $(\gamma) \neq \{s \mid L(s)$ is essentially complete$\}$. In other words, there may be a strategy tuple s which is not in CPNE (γ), though $L(s)$ is essentially complete.

We defined the Proportional Links Solution γ^P in section 2, and pointed out that it does not satisfy WLS. It also turns out that the conclusions of theorem 2 are no longer valid in the linking game $\Gamma(\gamma^P)$. While we do not have any general characterization results for $\Gamma(\gamma^P)$, we show below that complete structures will not necessarily be coalition-proof equilibria of $\Gamma(\gamma^P)$ by considering the special case of the 3-player *majority* game.[13]

Proposition 2. *Let N be a player set with $|N| = 3$, and let v be the majority game on $|N|$. Then, $s \in$ CPNE (γ^P) iff $L(s) = \{\{i,j\}\}$, i.e., only one pair of agents forms a link.*

Proof. Suppose only i and j form a link according to s. Then, $f_i^{\gamma^P}(s) = f_j^{\gamma^P}(s) = \frac{1}{2}$. Check that if i deviates and forms a link with k, then i's payoff remains at $\frac{1}{2}$. Also, clearly i and j together do not have any profitable deviation. Hence, s is coalition-proof.

Suppose $L(s) = \emptyset$. Then, $f_i^{\gamma^P}(s) = 0$ for all i. Suppose there are i and j such that $j \in s_i$. Then, s is not a Nash equilibrium since j can profitably deviate to $s_j' = \{i\}$. Note that $L(s_j', s_{-j}) = \{i,j\}$, and $f_j^{\gamma^P}(s_j', s_{-j}) = \frac{1}{2}$.

If $s_i = \emptyset$ for all i, then *any* two agents, say i and j, can deviate profitably to form the link $\{i,j\}$. Neither i nor j has a further deviation.

Now, suppose that N is a connected set according to s. There are two possibilities.

Case (i) : $L(s) = \bar{L}$. In that case, $f_i^{\gamma^P} = \frac{1}{3}$ for all $i \in N$. Let i and j deviate and break links with k. Then, both i and j get a payoff of $\frac{1}{2}$. Suppose i makes a further deviation. The only deviation which needs to be considered is if i re-establishes a link with k. Check that i's payoff *remains* at $\frac{1}{2}$. So, in this case s cannot be a coalition-proof equilibrium.

[13] v is a majority game if a majority coalition has worth 1, and all other coalitions have zero worth.

Case (ii) : $L(s) \neq \bar{L}$. Since N is a connected set in $L(s)$, the only possibility is that there exist i and j such that both are connected to k, but not to each other. Then, both i and j have a payoff of $\frac{1}{4}$. Let now i and j deviate, break links with k and form a link between each other. Then, their payoff increases to $\frac{1}{2}$. Check that neither player has any further profitable deviation. Again, this shows that s is not coalition-proof. □

Remark 4. The Proportional Links Solution γ^P satisfies CE and IP and is link monotonic in the case covered by Proposition 2. This observation shows that we cannot replace WLS by link monotonicity in Theorem 2.

The last equilibrium concept we discuss is strong Nash equilibrium. Since every strong Nash equilibrium is a coalition-proof Nash equilibrium, it follows immediately from Theorem 2 that for a solution satisfying CE, WLS, and IP it holds that if $s \in$ SNE (γ), then $L(s)$ is essentially complete for γ. However, strong Nash equilibria might not exist. One might think that for strong Nash equilibria to exist, some condition like *balancedness* of v is needed, but we have examples that show that balancedness of v is not necessary and even convexity of v is not sufficient for nonemptiness of the set of strong Nash equilibria of the linking game.

Conclusion

In this paper, we have studied the endogenous formation of *cooperation structures* in superadditive TU-games using a strategic game approach. In this strategic game, each player announces the set of players with whom he or she wants to form a link, and a link is formed if and only if both players want to form the link. Given the resulting cooperation structure, the payoffs are determined by some exogenous solution for cooperative games with cooperation structures. We have concentrated on the class of solutions satisfying three appealing properties. We have shown that in this setting both the undominated Nash equilibrium and the Coalition-Proof Nash Equilibrium of this strategic form game predict the formation of the full cooperation structure or some payoff equivalent structure. This also true for the concept of strong Nash equilibrium, although there are games for which the set of strong Nash equlibria may be empty.[14]

The results obtained in this paper all point in the direction of the formation of the full cooperation structure in a superadditive environment. However, as we have indicated earlier, these results are sensitive to the assumptions on solutions for cooperative game with cooperation structures. Further, the discussion in section 3 of Dutta et al. (1995) shows that in a context where links are formed sequentially rather than simultaneously other predictions may prevail.

[14] In a separate paper, Slikker et al. (2000), we show that another equilibrium for linking games, the *argmax sets of weighted potentials*, also predicts the formation of the full cooperation structure. See Monderer and Shapley (1996) for various properties of weighted potential games.

References

1. Aumann, R., Myerson, R. (1988) Endogenous formation of links between players and coalitions: an application of the Shapley value, in A. Roth (ed.) *The Shapley Value*, Cambridge University Press, Cambridge.
2. Bernheim, B., Peleg, B., Whinston, M. (1987) Coalition-Proof Nash equilibria I. Concepts, *Journal of Economic Theory* 42: 1–12.
3. Dutta, B., Nouweland, A. van den, Tijs, S. (1998) Link Formation in Cooperative Situations, *Int. J. Game Theory* 27: 245–256.
4. Dutta, B., Ray, D. (1989) A Concept of Egalitarianism under Participation Constraints, *Econometrica* 57: 615–636.
5. Hart, S., Kurz, M. (1983) Endogenous Formation of Coalitions, *Econometrica* 51: 1047–1064.
6. Jackson, M., and Wolinsky, A. (1996) A Strategic Model of Social and Economic Networks, *Journal of Economic Theory* 71: 44–74.
7. Kalai, E., and Samet, D. (1988) Weighted Shapley values. In A. Roth (ed.) *The Shapley Value*, Cambridge University Press, Cambridge.
8. Monderer, D. and Shapley, L. (1996) Potential games, *Games and Economic Behaviour* 14: 124–143.
9. Myerson, R. (1977) Graphs and cooperation in games, *Mathematics of Operations Research* 2: 225–229.
10. Myerson, R. (1991) *Game Theory: Analysis of Conflict.* Harvard University Press, Cambridge, Massachusetts.
11. Nouweland, A. van den (1993) *Games and Graphs in Economic Situations.* PhD Dissertation, Tilburg University, Tilburg, The Netherlands.
12. Qin, C. (1996) Endogenous formation of cooperation structures, *Journal of Economic Theory* 69: 218–226.
13. Slikker, M., Dutta, B., van den Nouweland, A., Tijs, S. (2000) Potential Maximizers and Network Formation. *Mathematical Social Sciences* 39: 55–70.

Network Formation Models With Costs for Establishing Links

Marco Slikker[1,*], **Anne van den Nouweland**[2,**]

[1] Department of Technology Management, Eindhoven University of Technology, P.O.Box 513, 5000 MB Eindhoven, The Netherlands (e-mail: M.Slikker@tm.tue.nl)
[2] Department of Economics, 435 PLC, 1285 University of Oregon, Eugene, OR 97403-1285, USA

Abstract. In this paper we study endogenous formation of communication networks in situations where the economic possibilities of groups of players can be described by a cooperative game. We concentrate on the influence that the existence of costs for establishing communication links has on the communication networks that are formed. The starting points in this paper are two game-theoretic models of the formation of communication links that were studied in the literature fairly recently, the extensive-form model by Aumann and Myerson (1988) and the strategic-form model that was studied by Dutta et al. (1998). We follow their analyses as closely as possible and use an extension of the Myerson value to determine the payoffs to the players in communication situations when forming links is not costless. We find that it is possible that as the costs of establishing links increase, more links are formed.

1 Introduction

In this paper we study endogenous formation of communication networks in situations where the economic possibilities of groups of players can be described by a cooperative game. We concentrate on the influence that the existence of costs for establishing communication links has on the communication networks that are formed. The starting points of this paper are two game-theoretic models of the formation of communication links that were studied in the literature fairly recently, the extensive-form model by Aumann and Myerson (1988) and the strategic-form model studied by Dutta et al. (1998).[1] In both of these papers

The authors thank an editor and an anonymous referee for useful suggestions and comments.

* This research was carried out while this author was a Ph.D. student at the Department of Econometrics and CentER, Tilburg University, Tilburg, The Netherlands.

** Support of the Department of Econometrics of Tilburg University and of the NSF under Grant Number SBR-9729568 is gratefully acknowledged.

[1] The model studied by Dutta et al. (1998) was actually first mentioned in Myerson (1991).

forming communication links is costless and, once a communication network has been formed, an external allocation rule is used to determine the payoffs to the players in different communication networks. The external allocation rule used by Aumann and Myerson (1988) is the Myerson value (cf. Myerson (1977)) and Dutta et al. (1998) considered a class of external allocation rules that contains the Myerson value. We follow their analyses as closely as possible and use a natural extension of the Myerson value to determine the payoffs to the players in communication situations with costs for establishing links.

The goal of this paper is to study the influence that costs of forming communication links have on the structures that are formed. In order to be able to clearly isolate the influence of the costs, we assume that costs are equal for all possible communication links. Starting from costs equal to zero, we increase the costs and see how these increasing costs induce different equilibrium communication structures. Throughout the paper, we initially restrict our analysis to situations in which the underlying cooperative games are 3-player symmetric games, and then extend our scope to games with more than three players.

In the extensive-form game of link formation we consider communication structures that are formed in subgame perfect Nash equilibria. We find for this game, with 3 symmetric players, that the pattern of structures formed as costs increase depends on whether the underlying coalitional game is superadditive and/or convex. We find that in case the underlying game is not superadditive or in case it is convex, increasing costs for forming communication links result in the formation of fewer links in equilibrium. However, if the underlying game is superadditive but not convex, then increasing costs initially lead to the formation of fewer links, then to the formation of more links, and finally lead to the formation of fewer links again.

For the strategic-form game of link formation we briefly discuss the inappropriateness of Nash equilibria and strong Nash equilibria and then consider undominated Nash equilibria and coalition-proof Nash equilibria. We find for this game, with 3 symmetric players, that the pattern of structures formed in undominated Nash equilibria and coalition-proof Nash equilibria as costs increase also depends on whether the underlying coalitional game is superadditive and/or convex. In contrast to the results for the extensive-from game of link formation, we find that in all cases increasing costs for forming communication links result in the formation of fewer links in equilibrium. We restrict our analysis of the formation of networks to symmetric 3-player games for reasons of clarity of exposition, but we prove the existence of coalition-proof Nash equilibria for 3-players games in general to show that the analysis using the coalition-proof Nash equilibrium concept can be extended to such games.

We then extend our scope to games with more than three players. We show that the relationship of costs and structures formed cannot be related back simply to superadditivity and/or convexity of the underlying game. Additionally, we show that the possibility that higher costs lead to more links being formed is still present for games with more than three players.

The importance of network structures in the organization of many economic relationships has been extensively documented (see e.g. the references in Goyal (1993) and Jackson and Wolinsky (1996)). The game-theoretic literature on communication networks was initiated by Myerson (1977), who studied the influence of restrictions in communication on the allocation of the coalitional values in TU-games. The influence of the presence of communication restrictions on cooperative games has been studied by many authors since and an extensive survey on this subject can be found in van den Nouweland (1993).

In the current paper we study the formation of communication networks. Formally, a communication network (cf. Myerson (1977)) is a graph in which the players are the nodes and in which two players are connected by a communication link (an edge in the graph) if and only if they can communicate with each other in a direct and meaningful way. The game theoretic literature on the formation of communication networks includes a number of papers, including Aumann and Myerson (1988), Goyal (1993), Dutta and Mutuswami (1997), Jackson and Wolinsky (1996), Watts (1997), Dutta et al. (1998), Bala and Goyal (2000), and Slikker and van den Nouweland (2001).

The current paper is most closely related to Aumann and Myerson (1988) and Dutta et al. (1998). Both of these two papers study the formation of communication links in situations where the economic possibilities of the players can be described by a cooperative game. It is the models in these two papers that we use to study the formation of communication links in the current paper. However, in these two papers establishing communication links is costless, whereas we impose costs for forming communication links.

To our knowledge, the formation of communication networks when there are costs for forming communication links has only been studied in specific parametric models, as is the case in Goyal (1993), Watts (1997), Bala and Goyal (2000), and some examples in Jackson and Wolinsky (1996). The first three of these papers study the formation of networks within the framework of a parametric model of information transmission. These papers employ different processes of network formation and study the efficiency and stability of networks. Jackson and Wolinsky (1996) do not specify a specific model of network formation, but they study the stability and efficiency of networks in situations where self-interested agents can form and sever links. In their paper, a value function gives the value of each possible network and an exogenously given allocation rule is used to determine the payoffs to individual players for each possible network structure. They show that for anonymous and component balanced allocation rules efficient graphs need not be stable. The value function used by Jackson and Wolinsky (1996) allows for costs of communication links to be incorporated in the model in an indirect way.

Our paper is less general than Jackson and Wolinsky (1996) because, following Aumann and Myerson (1988) and Dutta et al. (1998), we restrict ourselves to situations in which the economic possibilities of the players can be described by a coalitional game. However, we explicitly model the costs of establishing communication links, rather than having those implicit in the value function. This allows us to study the influence of these costs.

The outline of the paper is as follows. In Sect. 2 we provide general definitions concerning communication situations and allocation rules. In Sect. 3 we compute the payoffs allocated to the players in different communication situations according to the extension of the Myerson value that we use as the external allocation rule in this paper. We describe and study the linking game in extensive form in Sect. 4 and Sect. 5 contains our study of the linking game in strategic form. The models of Sects. 4 and 5 are compared in Sect. 6, in which we also reflect on the results obtained for the two models. In Sect. 7 we extend the scope of our analysis to games with more than 3 players. We conclude in Sect. 8.

2 Communication Situations

In this section we will describe communication situations and an allocation rule for these situations, the Myerson value. Additionally, we will introduce communication costs and describe how these costs will be divided between the players.

A communication situation (N,v,L) consists of a cooperative game (N,v), describing the economic possibilities of all coalitions of players, and a communication graph (N,L), which describes the communication channels between the players. The characteristic function v assigns a value $v(S)$ to all coalitions $S \subseteq N$, with $v(\emptyset) = 0$. We will restrict ourselves to zero-normalized non-negative games, i.e., $v(\{i\}) = 0$ for all $i \in N$ and $v(S) \geq 0$ for all $S \subseteq N$. Communication is two-way and is represented by an undirected communication graph, i.e., the set of links L is a subset of $\bar{L} := \{\{i,j\} \mid \{i,j\} \subseteq N,\ i \neq j\}$. The communication graph (N,L) should be interpreted as a way to model restricted cooperation between the players. Players can only cooperate with each other if they are directly connected with each other, i.e., there is a link between them, or if they are indirectly connected, i.e., there is a path via other players that connects them. Note that indirect communication between two players requires the cooperation of the players on a connecting path between them as well. The communication structure (N,L) gives rise to a partition of the player set into groups of players who can communicate with each other. Two players belong to the same partition element if and only if they are connected with each other, directly or indirectly. A partition element is called a *communication component* and the set of communication components is denoted N/L. The communication graph (N,L) also induces a partition of each coalition $S \subseteq N$.[2] This partition is denoted by S/L and it consists of the communication components of the subgraph $(S,L(S))$, where $L(S)$ contains the communication links within S, i.e., $L(S) := \{\{i,j\} \in L \mid \{i,j\} \subseteq S\}$.

The restricted communication within a coalition $S \subseteq N$ influences the economic possibilities of the coalition. Cooperation between players in different communication components is not possible, so benefits of cooperation can only

[2] $S \subseteq N$ denotes that S is a subset of N, $S \subset N$ denotes that S is a strict subset of N.

be realized within communication components. We define the value of coalition S in the communication situation (N, v, L) as the sum of the values of the communication components of S,

$$v^L(S) := \sum_{T \in S/L} v(T).$$

The game (N, v^L) is usually called the *graph-restricted game*. The Myerson value of the communication situation (N, v, L) coincides with the Shapley value Φ (see Shapley (1953)) of the graph-restricted game,

$$\mu(N, v, L) = \Phi(N, v^L).$$

Myerson (1977) characterizes this rule using two properties, component balancedness and fairness.[3] Component balancedness states that the players in a communication component C divide the value of this communication component, $v(C)$, between them. Fairness states that the addition (deletion) of a link in a communication situation should have the same cardinal effect on the two players that form this link.

In the description of the model above, it is assumed that there are no costs for establishing communication links. In the following we will introduce such costs and integrate these in the analysis of the communication situations described above.

We will assume that the formation of a communication link between any two players results in a fixed cost $c \geq 0$. Adding costs for establishing links has the effect that the value that connected players can obtain also depends on how many links they form and not just on whether they are connected or not. To determine the allocation of the costs and the benefits, we use the natural extension of the Myerson value that was introduced in Jackson and Wolinsky (1996). They prove that on the domain of networks whose values are described by means of a value function[4], there exists a unique fair and component balanced allocation rule[5]. For a value function w and a graph (N, L), this rule assigns to the players their Shapley values in the game $(N, U_{w,L})$ defined by $U_{w,L}(S) = \sum_{C \in S/L} w(L(C))$.[6] We apply this rule to the value function $w_{v,c}$ with $w_{v,c}(A) = \sum_{C \in N/A} v(C) - c|A|$ for all $A \subseteq \bar{L}$, which describes the worth obtainable by the players in network (N, A) minus the costs of the links in A if the cooperative game is (N, v) and the cost per link is c. Hence, we consider the Shapley value of the game $(N, U_{w_{v,c},L})$ with $U_{w_{v,c},L}(S) = \sum_{C \in S/L} w_{v,c}(L(C))$ for each $S \subseteq N$. We call this the *cost-extended Myerson value* of the situation (N, v, L, c) and denote it by $\nu(N, v, L, c)$.

[3] Myerson (1977) refers to component balancedness as component efficiency. We prefer to use component balancedness to avoid confusion with efficiency of graphs.

[4] A value function is a function that assigns a value to all possible sets of links $L \subseteq \bar{L}$.

[5] An allocation rule in this setting is a description of how the value associated with each network is distributed to the individual players.

[6] We are using the notation adopted in the current paper and not that used by Jackson and Wolinsky (1996).

Using the additivity property of the Shapley value, it is a straightforward exercise to show that

$$\nu_i(N, v, L, c) = \mu_i(N, v, L) - \frac{1}{2}|L_i|c \quad \text{for all } i \in N,$$

where $L_i := \{\{i,j\} \mid \{i,j\} \in L\}$ denotes the set of links in L in which player i is involved.

The discussion in the previous paragraph shows that we can interpret the cost-extended Myerson value in two inherently different ways. The first interpretation is that players bargain over the division of the benefits and the costs of link formation simultaneously. The second interpretation is that the players first incur the costs of link formation and divide these costs in a fair manner, and then, after these costs are sunk, bargain over the division of the benefits. Both points of view, while very different methodologically, lead to the same end result, namely the cost-extended Myerson value.

In the following sections we will introduce costs of establishing communication links in two well-known models of link formation, the model of link formation in extensive form introduced by Aumann and Myerson (1988) and the model of link formation in strategic form introduced by Myerson (1991). Throughout this paper we mostly restrict ourselves to 3-person cooperative games where the worth of a coalition only depends on how many members it has and not on the identities of these members.

3 The Cost-extended Myerson Value for Symmetric 3-Player Games

In this section we will compute the payoffs according to the cost-extended Myerson value for symmetric 3-player games and all possible communication structures between the three players of these games. Due to symmetry, we need to distinguish only 5 different positions a player can have in a communication graph. We will analyze the preferences of the players over these positions, depending on the underlying cooperative game and the costs of establishing communication links.

Let (N, v) be a symmetric 3-player game, i.e., there exist $w_1, w_2, w_3 \in \mathbf{R}$ such that $v(S) = w_{|S|}$ for all $S \subseteq N$ with $S \neq \emptyset$, and let c denote the non-negative costs for establishing a communication link.

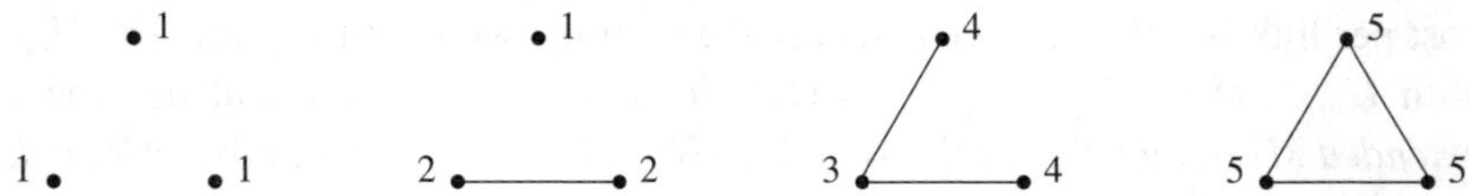

Fig. 1. Different positions

In Fig. 1 we distinguish 5 different positions in the set of graphs with 3 vertices. Position 1 is the isolated position. An isolated player receives zero payoff.[7] Note that in the graph with one link, one of the players is isolated:

$$\nu_i(N, v, \emptyset, c) = \nu_i(N, v, \{\{j,k\}\}, c) = 0 \quad \text{if } i \notin \{j,k\}. \tag{1}$$

Position 2 denotes the linked position in a graph with one link. The two linked players equally divide the value of a 2-person coalition and the costs,

$$\nu_j(N, v, \{\{j,k\}\}, c) = \frac{1}{2}w_2 - \frac{1}{2}c. \tag{2}$$

Position 3 is the central position in the graph with two links. A player in this position receives

$$\nu_i(N, v, \{\{i,j\},\{i,k\}\}, c) = \frac{1}{3}w_3 + \frac{1}{3}w_2 - c. \tag{3}$$

Position 4 is the non-central position in the graph with two links. The payoff a player in this position receives equals

$$\nu_j(N, v, \{\{i,j\},\{i,k\}\}, c) = \frac{1}{3}w_3 - \frac{1}{6}w_2 - \frac{1}{2}c. \tag{4}$$

Finally, position 5 represents a position in the graph with three links. In the graph with three links, every player receives

$$\nu_i(N, v, \bar{L}, c) = \frac{1}{3}w_3 - c. \tag{5}$$

Obviously, depending on (N, v) and c a player will have different preferences over positions 1 through 5. If the costs c are fairly high, then a player will find that in some cases the benefits from being linked to a player do not outweigh the costs for forming a link. The preferences of a player are described in Table 1.

Table 1. Preferences over different positions

Preference	Condition independent of c	Condition dependent on c
$1 \succ 2$		$c > w_2$
$1 \succ 3$		$c > \frac{1}{3}w_3 + \frac{1}{3}w_2$
$1 \succ 4$		$c > \frac{2}{3}w_3 - \frac{1}{3}w_2$
$1 \succ 5$		$c > \frac{1}{3}w_3$
$2 \succ 3$		$c > \frac{2}{3}w_3 - \frac{1}{3}w_2$
$2 \succ 4$	$2w_2 > w_3$	
$2 \succ 5$		$c > \frac{2}{3}w_3 - w_2$
$3 \succ 4$		$c < w_2$
$3 \succ 5$	$w_2 > 0$	
$4 \succ 5$		$c > \frac{1}{3}w_2$

If a condition in Table 1 holds with equality then a player is indifferent between the positions while a reverse preference holds if the reverse inequality holds. In the next sections we will consider the influence of costs of establishing communication links in link formation games.

[7] Recall that we restrict ourselves to zero-normalized games.

4 Linking Game in Extensive Form

In this section we will introduce a slightly modified version of the linking game in extensive form that was introduced and studied by Aumann and Myerson (1988). The modification consists of the incorporation of costs of establishing communication links. Subsequently, following Aumann and Myerson (1988), we will study the subgame perfect Nash equilibria (SPNE) in this model. We provide an example that illustrates some curiosities that can arise and we also provide a systematic analysis of 3-player symmetric games.

4.1 The Game

We will now describe the linking game in extensive form. This linking game is a slightly modified version of the game in extensive form as it was introduced by Aumann and Myerson (1988), the only difference being that we include costs for establishing links in the payoffs to the players.

A TU-cooperative game (N, v) and a cost per link c are exogenously given and initially there are no communication links between the players. The game consists of pairs of players being offered to form links, according to some exogenously given rule of order that is common knowledge to the players. A link is formed only if both potential partners agree on forming it. Once a link has been formed, it cannot be broken in a further stage of the game. The game is of perfect information: at any time, the entire history of offers, acceptances, and rejections is known to all players. After the last link has been formed, each of the pairs of players who have not yet formed a link, are given an opportunity to form an additional link. The process stops when, after the last link has been formed, all pairs of players that have not yet formed a link have had a final opportunity to do so and declined this offer. This process results in a set of links. We will denote this set by L. The payoff to the players is then determined by the cost-extended Myerson value, i.e., if (N, L) is formed player i receives

$$\nu_i(N, v, L, c) = \mu_i(N, v, L) - \frac{1}{2}|L_i|c.$$

In the original model of Aumann and Myerson (1988) there are no costs for links ($c = 0$) and player i receives $\mu_i(N, v, L)$.

Aumann and Myerson (1988) already noted that the order in which two players in a pair decide whether or not to form a link has no influence. Furthermore, since the game is of perfect information it has subgame perfect Nash equilibria (see Selten 1965).

4.2 An Example

In this section we will consider the 3-player symmetric game (N, v) with

$$v(S) := \begin{cases} 0 & \text{if } |S| \leq 1 \\ 60 & \text{if } |S| = 2 \\ 72 & \text{if } S = N \end{cases} . \tag{6}$$

This game was analyzed by Aumann and Myerson (1988), who showed that in the absence of costs of establishing communication links, every subgame perfect Nash equilibrium results in the formation of exactly one link. We will analyze the influence of link formation costs on the subgame perfect Nash equilibria of the model. The payoffs for the four classes of structures that can result follow directly from Sect. 3. A survey of these payoffs can be found in Table 2.

Table 2. Payoffs in different positions

Position	Payoff
1	0
2	$30 - \frac{1}{2}c$
3	$44 - c$
4	$14 - \frac{1}{2}c$
5	$24 - c$.

Aumann and Myerson (1988) study this example with $c = 0$. If two players, say i and j, form a link, they will each receive a payoff of 30. Certainly, both would prefer to form a link with the remaining player k, provided the other player does not form a link with player k, and receive 44. However, if player i forms a link with player k he can anticipate that subsequently players j and k will also form a link to get 24 rather than 14. So, both players i and j know that if one of them forms a link with player k they will end up with a payoff of 24, which is strictly less than 30, the payoff they receive if only the link between players i and j is formed. Hence, every subgame perfect Nash equilibrium results in the formation of exactly one link.

What will happen if establishing a communication link with another player is not free any more? One would expect that relatively small costs will not have very much influence and that larger costs will result in the formation of fewer links.

For small costs, say $c = 1$, we can repeat the discussion above and conclude that exactly one link will form. However, if the costs are larger the analysis changes. Assume for example that $c = 22$. Then, forming one link will result in a payoff of 19 for the two players forming the link, and the remaining player will receive 0. Forming two links will give the central player 22 and the other two players will receive 3 each. Finally, the full cooperation structure will give all players a payoff 2. We see that this changes the incentives of the players. Once two links are formed, the two players that are not linked with each other yet, prefer to stay in the current situation and receive 3 instead of forming a link and receive only 2. In case one link has been formed, a player who is already linked is now willing to form a link with the isolated player since this would increase his payoff (from 19 to 22) and the threat of ending up in the full cooperation structure has disappeared. Obviously, all players prefer forming some links to no link at all. Similar to the argument that in the absence of costs all three structures with one link are supported by a subgame perfect Nash equilibrium (see Aumann and Myerson (1988)), it follows that with communication costs equal to 22 all three structures with two links are supported by a subgame perfect Nash equilibrium.

The surprising result in this example is that an increase in the costs of establishing a communication link results in more communication between the players (2 links rather than 1). In the following subsection we will again see this result. We will also show that a further increase in the costs will result in a decrease in the number of links between the players.

4.3 Symmetric 3-Player Games

In this subsection we will describe the communication graphs that will result in symmetric 3-player games with various levels of costs for establishing links. To find which communication structures are formed in subgame perfect Nash equilibria, we simply use the general expressions for the payoffs that we provided in Sect. 3 and the preferences of the players over different positions that were analyzed in Table 1. It takes some tedious calculations, but eventually it turns out that we need to distinguish three classes of games that result in different communication structure patterns with changing costs of establishing communication links.

Firstly, assume that the game (N, v) satisfies $w_2 > w_3$. Then we find that the structures that are supported by subgame perfect Nash equilibria as a function of the costs of communication links are as summarized in Fig. 2 on page 242.

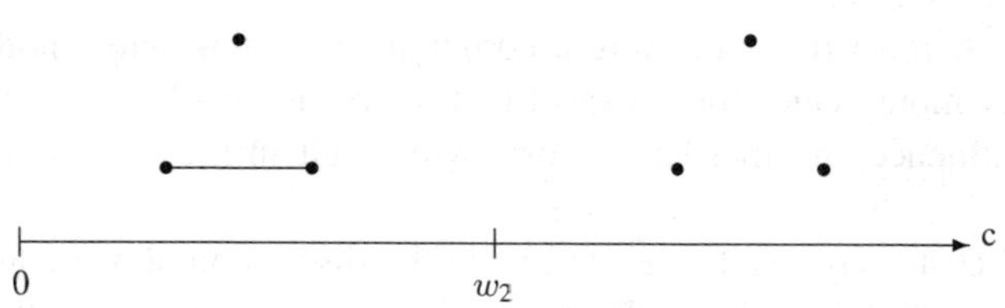

Fig. 2. Communication structures according to SPNE in case $w_2 > w_3$

We note that on the boundary, i.e., $c = w_2$, both structures that appear for $c < w_2$ and for $c > w_2$ are supported by a subgame perfect Nash equilibrium. If $w_2 > w_3$ the full communication structure, in which all players are connected directly, will never form. Checking the preferences of the players, we see that the full communication structure would be formed only if $c < \frac{2}{3}w_3 - w_2$. Since $\frac{2}{3}w_3 - w_2 < 0$ and since the costs of establishing a communication link are non-negative the full cooperation structure will not be formed.

Secondly, assume the game (N, v) satisfies $2w_2 > w_3 > w_2$. The structures resulting from subgame perfect Nash equilibria for this class of games are summarized in Fig. 3.

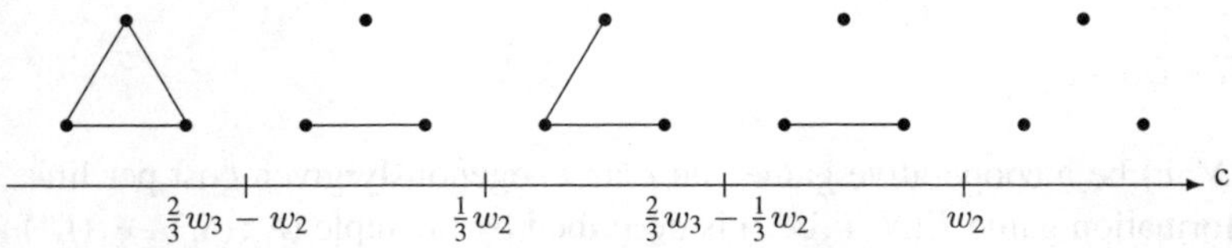

Fig. 3. Communication structures according to SPNE in case $2w_2 > w_3 > w_2$

The example in Sect. 4.2 belongs to this class of games. In that example $\frac{2}{3}w_3 - w_2 < 0$. Since the condition $2w_2 > w_3 > w_2$ can result in $\frac{2}{3}w_3 - w_2 < 0$ but also in $\frac{2}{3}w_3 - w_2 > 0$, we have not explicitly indicated $c = 0$ in Fig. 3.

Thirdly, consider the class of games satisfying $w_3 > 2w_2$. For these games the structures supported by subgame perfect Nash equilibria are summarized in Fig. 4.

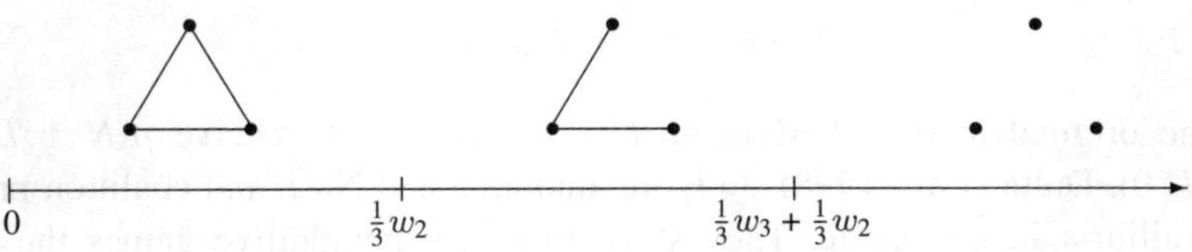

Fig. 4. Communication structures according to SPNE in case $w_3 > 2w_2$

The discussion above makes a distinction between three classes of games. Note that if $w_2 = w_3$ then Figs. 2 and 3 lead to the same results since some of the boundaries coincide. If $w_3 = 2w_2$ then Figs. 3 and 4 lead to the same results.

The communication structure patterns above result in three classes of games. The first class, with games satisfying $w_2 > w_3$, contains only non-superadditive games. The second class, defined by $2w_2 > w_3 > w_2$, contains only superadditive games that are not convex. The last class, with $w_3 > 2w_2$, contains only convex games.

We conclude that for non-superadditive games and for convex games increasing costs of establishing communication links results in a decreasing number of communication links, while for superadditive non-convex games increasing costs can result in more communication links.

5 Linking Game in Strategic Form

In this section we will introduce costs of establishing communication links in the link formation game in strategic form that was introduced by Myerson (1991) and subsequently studied by Qin (1996), Dutta et al. (1998), and Slikker (1998). We will analyze this model by means of the Nash equilibrium, strong Nash equilibrium, undominated Nash equilibrium, and coalition proof Nash equilibrium concepts.

5.1 The Game

Let (N, v) be a cooperative game and c an exogenously given cost per link. The link formation game $\Gamma(N, v, c, \nu)$ is described by the tuple $(N; (S_i)_{i\in N}; (f_i^{\nu})_{i\in N})$. For each player $i \in N$ the set $S_i = 2^{N\setminus\{i\}}$ is the strategy set of player i. A strategy of player i is an announcement of the set of players he wants to form communication links with. A communication link between two players will form if and only if both players want to form the link. The set of links that form according to strategy profile $s \in S = \prod_{i\in N} S_i$ will be denoted by

$$L(s) := \{\{i,j\} \subseteq N \mid i \in s_j,\ j \in s_i\}.$$

The payoff function $f^{\nu} = (f_i^{\nu})_{i\in N}$ is defined as the allocation rule ν, the cost-extended Myerson value, applied to $(N, v, L(s), c)$,

$$f^{\nu}(s) = \nu(N, v, L(s), c).$$

In the original model of Myerson (1991) the players receive $\mu(N, v, L) = \nu(N, v, L, 0)$. Dutta et al. (1998) study the undominated Nash and coalition proof Nash equilibria in this game. They show that in superadditive games the full communication structure will form or a structure that is payoff equivalent to it. Slikker (1998) shows a similar result for (strictly) perfect and (weakly/strictly) proper equilibria.

5.2 Nash Equilibria and Strong Nash Equilibria

In this section we consider Nash equilibria and strong Nash equilibria. We present an example showing that many communication structures can result from Nash equilibria, while strong Nash equilibria do not always exist.

Recall that a strategy profile is a Nash equilibrium if there is no player who can increase his payoff by unilaterally deviating from it. A strategy profile is called a strong Nash equilibrium if there is no coalition of players that can strictly increase the payoffs of all its members by a joint deviation (Aumann 1959).

Consider the following example. Let (N, v) be the symmetric 3-player game with

$$v(S) := \begin{cases} 0 & \text{if } |S| \leq 1 \\ 60 & \text{if } |S| = 2 \\ 72 & \text{if } S = N \end{cases} . \tag{7}$$

The payoffs to the players for the five positions we distinguished in Fig. 1 are summarized in Table 2 on page 241.

If $c = 0$ every structure can be supported by a Nash equilibrium, since nobody wants to break a link and two players are needed to form a link.[8] If costs rise, fewer structures are supported by Nash equilibria. For example, if $c > 20$ then a player prefers position 4 to position 5, and hence, the full cooperation structure is not supported by a Nash equilibrium. However, since a communication link can only be formed if two players want to do so, the communication structure with zero links is supported by a Nash equilibrium for any cost c.

For symmetric 3-player games, it is fairly easy to check for any of the four possible communication structures under what conditions on the costs they are supported by a Nash equilibrium. The results, which turn out to depend on whether the game is superadditive and/or convex, are represented in Figs. 5, 6, and 7.

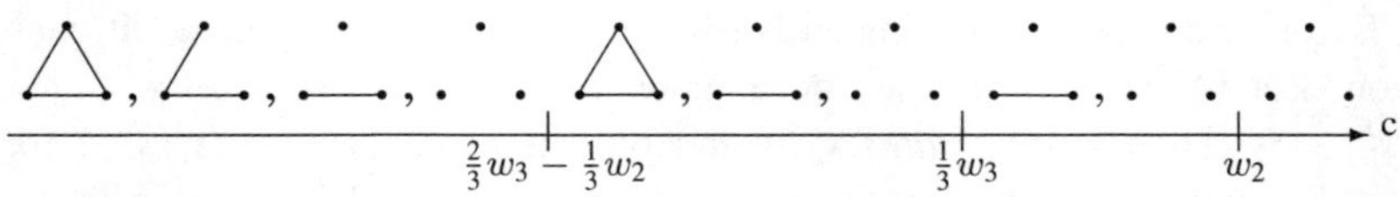

Fig. 5. Communication structures according to NE in case $w_2 > w_3$

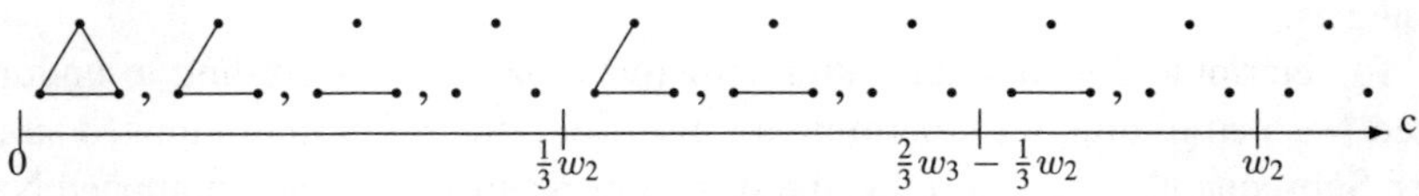

Fig. 6. Communication structures according to NE in case $2w_2 > w_3 > w_2$

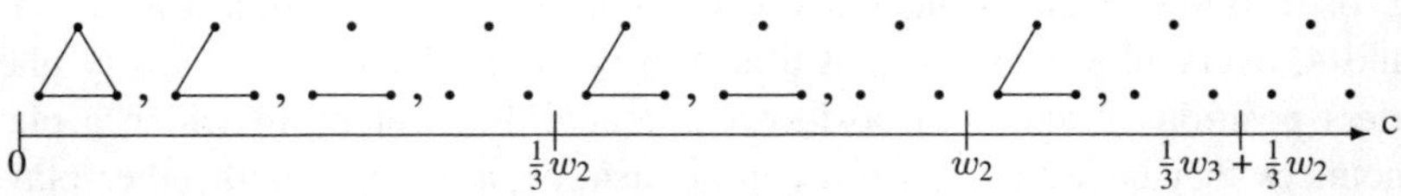

Fig. 7. Communication structures according to NE in case $w_3 > 2w_2$

Since Nash equilibria can result in a fairly large set of structures, we consider the refinement to strong Nash equilibria for the linking game in strategic form. Consider the game discussed earlier in this section and suppose that the costs per link are 20. We will show that no strong Nash equilibrium exists by considering all possible communication structures. Firstly, the communication structures with zero and three links cannot result from a strong Nash equilibrium since two players can deviate to a strategy profile resulting in only the link between them, improving their payoffs from 0 or 4 to 20. A structure with two links is not

[8] This was already proven for all superadditive games by Dutta et al. (1998).

supported by a strong Nash equilibrium since the two players in the non-central positions can deviate to a strategy profile resulting in only the link between them and improve their payoffs from 4 to 20. Finally, a communication structure with one link is not supported by a strong Nash equilibrium since one player in a linked position and the player in the non-linked position can deviate to a strategy profile resulting in an additional link between them, increasing both their payoffs by 4. We conclude that strong Nash equilibria do not always exist.[9]

5.3 Undominated Nash Equilibria

The multiplicity of structures resulting from Nash equilibria and the non-existence of strong Nash equilibria for several specifications of the underlying game and costs for establishing links, inspires us to study two alternative equilibrium refinements. The current section is devoted to undominated Nash equilibria and in Sect. 5.4 we analyze coalition proof Nash equilibria.

Before we define undominated Nash equilibria we need some additional notation. Let $(N, (S_i)_{i \in N}, (f_i)_{i \in N})$ be a game in strategic form. Let $i \in N$ and $s_i, s_i' \in S_i$. Then s_i *dominates* s_i' if and only if $f_i(s_i, s_{-i}) \geq f_i(s_i', s_{-i})$ for all $s_{-i} \in S_{-i}$ with strict equality for at least one $s_{-i} \in S_{-i}$. We will denote the set of undominated strategies of player i by S_i^u. Further, we define $S^u := \prod_{i \in N} S_i^u$. A strategy profile $s \in S$ is an *undominated Nash equilibrium* (UNE) if s is a Nash equilibrium and $s \in S^u$, i.e., if s is a Nash equilibrium in undominated strategies.

To determine the communication structures that result according to undominated Nash equilibria, we determine for any cost c the set of undominated strategies. Subsequently, we determine the structures resulting from undominated Nash equilibria.

For example, consider a symmetric 3-person game (N, v) with $2w_2 > w_3 > w_2$ and $c < \frac{1}{3}w_2$. The structures supported by Nash equilibria can be in found in Fig. 6. It follows from Table 1 that every player prefers position 5 to positions 1 and 4, every player prefers position 3 to positions 1 and 2, and every player prefers positions 2 and 4 to position 1. Hence, the strategy in which a player announces that he wants to form communication links with both other players dominates his other strategies. So, this strategy is the unique undominated strategy. If all three players choose this undominated strategy, then it is not profitable for any player to unilaterally deviate, implying that the unique undominated Nash equilibrium results in the full cooperation structure

The following example illustrates a tricky point that may arise when determining the undominated Nash equilibria. Consider a symmetric 3-person game (N, v) with $w_3 > 2w_2$ and $\frac{1}{3}w_2 < c < w_2$. The structures supported by Nash equilibria can be in found in Fig. 7. For every player i, strategy $s_i = \emptyset$ is dominated by a strategy in which the player announces that he wants to form a communication link with one other player since positions 2 and 4 are preferred

[9] Dutta et al. (1998) already note that in case $c = 0$ strong Nash equilibria might not exist.

to position 1. It is an undominated strategy for a player to announce that he wants to form a link with exactly one other player. Such a strategy is not dominated by $s_i = \emptyset$ because $2 \succ 1$ and it is not dominated by $s_i = N \backslash \{i\}$ because $4 \succ 5$. Since $3 \succ 1$ and $3 \succ 2$, it follows that $s_i = N \backslash \{i\}$ is an undominated strategy as well. Hence, the communication structures with 0, 1, 2, and 3 links can be supported by strategies that are undominated. In Fig. 7 we see that the structures with 0, 1, and 2 links are supported by Nash equilibria. However, this does not imply that all these structures are supported by undominated Nash equilibria. Consider a structure with one link, say link $\{i, j\}$. Let player k be the third player, who is isolated. Furthermore, let $s = (s_i, s_j, s_k)$ be a triple of undominated strategies such that $L(s) = \{\{i, j\}\}$. Since $\emptyset$ is a dominated strategy, we know that $i \in s_k$ or $j \in s_k$ or both. Suppose without loss of generality that $i \in s_k$. Since $L(s) = \{\{i, j\}\}$, it follows that $s_i = \{j\}$. However, since $4 \succ 2$, player i can strictly improve his payoff by deviating to $s_i' = \{j, k\}$. We conclude that s is not a Nash equilibrium. This shows that a structure with one link is not supported by an undominated Nash equilibrium. A similar argument shows that the structure with no links is not supported by an undominated Nash equilibrium. A structure with 2 links, however, is supported by an undominated Nash equilibrium: $t = (t_i, t_j, t_k) = (\{j\}, \{i, k\}, \{j\})$ is an undominated Nash equilibrium that results in communication structure $L(t) = \{\{i, j\}, \{j, k\}\}$. We conclude that all undominated Nash equilibria result in the formation of exactly two links.

Proceeding in the manner described above, we find all the structures that result according to undominated Nash equilibria. The results are schematically represented in Figs. 8, 9, and 10.

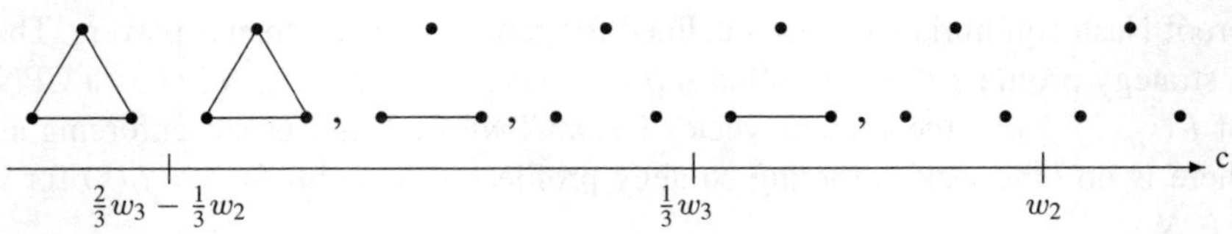

Fig. 8. Communication structures according to UNE in case $w_2 > w_3$

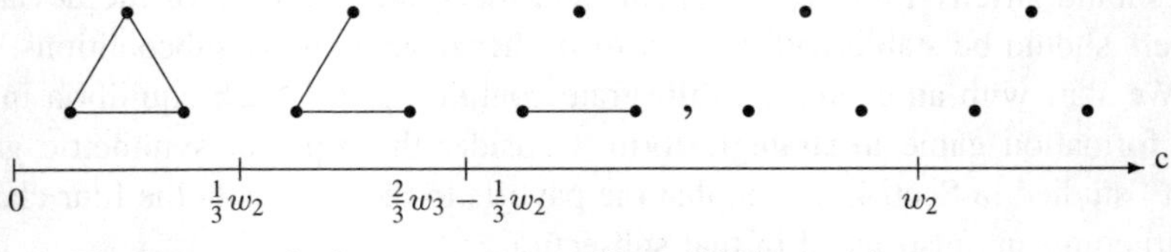

Fig. 9. Communication structures according to UNE in case $2w_2 > w_3 > w_2$

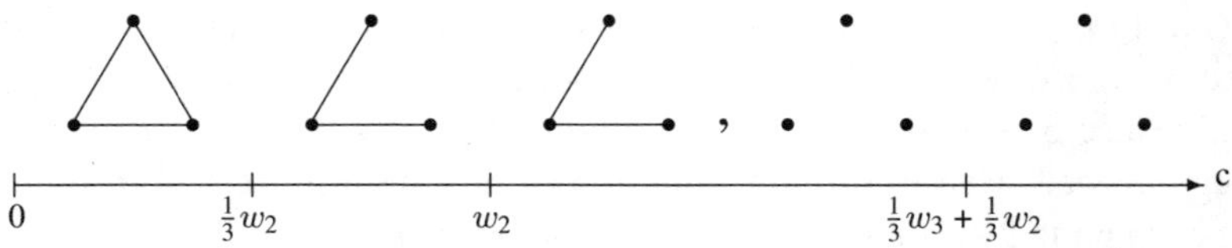

Fig. 10. Communication structures according to UNE in case $w_3 > 2w_2$

5.4 Coalition Proof Nash Equilibria

In this section we consider communication structures that result according to coalition proof Nash equilibria. We first give the definition of coalition proof Nash equilibria, then study an example and continue with general 3-player symmetric games. We also show that coalition proof Nash equilibria always exist for 3-player games.

Before we define the concept of *coalition proof Nash equilibrium* (CPNE) we will introduce some notation. Let $(N, (S_i)_{i\in N}, (f_i)_{i\in N})$ be a game in strategic form. For every $T \subset N$ and $\hat{s}_{N\setminus T} \in S_{N\setminus T}$, let $\Gamma(\hat{s}_{N\setminus T})$ be the game induced on the players of T by the strategies $\hat{s}_{N\setminus T}$, so

$$\Gamma(\hat{s}_{N\setminus T}) = (T, (S_i)_{i\in T}, (f_i^*)_{i\in T})$$

where for all $i \in T$, $f_i^* : S_T \to \mathbf{R}$ is given by $f_i^*(s_T) := f_i(s_T, \hat{s}_{N\setminus T})$ for all $s_T \in S_T$.

Now, coalition proof Nash equilibria are defined inductively. In a one-player game with player set $N = \{i\}$, $\hat{s}_i \in S = S_i$ is a CPNE of $\Gamma = (\{i\}, S_i, f_i)$ if $\hat{s}_i$ maximizes f_i over S_i. Let Γ be a game with $n > 1$ players. Assume that coalition proof Nash equilibria have been defined for games with less than n players. Then a strategy profile $\hat{s} \in S_N$ is called *self-enforcing* if for all $T \subset N$, $\hat{s}_T$ is a CPNE of $\Gamma(\hat{s}_{N\setminus T})$. Now, the strategy vector $\hat{s}$ is a CPNE of Γ if $\hat{s}$ is self-enforcing and there is no other self-enforcing strategy profile $s \in S_N$ with $f_i(s) > f_i(\hat{s})$ for all $i \in N$.

The set of coalition proof Nash equilibria is a superset of the set of strong Nash equilibria. The strong Nash equilibrium concept demands that no coalition can deviate to a profile that strictly improves the payoffs of all players in the coalition. The coalition proof Nash equilibrium concept has similar requirements, but the set of allowed deviations is restricted. Every player in the deviating coalition should strictly improve his payoff and the strategy profile of the deviating players should be stable with respect to further deviations by subcoalitions.

We start with an example to illustrate coalition proof Nash equilibria in the link formation game in strategic form. Consider the 3-player symmetric game (N, v) studied in Sect. 4.2. Note that the payoffs to the players in the four classes of structures are also listed in that subsection.

If $c = 0$, it follows from Dutta et al. (1998) that the full communication structure (with 3 links) is formed in the unique coalition-proof Nash equilibrium of the linking game in strategic form. To understand this, we consider the four classes of structures one-by-one. First note that the players would unilaterally like to form any additional links they can, which implies that in a Nash equilibrium s there can be no two players i and j such that $i \in s_j$ and $j \notin s_i$.

Hence, the structure with no links can only be formed in a Nash equilibrium if all 3 players state that they do not want to communicate with any of the other players, i.e., $s_i = s_j = s_k = \emptyset$. This strategy is not a CPNE, because two players i and j can deviate to $t_i = \{j\}$ and $t_j = \{i\}$ and form the link between them to get 30 rather than 0 and then neither one of these players wants to deviate further.

A structure with one link, say link $\{i,j\}$, can only be formed in a Nash equilibrium if $s_i = \{j\}$, $s_j = \{i\}$, and $s_k = \emptyset$. But players i and k have an incentive to deviate to the strategies $t_i = \{j,k\}$ and $t_k = \{i\}$ and form an additional link. This will give player i 44 rather than 30 and player k 14 rather than 0 and neither i nor k wants to deviate further because they do not want to break links and they cannot form new links. This shows that a structure with one link will not be formed in a CPNE.

In a Nash equilibrium, a structure with two links, say $\{i,j\}$ and $\{j,k\}$, can only be formed if $s_i = \{j\}$, $s_j = \{i,k\}$, and $s_k = \{j\}$. But players i and k have an incentive to deviate to the strategies $t_i = \{j,k\}$ and $t_k = \{i,j\}$ and form an additional link, so that they will each get 24 rather than 14. They will not want to deviate further, since this can only involve breaking links.

So, the only structure that can possibly be supported by a CPNE is the full communication structure. Suppose $s_i = \{j,k\}, s_j = \{i,k\}$, and $s_k = \{i,j\}$. The only deviations from these strategies that give all deviating players a higher payoff, are deviations by two players who break the links with the third player and induce the structure with only the link between themselves. Suppose players i and j deviate to the strategies $t_i = \{j\}$ and $t_j = \{i\}$ which will give both players 30 rather than 24. Then player i has an incentive to deviate further to $u_i = \{j,k\}$, in which case links $\{i,j\}$ and $\{i,k\}$ will be formed and player i will get 44 instead of 30. This shows that deviations from s by two players are not stable against further deviations by subcoalitions of the deviating coalition. Hence, s is a CPNE.

What will happen in this example if establishing communication links is not costless? Of course, for small costs, there will only be minor changes to the discussion above and the conclusion will be unchanged. But if the costs are larger, then some of the deviations that were previously taken into consideration will no longer be attractive. Suppose for instance that $c = 24$. Then *all* players will prefer a structure with two links above the structure with three links, in which they all get 0. In a structure with two links, no player wants to break any links, since this will reduce his or her payoff by 2. Hence, for these costs, exactly two links will be formed in a CPNE.

We now continue with the description of coalition proof Nash equilibria in symmetric 3-player games. Table 3 provides an overview of coalition proof Nash equilibria depending on the position a player prefers most.

Table 3. Coalition proof Nash equilibria depending on preferences of the players

Most preferred position	Additional condition	Structure resulting from CPNE
1		0 links
2		1 link
3	$5 \succ 4$	3 links
3	$4 \succ 5$	2 links
4	$\nu_3(N, v, L, c) > 0$	2 links
4	$\nu_3(N, v, L, c) < 0$	0 links
5		3 links

This table can be used to determine the coalition proof Nash equilibria for the three classes of games we distinguished in Sect. 4.3. The following figures describe the communication structures resulting from coalition proof Nash equilibria. Figure 11 describes the structures resulting from CPNE for the class of games containing only non-superadditive games, Fig. 12 is for the class of games containing only superadditive but non-convex games, and Fig. 13 deals with the class of games containing only convex games.

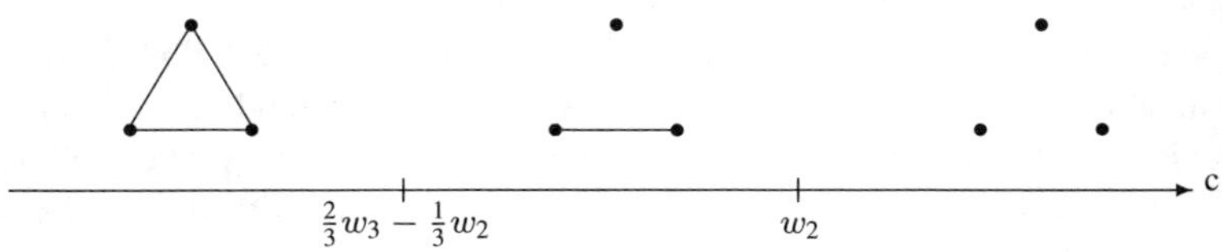

Fig. 11. Communication structures according to CPNE in case $w_2 > w_3$

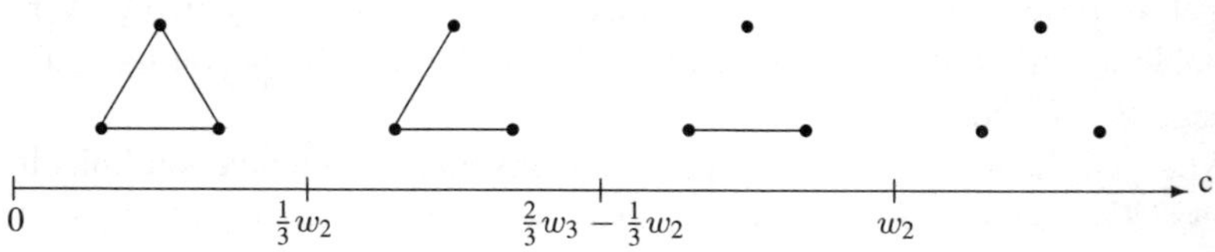

Fig. 12. Communication structures according to CPNE in case $2w_2 > w_3 > w_2$

In the discussion above we restricted our analysis of coalition proof Nash equilibria to symmetric 3-player games for clarity of exposition. However, the following theorem addresses the existence of CPNE for 3-player games in general to show that the analysis using coalition proof Nash equilibria can be extended to such games. The proof of this theorem can be found in the appendix.

Theorem 1. *Let (N, v) be a 3-player cooperative game and let $c \geq 0$ be the costs of establishing a communication link. Then there exists a coalition proof Nash equilibrium in the link formation game $\Gamma(N, v, c, \nu)$.*

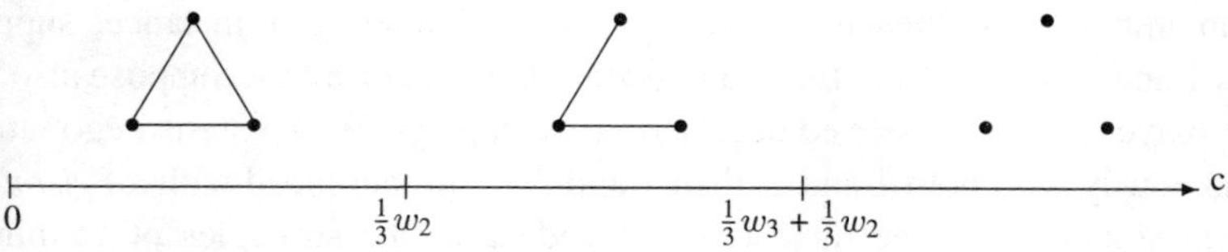

Fig. 13. Communication structures according to CPNE in case $w_3 > 2w_2$

6 Comparison of the Linking Games

In this section we compare the two models of link formation studied in the previous sections. We start with an illustration of the differences between these models in the absence of cooperation costs.[10] Subsequently, we analyze and compare some of the results of the previous sections.

To illustrate the differences between the model of link formation in extensive form and the model of link formation in strategic form, we assume $c = 0$ and we consider the 3-person game (N, v) with player set $N = \{1, 2, 3\}$ and

$$v(S) := \begin{cases} 0 & \text{if } |S| \leq 1 \\ 60 & \text{if } |S| = 2 \\ 72 & \text{if } S = N \end{cases} . \tag{8}$$

This game was also studied in Sects. 4.2 and 5.2. The prediction of the linking game in extensive form is that exactly one link will be formed. Suppose that, at some point in the game, link $\{1, 2\}$ is formed. Notice that either of 1 and 2 gain by forming an additional link with 3, provided that the other player does not form a link with 3. Two further points need to be noted. Firstly, if player i forms a link with 3, then it is in the interest of j $(j \neq i)$ to also link up with 3. Secondly, if *all* links are formed, then players 1 and 2 are worse-off compared to the graph in which they alone form a link. Hence, the structure $(N, \{\{1, 2\}\})$ is sustained as an 'equilibrium' by a pair of mutual threats of the kind :

"If you form a link with 3, then so will I."

Of course, this kind of threat makes sense only if i will come to know whether j has formed a link with 3. Moreover, i can acquire this information only if the negotiation process is public. If bilateral negotiations are conducted secretly, then it may be in the interest of some pair to conceal the fact that they have formed a link until the process of bilateral negotiations has come to an end. It is also clear that if different pairs can carry out negotiations simultaneously and if links once formed cannot be broken, then the mutual threats referred to earlier cannot be carried out.[11]

[10] Parts of the current section are taken from an unpublished preliminary version of Dutta et al. (1998)

[11] Aumann and Myerson (1988) also stress the importance of perfect information in deriving their results.

Thus, there are many contexts where considerations other than threats may have an important influence on the formation of links. For instance, suppose players 1 and 2 have already formed a link amongst themselves. Suppose also that neither player has as yet started negotiations with player 3. If 3 starts negotiations simultaneously with both 1 and 2, then 1 and 2 are in fact faced with a *Prisoners' Dilemma* situation. To see this, denote l and nl as the strategies of forming a link with 3 and not forming a link with 3, respectively. Then, the payoffs to 1 and 2 are described by the following matrix (the first entry in each box is 1's payoff, while the second entry is 2's payoff).

Player 2

		l	nl
Player 1	l	$(24, 24)$	$(44, 14)$
	nl	$(14, 44)$	$(30, 30)$

Note that l, that is forming a link with 3, is a dominant strategy for both players. Obviously, in the linking game in strategic form, the complete graph will form simply because players 1 and 2 cannot sign a binding agreement to abstain from forming a link with 3.

The rest of this section is devoted to a discussion of the cost-graph patterns as derived in the previous sections. For the linking game in extensive form, we considered subgame perfect Nash equilibria. The equilibrium concept for the linking game in strategic form that is most closely related to subgame perfection is that of undominated Nash equilibrium. However, it appears from Figs. 8 through 13 that in some cases there is still a multiplicity of structures resulting from undominated Nash equilibria and that the structures resulting from coalition proof Nash equilibria are a refinement of the structures resulting from undominated Nash equilibria.[12] Therefore, we compare the cost-graph patterns for subgame perfect Nash equilibria in the linking game in extensive form with those for coalition proof Nash equilibria in the linking game in strategic form.

Comparing Figs. 2, 3, and 4 to Figs. 11, 12, and 13, respectively, we find that the predictions according to SPNE in the extensive-form model and those according to CPNE in the strategic-form model are remarkably similar.

For a class containing only convex games ($w_3 > 2w_2$), both models generate exactly the same predictions (see Figs. 4 and 13).

For non-superadditive games, we get almost the same predictions. The only difference between Figs. 2 and 11 is that the level of costs that marks the transition from the full communication structure to a structure with one link is possibly positive ($\frac{2}{3}w_3 - \frac{1}{3}w_2$) in the extensive-form model, whereas it is negative ($\frac{2}{3}w_3 - w_2$) in the strategic-form model.[13] Note that considering undominated Nash equilibria instead of coalition proof Nash equilibria for the linking game in strategic form will only aggravate this difference.

[12] We remark that, even for the (3-person) linking game in strategic form, CPNE is not a refinement of UNE on the strategy level.

[13] See the discussion of Fig. 2 on page 242.

The predictions of both models are most dissimilar for the class containing only superadditive non-convex games ($2w_2 > w_3 > w_2$). In the extensive-form model we get a structure with one link in case $\frac{2}{3}w_3 - w_2 < c < \frac{1}{3}w_2$ (see Fig. 3), whereas in the strategic-form model for these costs we get the full communication structure (see Fig. 12). For lower costs we find the full communication structure for both models.

The discussion on mutual threats at the start of this section is applicable to all games in the class containing only superadditive non-convex games ($2w_2 > w_3 > w_2$). Not only is the difference between the predictions of both models of link formation a result of the validity of mutual threats in the extensive-form model, so is the remarkable result that higher costs may result in more links being formed in the linking game in extensive form. For high cost, the mutual threats will no longer be credible. Such a threat is not credible since executing it would permanently decrease the payoff of the player who executes it.

We conclude the section with a short discussion of the efficiency of the graphs formed in equilibrium. Jackson and Wolinsky (1996) establish that there is a conflict between efficiency and stability if the allocation rule used is component balanced. Indeed, we see many illustrations of this result in the current paper. For example, for the strategic-form model of link formation we find in Sect. 5.4 that for small costs all links will be formed.[14] This is clearly not efficient, because the (costly) third link does not allow the players to obtain higher economic profits. Rather, building this costly link diminishes the profits of the group of players as a whole. It is formed only because it influences the allocation of payoffs among the players. The formation of two links in case the game is superadditive (see Figs. 12 and 13) is promising in this respect. However, from an efficiency point of view these should be formed if $w_3 - 2c > w_2 - c$, or $c < w_3 - w_2$, and the cutoffs in Figs. 12 and 13 appear at different values for c.

7 Extensions

In this section we will extend our scope to games with more than three players. We study to what extend our results of the previous sections with respect to games with three players do or do not hold for games with more players.

The first point of interest is whether we will again find a division of games into three classes, non-superadditive games, superadditive but non-convex games, and convex games when studying network formation for games with more than three players. The following two examples of symmetric 4-player games illustrate that this is not the case. In these examples we consider two different superadditive games that are not convex. However, the patterns of structures formed according to subgame perfect Nash equilibria of the linking game in extensive from, are shown to be different for these games.

The first example we consider is the symmetric 4-player game (N, v_1) described by $w_1 = 0$, $w_2 = 60$, $w_3 = 180$, and $w_4 = 260$. Some tedious calculations,

[14] For costs equal to zero, this follows directly from the results obtained by Dutta et al. (1998).

to which we will not subject the reader, show that for this game, the structures that are formed according to subgame perfect Nash equilibria of the linking game in extensive form are as represented in Fig. 14.

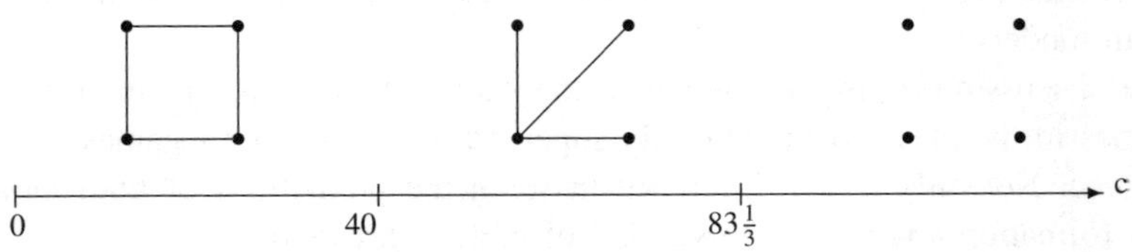

Fig. 14. Communication structures according to SPNE for the game (N, v_1)

Note that for the game (N, v_1), according to subgame perfect Nash equilibria, the number of links decreases as the cost per link c increases.

Different structures are formed for the second symmetric 4-player game we consider, (N, v_2) described by $w_1 = 0$, $w_2 = 12$, $w_3 = 180$, and $w_4 = 220$. Using backward induction, it is fairly easy to show that if $c = 10$ then all subgame perfect Nash equilibria result in the formation of exactly two links connecting three players with each other as represented in Fig. 15a.

Fig. 15. Communication structures according to SPNE for the game (N, v_2)

If $c = 40$, however, subgame perfect Nash equilibria result in the formation of a star graph with 3 links as represented in Fig. 15b.

Consideration of the two games (N, v_1) and (N, v_2), which are both superadditive and not convex, shows that for symmetric games with more than three players, the relationship between costs and number of links formed cannot be related back simply to superadditivity and/or convexity of the game[15]. This is not very surprising since, opposed to zero-normalized symmetric 3-player games, for zero-normalized symmetric games with more than three players, superadditivity and convexity cannot be described by a single inequality. For a zero-normalized symmetric 4-player game there are 3 conditions for a game to be superadditive ($w_3 > w_2$, $w_4 > w_3$, and $w_4 > 2w_2$) and 2 conditions for a game to be convex ($w_3 - w_2 > w_2$ and $w_4 - w_3 > w_3 - w_2$). For zero-normalized symmetric

[15] Note that we might have something similar to what we observed when comparing Figs. 2 and 11, and that the level of costs that would mark a transition from a structure like in Fig. 15a would be negative for the game (N, v_1). However, we can show that the patterns of structures formed in subgame perfect Nash equilibria as costs increase are different for the two games (N, v_1) and (N, v_2).

3-player games it is really the two conditions for superadditivity and convexity that are important. Following this line of thought, we are lead to consider the possibility that for zero-normalized symmetric 4-player games we will get patterns of communication structures formed that depend on which of the five superadditivity and convexity conditions are satisfied by the game. However, this turns out not to be true. A counterexample is provided by the games (N, v_1) and (N, v_2) discussed above. These games both satisfy all superadditivity conditions and exactly one convexity condition, namely $w_3 - w_2 > w_2$. Nevertheless, we already saw that the patterns of communication structures formed according to subgame perfect Nash equilibria differ. Relating back the relationship of costs and structures formed remains the subject of further research.

The most interesting result that we obtain for symmetric 3-player games is that in the linking game in extensive form it is possible that as the cost of establishing links increases, more links are formed. This result can be extended to games with more than 3 players. The game (N, v_2) that we saw earlier in this section is a symmetric 4-player game for which communication structures formed according to subgame perfect Nash equilibria have two links if $c = 10$ but have 3 links if $c = 40$. So, an increase in costs can result in an increase in the number of links formed according to subgame perfect Nash equilibria. By means of an example, we will show in the remainder of this section that for n-player games with n odd, it is possible that as the cost for establishing links increases, more links are formed according to subgame perfect Nash equilibria of the linking game in extensive form.

Let $n \geq 3$ be odd and let $N = \{1, \ldots, n\}$. Consider the symmetric n-player game (N, v_n) described by $w_1 = 0$, $w_2 = 60$, $w_3 = 72$, and $w_k = 0$ for all $k \in \{4, \ldots, n\}$. Let $c = 2$ and let s be a subgame perfect Nash equilibrium of the linking game in extensive form.[16] Denote by $L(s)$ the links that are formed if s is played. Firstly, note that $(N, L(s))$ does not contain a component with 4 or more players. This is true, because in such a component at least one player would get a negative payoff according to the cost-extended Myerson value.[17] Such a player would have a payoff of zero if he refused to form any link. Hence, for any $C \in N/L(s)$ it holds that $|C| \in \{1, 2, 3\}$. Suppose $C \in N/L(s)$ such that $|C| = 3$. Then the players in C are connected by 3 links such that they are all in position 5 (see Fig. 1) and each gets a payoff of 22. This follows because if two players in C are in position 4, then they both get 13 and they both prefer to form a link between them to get 22 instead. We conclude that for every $C \in N/L(s)$ either $|C| = 1$, or $|C| = 2$ and both players in C get 29 each, or $|C| = 3$ and each player in C gets 22. This, in turn, leads to the conclusion that there exists no $C \in N/L(s)$ with $|C| = 3$, because if this were the case, then at some point in the game tree (which may or may not be reached during actual play) a player who is connected to exactly one other player and would receive 29 if he makes no further links, chooses to make a link with a third

---

[16] Recall that subgame perfect Nash equilibria exist.

[17] Component balancedness of the cost-extended Myerson and the positive costs for links imply that the players in a component with more than 3 players divide a negative value amongst themselves.

player and then ends up getting only 22. This would clearly not be behavior that is consistent with subgame perfection. We also argue that there can be at most one $C \in N/L(s)$ with $|C| = 1$, because if there were at least two isolated players, then two of these players can increase their payoffs from 0 to 29 by forming a link.

Hence, there is at most one $C \in N/L(s)$ with $|C| = 1$ and for all other $C \in N/L(s)$ it holds that $|C| = 2$. Since n is odd, this means that exactly $\frac{n-1}{2}$ links are formed in a subgame perfect Nash equilibrium of the linking game in extensive form.[18]

Now, let $c = 22$ and let s be a subgame perfect Nash equilibrium of the linking game in extensive form with this higher cost and denote by $L(s)$ the links that are formed if s is played. As before, it easily follows that for every $C \in N/L(s)$ it holds that $|C| \in \{1, 2, 3\}$. The payoff to a player in position 5 would be 2, whereas the payoff to a player in position 4 is 3. Hence, there will be no $C \in N/L(s)$ consisting of 3 players who are connected by 3 links. Further, there can obviously be no more than one isolated player in $(N, L(s))$. Suppose that there is an isolated player, i.e., there is a $C \in N/L(s)$ with $|C| = 1$. Then there can be no $C \in N/L(s)$ with $|C| = 2$, since one of the players who is connected to exactly one other player could improve his payoff from 19 to 22 by forming a link with an isolated player, whose payoff would then increase from 0 to 3, and both improvements would be permanent. Since n is odd, it is not possible that $|C| = 2$ for all $C \in N/L(s)$. Then, we are left with two possibilities. The first possibility is that there is a $C \in N/L(s)$ with $|C| = 1$ and all other components of $(N, L(s))$ each consist of 3 players who are connected by 2 links. Note that this can only be the case if there exists a $k \in \mathbb{N}$ such that $n = 3k + 1$. Then, $|L(s)| = 2k = \frac{2(n-1)}{3} \geq \frac{n+1}{2}$. The second possibility is that there is no isolated player in $(N, L(s))$ and each component of $(N, L(s))$ consists either of 3 players who are connected by 2 links or it consists of 2 players who are connected by 1 link. Since n is odd, there must be at least one component consisting of three players. We conclude that also in this case $|L(s)| \geq \frac{n+1}{2}$.

Summarizing, we have that for the game (N, v_n) with $n \geq 3$, n odd, if $c = 2$, then in a subgame perfect Nash equilibrium $\frac{n-1}{2}$ links are formed and if $c = 22$, then $\frac{n+1}{2}$ or more links are formed. Hence, we have shown that for games with more than 3 players it is still possible that the number of links formed in a subgame perfect Nash equilibrium increases as the costs for establishing links increases.

8 Conclusions

In this paper, we explicitly studied the influence of costs for establishing communication links on the communication structures that are formed in situations where the underlying economic possibilities of the players are given by a cooperative

[18] If n were even, then $|L(s)| = \frac{n}{2}$.

game. To do so, we considered two existing models of the formation of communication networks, the extensive-form model of Aumann and Myerson (1988) and the strategic-form model studied by Dutta et al. (1998). For these models, we studied how the communication networks that are formed change as the costs for establishing links increase. In order to be able to isolate the influence of the costs, we assumed that costs are equal for all possible communication links. We mainly restricted our analysis to 3-player symmetric games because our proofs involve explicit computations and this of course puts severe restrictions on the type of situations that we can analyze while not loosing ourselves and the readers in complicated computations. The proof of the existence of coalition-proof Nash equilibria in the strategic-form game of link formation for general 3-player games provides a glimpse of the type of difficulties that we would have to deal with if we extended our analysis beyond symmetric games.

In the extensive-form game of link formation of Aumann and Myerson (1988), we considered communication structures that are formed in subgame perfect Nash equilibria. We find that for this game, with 3 symmetric players, the pattern of structures formed as costs increase depends on whether the underlying coalitional game is superadditive and/or convex. In case the underlying game is not superadditive or in case it is convex, increasing costs for forming communication links result in the formation of fewer links in equilibrium. However, if the underlying game is superadditive but not convex, then increasing costs initially lead to the formation of fewer links, then to the formation of more links, and finally lead to the formation of fewer links again. We show that the possibility that increasing costs for establishing links lead to more links being formed, is still present for games with more than 3 players. This is, in our view, the most surprising result of the paper. It shows that subsidizing the formation of links does not necessarily lead to more links being formed. Hence, authorities wishing to promote more cooperation cannot always rely on subsidies to accomplish this goal. In fact, such subsidies might have an adverse effect.

For the strategic-form game of link formation studied by Dutta et al. (1998) we briefly discussed the inappropriateness of Nash equilibria and strong Nash equilibria and went on to consider undominated Nash equilibria and coalition-proof Nash equilibria. We find that for this game, with 3 symmetric players, the pattern of structures formed as costs increase also depends on whether the underlying coalitional game is superadditive and/or convex. In contrast to the results for the extensive-form game of link formation, we find that in the strategic-form model in all cases increasing costs for forming communication links result in the formation of fewer links in equilibrium. The results we obtain for the two models are otherwise remarkably similar.

In order to follow the analyses in Aumann and Myerson (1988) and Dutta et al. (1998) as closely as possible, we extended the Myerson value to situations in which the formation of links is not costless. We did so in a manner that is consistent with the philosophy of the Myerson value. The Myerson value was introduced by Myerson (1977) as the unique allocation rule satisfying component balancedness and fairness. Myerson's analysis was restricted to situations in which the formation of communication links is costless. Jackson and Wolinsky (1996) note that Myerson's result can be extended to situations in which a value function describes the economic possibilities of the players in different networks (see Jackson and Wolinsky 1996, Theorem 4 on page 65). It seems reasonable to view the unique allocation rule for such situations that is component balanced and fair as the natural extension of the Myerson value. Since costs for forming links can be implicitly taken into account using value functions, this extension of the Myerson value can be used to determine allocations when an underlying cooperative game describes the economic possibilities of the players and in which there are costs for forming links. It is this allocation rule that we use.

Appendix

This appendix is devoted to the existence of CPNE for general 3-player games. Hence, we extend the scope of our investigation beyond symmetric games. We do, however, still restrict ourselves to zero-normalized non-negative games. For convenience, we will assume (without loss of generality) that

$$v(\{1,2\}) \geq v(\{1,3\}) \geq v(\{2,3\}).$$

Throughout the rest of this appendix we call a deviation by a coalition *profitable* if it strictly improves the payoffs of all deviating players. A deviation is called *stable* if the deviation is a coalition proof Nash equilibrium in the subgame induced on the coalition of deviating players by the strategies of the other players, i.e., a deviation from strategy profile s by coalition T is stable if it is a CPNE in the game $\Gamma(s_{N\setminus T})$ as defined on page 248. A deviation is called self-enforcing if this deviation is self-enforcing in the subgame induced on the coalition of deviating players by the strategies of the other players, i.e., a deviation from strategy profile s by coalition T is self-enforcing if it is self-enforcing in the game $\Gamma(s_{N\setminus T})$

The following lemmas will be used in the proof of existence of coalition proof Nash equilibria in 3-player games.

Lemma 1. *Let $\Gamma(N,v,c,\nu)$ be a 3-player link formation game with $c < \frac{2}{3}v(N) + \frac{1}{3}v(\{1,3\}) - \frac{2}{3}v(\{1,2\})$. Let $\bar{s}$ be the strategy profile with $\bar{s}_i = N\setminus\{i\}$ for all $i \in N$ which results in the full communication structure. Let $i,j \in N$. Then the deviation from $\bar{s}$ by $\{i,j\}$ given by $(s_i,s_j) = (\{j\},\{i\})$ is not stable.*

Proof. We will show that there exists a further deviation of (s_i,s_j) which is profitable and stable, implying that (s_i,s_j) is not stable. First, assume $\{i,j\} = \{1,2\}$. Consider a further deviation $t_1 = \{2,3\}$ by player 1. Then[19]

$$f_1^{\nu}(t_1,s_2,\bar{s}_3) = \nu_1(\{\{1,2\},\{1,3\}\}) = \frac{1}{3}v(N) + \frac{1}{6}v(\{1,3\}) + \frac{1}{6}v(\{1,2\}) - c$$
$$> \frac{1}{2}v(\{1,2\}) - \frac{1}{2}c = \nu_1(\{\{1,2\}\}) = f_1^{\nu}(s_1,s_2,\bar{s}_3),$$

[19] If there is no ambiguity about (N,v,c) we simply write $\nu(L)$ instead of $\nu(N,v,L,c)$.

where the inequality follows since $c < \frac{2}{3}v(N) + \frac{1}{3}v(\{1,3\}) - \frac{2}{3}v(\{1,2\})$. Since the strategy space of a player is finite there exists a strategy of player 1 that maximizes his payoff, given strategies $(s_2, \bar{s}_3)$ of players 2 and 3. This strategy is a profitable and stable deviation from (s_1, s_2). We conclude that (s_1, s_2) is not stable.

Similarly, by considering $t_1 = \{2,3\}$ we find that there exists a profitable and stable further deviation if $\{i,j\} = \{1,3\}$ and considering $t_2 = \{1,3\}$ implies that there exists a profitable and stable further deviation if $\{i,j\} = \{2,3\}$. In both cases we use that $v(\{1,2\}) \geq v(\{1,3\}) \geq v(\{2,3\})$. □

Lemma 2. *Let $\Gamma(N, v, c, \nu)$ be a link formation game. Let s be a strategy profile. If there exists a profitable and self-enforcing deviation from s by N, then the game has a CPNE.*

Proof. Suppose t^1 is a deviation from s by N that is profitable and self-enforcing. Since t^1 is a self-enforcing deviation by N, there exists no profitable and stable deviation from t^1 by any $S \subset N$. If there is no profitable and self-enforcing deviation from t^1 by N then t^1 is a CPNE. If t^2 is a profitable and self-enforcing deviation from t^1 by N, then

$$f^{\nu}(t^2) > f^{\nu}(t^1) > f^{\nu}(s).$$

Repeat the process above to find a sequence $(t^1, t^2, t^3, \ldots)$ such that t^k is a profitable and self-enforcing deviation from t^{k-1} for all $k \geq 2$. It holds that

$$f^{\nu}(t^1) < f^{\nu}(t^2) < f^{\nu}(t^3) < \ldots.$$

Since the strategy space of every player is finite this process has to end in finitely many steps. The last strategy profile in the sequence is a CPNE. □

We can now prove that coalition proof Nash equilibria exist in 3-player link formation games in strategic form.

Proof of Theorem 1. If $(\emptyset, \emptyset, \emptyset)$ is a CPNE we are done. From now on assume $(\emptyset, \emptyset, \emptyset)$ is not a CPNE.

Hence, there exists a profitable and stable deviation from $(\emptyset, \emptyset, \emptyset)$ by some $T \subset N$ or a profitable and self-enforcing deviation by N. If there exists a profitable and self-enforcing deviation by N it follows by Lemma 2 that we are done. So, from now on assume there exists no profitable and self-enforcing deviation from $(\emptyset, \emptyset, \emptyset)$ by N. Hence there exists a profitable and stable deviation from $(\emptyset, \emptyset, \emptyset)$ by some $T \subset N$. Since a player cannot unilaterally enforce the formation of a link, we conclude that there exists a profitable and stable deviation by a coalition with (exactly) two players.

So, there exists a profitable and stable deviation from $(\emptyset, \emptyset, \emptyset)$ by 2 players, say i and j. The structures players i and j can enforce are the structure with no links and the structure with link $\{i,j\}$. Since the structure with no links does not change their payoffs, it follows that this profitable and stable deviation results in link $\{i,j\}$. This deviation is profitable and stable iff $v(\{i,j\}) > c$.[20] Since, $v(\{1,2\}) \geq v(\{i,j\}) > c$ it follows that $(s_1, s_2) = (\{2\}, \{1\})$ is a profitable and stable deviation from $(\emptyset, \emptyset, \emptyset)$ and that $s = (\{2\}, \{1\}, \emptyset)$ is a Nash equilibrium. If s is a CPNE in the game $\Gamma(N, v, c, \nu)$ we are done. So, from now on assume that s is not a CPNE.

Hence, there exists a profitable and stable deviation from s by some $T \subset N$ or a profitable and self-enforcing deviation by N. However, no profitable and self-enforcing deviation by 3 players exists, since this would be a profitable and self-enforcing deviation from $(\emptyset, \emptyset, \emptyset)$. Since s is a Nash equilibrium, we derive that there exists a profitable and stable deviation by a coalition with (exactly) two players. Since coalition $\{1,2\}$ can only break link $\{1,2\}$, it follows that there exists a profitable and stable deviation from s by coalition $\{1,3\}$ or by coalition $\{2,3\}$. We will distinguish between these two cases.

CASE A: There exists a profitable and stable deviation from s by coalition $\{1,3\}$, say (t_1, t_3). Since $v(\{1,2\}) \geq v(\{1,3\})$ it follows that the deviation from s cannot result in link $\{1,3\}$ alone, since this would not improve the payoff of player 1. Hence, the deviation results in links $\{1,2\}$ and $\{1,3\}$, the only two links that can be enforced by players 1 and 3, given the strategy of player 2. Note that such a deviation is profitable if and only if

$$c < \frac{2}{3}v(N) - \frac{2}{3}v(\{1,2\}) + \frac{1}{3}v(\{1,3\}). \tag{9}$$

So, inequality (9) must hold. Since a further deviation by player 1 or player 3 can only result in breaking links, it follows that $(t_1, t_3) = (\{2,3\}, \{1\})$ is a profitable and stable deviation from s. Also, $\nu_2(\{\{1,2\},\{1,3\}\}) \geq \nu_3(\{\{1,2\},\{1,3\}\}) > 0$, where the weak inequality follows since $v(\{1,2\}) \geq v(\{1,3\})$ and the strict inequality follows by inequality (9). It follows that (t_1, s_2, t_3) is a Nash equilibrium, since unilaterally player 2 can only break link $\{1,2\}$. If (t_1, s_2, t_3) is a CPNE in the game $\Gamma(N, v, c, \nu)$ we are done. From now on assume (t_1, s_2, t_3) is not a CPNE.

Since coalitions $\{1,2\}$ and $\{1,3\}$ cannot enforce an additional link, they cannot make a profitable and stable deviation from (t_1, s_2, t_3). There exists no profitable and self-enforcing deviation by N from (t_1, s_2, t_3) since this would be a profitable and self-enforcing deviation from $(\emptyset, \emptyset, \emptyset)$. So, there exists a profitable and stable deviation from (t_1, s_2, t_3) by coalition $\{2,3\}$, say (u_2, u_3). Since both players receive a positive payoff according to (t_1, s_2, t_3), any profitable deviation results in at least the formation of link $\{2,3\}$. Since player 3 receives at least as much in the structure with links $\{1,2\}$ and $\{1,3\}$ as in the structure with links $\{1,2\}$ and $\{2,3\}$ this last structure will not form after deviation (u_2, u_3). Similarly, since player 2 receives at least as much in the structure with links $\{1,2\}$ and $\{1,3\}$ as in the structure with links $\{1,3\}$ and $\{2,3\}$ this last structure will not form after deviation (u_2, u_3). Finally, player 2 prefers the communication structure with links $\{1,2\}$ and $\{2,3\}$ above the communication structure with link $\{2,3\}$ since

[20] We remind the reader that we restrict ourselves to zero-normalized games.

$$c < \frac{2}{3}v(N) - \frac{2}{3}v(\{2,3\}) + \frac{1}{3}v(\{1,2\}),$$

where the inequality follows from inequality (9) and $v(\{1,2\}) \geq v(\{1,3\}) \geq v(\{2,3\})$. So, the deviation by players 2 and 3 to the communication structure with link $\{2,3\}$ alone will not be stable. We conclude that t_1 and deviation (u_2, u_3) together result in the full communication structure. We will show that (t_1, u_2, u_3) is a CPNE in the game $\Gamma(N, v, c, \nu)$. The deviation (u_2, u_3) from (t_1, s_2, t_3) is profitable iff $v(\{2,3\}) > 3c$. But, if $v(\{2,3\}) > 3c$ there is no profitable deviation from (t_1, u_2, u_3) to a structure with two links since $v(\{1,2\}) \geq v(\{1,3\}) \geq v(\{2,3\}) > 3c$. By Lemma 1 and inequality (9) it follows that there is no profitable and stable deviation from (t_1, u_2, u_3) to a structure with one link. Since $v(\{1,2\}) \geq v(\{1,3\}) \geq v(\{2,3\}) > 3c > 0$ it follows that a deviation to the communication structure with no links cannot be stable. We conclude that (t_1, u_2, u_3) is a CPNE, showing the existence of a CPNE in the game $\Gamma(N, v, c, \nu)$ in CASE A.

CASE B: There exists a profitable and stable deviation from s by coalition $\{2,3\}$, say (t_2, t_3). Since $v(\{1,2\}) \geq v(\{2,3\})$ it follows that the deviation from s cannot result in link $\{2,3\}$ alone. Hence, the deviation results in links $\{1,2\}$ and $\{2,3\}$, the only two links that can be enforced by players 2 and 3, given the strategy of player 1. Note that such a deviation is profitable if and only if

$$c < \frac{2}{3}v(N) - \frac{2}{3}v(\{1,2\}) + \frac{1}{3}v(\{2,3\}). \tag{10}$$

However, since $v(\{2,3\}) \leq v(\{1,3\})$ it follows that inequality (10) implies inequality (9). Hence, there exists a profitable and stable deviation from s by coalition $\{1,3\}$. Then CASE A applies and we conclude that a CPNE in the game $\Gamma(N, v, c, \nu)$ exists.

This completes the proof of the theorem. □

References

Aumann, R. (1959) Acceptable points in general cooperative n-person games. In: Tucker, A., Luce, R. (eds.) *Contributions to the theory of games IV.* Princeton University Press, pp. 287–324

Aumann, R., Myerson R. (1988). Endogenous formation of links between players and coalitions: an application of the Shapley value. In: Roth, A. *The Shapley value.* Cambridge University Press, Cambridge, United Kingdom, pp. 175–191

Bala, V., Goyal, S. (2000) A non-cooperative theory of network formation. *Econometrica* 68: 1181–1229

Dutta, B., van den Nouweland, A., Tijs, S. (1998) Link formation in cooperative situations. *International Journal of Game Theory* 27: 245–256

Dutta, B., Mutuswami, S. (1997) Stable Networks. *Journal of Economic Theory* 76: 322–344

Goyal, S. (1993) Sustainable Communication Networks. Discussion Paper TI 93-250, Tinbergen Institute, Erasmus University, Rotterdam

Jackson, M., Wolinsky, A. (1996) A Strategic Model of Social and Economic Networks. *Journal of Economic Theory* 71: 44–74

Myerson, R. (1977) Graphs and cooperation in games. *Mathematics of Operations Research* 2: 225–229

Myerson, R. (1991) *Game theory: Analysis of conflict.* Harvard University Press, Cambridge, Massachusetts

Qin, C. (1996) Endogenous formation of cooperation structures. *Journal of Economic Theory* 69: 218–226

Selten, R. (1965) Spieltheoretische Behandlung eines Oligopolmodells mit Nachfragetraegheit. *Zeitschrift für die gesamte Staatswissenschaft* 121: 301–324, 667–689

Shapley, L. (1953) A value for n-person games. In: Tucker, A., Kuhn, H. (eds.) *Contributions to the theory of games II* pp. 307–317

Slikker, M. (1998) A note on link formation. CentER Discussion Paper 9820, Tilburg University, Tilburg, The Netherlands

Slikker, M., van den Nouweland, A. (2001) A one-stage model of link formation and payoff division. *Games and Economic Behavior* 34: 153–175

van den Nouweland, A. (1993) Games and graphs in economic situations. Ph. D. Dissertation, Tilburg University Press, Tilburg, The Netherlands

Watts, A. (1997) A Dynamic Model of Network Formation. Working paper

Network Formation With Sequential Demands

Sergio Currarini[1], **Massimo Morelli**[2]

[1] Department of Economics, University of Venice, Cannaregio N° 873, 30121 Venezia, Italy (e-mail: s.currarini@rhbnc.ac.uk)
[2] Department of Economics, Ohio State University, 425 ARPS Hall, 1945 North High Street, Columbus, OH 43210, USA (e-mail: morelli@economics.sbs.ohio-state.edu)

Abstract. This paper introduces a non-cooperative game-theoretic model of sequential network formation, in which players propose links and demand payoffs. Payoff division is therefore endogenous. We show that if the value of networks satisfies size monotonicity, then each and every equilibrium network is efficient. The result holds not only when players make absolute participation demands, but also when they are allowed to make link-specific demands.

JEL Classification: C7

Key Words: Link formation, efficient networks, payoff division

1 Introduction

We analyze the formation process of a cooperation structure (or network) as a non-cooperative game, where players move sequentially. The main difference between this paper and the seminal work in this area by Aumann and Myerson (1988) is that we are interested in situations in which it is impossible to pre-assign a fixed imputation to each cooperation structure, i.e., situations in which the distribution of payoffs is *endogenous*.[1] Indeed, the formation of international cooperation networks, and, more generally, of any market network, occurs

We wish to thank Yossi Feinberg, Sanjeev Goyal, Andrew McLennan, Michael Mandler, Tomas Sjöström, Charles Zheng, an anonymous referee, and especially Matthew Jackson, for their useful comments. We thank John Miranowski for giving us the opportunity to work together on this project at ISU. We would also like to thank the workshop participants at Columbia, Penn State, Stanford, Berkeley, Minnesota, Ohio State, and the 1998 Spanish game theory meetings. The usual disclaimer applies.

[1] Slikker and Van Den Nouweland (2001) studied a link formation game with endogenous payoff division but with a simultaneous-move framework.

through a bargaining process, in which the *demand* of a payoff for participation is a crucial variable.

The most important theoretical debate stemming from Aumann and Myerson (1988) is about the potential conflict between efficiency and stability of networks. In the example of sequential network formation game studied by Aumann and Myerson the specific imputation rule that they consider (the Myerson value) determines an inefficient equilibrium network. The implication of their paper is therefore that *not all* fixed allocation rules are compatible with efficiency, even if the game is sequential. Jackson and Wolinsky (1996) consider value functions depending on the communication structure rather than on the set of connected players and demonstrate that efficiency and stability are indeed incompatible under fairly reasonable assumptions (anonymity and component balancedness) on the fixed imputation rules. Their approach is axiomatic, and hence their result does not have direct connections with the Aumann and Myerson result, which was obtained in a specific extensive form game. The strong conclusion of Jackson and Wolinsky is that *no* fixed allocation rule would ensure that at least one stable graph is efficient for every value function.[2] Dutta and Mutuswami (1997) show, on the other hand, that a mechanism design approach (where the allocation rules themselves are the mechanisms to play with) can help reconcile efficiency and stability. In particular, they solve the impossibility result highlighted by Jackson and Wolinsky by imposing the anonimity axiom only on the equilibrium network. With a similar mechanism design approach, one could probably find fixed allocation rules that lead to efficient network formation in sequential games like the one of Aumann and Myerson. However, since in many situations of market network formation there is no mechanism designer who can select the "right" allocation mechanism, we are here interested to ask what happens to the conflict between efficiency and stability discussed above when payoff division is endogenous.

The main result of this paper is that, if the value function satisfies size monotonicity (i.e., if the efficient networks connect all players in some way), then the sequential network formation process with endogenous payoff division leads *all* equilibria to be efficient (Theorem 2). As shown in Example 2, there exist value functions satisfying size monotonicity for which *no* allocation rule can eliminate inefficient equilibria when the game is simultaneous move, nor with the Jackson and Wolinsky concept of stability. So our efficiency result could not be obtained without the sequential structure of the game. We will also show (see Example 3) that the sequential structure alone, without endogenous payoff division, would not be sufficient.

In the game that we most extensively analyze, we assume that players propose links and formulate a single *absolute* demand, representing their final payoff demand. This is representative of situations such as the formation of economic unions, in which negotiations are multilateral in nature, and each player (country) makes an absolute claim on the total surplus from cooperation. We will show

[2] See also Jackson and Watts (2002) and Qin (1996).

that the result that all equilibria are efficient extends to the case in which players attach to each proposed link a separate payoff demand.

The next section describes the model and presents the link formation game. Section 3 contains the analysis of the Subgame Perfect Equilibria of the game, the main results, and a discussion of them. Section 4 presents the extension to link-specific demands, and Sect. 5 concludes.

2 The Model

2.1 Graphs and Values

Let $N = \{1, \ldots, n\}$ be a finite set of players. A graph g is a set L of links (non-directed segments) joining pairs of players in N (nodes). The graph containing a link for every pair of players is called complete graph, and is denoted by g^N. The set G of all possible graphs on N is then $\{g : g \subseteq g^N\}$. We denote by ij the link that joins players i and j, so that if $ij \in g$ we say that i and j are directly connected in the graph g. For technical reasons, we will say that each player is always connected to himself, i.e. that $ii \in g$ for all $i \in N$ and all $g \in G$. We will denote by $g + ij$ the graph obtained adding the link ij to the graph g, and by $g - ij$ the graph obtained removing the link ij from g.

Let $N(g) \equiv \{i : \exists j \in N \text{ s.t. } ij \in g\}$. Let $n(g)$ be the cardinality of $N(g)$. A *path* in g connecting i_1 and i_k is a set of nodes $\{i_1, i_2, \ldots, i_k\} \subseteq N(g)$ such that $i_p i_{p+1} \in g$ for all $p = 1, \ldots, k - 1$.

We say that the graph $g' \subset g$ is a *component* of g if

1. for all $i \in N(g')$ and $j \in N(g')$ there exists a path in g' connecting i and j;
2. for any $i \in N(g')$ and $j \in N(g)$, $ij \in g$ implies that $ij \in g'$.

So defined, a component of g is a maximal connected subgraph of g. In what follows we will use the letter h to denote a component of g (obviously, when all players are indirectly or directly connected in g the graph g itself is the unique component of g). Note that according to the above definition, each isolated player in the graph g represents a component of g. The set of components of g will be denoted by $C(g)$. Finally, $L(g)$ will denote the set of links in g.

To each graph $g \subseteq g^N$ we associate a value by means of the function $v : G \to R_+$. The real number $v(g)$ represents the aggregate utility produced by the set of agents N organized according to the graph (or network) g. We say that a graph g^* is *efficient* with respect to v if $v(g^*) \geq v(g)\ \forall g \subseteq g^N$. $G^*(v)$ will denote the set of efficient networks relative to v.

We restrict the analysis to *anonymous* and additive value functions, i.e., such that $v(g)$ does not depend on the identity of the players in $N(g)$ and such that the value of a graph is the sum of the values of its components.

2.2 The Link Formation Game

We will study a sequential game $\Gamma(v)$, in which agents form links and formulate payoff demands. In this section we consider the benchmark case in which each agent's demand consists of a positive real number, representing his demanded payoff in the game.

In the formulation of the game $\Gamma(v)$, it will be useful to refer to some additional definitions. A pre-graph on N is a set A of directed arcs (directed segments joining two players in N). The arc from player i to player j is denoted by a_i^j. The set of arcs A uniquely induces the graph

$$g(A) \equiv \left\{ ij \in g^N : a_i^j \in A \text{ and } a_j^i \in A \right\}.$$

2.2.1 Players, Aactions, and Histories

In the game $\Gamma(v)$ the set of players $N = \{1, \ldots, i, \ldots, n\}$ is exogenously ordered by the function $\rho : N \to N$. We use the notation $i \leq j$ as equivalent to $\rho(i) \leq \rho(j)$. Players sequentially choose actions according to the order ρ. An action x_i for player i is a pair (a_i, d_i), where a_i is a vector of arcs sent by i to some subset of players in $N \backslash i$ and $d_i \in [0, D]$ is i's payoff demand, where D is some positive finite real number.[3]

A history $x = (x_1, \ldots, x_n)$ is a vector of actions for each player in N. We will use the notation (borrowed from Harris 1985)

$$\lambda_i x \equiv (x_1, \ldots, x_{i-1})$$

to identify a subgame. We denote by X the set of possible histories, by $\lambda_i X$ the set of possible histories before player i and by X_i the set of possible actions for player i.

2.2.2 From Histories to Graphs

Players' actions induce graphs on the set N as follows. Firstly, we assume that at the beginning no links are formed, i.e., the game starts from the empty graph $g = \{\emptyset\}$. The history x generates the graph $g(x)$ according to the following rule. Let $A(x) \equiv (a_1, \ldots, a_n)$ be the arcs sent by the players in the history x.

- If h is a component of $g(A(x))$ and h is *feasible given* x, i.e., if

$$\sum_{i \in N(h)} d_i \leq v(h), \tag{1}$$

  then $h \in C(g(x))$;
- If h is a component of $g(A(x))$ and (1) is violated, then $h \notin C(g(x))$ and $i \in C(g(x))$ for all $i \in N(h)$;
- If h is not a component of $g(A(x))$, then $h \notin C(g(x))$.

[3] Assuming an upper bound on demands is without loss of generality, since one could always set $D = v(g^*)$ without affecting any of the equilibria of the game.

In words, the component h forms as the outcome of the history x if and only if the arcs sent in x generate h and the demands of the players in $N(h)$ are compatible, in the sense that they do not exceed the value produced by the component h.

2.2.3 Payoffs and Strategies

The payoff of player i is defined as a function of the history x. Letting $h_i(x) \in C(g(x))$ denote the component of $g(x)$ containing i, player i gets

$$P_i(x) = \begin{cases} d_i & \text{if } \sum_{j \in h_i(x)} d_j \leq v(h_i(x)) \\ 0 & \text{otherwise.} \end{cases} \tag{2}$$

This implies that we allow for free disposal.

A strategy for player i is a function $\sigma_i : \lambda_i X \to X_i$. A strategy profile for $\Gamma(v)$ is a vector of functions $\sigma = (\sigma_1, \ldots, \sigma_n)$. A Subgame Perfect Equilibrium (henceforth SPE) for $\Gamma(v)$ is defined as follows. For any subgame $\lambda_i x$, let $\sigma \,|\lambda_i x$ denote the restriction of the strategy profile σ to the subgame. A strategy profile σ^* is a SPE of $\Gamma(v)$ if for every subgame $\lambda_i x$ the profile $\sigma^* \,|\lambda_i x$ represents a Nash Equilibrium. We will denote by $f(\lambda_i x)$ a SPE path of the subgame $\lambda_i x$, i.e., equilibrium continuation histories after $\lambda_i x$. We will only consider equilibria in pure strategies.

3 Equilibrium

In this section we analyze the set of SPE of the game $\Gamma(v)$. We first show that SPE always exist. We then study the efficiency properties of SPE. Finally, we illustrate by example what is the role of the two main features of $\Gamma(v)$, namely the sequential structure and the endogeneity of payoff division, for the efficiency result.

3.1 Existence of Equilibrium

Since the game $\Gamma(v)$ is not finite in the choice of payoff demands, we need to establish existence of a SPE (see the Appendix for the proof).

Theorem 1. *The game $\Gamma(v)$ always admits Subgame Perfect Equilibria in pure strategies.*

3.2 Efficiency Properties of Equilibria

This section contains the main result of the paper: *all* the SPE of $\Gamma(v)$ induce an efficient network. We obtain this result for a wide class of value functions, satisfying a weak "superadditivity" condition, that we call *size monotonicity*. We first provide the definition and some discussion of this condition, then we prove our main result. We then analyze the role of each feature of our game (sequentiality and endogenous payoff division) and of size monotonicity in obtaining our result, and discuss the latter in the framework of the efficiency-stability debate related to Aumann and Myerson (1988) and Jackson and Wolinsky (1996) seminal contributions.

Definition 1. *The link ij is critical for the graph g if $ij \in g$ and $\#C(g) > \#C(g - ij)$.*

In words, a link is critical for a graph if by removing it we increase the number of components. Intuitively, a critical link is essential for the component it belongs to in the sense that without it that component would split in two different components.

Definition 2. *The value function v satisfies size monotonicity if and only if for all graphs g and critical link $ij \in g$*

$$v(g) > v(g - ij).$$

Size monotonicity requires that merging components in the "minimal" way strictly increases the value of the graph. By "minimal" we mean here that such merging occurs through a single additional link. This condition is trivially satisfied when additional links always increase the value of the graph, leading to an efficient fully connected graph. However, this condition is also compatible with cases in which "more" communication (more connected players) originates more value, but, for a fixed set of players that are communicating, this value decreases with the number of links used to communicate. Value functions exhibiting congestion in the number of links within components satisfy this assumption. The extreme case is represented by value functions such that the efficient graph consists of a single path connecting all players, or the star graph, with one player connected with all other players and no other pair of players directly linked (minimally connected graphs). One example that would originate such value functions is the symmetric connection model studied in Jackson and Wolinsky (1996), with a cost of maintaining links for each player, which is a strictly convex and increasing function of the number of maintained links.

The next lemma formally proves one immediate implication of size monotonicity, i.e., that all players are (directly or indirectly) connected.

Lemma 1. *Let v satisfy size monotonicity. All efficient graphs are connected, i.e., if g is efficient then $C(g) = \{g\}$ and $N(g) = N$.*

Proof. Consider a graph g such that $C(g) = \{h_1, \ldots, h_p\}$, with $p > 1$. Then let $i \in h_1$ and $j \in h_2$ ($ij \notin g$). The link ij is a critical link according to Definition 1, so that, by size monotonicity of v, we have that $v(g) < v(g + ij)$, implying that g is not efficient. **QED.**

We now state our main theorem, proving that size monotonicity is a sufficient condition for *all* SPE to be efficient.

Theorem 2. *Let v satisfy size monotonicity. Every SPE of $\Gamma(v)$ leads to an efficient network.*

We prove the theorem in two steps. We first prove by an induction argument in step 1 that if a given history is not efficient and satisfies a certain condition on payoff demands, then some player has a profitable deviation. Then, in step 2, we show that if some history x such that $g(x) \notin G^*$ is a SPE, then the condition on payoff demands introduced in step 1 would be satisfied, which implies that there exists a profitable deviation from *any* history that leads to an inefficient network.

The proof relies on two lemmas, the first characterizing equilibrium payoffs and the second characterizing equilibrium graphs.

Lemma 2. *Let v satisfy size monotonicity. For any arbitrary history of $\Gamma(v)$, $\lambda_m x$, the continuation equilibrium payoff for player m, $P_m(f(\lambda_m x))$, is strictly positive, for all $m = 1, \ldots, n-1$.*

Proof. Recall that n is the last player in the order of play ρ, and let $m < n$ be any player moving before n. Consider an arbitrary history $\lambda_m x$. In order to prove that the continuation equilibrium payoff is strictly positive for player m, let us show that there exists $\varepsilon > 0$ such that if player m plays the action $x_m = \left(a_m^n, \varepsilon\right)$, then it is a dominant strategy for player n to reciprocate m's arc and form some feasible component h with $mn \in h$.

Suppose first that $\varepsilon = 0$, so that, at the arbitrary history $\lambda_m x$, player m chooses $x_m = (a_m^n, 0)$.

We want to show that there cannot be an equilibrium continuation history $f(\lambda_m x, x_m)$ such that, denoting the history $(\lambda_m x, x_m, f(\lambda_m x, x_m))$ by $\hat{x}$, $h_m(\hat{x}) = mm$ (i.e., where m is alone even though she demands 0). Suppose this is the case, and let $\hat{x}_n = (a_n, d_n)$ be a strategy for player n such that $a_n^m \notin a_n$. Let $h_n(\hat{x})$ be the component including n if this continuation history is played. Denote by h_n' the component obtained by adding the link mn to $h_n(\hat{x})$. By size monotonicity,

$$v(h_n(\hat{x})) < v(h_n').$$

If the component $h_n(\hat{x})$ is feasible, the component h_n' is feasible too, for some demand $d_n + \delta > d_n$ of player n.[4] It follows that it is dominant for n to reciprocate m's arc and get a strictly greater payoff. So $\hat{x}$ cannot be an equilibrium continuation payoff.

Consider then $x_m(\varepsilon) \equiv (a_m^n, \varepsilon)$ with $\varepsilon > 0$.

Consider the continuation history $\hat{x}(\varepsilon) = f(\lambda_m x, x_m(\varepsilon))$, with

$$\hat{x}(\varepsilon) = (\lambda_m x, x_m(\varepsilon), f(\lambda_m x, x_m(\varepsilon))),$$

[4] If $h_n(\lambda_m x, x_m, \hat{x})$ is not feasible, then either there exists some positive demand d_n' for player n such that $\sum_{i \in N(h_n') \backslash n} d_i + d_n' = v(h_n')$ or player n could just reciprocate player m's arc and demand $d_n' = v(mn) > 0$ (this last inequality by size monotonicity).

and $\hat{x}_n = (a_n, d_n)$ such that $a_n^m \notin a_n$. Let $h_n(\hat{x}(\varepsilon))$ be the component that includes n given $\hat{x}(\varepsilon)$. Let again $h_n'(\varepsilon) \equiv mn \cup h_n(\hat{x}(\varepsilon))$. Define

$$\delta_{\min} \equiv \min_{\varepsilon \geq 0} v\left(h_n'(\varepsilon)\right) - v\left(h_n(\hat{x}(\varepsilon)\right) > 0$$

where the strict inequality comes form size monotonicity.
Let $0 < \varepsilon < \delta_{\min}$.
If $h_n(\hat{x}(\varepsilon))$ is feasible, then $h_n'(\varepsilon)$ is feasible too, for some positive additional demand of player n. Thus, it is possible for player n to demand a strictly higher payoff than d_n (this because $\varepsilon < \delta_{\min}$).[5] Therefore a positive payoff is always attainable by any player $m < n$, at any history. **QED.**

Lemma 3. *Let v satisfy size monotonicity. Let x be a SPE history of the game $\Gamma(v)$. In the induced graph $g(x)$ all players are connected, i.e., $C(g(x)) = \{g(x)\}$ and $N(g(x)) = N$.*

Proof. Suppose that $C(g(x)) = \{h_1, \ldots, h_k\}$ with $k > 1$. Let again n be the last player in the ordering ρ. Note first that there must be some component h_p such that $n \notin h_p$, since otherwise the assumption that $k > 1$ would be contradicted. Also, note that by Lemma 2, x being an equilibrium implies that[6]

$$\sum_{i \in N(h_p)} d_i = v(h_p) \quad \forall p \in \{1, \ldots, k\}.$$

Let us then consider h_p and the last player m in $N(h_p)$ according to the ordering ρ. Let $\hat{x}_m(\varepsilon) = \left(a_m \cup a_m^n, d_m + \varepsilon\right)$, with continuation history $f(\lambda_m x, \hat{x}_m(\varepsilon))$. Let

$$\hat{x}(\varepsilon) = (\lambda_m x, \hat{x}_m(\varepsilon), f(\lambda_m x, \hat{x}_m(\varepsilon)))$$

and let $h_n(\hat{x}(\varepsilon))$ be the component including n in $g(\hat{x}(\varepsilon))$. Suppose first that $mn \notin h_n(\hat{x}(\varepsilon))$ and $in \in h_n(\hat{x}(\varepsilon))$ for some $i \in N(h_p)$. Note first that if some player $j > m$ is in $h_n(\hat{x}(\varepsilon))$, then by Lemma 2 $h_n(\hat{x}(\varepsilon))$ is feasible given x_n, and since player m is getting a higher payoff than under x, the action $\hat{x}_m(\varepsilon)$ is a profitable deviation for him. We therefore consider the case in which no player $j > m$ is in $h_n(\hat{x}(\varepsilon))$, and $h_n(\hat{x}(\varepsilon))$ is not feasible. In this case, it is a feasible strategy for player n, who is getting a zero payoff under x_n, to reciprocate only player m's arc and form the component h_n' such that, by size monotonicity,

$$v(h_n'(\varepsilon)) > v(h_p).$$

[5] If instead $h_n(\lambda_m x, x_m(\epsilon_m), \hat{x}_m(\epsilon_m))$ is not feasible, then either there exists some positive demand d_n' such that $\sum_{i \in N(h_n'(\varepsilon_m)) \setminus n} d_i + d_n' = v(h_n'(\varepsilon_m))$ or player n could just reciprocate player m's arc and demand $d_n' = v(mn) - \varepsilon_m > 0$ (this last inequality again by size monotonicity).

[6] Note that there cannot be any equilibrium where the last player demands something unfeasible: since in every equilibrium the last player obtains a zero payoff, one could think that she could then demand anything, making the complete graph unfeasible, but this would entail a deviation by one of the previous players, who would demand ϵ less, in order to make n join in the continuation equilibrium. Thus, the unique equilibrium demand of player n is 0.

If ε is small enough we get

$$v(h_n'(\varepsilon)) - v(h_p) > \varepsilon$$

which implies that reciprocating only player m's arc and demanding $d_n = v(h_n'(\varepsilon)) - v(h_p) - \varepsilon > 0$ is a profitable deviation for player n.

Thus, we can restrict ourselves to the case in which $in \notin h_n(\hat{x}(\varepsilon))$ for all $i \in N(h_p)$. Let $h_n'(\varepsilon)$ be obtained by adding the link mn to $h_n(\hat{x}(\varepsilon))$. By size monotonicity

$$v\left(h_n'(\varepsilon)\right) - v\left(h_n(\hat{x}(\varepsilon))\right) > 0.$$

Let also

$$\delta_{\min} \equiv \min_{\varepsilon \geq 0} \left[v\left(h_n'(\varepsilon)\right) - v\left(h_n(\hat{x}(\varepsilon))\right)\right] > 0.$$

Consider a demand ε such that $0 < \varepsilon < \delta_{\min}$. As in the proof of Lemma 2, we claim that if player m demands ε, then it is dominant for player n to reciprocate player m's link and form the component $h_n'(\varepsilon)$. Note first that, given that $0 < \varepsilon_m < \delta_{\min}$, if $h_n(\hat{x}(\varepsilon)$ is feasible, then $h_n'(\varepsilon)$ is feasible for some positive additional demand (w.r.t. d_n) of player n. If instead $h_n(\hat{x}(\varepsilon)$ was not feasible, then player n would be getting a zero payoff, and this would be strictly dominated by reciprocating m's arc and getting a payoff of

$$\left[v\left(h_n'(\varepsilon)\right) - v\left(h_n(\hat{x}(\varepsilon))\right)\right] - \varepsilon$$

which, again by the fact that $\varepsilon < \delta_{\min}$, is strictly positive. **QED.**

Proof of Theorem 2.

Step 1. Induction argument.

Induction Hypothesis (H): Let x be an arbitrary history such that $g(x) \notin G^*$. Let m be the first player in the ordering ρ such that there is no x^* such that (1) $\lambda_{m+1}x^* = \lambda_{m+1}x$ and (2) $g(x^*) \in G^*$. Let x be such that

$$\sum_{i=1}^{m} d_i \leq v\left(g(x)\right) - \sum_{i=m+1}^{n} d_i.$$

Then there exists some $\varepsilon > 0$ and action $x_m^* = \left(a_m^*, d_m + \varepsilon\right)$ that induce a continuation history $f\left(\lambda_m x, x_m^*\right)$ such that, denoting by x^* the history $(\lambda_m x, x_m^*, f(\lambda_m x, x_m^*))$, $g(x^*) \in G^*$ and $\sum_{i=1}^{n} \hat{d}_i = v(g(x^*))$.

(H) true for player n: Let $x_n = (a_n, d_n)$. Let player m, as defined in (H), be n. In words, this means that n could still induce the efficient graph by deviating to some other action. Formally, there exist some arcs a_n^* and a demand d_n' such that $g\left(x_1, \ldots, x_{n-1}, a_n^*, d_n'\right) \in G^*$ and, therefore, such that $v\left(g\left(x_1, \ldots, x_{n-1}, a_n^*, d_n'\right)\right) > v(g(x))$. By (H)

$$\sum_{i=1}^{n} d_i \leq v\left(g(x)\right)$$

and by size monotonicity all players are connected in $g\left(x_1,\ldots,x_{n-1},a_n^*,d_n^*\right)$. These two facts imply that player n can induce the efficient graph and demand $d_n' = d_n + \varepsilon_n$ with

$$\varepsilon_n = \left[v\left(g^*\right) - v\left(g\left(x\right)\right)\right] > 0.$$

(H) true for player $m+1$ implies (H) true for player m: Suppose again that x is an inefficient history and that m is the first player in x such that the action a_m is not compatible with efficiency in the sense of assumption (H). Let a_m^* be some action compatible with efficiency and let $x_m^*(\varepsilon) = \left(a_m^*, d_m + \varepsilon\right)$. Let also $f\left(\lambda_m x, x_m^*(\varepsilon)\right)$ represent the corresponding continuation history, and $x^*(\varepsilon) = \left(\lambda_m x, x_m^*(\varepsilon), f\left(\lambda_m x, x_m^*(\varepsilon)\right)\right)$. We need to show that there exists $\varepsilon > 0$ such that $g\left(x^*(\varepsilon)\right) \in G^*$. Note first that in the history $x^*(\varepsilon)$, the first player k such that a_k is not compatible with efficiency must be such that $k > m$. Since by (H)

$$\sum_{i=1}^{m} d_i \leq v\left(g\left(x\right)\right) - \sum_{i=m+1}^{n} d_i$$

there exists an $\varepsilon > 0$ such that

$$\sum_{i=1}^{m-1} d_i + d_m + \varepsilon < v\left(g^*\right) - \sum_{i=m+1}^{n} d_i.$$

Thus, if player m plays $x_m^*(\varepsilon)$, player $(m+1)$ faces a history $\left(\lambda_m x, x_m^*(\varepsilon)\right)$ that satisfies the inductive assumption (H). Suppose now that player $(m+1)$ optimally plays some action x_{m+1} such that no efficient graph is compatible (in the sense of assumption (H)) with the history $\left(\lambda_m x, x_m(\varepsilon), x_{m+1}\right)$. Then, by (H) we know there would be a deviation for player $(m+1)$, contradicting the assumption that x_{m+1} is part of the continuation history at $\left(\lambda_m x, x_m(\varepsilon)\right)$. Thus, we know that player $(m+1)$ will optimally play some strategy x_{m+1}^* such that the continuation history $f\left(\left(\lambda_m x, x_m(\varepsilon), x_{m+1}^*\right)\right)$ induces a feasible efficient graph.

Step 2. We now show that the induction argument can be applied to each candidate SPE history x of $\Gamma(v)$ such that $v(g(x)) < v(g^*)$ (which we want to rule out). This is shown to imply that the first player m (such that there does not exist x^* such that $\lambda_{m+1}x^* = \lambda_{m+1}x$ and $v\left(g\left(x^*\right)\right) = v(g^*)$) has a profitable deviation.

Note first that by Lemma 3 if x is a SPE history then all players are connected. This, together with Lemma 2, directly implies that

$$\sum_{i=1}^{n} d_i = v\left(g\left(x\right)\right)$$

or, equivalently, that

$$\sum_{i=1}^{m} d_i = v\left(g\left(x\right)\right) - \sum_{i=m+1}^{n} d_i$$

for all $m = 1,\ldots,n$. It follows that the induction argument can be applied to all inefficient SPE histories to conclude that the first player whose action is not compatible with efficiency in the sense of assumption (H) has some action $x_m^*(\varepsilon) = \left(a_m^*, d_m + \varepsilon\right)$ such that $\varepsilon > 0$ and such that the induced graph $g(x^*(\varepsilon)) \in G^*$ is feasible, where, as usual, $x^*(\varepsilon) = \left(\lambda_m x, x_m^*(\varepsilon), f\left(\lambda_m x, x_m^*(\varepsilon)\right)\right)$. Since $g(x^*(\varepsilon))$ is feasible, then the action $x_m^*(\varepsilon)$ represents a deviation for player m, proving the theorem. **QED.**

The efficiency theorem extends to the case in which the order of play is random, i.e., in which each mover only knows a probability distribution over the identity of the subsequent mover. This is true because the value function is assumed to satisfy anonymity. Another important remark about the role of the order of play regards the asymmetry of equilibrium payoffs: for any given order of play the equilibrium payoffs are clearly asymmetric, since the last mover always obtains 0. However, if *ex ante* all orders of play have the same probability, then the expected equilibrium payoff is $E(P_i(g(x(\rho)))) = \frac{v(g^*)}{n} \; \forall i$.

3.3 Discussion

In this section we want to discuss our result in the framework of the recent literature debate on the possibility of reconciling efficiency and stability in the process of formation of networks. As we pointed out in the introduction, this debate has been initiated by two seminal papers: Aumann and Myerson (1988) have shown that if the Myerson value is imposed as a fixed imputation rule, then forward looking players forming a networks through sequential link formation can induce inefficient networks. The value function they consider is obtained from a traditional coalitional form game. Jackson and Wolinksy (1996) obtained a general impossibility result considering value functions that depend on the communication structure rather than only on the set of connected players. This incompatibility has been partially overcome by Dutta and Mutuswami (1997) who show that it disappears if component balancedness and anonymity are required only on stable networks.

We first note that the *size monotonicity* requirement of Theorem 2 in the present paper is compatible with the specific value function for which Jackson and Wolinsky show that no anonymous and component balanced imputation rule exists such that at least one stable graph is efficient. In this sense, we can conclude that in our game the aforementioned conflict between efficiency and stability does not appear. Since however imputation rules of the type considered by Dutta and Mutuswami allow for efficient and stable networks, our game can be considered as another way to overcome that conflict.

The real novelty of our efficiency result is therefore the fact that *all subgame perfect equilibria of our game are efficient*. In the rest of this section we will show that both the sequential structure of the game and the endogeneity of the final imputation rule are "tight" conditions for the result, as well as the size monotonicity requirement. Indeed, we first show that relaxing size monotonicity generates inefficient equilibria. We then construct a value function for which all fixed component balanced and anonymous imputation rules generate at least one inefficient stable graph in the sense of Jackson and Wolinsky. The same is shown for a game of endogenous payoff division in which agents move simultaneously. We finally show that sequentiality alone does not generate our result, since no fixed component balanced and anonymous imputation rule exists such that all subgame perfect equilibria are efficient.

3.3.1 Eliminating Size Monotonicity

The next example shows that if a value function v does not satisfy size monotonicity, then the SPE of $\Gamma(v)$ may induce an inefficient network.

Example 1. Consider a four-player game with the following value function:

$$\begin{aligned} v(h) &= 9 \textit{ if } N(h) = N \\ v(h) &= 8 \textit{ if } \#N(h) = 3 \textit{ and } \#L(h) = 2; \\ v(h) &= 5 \textit{ if } \#N(h) = 2; \\ v(h) &= 0 \textit{ otherwise}. \end{aligned}$$

The efficient network is one with two separate links. We show that the history x such that

$$\begin{aligned} x_1 &= \left((a_1^2, a_1^3, a_1^4), 3\right) \\ x_2 &= \left((a_2^1, a_2^3, a_2^4), 3\right) \\ x_3 &= \left((a_3^2, a_3^4), 3\right) \\ x_4 &= (a_4^3, 0) \end{aligned}$$

is a SPE of the game $\Gamma(v)$, leading to the inefficient graph $(12, 23, 34)$.

1. **Player 4:** given that at the history $\lambda_4 x$ we have $d_1 + d_2 + d_3 = 9$, player 4 optimally reciprocates the arc of player 3.
2. **Player 3:** sending just a_3^2 or a_3^1 or both, would let player 3 demand at most $d_3 = 2$; forming a link just with player 4 would allow player 3 to demand at most $d_3 = 3$, since player 4 would have at that node the outside option of going with the first two movers.
3. **Player 2:** If $d_2 > d_1 = 3$, then player 3 has the outside option of just reciprocating the arc of player 1 and demand $d_3 = 3$. Thus, $d_2 > 3$ is not a profitable deviation for player 2. In terms of arcs, note first that if player 2 sends just a_2^1 then $d_2 \leq 2$, given that $d_1 = 3$. Suppose now that player 2 sends arcs only to 1 and 4 demanding $d_2 = 3 + \epsilon$. In this case player 3 would react by sending an arc just to player 4, demanding $3 + \epsilon - \delta$ $(\epsilon > \delta > 0)$, which 4 would optimally reciprocate.
4. **Player 1:** We just check that player 1 could not demand $d_1 = 3 + \epsilon > 3$. If he does, then player 2 can "underbid" by a small δ, as in the argument above, so that player 3 and/or 4 would always prefer to reciprocate links with player 2.

This example has shown that when size monotonicity is violated then inefficient equilibria may exist. The intuition for the failure of Theorem 2 when v is not size monotonic can be given as follows. By Lemma 1, under size monotonicity all efficient graphs are connected (though not necessarily fully connected). It follows that the gains from efficiency can be shared among all players in equilibrium (since efficiency requires all players to belong to the same component). When size monotonicity fails, however, the efficient graph may consist of more than one component. It becomes then impossible to share the gains from efficiency among all players, since side payments across components are not allowed in the game $\Gamma(v)$. It seems reasonable to conjecture that it would be possible to conceive a game form allowing for such side payments and such that all equilibria are efficient even when size monotonicity fails.

3.3.2 The Role of Sequentiality

The next example displays a value function satisfying size monotonicity, and serves the purpose of demonstrating the crucial role of the sequential structure of our game for the result that *all* equilibria are efficient. In fact, neither using the stability concept of Jackson and Wolinsky, nor with a simultaneous move game, it is possible to eliminate all inefficient equilibria.

Example 2. Consider a four-player game with the following value function:

$$
v(h) = \begin{cases} 1 & \textit{if } \#N(h) = 2; \\ 2 & \textit{if } \#N(h) = 3; \\ 20 & \textit{if } \#N(h) = 4 \textit{ and } \#L_i = 2\ \forall i; \\ 24 & \textit{if } h = g^N; \\ 4 & \textit{otherwise}. \end{cases}
$$

This value function satisfies size monotonicity, and the only two connected networks with value greater than 4 are the complete graph and the one where each player has two links.

Let us first show that the inefficient network with value equal to 20 is stable, in the sense of Jackson and Wolinsky (1996), *for every allocation rule* satisfying anonymity and component balancedness. To see this, note that in such network anonymity implies that each player would receive 5, which is greater than anything achievable by either adding a new link or severing one ($5 > 4$). Along the same line it can be proved that the complete (efficient) graph is stable.

Similarly, even if we allow payoff division to be endogenous, a simultaneous move game would always have an equilibrium profile leading to the inefficient network with value equal to 20. To see this, consider a simultaneous move game where every player announces at the same time a set of arcs and a demand (keeping all the other features of the game as in $\Gamma(v)$). Consider a strategy profile in which every player demands 5 and sends only two arcs, in a way that every arc is reciprocated. It is clear that any deviation in terms of arcs (less or more) induces a network with value 4, and hence the deviation cannot be profitable.

On the other hand, given the sequential structure of $\Gamma(v)$, the inefficient networks are never equilibria, and the intuition can be easily obtained through the example above: calling σ the strategy profile leading to the inefficient network discussed above, the first mover can deviate by sending all arcs and demanding more than 5, since in the continuation game he expects the third arc will be reciprocated and the complete graph will be formed.

3.3.3 The Role of Endogenous Payoff Division

Having shown the crucial role of sequentiality, the next task is to show the relevance of the other innovative aspect of $\Gamma(v)$, namely, endogenous payoff division. Consider a game $\Gamma(v, Y)$ that is like $\Gamma(v)$ but for the fact that the action space of each player only includes the set of possible arcs he could send, and no payoff demand can be made. The imputation rule Y (of the type considered in Jackson and Wolinsky 1996) determines payoffs for each network. We can now show by example that there are some value functions that satisfy size monotonicity for which no allocation rule satisfying anonymity and component balancedness can eliminate all inefficient networks from the set of equilibrium outcomes of $\Gamma(v, Y)$.

Proposition 1. *There exists value functions satisfying size monotonicity and such that every fixed imputation rule Y satisfying anonymity and component balancedness induces at least one inefficient equilibrium in the associated sequential game $\Gamma(v, Y)$.*

Proof. By Example.

Example 3. Consider a three-player game $\Gamma(v, Y)$ with the following value function:[7]

$$\begin{aligned} v(12) &= v(23) = v(13) = 1; \\ v(12, 23) &= v(13, 12) = v(13, 23) = 1 + \varepsilon > 1; \\ v(12, 13, 23) &= 1. \end{aligned}$$

Given anonymity of Y, the only payoff distribution if the complete graph forms is $P_i(g^N) = \frac{1}{3}$. Similarly, if $h = ij$, then both i and j must receive $\frac{1}{2}$. If $h = (ij, jk)$, then let us call x the payoff to i and k and y the payoff to the pivotal player, j, with $(2x + y = 1 + \epsilon)$. Let ϵ be small, so that $\frac{1+\epsilon}{3} < \frac{1}{2}$.

1. If $y \geq \frac{1}{2}$, the first mover cannot send one arc only. If he sends an arc only to the second mover, then player 2's best response is to send two arcs and get y; if he sends an arc to the third mover only, the second mover does the

[7] This value function was used in Jackson and Wolinsky (1996) to get their impossibility result under the axiomatic approach discussed in the previous section.

same, and the third mover gets y. So, if the first mover sends only one arc his payoff is $\frac{1+\epsilon-y}{2} < \frac{1}{3}$. By sending both arcs, player 1 would end up forming the complete graph and obtaining $\frac{1}{3}$, which makes the complete graph an equilibrium network.

2. If $y < \frac{1}{2}$, note that there always exists an equilibrium continuation history leading to the graph (12) if player 1 sends the arc only to player 2. Thus, if $x < \frac{1}{2}$, player 1 cannot get as much as $\frac{1}{2}$ on any other network, and sending an arc only to player 2 will therefore be an equilibrium strategy. If on the contrary $x \geq \frac{1}{2}$, there could be an incentive for player 1 to form the efficient graph and get x. However, it can be easily checked that in this case, the following strategy profile is an equilibrium:

$$\sigma_2 = a_1^2$$

$$\sigma_2 = \begin{cases} \sigma_2\left(a_1^2, a_1^3\right) = \left(a_2^1, a_2^3\right) \\ \sigma_2\left(a_1^3\right) = \left(a_2^1, a_2^3\right) \\ \sigma_2\left(a_1^2\right) = \left(a_2^1\right) \end{cases}$$

$$\sigma_3 = \begin{cases} \sigma_3\left(a_1^2, a_1^3, a_2^1, a_2^3\right) = \sigma_3\left(a_1^2, a_1^3, a_2^1\right) = \sigma_3\left(a_1^2, a_2^1\right) = a_3^1 \\ \sigma_3\left(a_1^2, a_1^3, a_2^3\right) = a_3^2 \\ \sigma_3\left(a_1^3, a_2^3, a_2^1\right) = \sigma_3\left(a_1^3, a_2^3\right) = a_3^2 \\ \sigma_3\left(a_1^3, a_2^1\right) = \sigma_3\left(a_1^2, a_2^1, a_2^3\right) = \sigma_3\left(a_1^2, a_2^3\right) = \left(a_3^1, a_3^2\right). \end{cases}$$

In words, there are optimal strategies that support the pair (12) as a SPE equilibrium. **QED.**

4 Link-Specific Demands

Consider now a variation of the game, $\Gamma_1(v)$, which differs from $\Gamma(v)$ in that players can attach payoff demands on each arc they send, rather than demanding just one aggregate payoff from the whole component. Player i's demand d_i is a vector of real positive numbers, one for each arc sent in the vector a_i. We describe how payoffs depend on histories in $\Gamma_1(v)$ on the basis of the formal description of the game $\Gamma(v)$:

1. The feasibility condition given in (1) is replaced by:

$$\sum_{i \in N(h)} \sum_{j: ij \in h} d_i^j \leq v(h); \tag{3}$$

2. The payoff for player i in the component $h \in C(g(x))$ is given by

$$P_i(x) = \begin{matrix} \sum_{j \neq i: ij \in h} d_i^j & \textit{if } L(h(x)) \neq \emptyset \\ 0 & \textit{otherwise} \end{matrix} \tag{4}$$

(instead of (2)). In words, the payoff for player i from history x would be equal to the sum of the link-specific demands made by i to the members of her component whom she is directly linked to.

The same efficiency result as the one obtained in Theorem 2 can be obtained for the game $\Gamma_1(v)$. Proofs are found in the appendix.

Lemma 4. *Let v satisfy size monotonicity. Let $\lambda_m x$ be an arbitrary history of the game $\Gamma_1(v)$. Then $P_i(f(\lambda_m x)) > 0$ for all $i = 1, ..., n-1$.*

Lemma 5. *Let v satisfy size monotonicity. Let x be a SPE history of the game $\Gamma_1(v)$. In the induced graph $g(x)$ all players are connected, i.e., $C(g(x)) = \{g(x)\}$ and $N(g(x)) = N$.*

Theorem 3. *Let v satisfy size monotonicity. Every SPE of $\Gamma_1(v)$ leads to an efficient network.*

5 Conclusions

This paper provides an important result for all the situations in which a communication network forms in the absence of a mechanism designer: if players sequentially form links and bargain over payoffs, the outcome is an efficient network. This result holds as long as disaggregating components *via* the removal of "critical" links lowers the aggregate value of the network. In other words, efficiency arises whenever more communication is good, at least when it is obtained with the minimal set of links. We have shown this result by proving that all the subgame perfect equilibria of a sequential link formation game, in which the relevant players demand absolute payoffs, lead to efficient networks. On the other hand, endogenous payoff division is not sufficient to obtain optimality when the optimal network has more than one component. Allowing for link-specific demands we obtain identical results.

Appendix

Proof of Theorem 1. We prove the theorem by showing that every player's maximization problem at each subgame has a solution. Using the notation introduced in the previous sections, we show that for each player m and history x, there exists an element $x_m \in X_m$ maximizing m's payoff given the continuation histories originating at $(\lambda_m x, x_m)$. Since the choice set X_m is given by the product set $A_m \times [0, D]$, where the finite set A_m is the set of vectors of arcs that player m can choose to send to other players in the game, it suffices to show that we can associate with each vector of arcs $a_m \in A_m$ a maximal feasible demand $d_m(a_m)$.

Suppose not. Then, given a_m, $\forall d_m \exists \varepsilon > 0$ such that $(d_m + \varepsilon)$ is feasible. This, together with the fact that the set $[0, D]$ is compact, imply that there exists some demand $\overline{d}_m(a_m)$ which is not feasible given a_m and which is the limit of some sequence of feasible demands $(d_m^p)_{p=1,...,\infty}$. We prove the theorem by contradicting this conclusion.

First, we denote by $\overline{x}$ a continuation history given $(a_m, \overline{d}_m(a_m))$, and, for all p, we denote by $\overline{x}(p)$ a continuation history given $(a_m, (d_m^p))$. For all p, feasibility of d_m^p implies that player m belongs to some component h_m^p such that

$$v(h_m^p) \geq \sum_{\substack{i \in N(h_m^p) \\ i = m}} d_i(p) + d_m^p. \tag{5}$$

We claim that as $d_m^p \to \overline{d}_m(a_m)$ (5) remains satisfied for some component $\overline{h}_m$. Suppose first that there exists p such that the component h_m^p is the same for all $p \geq p$. We proceed by induction.

Induction Hypothesis: Consider the history x and the histories $x(p)$, $p \geq \overline{p}$, the history identical to x but for player m's demand which is d_m^p. If x_i is the best response of player n at the subgame $\lambda_n x(p)$ for all $p \geq \overline{p}$ then x_i is a best response of player n at $\lambda_n x$.

Player n: At the subgame $\lambda_n x(p)$ player n can either optimally join a component including m or not join any component including m. In the first case, his payoff by not joining m's component with action $\overline{x}_n(p)$ is weakly greater than the one he gets by joining with any action $x_n(p)$:

$$P_n(\lambda_n x(p), \overline{x}_n(p)) \geq P_n(\lambda_n x(p), x_n(p))$$

Bringing m's demand to the limit does not change the above inequality.

In the second case, player n's payoff is maximized by joining a component including m with action $\overline{x}_n(p)$:

$$P_n(\lambda_n x(p), \overline{x}_n(p)) \geq P_n(\lambda_n x(p), x_n(p))$$

We can apply the same limit argument in this case, by noting that at the limit condition (5) remains satisfied.

True for player k + 1 implies true for player k: Assume that the induction hypothesis is satisfied for all players $k+1, \ldots, n$. Then, the continuation histories after the subgames $\lambda_{k+1} x(p)$ and $\lambda_{k+1} x$ are the same. Player k's optimal choice $\overline{x}_k(p)$ at $\lambda_k x(p)$ satisfies the following condition for all $x_k \in X_k$:

$$P_k(\lambda_k x(p), \overline{x}_k(p), f(\lambda_k x(p), \overline{x}_k(p))) \geq P_k(\lambda_k x(p), x_k(p), f(\lambda_k x(p), x_k(p))).$$

Since we have argued that by the induction hypothesis that

$$f(\lambda_k x(p), \overline{x}_k(p)) = f(\lambda_k x, \overline{x}_k(p));$$
$$f(\lambda_k x(p), x_k(p)) = f(\lambda_k x(p), x_k(p)),$$

we conclude that at the limit

$$P_k(\lambda_k x, \overline{x}_k(p), f(\lambda_k x, \overline{x}_k(p))) \geq P_k(\lambda_k x, x_k(p), f(\lambda_k x(p), x_k(p))).$$

This means that $\overline{x}_k(p)$ is still optimal at $\lambda_k x$. Moreover, the feasibility condition (5) still holds whenever player k was joining a component including m. This concludes the induction argument.

The above argument directly implies that if component h_m^p is still feasible at the limit, so that the demand $\overline{d}_m(a_m)$ is itself feasible.

Finally, suppose that there exists no p such that the component $\overline{h}_m^p$ is the same for all $p \geq \overline{p}$. In this case, since the set of possible components to which

m can belong to given a_m is finite, for each possible such component h we can associate a subsequence $\{d_m(h)\}_{p=1,\dots,\infty} \to \overline{d}_m(a_m)$. The feasibility condition applied to each component h implies that for all h:

$$v(h) \geq \sum_{\substack{i \in N(h) \\ i = m}} d_i + d_m(h).$$

We can apply the above induction argument to this case by considering some converging subsequence, thereby showing that there exists some feasible component $\overline{h}_m$ induced by the demand $\overline{d}_m(a_m)$. **QED.**

Proof of Lemma 4. Let n be the last player in the ordering ρ and let $m < n$. Consider an arbitrary history $\lambda_m x$. We show that there exists a demand $d_m^n > 0$ such that if player m plays the action $x_m = \left(a_m^n, d_m^n\right)$ then it is a dominating strategy for player n to reciprocate m's arc and form some feasible component h with $mn \in h$.

For a given $d_m^n > 0$, let $x_m\left(d_m^n\right) = \left(a_m^n, d_m^n\right)$, and consider again the continuation history $\hat{x}\left(d_m^n\right) = f\left(\lambda_m x, x_m\left(d_m^n\right)\right)$. Let also $x_n = (a_n, d_n)$[8] be a strategy for player n such that $a_n^m \notin a_n$. Let $h(n, d_m^n)$ be the component that includes n if x_n is played at the history $\lambda_n \hat{x}\left(d_m^n\right)$ and $h'(n, d_m^n)$ be the component obtained by adding the link mn to $h(n, d_m^n)$. Define

$$\delta_{\min} \equiv \min_{d_m^n > 0} \left\{v\left(h'(n, d_m^n)\right) - v\left(h(n, d_m^n)\right)\right\} > 0,$$

where the last inequality comes form size monotonicity. Let now $0 < d_m^n < \delta_{\min}$. Note first that if $h(n, d_m^n)$ is feasible, then $h'(n, d_m^n)$ is feasible for some positive demand d_n^m of player n. Thus, player n can get a strictly higher payoff than under x_n (this because $\varepsilon < \delta_{\min}$). If instead $h(n, d_m^n)$ is not feasible, then either there exists some positive demand d_n^m for player n such that

$$\sum_{i \in N(h'(n, d_m^n)) \setminus n} \sum_{j : ij \in h'(n, d_m^n)} d_i^j + d_n^m = v(h'(n, d_m^n))$$

or player n could just reciprocate player m's arc and demand her $d_n^m = v(mn) - d_m^n > 0$ (this last inequality again follows from size monotonicity). It follows that it is dominant for n to reciprocate m's arc and get a strictly positive payoff. **QED.**

Proof of Lemma 5. Suppose that $C(g(x)) = \{h_1, \dots, h_k\}$ with $k > 1$. Let again n be the last player in the ordering ρ. Note first that there must be some component h_p such that $n \notin h_p$, since otherwise the assumption that $k > 1$ would be contradicted. Also, note that by Lemma 4, x being an equilibrium implies that for all $p = 1, \dots, k$

$$\sum_{i \in N(h_p)} \sum_{j : ij \in h_p} d_i^j = v(h_p).$$

[8] Recall that in game $\Gamma_2(v)$ d_n is a vector, with as many dimensions as the number of arcs sent by n.

Let us then consider h_p and the last player m in $N(h_p)$ according to the ordering ρ. Let $\hat{x}_m\left(d_m^n\right) = \left(a_m \cup a_m^n, d_m \cup d_m^n\right)$, with continuation history $\hat{x}\left(d_m^n\right) = f\left(\lambda_m x, \hat{x}_m\left(d_m^n\right)\right)$. Let $h(n, d_m^n)$ be the component including n in $g(\hat{x}\left(d_m^n\right))$. Suppose first that $mn \notin h(n, d_m^n)$ and $in \in h(n, d_m^n)$ for some $i \in N(h_p)$. Consider then the demand

$$\hat{d}_m^n < \min_{j \in N(h_p)} \left\{d_j^n\right\}.$$

Let now player m play $\hat{d}_m^n$. Suppose that still $in \in N(h(n, d_m^n))$ for some $i \in N(h_p)$. Then it would be a profitable deviation for player n to reciprocate the arc sent by m instead of the arc sent by some other player $i \in N(h_p)$, to which a demand $d_i^n > d_m^n$ is attached.

Suppose now that $in \notin N(h(n, d_m^n))$ for all $i \in N(h_p)$. Let $h'(n, d_m^n)$ be obtained by adding the link mn to $h(n, d_m^n)$. By size monotonicity

$$v\left(h'(n, d_m^n)\right) - v\left(h(n, d_m^n)\right) > 0.$$

Now let

$$\delta_{\min} \equiv \min_{d_m^n \geq 0} \left[v\left(h'(n, d_m^n)\right) - v\left(h(n, d_m^n)\right)\right] > 0.$$

Consider now a demand $0 < d_m^n < \delta_{\min}$. As in the proof of Lemma 4, we claim that it is dominant for player n to reciprocate player m's link and form a feasible component. Note first that, given that $0 < d_m^n < \delta_{\min}$, if $h(n, d_m^n)$ is feasible, then $h(n, d_m^n)$ is feasible for some positive demand d_n^m of player n. If instead $h(n, d_m^n)$ was not feasible, then player n would be getting a zero payoff, and this would be strictly dominated by reciprocating m's arc and getting a payoff of $\left[v\left(h'(n, d_m^n)\right) - v\left(h(n, d_m^n)\right)\right] - d_m^n$, which, again by the fact that $d_m^n < \delta_{\min}$, is strictly positive. **QED.**

Proof of Theorem 3. We proceed by first showing by induction, in step 1, that if a given history is not efficient and satisfies a certain condition on payoff demands, then some player has a profitable deviation. In step 2 we establish that if a history x, leading to an inefficient graph, was SPE, then it would have to satisfy the condition on payoff demands described in step 1, which implies that there exists a profitable deviation from *any* such history x leading to an inefficient graph.

Step 1. Induction Argument.

Induction Hypothesis (H): Let x be an arbitrary history such that $g(x) \notin G^*$. Let m be the first player in the ordering ρ such that there is no x^* such that (1) $\lambda_{m+1} x^* = \lambda_{m+1} x$ and (2) $g(x^*) \in G^*$. Let x be such that

$$\sum_{i=1}^{m} \sum_{j:ij \in N(h(i))} d_i^j \leq v\left(g\left(x\right)\right) - \sum_{i=m+1}^{n} \sum_{j:ij \in N(h(i))} d_i^j.$$

Then there exists some $\varepsilon_m > 0$ such that the action $x_m^* = \left(a_m^*, d_m + \varepsilon_m\right)$ induces a history $\hat{x} = f\left(\lambda_m x, x_m^*\right)$ such that $g(\hat{x}) \in G^*$ and $\sum_{i=1}^{n} \sum_{j:ij \in N(h(i))} \hat{d}_i^j = v(g(\hat{x})$.

(H) true for player n: Let $x_n = (a_n, d_n)$. By assumption (H), there exists some arcs a_n^* such that $g(\lambda_n a, a_n^*) \in G^*$ and, therefore, such that $v(g(\lambda_n a, a_n^*)) > v(g(x))$. By (H)

$$\sum_{i=1}^{n} \sum_{j:ij \in N(h(i))} d_i^j \leq v(g(x));$$

Moreover, by size monotonicity all players are connected in $g(\lambda_n a, a_n^*)$.[9] These two facts imply that player n can induce the efficient graph and demand the vector $d_n + \varepsilon_n$, where

$$\sum_{i \in N(g^*):in \in g^*} \varepsilon_n^i = [v(g(\lambda_n a, a_n^*)) - v(g(x))] > 0.$$

(H) true for player m + 1 implies (H) true for player m: Suppose again that x is an inefficient history and that m is the first player in x such that the action a_m is not compatible with efficiency (in the sense of assumption (H)). Let a_m^* be some vector of arcs compatible with efficiency and let $x_m^*(\varepsilon) = (a_m^*, d_m + \varepsilon)$. Let $x^*(\varepsilon) \equiv f(\lambda_m x, x_m^*(\varepsilon))$ represent the relative continuation history. We need to show that there exists $\varepsilon > 0$ such that $g(x^*(\varepsilon)) \in G^*$. Note first that in the history $x^*(\varepsilon)$ the first player k such that a_k is not compatible with efficiency must be such that $k > m$. Also, since by (H)

$$\sum_{i=1}^{m} \sum_{j:ij \in N(h(i))} d_i^j \leq v(g(x)) - \sum_{i=m+1}^{n} \sum_{j:ij \in N(h(i))} d_i^j$$

there exists an $\varepsilon_m > 0$ such that

$$\sum_{i=1}^{m-1} \sum_{j:ij \in N(h(i))} d_i^j + \sum_{j:mj \in N(h(m))} (d_m^j + \varepsilon_m) < v(g^*) - \sum_{i=m+1}^{n} \sum_{j:ij \in N(h(i))} d_i^j.$$

Thus, if player m plays $x_m^*(\varepsilon_m)$, player $m+1$ faces a history $(\lambda_m x, x_m^*(\varepsilon_m))$ that satisfies the inductive assumption (H). Suppose now that player $m+1$ optimally plays some action x_{m+1} such that no efficient graph is compatible (in the sense of assumption (H)) with the history $(\lambda_m x, x_m^*(\varepsilon_m), x_{m+1})$. Then, by (H) we know there would be a deviation for player $m + 1$, contradicting the assumption that x_{m+1} is part of the continuation history at $(\lambda_m x, x_m^*(\varepsilon_m))$. Thus, we know that player $m + 1$ will optimally play some strategy x_{m+1}^* such that the continuation history $f((\lambda_m x, x_m^*(\varepsilon_m), x_{m+1}^*))$ induces a feasible efficient graph.

Step 2. We now show that the induction argument can be applied to each SPE history x of $\Gamma_2(v)$ such that $g(x) \notin G^*$. This is shown to imply that the first player m such that there is no x^* such that $\lambda_{m+1}x^* = \lambda_{m+1}x$ and $g(x^*) \in G^*$ has a profitable deviation.

[9] $\lambda_i a$ constitutes a slight abuse of notation, describing the history of arcs sent before the turn of player i.

Note first that by Lemma 5 if x is a SPE history then all players are connected. This, together with Lemma 4, directly implies that

$$\sum_{i=1}^{n} \sum_{j:ij \in N(h(i))} d_i^j = v(g(x))$$

or, equivalently, that

$$\sum_{i=1}^{m} \sum_{j:ij \in N(h(i))} d_i^j = v(g(x)) - \sum_{i=m+1}^{n} \sum_{j:ij \in N(h(i))} d_i^j$$

for all $m = 1, \ldots, n$. It follows that the induction argument can be applied to all inefficient SPE histories, to conclude that the first player whose action is not compatible with efficiency in the sense of (H), has some action $x_m^*(\varepsilon_m) = \left(a_m^*, d_m + \varepsilon_m\right)$ such that $\varepsilon_m > 0$ and such that the induced graph $g\left(f\left(\lambda_m x, x_m^*(\varepsilon_m)\right)\right) \in G^*$ is feasible. Since $g\left(f\left(\lambda_m x, x_m^*\right)\right)$ is feasible, then the action $x_m^*(\varepsilon_m)$ represents a deviation for player m, proving the theorem. **QED.**

References

Aumann, R., Myerson, R. (1988) Endogenous formation of links between players and coalitions: an application of the Shapley value. In: Roth, A. (ed.) *The Shapley Value*. Cambridge University Press, Cambridge

Bala, V., Goyal, S. (1998) Self Organization in Communication Networks. Working Paper at Erasmus University, Rotterdam

Dutta, B., Mutuswami, S. (1997) Stable networks. *Journal of Economic Theory* 76: 322–344

Harris, C. (1985) Existence and characterization of perfect equilibrium in games of perfect information. *Econometrica* 53: 613–628

Jackson, M.O., Watts, A. (2002) The evolution of social and economic networks. *Journal of Economic Theory* (forthcoming)

Jackson, M.O., Wolinsky, A. (1996) A strategic model of social and economic networks. *Journal of Economic Theory* 71: 44–74

Slikker, M., Van Den Nouweland, A. (2001) A one-stage model of link formation and payoff division. *Games and Economic Behavior* 34: 153–175

Quin. (1996) Endogenous formation of cooperation structures. *Journal of Economic Theory* 69: 218–226

Coalition Formation in General NTU Games

Anke Gerber

Institute for Empirical Research in Economics, University of Zurich, Blümlisalpstrasse 10, CH-8006 Zurich, Switzerland; (e-mail: agerber@iew.unizh.ch)

Abstract. A general nontransferable utility (NTU) game is interpreted as a collection of pure bargaining games that can be played by individual coalitions. The threatpoints or claims points respectively, in these pure bargaining games reflect the players' opportunities outside a given coalition. We develop a solution concept for general NTU games that is consistent in the sense that the players' outside opportunities are determined by the solution to a suitably defined reduced game. For any general NTU game the solution predicts which coalitions are formed and how the payoffs are distributed among the players.

Key Words: Endogenous coalition formation, bargaining, outside opportunities.

JEL Classification: C71, C78

1 Introduction

There are many economic situations in which coalition formation and bargaining over the gains from cooperation play a central role. Examples include the problem of firm formation and profit distribution in a coalition production economy, decisions about the provision of public goods in a local public goods economy or the question of formation of government. Common to these problems is that also coalitions different from the grand and single player coalitions play a role, which is an extension of the pure bargaining situation that was first analysed by

This paper is part of the author's dissertation at Bielefeld University, Germany. The author is grateful to Bhaskar Dutta and an anonymous referee for useful comments. Financial support through a scholarship of the Deutsche Forschungsgemeinschaft (DFG) at the graduate college "Mathematical Economics" at Bielefeld University is gratefully acknowledged.

Nash (1950). We can formulate these problems as general nontransferable utility games and, as we will see, analyse them by using tools from bargaining theory.[1]

Given an NTU game the main questions that arise are: 1. which coalitions are formed; and 2. which payoff vector is chosen by the coalitions that are actually formed? Simple and natural as these questions appear to be we find that the literature has mainly provided an answer to the second question while assuming that the coalition structure is exogenously given. All classic solutions for NTU games rest on this assumption: the *core* (Scarf 1967 and Aumann and Drèze 1974), the *Shapley NTU value* (Shapley 1969), the *Harsanyi solution* (Harsanyi 1959, Harsanyi 1963) and the *bargaining set* (Aumann and Maschler 1964 in the TU case, Asscher 1976, Asscher 1977 in the NTU case). Of course, there are situations in which we can justify the simplifying assumption of an exogenously given coalition structure. For example, if the underlying game is superadditive so that there are increasing returns to cooperation, then there are good reasons to expect the formation of the grand coalition. But even in the superadditive case we get counterintuitive results for solutions that take the formation of the grand coalition for granted (see the discussion on the Shapley NTU value in Aumann 1985, Aumann 1986, Roth 1980, Roth 1986 and Shafer 1980).

Realizing that in general the coalition structure will not be given exogenously other approaches model coalition formation and payoff distribution as a two-stage process. In the first stage the players form coalitions and in the second stage the payoffs are determined according to some solution concept that is defined with respect to an exogenously given coalition structure (see Hart and Kurz 1983 and Shenoy 1979). What is critical here is that players in the first stage make commitments to form certain coalitions although this restricts their bargaining possibilities in the second stage.

There are only few (cooperative) approaches that simultaneously address the questions of coalition formation and payoff distribution. Among these is the concept of a *bargaining aspiration outcome* (Bennett and Zame 1988). A bargaining aspiration (see Albers 1974, Albers 1979) is a vector of prices that players demand for their participation in any coalition. These prices are maximal and achievable and there must not be a one-sided dependence between any two players. A bargaining aspiration defines an outcome of the game if there exists a partition of the set of players into coalitions that can afford the prices of their members. Unfortunately, the existence of such a bargaining aspiration outcome is not guaranteed in general. The endogenous formation of coalitions is also achieved by the bargaining set defined in Zhou (1994), which comprises all payoff vectors that are feasible for *some* coalition structure and for which there exists no justified objection. In contrast to the Aumann–Maschler bargaining set coalitions rather than single players are the initiators of objections and counter-objections. However, nonemptiness can only be proved for a very restricted class of NTU games.

[1] Of course, our analysis will include as a special case all situations in which utility is transferable between the players (TU games).

Our approach to the solution of NTU games is based upon the fact that there will naturally be a mutual relation between payoffs and coalition structures. On the one hand the payoffs which the players expect to achieve in different coalitions determine with whom they will cooperate in the end. On the other hand the "bargaining power" of the members of some coalition S and thereby their payoffs clearly depend on what these players expect to achieve outside coalition S. That is, the payoffs in S depend in particular on the coalition structure that would emerge if coalition S were not formed. Thus, the payoffs influence the coalition structure and vice versa. The main idea of our solution concept is the following. We interpret an NTU game as a collection of pure bargaining games that can be played by single coalitions. For each coalition we take as exogenously given a solution concept for pure bargaining games which is meant to reflect a common notion of fairness in this coalition. Given these bargaining solutions the players can determine their payoffs in the various coalitions and decide which coalitions to form. Since the feasible set for each coalition is well defined in an NTU game the main issue will be to choose an appropriate disagreement point and possibly claims point for each bargaining game. Naturally these points should depend on the players' opportunities outside the given coalition. We will see that under this requirement the disagreement and claims points link the otherwise isolated bargaining games. Given the players' payoffs in each coalition we will apply the *dynamic solution* (Shenoy 1979, Shenoy 1980) in order to determine stable coalition structures.

The $\mathscr{C}$-solution we define is consistent in the sense that the outside opportunities in each coalition S are determined by the players' expected payoffs in the $\mathscr{C}$-solution of the game that is reduced by coalition S.[2] In this way we ensure credibility of the outside opportunities. By definition the $\mathscr{C}$-solution exists for all NTU games which is an important property.

The paper is organized as follows. In Sect. 2 we review solution concepts for abstract games. The $\mathscr{C}$-solution is defined in Sect. 3. Section 4 is devoted to the discussion of some properties of the new solution concept. We also consider special classes and several examples of NTU games. Finally, we close the paper with some concluding remarks in Sect. 5.

2 The Dynamic Solution for Abstract Games

In this section we recall the definition of the dynamic solution (Shenoy 1979, Shenoy 1980) which we will use later to select stable coalition structures.

Let X be an arbitrary set and let $\text{dom} \subset X \times X$ be a binary relation on X called *domination*.[3] Then (X, dom) is called an *abstract game*. An element $x \in X$ is said to be *accessible* from $y \in X$, denoted $y \to x$, if either $x = y$ or if there exist $z_0, z_1, \ldots, z_m \in X$ such that $z_0 = x$, $z_m = y$, and

[2] Guesnerie and Oddou 1979 introduce the term *C-stable solution* for the core defined with respect to an arbitrary coalition structure. We thank Shlomo Weber for pointing the similarity of terms out to us and hope the reader will not confuse the two concepts.

[3] Weak and strong set inclusion is denoted by $\subset$ and $\subsetneq$, respectively.

z_0 dom z_1 dom z_2 dom ... dom z_{m-1} dom z_m. The binary relation accessible is the transitive and reflexive closure of dom. The *core* of the abstract game (X, dom) is the set

$$\text{Core} = \{x \in X \mid \nexists\, y \in X \text{ such that } y \text{ dom } x\}.$$

Since the core is empty for a large class of games we aim at a solution concept with weaker stability requirements.

Definition 1. *The set $S \subset X$ is an elementary dynamic solution of an abstract game (X, dom) if*

1. $x \not\rightarrow y$ *for all* $x \in S$ *and* $y \in X \setminus S$,
2. $x \rightarrow y$ *and* $y \rightarrow x$ *for all* $x, y \in S$.

P is the dynamic solution of an abstract game (X, dom) if

$$P = \bigcup \{S \mid S \text{ is an elementary dynamic solution of } (X, \text{dom})\}.$$

Observe that the dynamic solution of an abstract game always exists and is unique, though it may be empty. The concept of the dynamic solution was introduced by Shenoy (1979), Shenoy (1980) and is closely related to the notion of an *R-admissible set* in the context of social choice correspondences (Kalai et al. 1976). It can easily be shown that the core is a subset of the dynamic solution by regarding each core element as an elementary dynamic solution.

If X is finite the dynamic solution can be characterized as follows.

Lemma 1. *If X is finite, then $P \subset X$ is the dynamic solution of an abstract game (X, dom) if and only if P satisfies the following conditions.*

1. *(Internal Stability)*
   For all $x, y \in P$, it is true that $x \rightarrow y$ if and only if $y \rightarrow x$.
2. *(External Stability)*
   a) *For all $x \in P$, $y \in X \setminus P$, it is true that $x \not\rightarrow y$.*
   b) *For all $y \in X \setminus P$ there exists $x \in P$ such that $y \rightarrow x$.*

We give a brief sketch of the proof and refer to Shenoy (1980) for the details: The sufficiency of the conditions as well as the necessity of conditions 1. and 2.(a) is obvious. In order to prove the necessity of condition 2.(b) assume by way of contradiction that there exists $y_1 \in X \setminus P$ such that $y_1 \not\rightarrow x$ for all $x \in P$. Let $S(y_1)$ be the equivalence class (with respect to the relation $\rightarrow$) containing y_1. If $x \not\rightarrow y$ for all $x \in S(y_1)$ and $y \in X \setminus S(y_1)$, then $S(y_1)$ is an elementary dynamic solution and we get a contradiction since $S(y_1) \subset X \setminus P$. Hence there exists $y_2 \in X \setminus (P \cup S(y_1))$ such that $x \rightarrow y_2$ for some $x \in S(y_1)$. Let $S(y_2)$ be the equivalence class containing y_2 and repeat the argument above. Since X is finite we get a contradiction after a finite number of steps.

From condition 2.(b) in Lemma 1 it immediately follows that the dynamic solution is always nonempty if X is finite.

Theorem 1. *Let* (X, dom) *be an abstract game. If* X *is finite, then the dynamic solution is nonempty.*

There is a clear dynamic interpretation of the stability notion inherent in the definition of the dynamic solution. Let us assume that for all $x, y \in X$ there exists a transition probability for moving from x to y. Assume that for $x \neq y$ this transition probability is strictly positive if and only if y dom x. Then, any elementary dynamic solution is an irreducible closed set of the Markov process generated by these transition probabilities and the dynamic solution P is the union of all the irreducible closed sets.[4] If X is finite, the elements $x \in P$ are the persistent states whereas the elements $x \notin P$ are the transient states of the Markov process.[5] Moreover, the probability of staying forever outside the dynamic solution is zero which implies that any process that starts with an arbitrary element $x \in X$ will enter the dynamic solution after a finite number of steps with probability one. Therefore, the dynamic solution is stable in a very natural sense.

3 Endogenous Coalition Formation

Let us first introduce some notation. In the following $\mathbb{N}$ will denote the set of positive integers and $\mathbb{R}$ will denote the set of real numbers. By $|A|$ we denote the cardinality of a set A. The set $N = \{1, \ldots, n\}$, $n \in \mathbb{N}$, will denote the *player set*. By $\mathscr{P}(N)$ we denote the set of nonempty subsets (*coalitions*) of N. Let $\mathbb{R}^N$ be the cartesian product of $|N| = n$ copies of $\mathbb{R}$, indexed by the elements of N. No confusion should arise from the fact that by 0 we will also denote the vector $(0, \ldots, 0) \in \mathbb{R}^N$. Vector inequalities in $\mathbb{R}^N$ are denoted by $\geq, >, \gg$, i.e. for $x, y \in \mathbb{R}^N$, $x \geq y$ means $x_i \geq y_i$ for all $i \in N$, $x > y$ means $x \geq y$ and $x \neq y$, and $x \gg y$ means $x_i > y_i$ for all $i \in N$. For $S \in \mathscr{P}(N)$ and $x \in \mathbb{R}^N$, we denote by x_S the *projection* of x to the subspace $\mathbb{R}^N_S$ that is spanned by the vectors $\left(e^i\right)_{i \in S}$, where e^i denotes the ith unit vector in $\mathbb{R}^N$. A set $A \subset \mathbb{R}^N_S$ is called *comprehensive in* $\mathbb{R}^N_S$ if $x \in A$ implies that $y \in A$ for all $y \in \mathbb{R}^N_S$, $y \leq x$. For $A \subset \mathbb{R}^N_S$ let

$$\mathrm{WPO(A)} = \{\mathrm{x} \in \mathrm{A} \mid \mathrm{y} \notin \mathrm{A} \text{ for all } \mathrm{y} \in \mathbb{R}^{\mathrm{N}}_{\mathrm{S}} \text{ with } \mathrm{y_i} > \mathrm{x_i} \text{ for all } \mathrm{i} \in \mathrm{S}\}$$

be the set of *weakly Pareto optimal* points in A and let

$$\mathrm{PO(A)} = \{\mathrm{x} \in \mathrm{A} \mid \mathrm{y} \notin \mathrm{A} \text{ for all } \mathrm{y} \in \mathbb{R}^{\mathrm{N}}_{\mathrm{S}} \text{ with } \mathrm{y} \neq \mathrm{x} \text{ and } \mathrm{y} \geq \mathrm{x}\}$$

denote the set of *Pareto optimal* points in A.

We will study nontransferable utility games defined as follows.

---

[4] A set of states of a Markov process is closed if the process never leaves the set after entering it. A closed set is irreducible if no proper subset is closed.

[5] We remark that in the following we only have to deal with finite sets X.

Definition 2. *A correspondence* $V : \mathscr{P}(N) \twoheadrightarrow \mathbb{R}^N$ *is called a nontransferable utility (NTU) game if*

1. $V(S) \subset \mathbb{R}^N_S$ *for all* $S \in \mathscr{P}(N)$,
2. $V(S)$ *is nonempty, convex and closed in the relative topology of* $\mathbb{R}^N_S$ *for all* $S \in \mathscr{P}(N)$,
3. $V(S)$ *is comprehensive in* $\mathbb{R}^N_S$ *for all* $S \in \mathscr{P}(N)$,
4. $V(\{i\})$ *is bounded from above for all* $i \in N$,
5. $\{x \in V(S) \,|\, x \geq \underline{x}_S\}$ *is nonempty and bounded from above for all* $S \in \mathscr{P}(N)$, $|S| \geq 2$, *where* $\underline{x} \in \mathbb{R}^N$ *is given by*

$$\underline{x}_i = \sup\{t \in \mathbb{R} \,|\, te^i \in V(\{i\})\} \quad \textit{for all } i \in N.$$[6]

Our definition of an NTU game is fairly standard. Observe that no loss of generality is incurred by imposing the requirement that for all coalitions the feasible set contains some individually rational utility allocations (5. in Definition 2). Any coalition S for which this is not the case is irrelevant for the determination of the outcome of the game and the feasible set could be replaced by the degenerate set $\{x \in \mathbb{R}^N_S \,|\, x \leq \underline{x}_S\}$ to be in accordance with our definition.

Let $\mathscr{G}$ be the class of NTU games and let Π be the set of all *coalition structures* on N, i.e.

$$\Pi = \left\{ \{S_1, \ldots, S_m\} \;\middle|\; S_i \in \mathscr{P}(N) \text{ for all } i,\ S_i \cap S_j = \emptyset \text{ for } i \neq j,\ \bigcup_{i=1}^{m} S_i = N \right\}.$$

Then, for $V \in \mathscr{G}$ and $P \in \Pi$ we denote by $\mathscr{F}_V(P)$ the set of payoff vectors that are feasible given coalition structure P, i.e.

$$\mathscr{F}_V(P) = \{x \in \mathbb{R}^N \,|\, x_S \in V(S) \text{ for all } S \in P\}.$$

An element $(Q,x) \in \bigcup_{P\in\Pi}(\{P\} \times \mathscr{F}_V(P))$ is called a *payoff configuration*. The set of payoff configurations will be the outcome space of our solution, which therefore predicts which coalitions are formed and which utility distribution is chosen in these coalitions.

We will first define a dominance relation on the set $\bigcup_{P\in\Pi}(\{P\} \times \mathscr{F}_V(P))$. To this end let $P \in \Pi$ and $R \in \mathscr{P}(N)$. The set of *partners* of R in coalition structure P is the set

$$C^P(R) = \{i \in N \,|\, \text{there exists } T \in P \text{ such that } T \cap R \neq \emptyset \text{ and } i \in T \setminus R\}.$$

Thus, i is a partner of coalition R in P if i himself is not a member of R but forms a coalition with some member of R in coalition structure P. Observe that $C^P(R) = \emptyset$ if and only if R is the union of coalitions $S \in P$.

[6] For the sake of keeping notation as simple as possible we omit indexing $\underline{x}$ with V.

Definition 3. *Let $V \in \mathscr{G}$ and let $(P,x),(Q,y) \in \bigcup_{P\in\Pi}(\{P\} \times \mathscr{F}_V(P))$. Then coalition R can induce (P,x) from (Q,y) if*

$$P = \{R\} \cup \{\{T\} \,|\, T \in Q, T \cap R = \emptyset\} \cup \{\{i\} \,|\, i \in C^Q(R)\},$$

and x is such that $x_R \in V(R)$ and

$$x_T = \begin{cases} y_T \text{ , if } T \in P \cap Q \\ \underline{x}_{\{i\}} \text{ , if } T = \{i\} \text{ and } i \in C^Q(R) \end{cases} .$$

Thus, coalition R can induce a movement from (Q,y) to (P,x) if (P,x) results from the formation of R and the consequent breaking of coalitions with the partners of R in Q.[7]

Definition 4. *Let $V \in \mathscr{G}$ and let $(P,x),(Q,y) \in \bigcup_{P\in\Pi}(\{P\} \times \mathscr{F}_V(P))$. Then (P,x) dominates (Q,y), shortly (P,x)* dom *(Q,y), if there exists a coalition R which can induce (P,x) from (Q,y) and*

$$x_i > y_i \text{ for all } i \in R.$$

We believe that the dominance relation given in Definition 4 is very natural, if one views coalition formation as a dynamic process, where players form coalitions and break them up again in favor of more profitable coalitions. Of course, our definition of dominance imputes a myopic behavior on the part of the players which is justified if, for example, coalition formation is time consuming and the players are impatient.[8]

In the following we will describe how the players' payoffs are determined in each coalition. The main idea of our paper is to interpret an NTU game as a family of interdependent bargaining games (with or without claims) for individual coalitions. These games are defined as follows.

Definition 5. *Let $S \in \mathscr{P}(N)$. Then (A,d) is a (pure) bargaining game for coalition S if*

1. *$d \in A \subset \mathbb{R}^N_S$,*
2. *A is convex and closed in the relative topology of $\mathbb{R}^N_S$,*
3. *$\{x \in A \,|\, x \geq d\}$ is bounded,*
4. *A is comprehensive in $\mathbb{R}^N_S$.*

---

[7] We interpret the formation of a coalition T not only as an agreement to cooperate but also as an agreement about the payoff distribution in T. Therefore, as soon as some members of T decide to leave the coalition, the former agreement is void. This is true in particular since $x_T \in V(T)$ does not imply that $x_{T\setminus R} \in V(T \setminus R)$, so that in general the members of $T \setminus R$ need a new agreement after deviation of R if they want to stay together.

[8] See Chwe 1994 and Ray and Vohra 1997 for different approaches where players are assumed to be farsighted.

Any bargaining game is characterized by a set of feasible utility allocations A, measured in von Neumann–Morgenstern utility scales, and a point d, called *disagreement point* or *threatpoint* which marks the outcome of the game if the players do not agree on a utility allocation in the feasible set. For $S \in \mathscr{P}(N)$ let

$$\Sigma^{\mathbf{S}} = \{(A,d) \mid (A,d) \text{ is a bargaining game for coalition } S\}.$$

In many "real life" negotiations the resolution of the conflict depends not only on the threatpoint but also on the claims with which the players come to the bargaining table, given these claims are credible or verifiable. For an example the reader may think of wage negotiations between labor and management. If the claims of the players are feasible they will naturally serve as a disagreement point. If they are not feasible, like in a bankruptcy problem, they give rise to a new class of bargaining problems which were formally introduced by Chun and Thompson (1992).

Definition 6. *Let $S \in \mathscr{P}(N)$. Then (A,d,c) is a bargaining problem with claims for coalition S if*

1. $(A,d) \in \Sigma^{\mathbf{S}}$,
2. $c \in \mathbb{R}^N_S \setminus A$, $c > d$.

For $S \in \mathscr{P}(N)$ let

$$\Sigma^{\mathbf{S}}_{\mathbf{c}} = \{(A,d,c) \mid (A,d,c) \text{ is a bargaining problem with claims for coalition } S\}.$$

Bargaining problems with and without claims naturally arise in the context of nontransferable utility games if one formalizes the idea that the players' payoffs in each coalition should depend on their opportunities in other coalitions. We assume that within each coalition the payoffs are determined according to some bargaining solution that reflects the normative conceptions of its members. Thus, for each coalition $S \in \mathscr{P}(N)$, $|S| \geq 2$, we take as exogenously given a *bargaining solution* $\varphi^S : \Sigma^{\mathbf{S}} \cup \Sigma^{\mathbf{S}}_{\mathbf{c}} \to \mathbb{R}^N_S$ which assigns to each bargaining game with or without claims for coalition S a payoff vector that is considered "fair" by the members of S. We explicitly allow for different coalitions to use different bargaining solutions, i.e. the normative conceptions of a player may (but need not) depend on the respective coalition she is a member of, which we believe is a realistic assumption.

In an attempt to provide a very general approach we only impose a minimum set of conditions upon φ^S. For all $(A,d) \in \Sigma^{\mathbf{S}}$ and for all $(A,d,c) \in \Sigma^{\mathbf{S}}_{\mathbf{c}}$,

1. **(Feasibility)** $\varphi^S(A,d) \in A$ and $\varphi^S(A,d,c) \in A$,
2. **(Individual Rationality)** $\varphi^S(A,d) \geq d$ and $\varphi^S(A,d,c) \geq d$,
3. **(Weak Pareto Optimality)** $\varphi^S(A,d) \in$ WPO(A) and $\varphi^S(A,d,c) \in$ WPO(A).

Weak Pareto optimality of the bargaining solutions φ^S is not needed in the following. Without this requirement, however, the rationality of our solution concept would be obscure. The bargaining solution φ^S could be given by any combination of solutions to bargaining problems with and without claims, e.g. the *Nash solution* (Nash 1950) or the *Kalai–Smorodinsky solution* (Kalai and Smorodinsky 1975) for bargaining problems without claims and the *proportional solution* (Chun and Thomson 1992) for bargaining problems with claims.

Given an NTU game V and bargaining solutions φ^S ($S \in \mathscr{P}(N)$, $|S| \geq 2$), in each coalition the main issue will be to determine the players' outside opportunities as reflected by the disagreement point (and the claims point, in case the outside opportunities are not feasible). Of course, a player can claim to get an arbitrarily high payoff outside a given coalition, but her claim will increase her bargaining power only if this threat is credible. In the context of pure bargaining games it is often assumed that the disagreement point is given by the Nash equilibrium outcome of some underlying noncooperative game. For an NTU game credibility requires that player i's outside opportunity as a member of coalition S should be given by her expected payoff if negotiations in S break down and i settles for her alternatives in the remaining coalitions. Thus, we have to look at reduced games defined as follows. For $V \in \mathscr{G}$ and $S \in \mathscr{P}(N)$ the *reduced game*[9] $V^{-S} \in \mathscr{G}$ is given by

$$V^{-S}(T) = \begin{cases} V(T) & , T \neq S \\ \{y \in \mathbb{R}^N_T \mid y \leq \underline{x}_T\}, & T = S \end{cases} .$$

Indeed, if the members of S break off negotiations, this is equivalent to assigning to S a feasible set which is degenerate. Observe that with our definiton of reduced games we do not allow for renegotiations. In the game V^{-S} the members of S cannot use S as a threat any more since the feasible set is degenerate. In other words, V^{-S} truly reflects the *outside* opportunities of S.

The solution we propose in this paper is consistent in the sense that the players' outside opportunities in coalition S are determined by the *solution* to V^{-S}. If the expected payoffs for the members of S in the solution to V^{-S} are feasible they define a natural disagreement point for the bargaining game played by coalition S. If the outside opportunities are not feasible they define a natural claims point. Therefore, given a game $V \in \mathscr{G}$, a coalition $S \in \mathscr{P}(N)$, $|S| \geq 2$, and the bargaining solution φ^S we obtain a function ϕ^S_V which maps outside opportunities into feasible utility allocations for coalition S. Formally, define $\phi^S_V : \{y \in \mathbb{R}^N_S \mid y \geq \underline{x}_S\} \to \mathbb{R}^N_S$ by

$$\phi^S_V(y) = \begin{cases} \varphi^S(V(S), y) & , y \in V(S) \\ \varphi^S(V(S), \underline{x}_S, y) & , \text{else} \end{cases} ,$$

$y \in \mathbb{R}^N_S$, $y \geq \underline{x}_S$.[10] Observe that ϕ^S_V is well defined since $(V(S), y) \in \Sigma^{\mathbf{S}}$ if $y \in V(S)$, and $(V(S), \underline{x}_S, y) \in \Sigma^{\mathbf{S}}_{\mathbf{c}}$ if $y \notin V(S)$, $y \in \mathbb{R}^N_S$, $y \geq \underline{x}_S$.

[9] We are aware of the fact that the term *reduced game* has been used at other places in the literature with a different meaning. However, we could not think of a different term for the games we consider here which expresses equally well the fact that we *reduce* the original game by one coalition.

[10] As we will see in the following we only have to deal with individually rational outside opportunities.

Definition 7. *A coalition $S \in \mathscr{P}(N)$, $|S| \geq 2$, is called relevant in game $V \in \mathscr{G}$ if there exists $y \in V(S)$ such that $y > \underline{x}_S$. By $\mathscr{R}^V$ we denote the set of relevant coalitions in V.*

For example, there is at most one relevant coalition in a pure bargaining game for coalition S (namely S), if we interpret the bargaining game as an NTU game, which can be done in an obvious way. Rational players will either form relevant coalitions or they will stay on their own so that we can restrict ourselves to coalition structures generated by relevant and single player coalitions: Let

$$\Pi(\mathscr{R}^V) = \{P \in \Pi \mid S \in P \Rightarrow \ S \in \mathscr{R}^V \text{ or } S = \{i\} \text{ for some } i \in N\}.$$

Now the definition of our solution on the class $\mathscr{G}$ is straightforward: With each $V \in \mathscr{G}$ and each relevant coalition $S \in \mathscr{R}^V$ we associate a bargaining game with or without claims. As argued above credibility of the disagreement point and claims point requires these points to be determined by the players' opportunities outside S. In order to obtain a consistent solution concept we compute a player's outside opportunity in S as the average payoff he gets in the solution to the reduced game V^{-S}. The payoffs for the members of S are then determined by applying the bargaining solution φ^S. Finally, the dynamic solution picks the stable payoff configurations among those that are generated by $\mathscr{R}^V$ and the respective payoffs in these coalitions.

Definition 8. *Let $V \in \mathscr{G}$. The $\mathscr{C}$-solution is a set of payoff configurations which is inductively defined over $|\mathscr{R}^V|$ as follows.*

1. $|\mathscr{R}^V| = 0$:
   For $V \in \mathscr{G}$, $\mathscr{R}^V = \emptyset$, the $\mathscr{C}$-solution is given by

$$\{(\{\{1\},\{2\},\ldots,\{n\}\},(\underline{x}_1,\ldots,\underline{x}_n))\}.$$

2. $|\mathscr{R}^V| = m, m \geq 1$:
   Let $V \in \mathscr{G}$ with $|\mathscr{R}^V| = m$. For $S \in \mathscr{R}^V$ let

$$\{(P^1,x^1),(P^2,x^2),\ldots,(P^{k(S)},x^{k(S)})\} \subset \bigcup_{P\in\Pi} (\{P\} \times \mathscr{F}_{V^{-S}}(P))$$

be the $\mathscr{C}$-solution for V^{-S} and let $y^S = \frac{1}{k(S)} \sum_{l=1}^{k(S)} x_S^l \in \mathbb{R}_S^N$ be the average payoff distribution of the members of S in V^{-S}. The $\mathscr{C}$-solution for V is given by the dynamic solution to the abstract game (X, dom), where X is the set of payoff configurations (P,x) such that $P \in \Pi(\mathscr{R}^V)$ and for all $S \in P$,

$$x_S = \begin{cases} \phi_V^S\left(y^S\right), & |S| \geq 2 \\ \underline{x}_{\{i\}}, & \text{if } S = \{i\} \end{cases}.$$

Given the individual rationality of the bargaining solutions φ^S and given the uniqueness and nonemptiness of the dynamic solution (see Theorem 1) it is straightforward to see that the $\mathscr{C}$-solution is well defined, nonempty and unique.

Before discussing some properties of the $\mathscr{C}$-solution let us illustrate the definition with the following example.

Example 1 (Piano Mover Game). *Let* $N = \{1,2,3\}$ *and let* $V : \mathscr{P}(N) \twoheadrightarrow \mathbb{R}^N$ *be defined by*

$$V(\{1,2\}) = \{x \in \mathbb{R}^N_{\{1,2\}} \,|\, x_1 + x_2 \leq 50\},$$

$$V(\{1,3\}) = \{x \in \mathbb{R}^N_{\{1,3\}} \,|\, x_1 + x_3 \leq 50\},$$

$$V(S) = \{x \in \mathbb{R}^N_S \,|\, x \leq 0\}, \text{ else.}$$

In order to compute the $\mathscr{C}$*-solution for this game it turns out that we only have to specify the bargaining solution on the class of bargaining games without claims* $\Sigma^{\mathbf{S}}$*. For all* $S \in \mathscr{P}(N)$, $|S| \geq 2$, *let* $\varphi^S(A,d) = \nu^S(A,d)$ *for all* $(A,d) \in \Sigma^{\mathbf{S}}$, *where* $\nu^S : \Sigma^{\mathbf{S}} \to \mathbb{R}^N_S$ *is the Nash solution for coalition* S.[11] *In the following we will use a simplified notation for coalition structures, e.g. we write* $[\,12\,|\,3\,]$ *instead of* $\{\{1,2\},\{3\}\}$.

In the piano mover game the set of relevant coalitions is given by $\mathscr{R}^V = \{\{1,2\},\{1,3\}\}$*. We first consider the reduced game* $\left(V^{-\{1,2\}}\right)^{-\{1,3\}}$*, which, by definition, has no relevant coalitions. Thus, by Definition 8,*

$$\mathscr{C}\text{-solution for } \left(V^{-\{1,2\}}\right)^{-\{1,3\}} = \{([\,1\,|\,2\,|\,3\,],(0,0,0))\}\,. \tag{1}$$

Next, consider the reduced game $V^{-\{1,2\}}$ *where coalition* $\{1,3\}$ *is the only relevant coalition. By (1) and using the notation of Definition 8 we find that* $y^{\{1,3\}} = (0,0,0)$ *is the average payoff distribution of the members of coalition* $\{1,3\}$ *in the reduced game* $\left(V^{-\{1,2\}}\right)^{-\{1,3\}}$*. Since* $(0,0,0) \in V^{-\{1,2\}}(\{1,3\}) = V(\{1,3\})$, *we get*

$$\phi^{\{1,3\}}_{V^{-\{1,3\}}}(0,0,0) = \nu^{\{1,3\}}(V(\{1,3\}),(0,0,0)) = (25,0,25). \tag{2}$$

Obviously, $\Pi(\mathscr{R}^{V^{-\{1,2\}}}) = \{[\,1\,|\,2\,|\,3\,],[\,13\,|\,2\,]\}$, *so that the* $\mathscr{C}$*-solution for* $V^{-\{1,2\}}$ *is given by the dynamic solution to* (X, dom)*, where*

$$X = \{([\,1\,|\,2\,|\,3\,],(0,0,0)),([\,13\,|\,2\,],(25,0,25))\}\,. \tag{3}$$

The latter payoff configuration dominates the first but not vice versa. Hence,

$$\mathscr{C}\text{-solution for } V^{-\{1,2\}} = \{([\,13\,|\,2\,],(25,0,25))\}\,. \tag{4}$$

Similarly, we compute

---

[11] The *Nash solution for coalition* S, $\nu^S : \Sigma^{\mathbf{S}} \to \mathbb{R}^N_S$ is given by $\nu^S(A,d) = \operatorname{argmax}\{\prod_{i \in M}(x_i - d_i) \,|\, x \in A,\ x \geq d\}$, if $M \neq \emptyset$, and $\nu^S(A,d) = d$, else, where $M = \{i \in S \,|\, \text{there exists } x \in A,\ x \geq d, \text{ such that } x_i > d_i\}$.

$$\mathscr{C}\text{-solution for } V^{-\{1,3\}} = \{([\,12\,|\,3\,], (25,25,0))\}\,. \tag{5}$$

Finally, we consider the original game V. By (4) and again using the notation of Definition 8 the average payoff for players 1 *and* 2 *outside coalition* $\{1,2\}$ *is given by* $y^{\{1,2\}} = (25,0,0) \in V(\{1,2\})$ *and we get*

$$\phi_V^{\{1,2\}}(25,0,0) = \nu^{\{1,2\}}\left(V(\{1,2\}),(25,0,0)\right) = (37.5,12.5,0). \tag{6}$$

Also, by (5), $y^{\{1,3\}} = (25,0,0) \in V(\{1,3\})$ *is the outside option vector for coalition* $\{1,3\}$ *and therefore,*

$$\phi_V^{\{1,3\}}(25,0,0) = \nu^{\{1,3\}}\left(V(\{1,3\}),(25,0,0)\right) = (37.5,0,12.5). \tag{7}$$

Thus, since $\Pi(\mathscr{R}^V) = \{[\,1\,|\,2\,|\,3\,],[\,12\,|\,3\,],[\,13\,|\,2\,]\}$*, the $\mathscr{C}$-solution for V is given by the dynamic solution to* (X, dom)*, where*

$$X = \{([\,1\,|\,2\,|\,3\,],(0,0,0)),\ ([\,12\,|\,3\,],(37.5,12.5,0)),\ ([\,13\,|\,2\,],(37.5,0,12.5))\}\,. \tag{8}$$

Since the latter two payoff configurations do not dominate each other but both dominate the first, we obtain:

$$\mathscr{C}\text{-solution for } V = \{([\,12\,|\,3\,],(37.5,12.5,0)),\ ([\,13\,|\,2\,],(37.5,0,12.5))\}\,.[12] \tag{9}$$

In this game our intuition might suggest that player 1 (the strong piano mover) should be able to get the whole surplus i.e., one would expect a final payoff distribution of $(50,0,0)$*. This is also the payoff distribution predicted by the set of bargaining aspiration outcomes (Bennett and Zame 1988) and the core of the superadditive cover of V. The support for the payoff distribution* $(50,0,0)$ *comes from the fact that, given his exceptional role, player 1 should be able to play 2 and 3 (the weak piano movers) off against each other. However, empirical results provide more support for the prediction of the $\mathscr{C}$-solution. Maschler (1978, p. 253) conducted an experiment with high school children and found that the average payoff to player* 1 *was approximately* 37.82*, which is almost exactly the payoff predicted by the $\mathscr{C}$-solution and is also close to the payoff predicted by the Shapley NTU value and the Harsanyi solution for the superadditive cover of V. The latter are both given by* $\{1/3(100,25,25)\}$*.*

Coming back to the definition of the $\mathscr{C}$-solution there are, of course, many other possibilities to define an outside opportunity payoff vector for each relevant coalition. For example, one could select *one* element of the $\mathscr{C}$-solution for any reduced game to support the $\mathscr{C}$-solution in the higher level game.[13] But this

---

[12] We remark that the $\mathscr{C}$-solution for the superadditive cover of V is given by

$$\{([\,12\,|\,3\,],(37.5,12.5,0)),\ ([\,13\,|\,2\,],(37.5,0,12.5)),\ ([\,123\,],(37.5,6.25,6.25))\}\,,$$

i.e. the average predicted payoffs remain the same.

[13] Compare the definition of a *bargaining equilibrium* for an assignment game in Crawford and Rochford (1986).

would mean imputing to all agents that they expect the same equilibrium in any reduced game which seems to be a very unrealistic assumption. Another possibility would be to take the minimally (or maximally) attainable outside opportunity level for each player. However, both choices do not reflect the true outside opportunities and would be subject to a debate among the players. For example, the minimally or maximally attainable level might be the same for all members of a coalition although the reduced game is highly asymmetric with respect to these players. On the contrary we believe that the average over the payoffs a player can obtain in the solution to the reduced game truly reflects his outside opportunities and cannot be attacked for not being credible.[14]

4 Properties of the $\mathscr{C}$-Solution

In the following we will discuss some characteristics of the $\mathscr{C}$-solution. Obviously, all properties will depend on the characteristics of the bargaining solutions chosen by each coalition. First, we show that the $\mathscr{C}$-solution is symmetric or, more general, anonymous if the bargaining solutions φ^S have this property. Let $\pi : N \to N$ be a permutation. With a slight abuse of notation we define $\pi(x) \in \mathbb{R}^N$ by $(\pi(x))_i = x_{\pi^{-1}(i)}$ $(i \in N, x \in \mathbb{R}^N)$, $\pi(A) \subset \mathbb{R}^N$ by $\pi(A) = \{z \mid \exists x \in A \text{ such that } \pi(x) = z\}$ $(A \subset \mathbb{R}^N)$, $\pi(S) \in \mathscr{P}(N)$ by $\pi(S) = \{j \mid \exists i \in S \text{ such that } \pi(i) = j\}$ $(S \in \mathscr{P}(N))$, $\pi(P) \in \Pi$ by $\pi(P) = \{T \mid \exists S \in P \text{ such that } \pi(S) = T\}$ $(P \in \Pi)$, and $\pi(V) \in \mathscr{G}$ by $\pi(V)(S) = \pi\left(V\left(\pi^{-1}(S)\right)\right)$ $(S \in \mathscr{P}(N), V \in \mathscr{G})$.

Definition 9. *Let $\varphi = \{\varphi^S \mid S \in \mathscr{P}(N),\ |S| \geq 2\}$. Then φ is anonymous if for all permutations $\pi : N \to N$ and for all $S \in \mathscr{P}(N)$, $|S| \geq 2$, it is true that*

$$\begin{aligned} \pi\left(\varphi^S(A,d)\right) &= \varphi^{\pi(S)}(\pi(A),\pi(d)) && \textit{for all } (A,d) \in \Sigma^{\mathbf{S}}, \\ \pi\left(\varphi^S(A,d,c)\right) &= \varphi^{\pi(S)}(\pi(A),\pi(d),\pi(c)) && \textit{for all } (A,d,c) \in \Sigma^{\mathbf{S}}_{\mathbf{c}}. \end{aligned}$$

In other words φ is anonymous if each φ^S is anonymous and only depends on $|S|$. The following theorem shows that the property of anonymity carries over to the $\mathscr{C}$-solution. The proof of the theorem and its corollary is straightforward.

Theorem 2. *Let φ be anonymous. Then for all permutations $\pi : N \to N$ and all $V \in \mathscr{G}$ it is true that*

$$(P,x) \in \mathscr{C}\textit{-solution for } V \iff (\pi(P),\pi(x)) \in \mathscr{C}\textit{-solution for } \pi(V).$$

Corollary 1. *Let $\pi : N \to N$ be a permutation and let $V \in \mathscr{G}$ be such that $\pi(V) = V$. If φ is anonymous, then*

$$(P,x) \in \mathscr{C}\textit{-solution for } V \iff (\pi(P),\pi(x)) \in \mathscr{C}\textit{-solution for } V.$$

[14] The fact that we take equal weights for all elements of the solution to the reduced games may be subject to criticism but is indeed justified since all elements of the $\mathscr{C}$-solution exhibit the same stability properties and there is nothing in our model which allows us to discriminate between them. Endogenous weights for the elements of the $\mathscr{C}$-solution can only be determined if we are willing to make some (strong) assumptions about the bargaining process so that we can appeal to the dynamic properties of the dynamic solution (recall the discussion of the dynamic solution after Theorem 1).

Keeping in mind that the players' payoffs are given in terms of von Neumann–Morgenstern utility scales we will check in the following how the $\mathscr{C}$-solution behaves under positive affine transformations of utility. Let $a, b \in \mathbb{R}^N$, $a \gg 0$. Then the mapping $L^{a,b} : \mathbb{R}^N \to \mathbb{R}^N$ is called a *positive affine transformation* if $\left(L^{a,b}(x)\right)_i = a_i x_i + b_i$ for all $i \in N$ and $x \in \mathbb{R}^N$. For $A \subset \mathbb{R}^N$ let $L^{a,b}(A) = \{z \mid \exists x \in A$ such that $z = L^{a,b}(x)\}$. With a slight abuse of notation we define $L^{a,b} : \mathscr{G} \to \mathscr{G}$ by

$$L^{a,b}(V)(S) = L^{a,b_S}(V(S)) \text{ for all } S \in \mathscr{P}(N),\ V \in \mathscr{G}.$$

Definition 10. *Let* $\varphi = \{\varphi^S \mid S \in \mathscr{P}(N), |S| \geq 2\}$. *Then* φ *is covariant under positive affine transformations of utility if for all positive affine transformations* $L^{a,b} : \mathbb{R}^N \to \mathbb{R}^N$ *and for all* $S \in \mathscr{P}(N)$, $|S| \geq 2$, *it is true that*

$$\begin{aligned} L^{a,b_S}\left(\varphi^S(A,d)\right) &= \varphi^S\left(L^{a,b_S}(A), L^{a,b_S}(d)\right) && \text{for all } (A,d) \in \Sigma^{\mathbf{S}}, \\ L^{a,b_S}\left(\varphi^S(A,d,c)\right) &= \varphi^S\left(L^{a,b_S}(A), L^{a,b_S}(d), L^{a,b_S}(c)\right) && \text{for all } (A,d,c) \in \Sigma^{\mathbf{S}}_{\mathbf{c}}. \end{aligned}$$

The proof of the following theorem is again straightforward.

Theorem 3. *Let* φ *be covariant under positive affine transformations of utility. Then for all positive affine transformations* $L^{a,b} : \mathbb{R}^N \to \mathbb{R}^N$ *and all* $V \in \mathscr{G}$ *it is true that*

$$(P, x) \in \mathscr{C}\text{-solution for } V \iff (P, L^{a,b}(x)) \in \mathscr{C}\text{-solution for } L^{a,b}(V).$$

Next we will study whether the payoff configurations predicted by the $\mathscr{C}$-solution are weakly efficient .[15]

Definition 11. *Let* $V \in \mathscr{G}$ *and let* $(P, x) \in \bigcup_{P \in \Pi}(\{P\} \times \mathscr{F}_V(P))$ *be a payoff configuration. Then* (P, x) *is called weakly efficient if there does not exist* $(Q, y) \in \bigcup_{P \in \Pi}(\{P\} \times \mathscr{F}_V(P))$ *with* $y \gg x$.

It turns out that the payoff configurations in the $\mathscr{C}$-solution are not weakly efficient, in general. One reason is the well known conflict between equity and efficiency which we illustrate with the following example of a transferable utility game.

Example 2. *Let* $N = \{1, 2, 3\}$ *and let* $V : \mathscr{P}(N) \twoheadrightarrow \mathbb{R}^N$ *be defined by*

$$\begin{aligned} V(\{i\}) &= \{x \in \mathbb{R}^N_{\{i\}} \mid x_i \leq 0\}, \text{ for all } i = 1, 2, 3, \\ V(\{1,2\}) &= \{x \in \mathbb{R}^N_{\{1,2\}} \mid x_1 + x_2 \leq 18\}, \\ V(\{1,3\}) &= \{x \in \mathbb{R}^N_{\{1,3\}} \mid x_1 + x_3 \leq 12\}, \\ V(\{2,3\}) &= \{x \in \mathbb{R}^N_{\{2,3\}} \mid x_2 + x_3 \leq 12\}, \\ V(N) &= \{x \in \mathbb{R}^N \mid x_1 + x_2 + x_3 \leq 22\}. \end{aligned}$$

[15] We confine to weak efficiency because strong efficiency cannot be expected with our notion of strong dominance.

Obviously, in this game only the formation of the grand coalition can lead to an efficient payoff configuration. Assume that φ is anonymous and covariant under positive affine transformations of utility. A straightforward computation, which we do not want to present here, then shows that the payoffs in the relevant coalitions are as follows:

$$\{1,2\}:(9,9,0),\ \{1,3\}:(9,0,3),\ \{2,3\}:(0,9,3),\ \{1,2,3\}:(26/3,26/3,14/3).$$

Hence, the $\mathscr{C}$-solution is given by

$$\{([\,12\,|\,3\,],(9,9,0)),([\,13\,|\,2\,],(9,0,3)),([\,1\,|\,23\,],(0,9,3))\}\,.$$

Since the grand coalition is not formed, none of the payoff configurations in the $\mathscr{C}$-solution is weakly efficient.

The example shows that if players are restricted to distribute payoffs in an equitable way, namely by applying a bargaining solution which takes into account each player's outside opportunities, then the resulting payoff configuration may turn out to be inefficient. The loss of efficiency is the "price" we have to pay for achieving fairness in our sense. The next question then is whether the payoff configurations in the $\mathscr{C}$-solution are constrained efficient. I.e., given that the payoffs within each relevant coalition have to be distributed according to a bargaining solution, is it possible to improve all players over a payoff configuration in the $\mathscr{C}$-solution? Again, the answer is positive as we can see from the following example.

Example 3. *Let $N=\{1,2,3,4\}$ and let $V:\mathscr{P}(N)\twoheadrightarrow\mathbb{R}^N$ be defined by*

$$\begin{aligned}
V(\{2,3\}) &= \{x\in\mathbb{R}^N_{\{1,2\}}\,|\,(x_2,x_3)\le(8,5)\},\\
V(\{1,2,3\}) &= \{x\in\mathbb{R}^N_{\{1,2,3\}}\,|\,(x_1,x_2,x_3)\le(10,7,4)\},\\
V(\{2,3,4\}) &= \{x\in\mathbb{R}^N_{\{2,3,4\}}\,|\,(x_2,x_3,x_4)\le(10,7,4)\},\\
V(\{1,3,4\}) &= \{x\in\mathbb{R}^N_{\{1,3,4\}}\,|\,(x_1,x_3,x_4)\le(4,10,7)\},\\
V(\{1,2,4\}) &= \{x\in\mathbb{R}^N_{\{1,2,4\}}\,|\,(x_1,x_2,x_4)\le(7,4,10)\},\\
V(N) &= \{x\in\mathbb{R}^N\,|\,(x_1,x_2,x_3,x_3)\le(1,9,6,1)\},\\
V(S) &= \{x\in\mathbb{R}^N_S\,|\,x\le 0\},\ \textit{else}.
\end{aligned}$$

Let the bargaining solutions φ be Pareto optimal. Then the payoffs within each relevant coalition are given by the unique Pareto optimal utility allocation in the feasible set and it is easily seen that the $\mathscr{C}$-solution is given by

$$\begin{gathered}
\{([\,123\,|\,4\,],(10,7,4,0)),([\,234\,|\,1\,],(0,10,7,4)),([\,134\,|\,2\,],(4,0,10,7))\\
([\,124\,|\,3\,],(7,4,0,10)),([\,1\,|\,23\,|\,4\,],(0,8,5,0)),([\,1234\,],(1,9,6,1))\,\}
\end{gathered}$$

Observe that the payoff configuration $([\,1\,|\,23\,|\,4\,],(0,8,5,0))$ is not weakly (constrained) efficient but nevertheless belongs to the $\mathscr{C}$-solution since it dominates the (weakly efficient) payoff configurations $([\,123\,|\,4\,],(10,7,4,0))$ and $([\,124\,|\,3\,],(7,4,0,10))$ that belong to the $\mathscr{C}$-solution.

However, contrary to the case of unconstrained efficiency, there always exist payoff configurations in the $\mathscr{C}$-solution which are constrained efficient since any inefficient payoff configuration necessarily leads to a weakly efficient one via the dominance relation we have defined.

We do not think that the lack of constrained efficiency that we may find for some payoff configurations in our solution is a serious drawback since there is no reason to believe that during a coalition formation process coalitions will only deviate to weakly efficient payoff configurations, in particular if we assume players to be myopic as it is reflected in our dominance relation.

4.1 Special Cases

It is difficult to obtain general statements about the behavior of the $\mathscr{C}$-solution on large classes of games since most properties of the original game are not inherited by the reduced games. We usually have to enter a case by case study, even for games that have a nice structure, like e.g. balanced transferable utility games. Whether a coalition is formed or not critically depends on the payoffs that are obtained in all relevant coalitions and these payoffs, in general, depend in a highly discontinuous way on the parameters of the game.[16] In this section we will consider some special cases for which we are able to obtain clear-cut predictions. We also present examples of games for which the $\mathscr{C}$-solution gives intuitive results that are different from the predictions of other solution concepts.

4.1.1 Symmetric Games

Corollary 1 can be used to determine the $\mathscr{C}$-solution for any symmetric game if φ is anonymous. Let $V \in \mathscr{G}$ be symmetric, i.e. $\pi(V) = V$ for all permutations $\pi : N \to N$, and let φ be anonymous. For $k = 1, \ldots, n$, let $\alpha_k = \max\{t \in \mathbb{R} \mid t\, e_S \in V(S)\}$ where $S \in \mathscr{P}(N)$ is such that $|S| = k$ and $e \in \mathbb{R}^N$ is defined by $e_i = 1$ for all $i \in N$. The numbers α_k are well defined since V is symmetric, $V(\{i\})$ is bounded from above for all $i \in N$ and $V(S)$ is bounded from above on the set of individually rational payoffs for all $S \in \mathscr{P}(N)$, $|S| \geq 2$.

Consider a coalition $S \in \mathscr{R}^V$ and a permutation $\pi : N \to N$ such that $\pi(S) = S$. Then $\pi(V^{-S}) = (\pi(V))^{-\pi(S)} = V^{-S}$. By Corollary 1 $(P,x) \in \mathscr{C}$-solution for V^{-S} if and only if $(\pi(P), \pi(x)) \in \mathscr{C}$-solution for V^{-S}. This implies $y^S = \pi(y^S)$ for all permutations $\pi : N \to N$ such that $\pi(S) = S$, where y^S is the outside option vector for coalition S. Since φ is anonymous we get $\phi_V^S(y^S) = \pi\left(\phi_V^S(y^S)\right)$ for all $S \in \mathscr{R}^V$ and for all permutations π such that $\pi(S) = S$. Therefore, $\phi_V^S(y^S) = \alpha_{|S|}\, e_S$ for all $S \in \mathscr{R}^V$ by weak Pareto optimality of φ^S.

Let X be the set of payoff configurations (P, x) such that $P \in \Pi\left(\mathscr{R}^V\right)$ and $x_S = \alpha_{|S|} e_S$ for all $S \in P$. By definition the $\mathscr{C}$-solution for V is the dynamic solution of the abstract game (X,dom). Let $\overline{X}$ be the set of all payoff

[16] We hope, though, that our solution will not be judged by the complexity of its computation.

configurations $(P, x) \in X$ defined as follows. If $P = \{S_1, \ldots, S_m\}$, where the coalitions are numbered so that $\alpha_{|S_1|} \geq \alpha_{|S_2|} \geq \ldots \geq \alpha_{|S_m|}$, then

$$\alpha_{|S_l|} = \max\left\{ \alpha_k \;\middle|\; k \leq |N \setminus \bigcup_{h=1}^{l-1} S_h| \right\}$$

for all $l = 1, \ldots, m$.

Theorem 4. *If $V \in \mathscr{G}$ is symmetric and φ is anonymous, then the set $\overline{X}$ is the $\mathscr{C}$-solution for V.*

Proof. First observe that no $(P, x) \in \overline{X}$ is dominated by some $(Q, y) \in X$: Let $(P, x) \in \overline{X}$ be given with $P = \{S_1, \ldots, S_m\}$ and $\alpha_{|S_1|} \geq \alpha_{|S_2|} \geq \ldots \geq \alpha_{|S_m|}$, and assume that there exists $S \in \mathscr{P}(N)$ such that $\alpha_{|S|} > x_i$ for all $i \in S$. Hence, there exists $1 \leq k \leq m$ such that k is minimal with the property that

$$\alpha_{|S|} > \alpha_{|S_k|}. \tag{10}$$

By definition of $(P, x) \in \overline{X}$ this implies that $|S| > |N \setminus \bigcup_{l=1}^{k-1} S_l|$. On the other hand from (10) and the fact that $\alpha_{|S|} > x_i$ for all $i \in S$ we conclude that $S \subset N \setminus \bigcup_{l=1}^{k-1} S_l$ which is a contradiction.

It remains to be shown that for any payoff configuration $(Q, y) \in X \setminus \overline{X}$ there exists $(P, x) \in \overline{X}$ which is accessible from (Q, y) with respect to the dominance relation dom. Let $(Q, y) \in X \setminus \overline{X}$. We will inductively define payoff configurations $(P^k, x^k) \in X$ and sets $A^k \subset N$ such that $(P^1, x^1) = (Q, y)$ and for all $k \geq 1$,

1. (P^{k+1}, x^{k+1}) dom (P^k, x^k),
2. $A^k \subsetneq A^{k+1}$,
3. if for some $S \in \mathscr{P}(N)$, $\alpha_{|S|} > x_i^k$ for all $i \in S$, then $S \subset N \setminus A^k$.

We stop this procedure if we have reached k so that there exists no $S \in \mathscr{P}(N)$ with $\alpha_{|S|} > x_i^k$ for all $i \in S$. Observe that the procedure indeed stops after a finite number of steps because of condition 3. and the fact that N is finite and the sets A^k are strictly increasing. If the procedure stops with a payoff configuration (P^k, x^k) then it is straightforward to see that $(P^k, x^k) \in \overline{X}$. Also, (P^k, x^k) is accessible from $(P^1, x^1) = (Q, y)$ which proves the theorem.

We inductively construct the sequence of payoff configurations as follows. Set $(P^1, x^1) = (Q, y)$ and choose some numbering of the coalitions in P^1 such that $P^1 = \{S_1, \ldots, S_{m_1}\}$ with

$$\alpha_{|S_1|} \geq \alpha_{|S_2|} \geq \ldots \geq \alpha_{|S_{m_1}|}.$$

Since $(P^1, x^1) \in X \setminus \overline{X}$ there exists $1 \leq l \leq m_1$ such that

$$\alpha_{|S_l|} < \max\left\{ \alpha_k \;\middle|\; k \leq |N \setminus \bigcup_{h=1}^{l-1} S_h| \right\}.$$

Let l_1 be minimal with this property and set $A^1 = \bigcup_{h=1}^{l_1-1} S_h$.

Assume that for $k \geq 1$, (P^l, x^l) and A^l, $l = 1, \ldots, k$, have already been defined so that conditions 1.–3. are fulfilled. If there exists no $S \in \mathscr{P}(N)$ such that $\alpha_{|S|} > x_i^k$ for all $i \in S$ we are done. Otherwise, choose some numbering of the coalitions in P^k such that $P^k = \{T_1, \ldots, T_{m_k}\}$ with

$$\alpha_{|T_1|} \geq \alpha_{|T_2|} \geq \ldots \geq \alpha_{|T_{m_k}|}.$$

Then there exists $1 \leq l \leq m_k$ such that

$$\alpha_{|T_l|} < \max \left\{ \alpha_k \;\middle|\; k \leq |N \setminus \bigcup_{h=1}^{l-1} T_h| \right\}.$$

Let l_k be minimal with this property. Observe that for all $i \in N \setminus \bigcup_{h=1}^{l_k-1} T_h$ there exists $T \subset N \setminus \bigcup_{h=1}^{l_k-1} T_h$ such that $i \in T$ and $\alpha_{|T|} > x_j^k$ for all $j \in T$. Thus, by 3., $i \notin A^k$, i.e. $A^k \subset \bigcup_{h=1}^{l_k-1} T_h$. Choose $T \subset N \setminus \bigcup_{h=1}^{l_k-1} T_h$ such that

$$\alpha_{|T|} = \max \left\{ \alpha_k \;\middle|\; k \leq |N \setminus \bigcup_{h=1}^{l_k-1} T_h| \right\}.$$

Then $\alpha_{|T|} > x_i^k$ for all $i \in T$ and by 3., $T \subset N \setminus A^k$. Set $A^{k+1} = \bigcup_{h=1}^{l_k-1} T_h \cup T$ and let (P^{k+1}, x^{k+1}) be the payoff configuration induced by T from (P^k, x^k) with $x_T^{k+1} = \alpha_{|T|} e_T$. Then (P^{k+1}, x^{k+1}) dom (P^k, x^k) and $A^k \subsetneq A^{k+1}$. Moreover, by construction, if there exists $S \in \mathscr{P}(N)$ such that $\alpha_{|S|} > x_i^{k+1}$ for all $i \in S$ then $S \subset N \setminus A^{k+1}$.

□

Remark 1. Under the notation used above, the $\mathscr{C}$-solution is identical to the core of the abstract game (X, dom). This follows from the fact that a payoff configuration $(P, x) \in X$ is dominated by some payoff configuration $(Q, y) \in X$ if and only if $(P, x) \in X \setminus \overline{X}$ (see the proof of Theorem 4).

4.1.2 Marriage Markets

It is interesting to note that the elements of the $\mathscr{C}$-solution for an NTU game which represents a marriage market (Gale and Shapley 1962) correspond to the set of stable matchings if the bargaining solutions are Pareto optimal. Thus, the predictions of the $\mathscr{C}$-solution are in accordance with the common perception of what should be the outcome for this class of games. Let the set of agents be given by $N = W \cup M$, where W and M are disjoint finite sets consisting of "women" and "men", respectively. Each $w \in W$ has preferences over the set $M \cup \{w\}$ that are represented by a utility function $u_w : M \cup \{w\} \to \mathbb{R}$. Thus, $u_w(m) > u_w(m')$ means that w prefers being matched (married) with m over being matched with m'. Similarly, each $m \in M$ has preferences over the set $W \cup \{m\}$ represented by $u_m : W \cup \{m\} \to \mathbb{R}$. W.l.o.g. normalize the utility of remaining single to be zero for all agents, i.e. $u_i(i) = 0$ for all $i \in N$. A *matching market* $\mathscr{M}$ then is given by the triple (W, M, u) where $u = (u_i)_{i \in N}$.

A *matching* is a one-to-one function $\mu : W \cup M \to W \cup M$ such that $\mu(\mu(i)) = i$ for all $i \in W \cup M$, and $\mu(w) \notin M$ ($\mu(m) \notin W$) implies $\mu(w) = w$ ($\mu(m) = m$). A matching $\mu : W \cup M \to W \cup M$ is *stable* if there exists no $i \in N$ such that $u_i(i) > u_i(\mu(i))$ (*individual rationality*), and if there exists no pair $(w, m) \in W \times M$ such that $u_w(m) > u_w(\mu(w))$ and $u_m(w) > u_m(\mu(m))$. Gale an Shapley (1962) proved that the set of stable matchings is nonempty for all matching markets (W, M, u).[17]

We can represent a matching market $\mathscr{M}$ as an NTU game $V^{\mathscr{M}} : \mathscr{P}(N) \twoheadrightarrow \mathbb{R}^N$ defined by

$$\begin{aligned} V^{\mathscr{M}}(\{w, m\}) &= \{x \in \mathbb{R}^N_{\{w,m\}} \mid (x_w, x_m) \leq (u_w(m), u_m(w))\}, \\ &\quad \text{for all } w \in W,\ m \in M,\ \text{ such that } u_w(m) \geq 0,\ u_m(w) \geq 0, \\ V^{\mathscr{M}}(S) &= \{x \in \mathbb{R}^N_S \mid x \leq 0\},\ \text{else}. \end{aligned}$$

Note that we defined the feasible sets for pairs to be degenerate if there is at least one agent for whom being matched with the other is worse than remaining single. Since we are only interested in individually rational matchings this definition imposes no loss of generality and guarantees that $V^{\mathscr{M}}$ is an NTU game in the sense of Definition 2. Obviously, $\mathscr{R}^{V^{\mathscr{M}}}$ is the set of all pairs $\{w, m\}$ such that $u_w(m) \geq 0$ and $u_m(w) \geq 0$ with strict inequality for at least one $i \in \{w, m\}$. Each individually rational matching μ corresponds to exactly one coalition structure $P^\mu \in \Pi\left(\mathscr{R}^{V^{\mathscr{M}}}\right)$.[18] The following theorem is a corollary of a result by Roth and Vande Vate (1990).

Theorem 5. *Let $\mathscr{M} = (W, M, u)$ be a marriage market and assume that the bargaining solutions φ^S are Pareto optimal for all S with $|S| = 2$, i.e. $\varphi^S(A, d) \in$ PO(A) for all $(A, d) \in \Sigma^{\mathbf{S}}$, and $\varphi^S(A, d, c) \in$ PO(A) for all $(A, d, c) \in \Sigma^{\mathbf{S}}_{\mathbf{c}}$. Then $\mu : W \cup M \to W \cup M$ is stable if and only if $(P^\mu, x^\mu) \in \mathscr{C}$-solution for $V^{\mathscr{M}}$, where P^μ is the coalition structure corresponding to μ and $x_i^\mu = u_i(\mu(i))$ for all $i \in N$.*

Proof. First of all observe that Pareto optimality of the bargaining solutions φ^S for $|S| = 2$ implies that the outside opportunities are irrelevant for the determination of the payoffs for each pair $\{w, m\} \in \mathscr{R}^{V^{\mathscr{M}}}$ since $x \in \mathrm{PO}(V^{\mathscr{M}}(\{w, m\}))$ if and only if $x_w = u_w(m)$, $x_m = u_m(w)$ and $x_i = 0$ for all $i \in N \setminus \{w, m\}$. Thus, the $\mathscr{C}$-solution for $V^{\mathscr{M}}$ is the dynamic solution to $(X^{\mathscr{M}}$,dom), where $X^{\mathscr{M}}$ is the set

[17] Gale and Shapley (1962) ruled out indifferences in the preferences of the agents. However, their existence proof is valid also in the general case. Also, Roth and Vande Vate (1990) provide a different existence proof which does not rely on preferences to be strict.

[18] Let μ be an individually rational matching. Then $P^\mu \in \Pi\left(\mathscr{R}^{V^{\mathscr{M}}}\right)$ is defined as follows. If $\mu(i) = i$, then $\{i\} \in P^\mu$. If $\mu(w) = m$ and $\{w, m\} \in \mathscr{R}^{V^{\mathscr{M}}}$ for some $w \in W$, $m \in M$, then $\{w, m\} \in P^\mu$. If $\mu(w) = m$ but $\{w, m\} \notin \mathscr{R}^{V^{\mathscr{M}}}$ for some $w \in W$, $m \in M$, then $\{w\}, \{m\} \in P^\mu$.

of all payoff configurations (P,x), such that $P \in \Pi\left(\mathscr{R}^{V^{\mathscr{M}}}\right)$ and x is such that whenever $\{w,m\} \in P$, then $x_w = u_w(m)$, $x_m = u_m(w)$, and whenever $\{i\} \in P$, then $x_i = 0$. If μ is stable, then, by definition, $(P^\mu, x^\mu) \in \mathscr{C}$-solution for $V^{\mathscr{M}}$ if P^μ is the coalition structure corresponding to μ and $x_i^\mu = u_i(\mu(i))$ for all $i \in N$. On the other hand, if μ is not stable (and w.l.o.g. μ individually rational), then Roth and Vande Vate (1990) have shown that there exists a finite sequence of matchings $\mu^1, \ldots, \mu^k$ such that $\mu^1 = \mu$, μ^k is stable and which has the following property. If $(P^{\mu^i}, x^{\mu^i}) \in X^{\mathscr{M}}$ is the payoff configuration corresponding to μ^i, then (P^{μ^k}, x^{μ^k}) dom $(P^{\mu^{k-1}}, x^{\mu^{k-1}})$ dom ... dom $(P^{\mu^1}, x^{\mu^1}) = (P^\mu, x^\mu)$. Thus, $(P^{\mu^k}, x^{\mu^k}) \in \mathscr{C}$-solution for $V^{\mathscr{M}}$ and (P^{μ^k}, x^{μ^k}) is accessible from (P^μ, x^μ) with respect to dom but not vice versa. This proves that $(P^\mu, x^\mu) \notin \mathscr{C}$-solution.

□

4.1.3 Transferable Utility Games

In the following we will determine the $\mathscr{C}$-solution for 3-person superadditive transferable utility (TU) games.[19] It will turn out that the $\mathscr{C}$-solution does not always predict the formation of the grand coalition. Even if the game is balanced, which is equivalent to the core being nonempty, there are cases where the grand coalition is never formed in the $\mathscr{C}$-solution.

A mapping $v : \mathscr{P}(N) \to \mathbb{R}$ is called a *transferable utility (TU) game*. A TU game $v : \mathscr{P}(N) \to \mathbb{R}$ is *superadditive*, if $v(S) + v(T) \leq v(S \cup T)$ for all $S, T \in \mathscr{P}(N)$, $S \cap T = \emptyset$. In particular, if v is superadditive, then v can equivalently be represented as an NTU game V in the class $\mathscr{G}$ by defining $V(S) = \{x \in \mathbb{R}^N_S \mid \sum_{i \in S} x_i \leq v(S)\}$ for all $S \in \mathscr{P}(N)$. In the following let $N = \{1,2,3\}$ and let φ be anonymous and covariant under positive affine transformations of utility. Then, w.l.o.g. we can confine to TU games that are normalized so that $v(\{i\}) = 0$ for all $i \in N$. For all coalitions S let $\varphi^S|_{\Sigma^S_{\mathbf{c}}}$ be the *proportional solution* (see Chun and Thomson 1992), i.e. $\varphi^S(A,d,c) = d + \bar{\lambda}(c-d)$, where $\bar{\lambda} = \max\{\lambda \in \mathbb{R} \mid d + \lambda(c-d) \in A\}$, for all $(A,d,c) \in \Sigma^S_{\mathbf{c}}$.[20]

1 Relevant Coalition. The case of one relevant coalition is almost trivial. Let $c > 0$ and let $v : \mathscr{P}(N) \to \mathbb{R}$ be given by $v(\{1,2,3\}) = c, v(S) = 0$, else. Then,

$$\mathscr{C}\text{-solution} = \left\{\left([\,123\,], \left(\frac{c}{3}, \frac{c}{3}, \frac{c}{3}\right)\right)\right\}.$$

2 Relevant Coalitions. Let $c \geq b > 0$ and let $v : \mathscr{P}(N) \to \mathbb{R}$ be given by $v(\{1,2,3\}) = c$, $v(\{1,2\}) = b$, $v(S) = 0$, else. This game is balanced for all choices of c and b with $c \geq b > 0$. The $\mathscr{C}$-solution of v (if represented as an NTU game) is as follows. If $c > b$, then

[19] We do not present the straightforward but tedious computation.

[20] Observe that it is not enough to assume that φ is anonymous and covariant under permutations in order to determine the solution of a bargaining game with claims in the transferable utility case.

$$\mathscr{C}\text{-solution} = \left\{ \left([\,123\,], \left(\frac{c}{3} + \frac{b}{6}, \frac{c}{3} + \frac{b}{6}, \frac{c}{3} - \frac{b}{3} \right) \right) \right\}.$$

If $c = b$, then

$$\mathscr{C}\text{-solution} = \left\{ \left([\,12\,|\,3\,], \left(\frac{c}{2}, \frac{c}{2}, 0 \right) \right), \left([\,123\,], \left(\frac{c}{2}, \frac{c}{2}, 0 \right) \right) \right\}.$$

Hence, in both cases the average predicted payoffs are the same and belong to the core of the game. If $c = b$, then in addition to the formation of the grand coalition the $\mathscr{C}$-solution predicts the formation of coalition $\{1, 2\}$.

3 Relevant Coalitions. Let $c \geq b \geq a > 0$ and let $v : \mathscr{P}(N) \to \mathbb{R}$ be given by $v(\{1,2,3\}) = c$, $v(\{1,2\}) = b$, $v(\{1,3\}) = a$, $v(S) = 0$, else. The game is balanced for all choices of a, b, c, with $c \geq b \geq a > 0$. The $\mathscr{C}$-solution of v is as follows. If $c > b > a$, then

$$\mathscr{C}\text{-solution} = \left\{ \left([\,123\,], \left(\frac{c}{3} + \frac{b}{6} + \frac{a}{4}, \frac{c}{3} + \frac{b}{6} - \frac{a}{4}, \frac{c}{3} - \frac{b}{3} \right) \right) \right\}.$$

If $c = b > a$, then

$$\begin{aligned} \mathscr{C}\text{-solution} \quad = \quad & \left\{ \left([\,12\,|\,3\,], \left(\frac{c}{2} + \frac{a}{4}, \frac{c}{2} - \frac{a}{4}, 0 \right) \right), \right. \\ & \left. \left([\,123\,], \left(\frac{c}{2} + \frac{a}{4}, \frac{c}{2} - \frac{a}{4}, 0 \right) \right) \right\}. \end{aligned}$$

If $a = b \geq \frac{8}{11}c$, then

$$\begin{aligned} \mathscr{C}\text{-solution} \quad = \quad & \left\{ \left([\,12\,|\,3\,], \left(\frac{3}{4}b, \frac{1}{4}b, 0 \right) \right), \right. \\ & \left([\,13\,|\,2\,], \left(\frac{3}{4}b, 0, \frac{1}{4}b \right) \right), \\ & \left. \left([\,123\,], \left(\frac{c}{3} + \frac{5}{12}b, \frac{c}{3} - \frac{5}{24}b, \frac{c}{3} - \frac{5}{24}b \right) \right) \right\}. \end{aligned}$$ [21]

If $a = b < \frac{8}{11}c$, then

$$\mathscr{C}\text{-solution} = \left\{ \left([\,123\,], \left(\frac{c}{3} + \frac{5}{12}b, \frac{c}{3} - \frac{5}{24}b, \frac{c}{3} - \frac{5}{24}b \right) \right) \right\}.$$

Again, in all cases there exists a payoff configuration in the $\mathscr{C}$-solution, in which the grand coalition is formed, but now the predicted payoffs do not necessarily belong to the core.

Table 1. The $\mathscr{C}$-solution for v with 4 relevant coalitions

		$\mathscr{C}$-solution
$a > \frac{2}{3}b$	$c < \frac{7}{16}a + \frac{9}{8}b$	$([\,13 \mid 2\,], (\frac{3}{8}a + \frac{b}{4}, 0, \frac{5}{8}a - \frac{b}{4}))$ $([\,1 \mid 23\,], (0, \frac{3}{8}a + \frac{b}{4}, \frac{5}{8}a - \frac{b}{4}))$
	$\frac{7}{16}a + \frac{9}{8}b \leq c \leq \frac{25}{16}a + \frac{3}{8}b$	$([\,13 \mid 2\,], (\frac{3}{8}a + \frac{b}{4}, 0, \frac{5}{8}a - \frac{b}{4}))$ $([\,1 \mid 23\,], (0, \frac{3}{8}a + \frac{b}{4}, \frac{5}{8}a - \frac{b}{4}))$ $([\,123\,], (\frac{c}{3} + \frac{b}{8} - \frac{7}{48}a, \frac{c}{3} + \frac{b}{8} - \frac{7}{48}a, \frac{c}{3} - \frac{b}{4} + \frac{7}{24}a))$
	$\frac{25}{16}a + \frac{3}{8}b < c$	$([\,123\,], (\frac{c}{3} + \frac{b}{8} - \frac{7}{48}a, \frac{c}{3} + \frac{b}{8} - \frac{7}{48}a, \frac{c}{3} - \frac{b}{4} + \frac{7}{24}a))$
$a = \frac{2}{3}b$	$c < \frac{23}{18}b$	$([\,12 \mid 3\,], (\frac{b}{2}, \frac{b}{2}, 0))$ $([\,13 \mid 2\,], (\frac{b}{2}, 0, \frac{b}{6}))$ $([\,1 \mid 23\,], (0, \frac{b}{2}, \frac{b}{6}))$
	$c = \frac{23}{18}b$	$([\,12 \mid 3\,], (\frac{b}{2}, \frac{b}{2}, 0))$ $([\,13 \mid 2\,], (\frac{b}{2}, 0, \frac{b}{6}))$ $([\,1 \mid 23\,], (0, \frac{b}{2}, \frac{b}{6}))$ $([\,123\,], (\frac{b}{2}, \frac{b}{2}, \frac{5}{18}b))$
	$c > \frac{23}{18}b$	$([\,123\,], (\frac{c}{3} + \frac{2}{27}b, \frac{c}{3} + \frac{2}{27}b, \frac{c}{3} - \frac{4}{27}b))$
$a < \frac{2}{3}b$	$c > b$	$([\,123\,], (\frac{c}{3} + \frac{b}{6}, \frac{c}{3} + \frac{b}{6}, \frac{c}{3} - \frac{b}{3}))$
	$c = b$	$([\,12 \mid 3\,], (\frac{b}{2}, \frac{b}{2}, 0))$ $([\,123\,], (\frac{b}{2}, \frac{b}{2}, 0))$

4 Relevant Coalitions. Since there are too many cases to consider when there are 4 relevant coalitions we restrict ourselves to the following one. Let $c \geq b \geq a > 0$ and let $v : \mathscr{P}(N) \to \mathbb{R}$ be given by $v(\{1,2,3\}) = c$, $v(\{1,2\}) = b$, $v(\{1,3\}) = v(\{2,3\}) = a, v(S) = 0$, else. Observe that v is balanced if and only if $a + b/2 \leq c$. The $\mathscr{C}$-solution for v is given in Table 1.

From Table 1 we see that there are 2 cases in which the $\mathscr{C}$-solution does not predict the formation of the grand coalition although the game is balanced, namely if $a > 2b/3$ and $a + b/2 \leq c < 7a/16 + 9b/8$ or if $a = 2/3b$ and $a + b/2 \leq c < 23b/18$. (Recall that Example 4 belongs to the latter case.) In these cases equity requires the players to distribute the payoffs in such a way that the formation of the grand coalition is not the best choice. We observe, however, that the $\mathscr{C}$-solution uniquely predicts the formation of the grand coalition if c becomes large enough. This fact holds true in general and is proved in the following lemma.

Lemma 2. *Let* $v^c : \mathscr{P}(N) \to \mathbb{R}$ *be a TU game (N arbitrary) such that* $\sum_{i \in S} v^c(\{i\}) \leq v^c(S)$ *for all* $S \in \mathscr{P}(N)$. *Let* T *be an arbitrary coalition with*

[21] This is the case of the superadditive cover of the piano mover game in Example 3.

$|T| \geq 2$, and let $v^c(T) = c$ and $v^c(S)$ be independent of c for all $S \neq T$. If φ is anonymous and covariant under positive affine transformations of utility, then there exists $\bar{c}$ such that $T \in P$ for all $(P, x) \in \mathscr{C}$-solution of v^c if $c \geq \bar{c}$. In particular, the $\mathscr{C}$-solution uniquely predicts the formation of the grand coalition if $v^c(N) = c$ is large enough.

Proof. Let v^c be a TU game as in the statement of the lemma and let V^c be its equivalent representation as an NTU game. Let T, $|T| \geq 2$, be a coalition with $v^c(T) = c$ and let $c > \sum_{i \in T} v^c(\{i\})$, so that $T \in \mathscr{R}^{V^c}$. Observe that $(V^c)^{-T}$ is independent of c. In particular, the outside option vector $y^T \in \mathbb{R}^N_T$ for coalition T is independent of c. Let $c \geq y = \sum_{i \in T} y_i^T$. Then $x_i(c) = \left(\phi^T_{V^c}(y^T)\right)_i = \left(\varphi^T(V^c(T), y^T)\right)_i = y_i^T + (c - y)/|T|$ which is strictly increasing in c for all $i \in T$. The maximum payoff a player can obtain in any relevant coalition $S \neq T$ is $c_1 = \max\{v^c(S) \mid S \in \mathscr{P}(N), S \neq T\}$. Thus, there exists $\bar{c} \geq c_1$ such that $x_i(c) > c_1$ for all $i \in T$ and for all $c \geq \bar{c}$ which implies the formation of coalition T in any payoff configuration in the $\mathscr{C}$-solution of v^c.

□

We close this section with two examples of 3-person NTU games for which the $\mathscr{C}$-solution gives very intuitive predictions. The first example was heavily discussed in the literature.

Example 4 (Roth 1980). *Let $N = \{1, 2, 3\}$ and $0 \leq p < 1/2$. Consider $V : \mathscr{P}(N) \twoheadrightarrow \mathbb{R}^N$, given by*

$$\begin{aligned}
V(\{i\}) &= \{x \in \mathbb{R}^N_{\{i\}} \mid x_i \leq 0\} \text{ for all } i \in N, \\
V(\{1,2\}) &= \{x \in \mathbb{R}^N_{\{1,2\}} \mid (x_1, x_2) \leq (1/2, 1/2)\}, \\
V(\{1,3\}) &= \{x \in \mathbb{R}^N_{\{1,3\}} \mid (x_1, x_3) \leq (p, 1-p)\}, \\
V(\{2,3\}) &= \{x \in \mathbb{R}^N_{\{2,3\}} \mid (x_2, x_3) \leq (p, 1-p)\}, \\
V(N) &= \text{cch}\left(\{(1/2, 1/2, 0), (p, 0, 1-p), (0, p, 1-p)\}\right),
\end{aligned}$$

where cch(A) *denotes the convex and comprehensive hull of $A \subset \mathbb{R}^N$, i.e. the smallest convex and comprehensive set in $\mathbb{R}^N$ containing A. The $\mathscr{C}$-solution for V is given by*

$$\left\{\left([\,12\,|\,3\,], (1/2, 1/2, 0)\right), \left([\,123\,], (1/2, 1/2, 0)\right)\right\}.$$

As Roth argues, the payoff distribution $(1/2, 1/2, 0)$ is the unique outcome of the game that is consistent with the hypothesis that the players are rational utility maximizers. This is exactly the payoff distribution predicted by the $\mathscr{C}$-solution, the core and the set of bargaining aspiration outcomes. The reason for expecting $(1/2, 1/2, 0)$ to be the outcome of the game is that $1/2$ is the maximum payoff players 1 and 2 can achieve in this game and they can realize it without cooperation of 3. Especially for p small it seems very unlikely that player 3 should be able to persuade 1 or 2 to form a coalition with him.

Despite the intuitive support for the payoff distribution $(1/2, 1/2, 0)$ *we can imagine a scenario in which it is not a priori clear that none of the coalitions* $\{1,3\}$ *and* $\{2,3\}$ *will form, so that the prediction of the Shapley NTU value* ($(1/3, 1/3, 1/3)$ *if* $p > 0$ *and additionally* $(1/2, 1/2, 0)$ *if* $p = 0$*) does not appear to be absurd any more. The unit of measurement might play a role here (are we talking about \$100, \$1000 or even one million dollar as a unit?), as well as the way the players are bargaining with each other (are they all bargaining openly in the same room or are they talking to each other on the phone and the left out player cannot interfere?). An extensive discussion of this example along these lines can be found in Aumann (1985), Aumann (1986) and Roth (1980), Roth (1986) and Shafer (1980).*

The following example is due to Owen (1972).

Example 5. *Let* $N = \{1, 2, 3\}$, *and let* $V : \mathscr{P}(N) \twoheadrightarrow \mathbb{R}^N$ *be given by*

$$V(\{1,2\}) = \left\{x \in \mathbb{R}^N_{\{1,2\}} \mid \exists y \in \mathbb{R}^N_+,\ y \geq x, \textit{such that } y_1 + 4y_2 \leq 100\right\},$$

$$V(N) = \left\{x \in \mathbb{R}^N \mid \exists y \in \mathbb{R}^N_+,\ y \geq x, \textit{such that } \textstyle\sum_{i=1}^3 y_i \leq 100\right\},$$

$$V(S) = \left\{x \in \mathbb{R}^N_S \mid x \leq 0\right\}, \textit{ else.}$$

Underlying the formal definition of the game might be the following scenario. The gains from a joint venture between players 1 and 2 are paid to player 1 who then has to transfer some share to player 2. If she mails the money to player 2 it is stolen with a very high probability of 3/4. If she transfers the money by her bank (player 3) nothing is lost, but surely player 3 will demand a fee for his service.

The $\mathscr{C}$*-solution for* V *is given by*

$$\{([\,123\,], (62.5, 25, 12.5))\}.$$

Let us compare the $\mathscr{C}$*-solution to the predictions of other solution concepts. The set of bargaining aspiration outcomes* ($\{([\,123\,], x)\}$, *where* $x \in \mathbb{R}^3_{++}$ *is such that* $\sum_{i=1}^3 x_i = 100$ *and* $x_2 > \frac{1}{3}x_3$*) and the core* ($\{x \in \mathbb{R}^3_+ \mid \sum_{i=1}^3 = 100,\ x_2 \geq \frac{1}{3}x_3\}$*) are both too large to give a good prediction for the outcome of the game. Both include extreme payoff distributions in which either player 1 or player 2 receive almost the whole surplus of 100. It seems that player 2 is in a weaker position than player 1 and we would expect the outcome to reflect this asymmetry of the game. However, the Shapley NTU value* ($\{(50, 50, 0)\}$*) and the Harsanyi solution* ($\{(40, 40, 20)\}$*) both assign equal payoffs to these players. Moreover, the Shapley NTU value predicts player 3 to offer his service for nothing. By comparison, the* $\mathscr{C}$*-solution predicts a payoff distribution which we would intuitively expect (at least in relative, not absolute terms). Player 1 keeps about 2/3 of the money for herself. The rest is transferred to player 2, where player 3 gets a fee to the amount of 12.5 for his service to transfer the money. At first sight the fee might appear to be large (1/3 of the transferred money). However, it naturally reflects the high risk to transfer the money by mail.*

5 Conclusion

The questions of coalition formation and payoff distribution are central to the theory of general NTU games. Nevertheless, there are only few approaches that simultaneously address both points. Often it is assumed that players will form the grand coalition or some other exogenously given coalition structure while smaller coalitions are only used as a threat to enforce certain payoffs. It is obvious that this approach to a solution for general NTU games is not appropriate in general, especially for games that are not superadditive.

We have provided a model of coalition formation which relies on the interpretation of an NTU game as a family of interdependent bargaining games. The disagreement points or claims points which link these games are determined by the players' expected payoffs if bargaining in the respective coalition breaks down. In bargaining theory the disagreement point and claims point are exogenously given. In our context these points arise endogenously as an aggregate of the players' outside opportunities in an NTU game. Observe, however, that the disagreement point and claims point are still exogenous in the bargaining problem of each coalition since the outside opportunities are independent of the agreement within the coalition (no renegotiations). This is due to the consistency property of the $\mathscr{C}$-solution: the opportunities outside a coalition are determined by the $\mathscr{C}$-solution to the reduced game where the respective coalition is not relevant any more.

Bennett (1991), Bennett (1997) presents an approach that is similar to ours in the sense that an NTU game is interpreted as a set of interrelated bargaining games. Given that each coalition has a conjecture about the agreements in other coalitions the disagreement point in each coalition is determined by the maximum amount each member can achieve in alternative coalitions. Then, as in our model, the payoffs in each coalition are computed according to a bargaining solution. These payoffs in turn serve as a conjecture for other coalitions and so on. A consistent conjecture is a fixed point of the mapping described above. Bennett (1991), Bennett (1997) proves that each consistent conjecture generates an aspiration and vice versa (for some choice of bargaining solutions). Bennett's multilateral bargaining approach is opposite to our model in two respects. First, it allows for renegotiations, which means that outside opportunities cannot be interpreted as disagreement payoffs as in our case. Thus, the application of a bargaining solution is questionable since players know about the indirect influence of any agreement on their outside opportunities and therefore on their disagreement point. Second, outside options in the multilateral bargaining approach are not credible in general. In order to obtain their maximum payoff outside a coalition two players might rely on the formation of two coalitions which cannot be formed simultaneously, i.e. outside options might not be overall feasible. Moreover, it is not analysed whether the members of a player's best alternative coalition really want to cooperate. They might as well have better alternatives. This criticism, of course, only applies out of equilibrium. Nevertheless, if we interpret a consistent conjecture as the limit outcome of a process in which players constantly adjust their conjectures, then any form of inconsistency before the limit is reached is all but harmless.

Unfortunately, experiments mostly deal with small TU games, where the number of players often does not exceed four, so that we cannot make a general statement about the suitability of the $\mathscr{C}$-solution as a predictor of "real" outcomes of coalitional games. Of course, the predictive power of the $\mathscr{C}$-solution depends on the appropriate choice of the bargaining solutions which in turn depends on the situation that is modelled by a game. We believe that the generality of our approach is advantageous and our examples underline that the $\mathscr{C}$-solution captures many important aspects that determine which coalitions are formed and which payoff vector is chosen in a general NTU game.

References

Albers, W. (1974) Zwei Lösungskonzepte für kooperative Mehrpersonenspiele, die auf Anspruchsniveaus der Spieler basieren. *Operations Research Verfahren* 21: 1–13

Albers, W. (1979) Core- and Kernel-variants based on imputations and demand profiles. In: Moeschlin, O., Pallaschke, D. (eds.) *Game Theory and Related Topics*. North-Holland Publishing Company, Amsterdam

Asscher, N. (1976) An ordinal bargaining set for games without side payments. *Mathematics of Operations Research* 1(4): 381–389

Asscher, N. (1977) A cardinal bargaining set for games without side payments. *International Journal of Game Theory* 6(2): 87–114

Aumann, R.J. (1985) On the non-transferable utility value: A comment on the Roth–Shafer examples. *Econometrica* 53(3): 667–677

Aumann, R.J. (1986) Rejoinder. *Econometrica* 54(4): 985–989

Aumann, R.J., Drèze, J.H. (1974) Cooperative games with coalition structures. *International Journal of Game Theory* 3(4): 217–237

Aumann, R.J., Maschler, M. (1964) The bargaining set for cooperative games. In: Dresher, M., Shapley, L.S., Tucker, A.W. (eds.) *Advances in Game Theory (Annals of Mathematics Studies 52)*. Princeton University Press, Princeton

Bennett, E. (1991) Three approaches to bargaining in NTU games. In: Selten, R. (ed.) *Game Equilibrium Models III*. Springer, Berlin, Heidelberg, New York

Bennett, E. (1997) Multilateral bargaining problems. *Games and Economic Behavior* 19(2): 151–179

Bennett, E., Zame, W.R. (1988) Bargaining in cooperative games. *International Journal of Game Theory* 17(4): 279–300

Chun, Y., Thomson, W. (1992) Bargaining problems with claims. *Mathematical Social Sciences* 24: 19–33

Chwe, M. S.-Y. (1994) Farsighted coalitional stability. *Journal of Economic Theory* 63(2): 299–325

Crawford, V.P., Rochford, S.C. (1986) Bargaining and competition in matching markets. *International Economic Review* 27(2): 329–348

Gale, D., Shapley, L.S. (1962) College admissions and the stability of marriage. *American Mathematical Monthly* 69(1): 9–15

Guesnerie, R., Oddou, C. (1979) On economic games which are not necessarily superadditve. *Economics Letters* 3: 301–306

Harsanyi, J.C. (1959) A bargaining model for the cooperative n-person game. In: Tucker, A.W., Luce, R.D. (eds.) *Contributions to the Theory of Games IV (Annals of Mathematics Studies 40)*. Princeton University Press, Princeton, New Jersey

Harsanyi, J.C. (1963) A simplified bargaining model for the n-person cooperative game. *International Economic Review* 4(2): 194–220

Hart, S., Kurz, M. (1983) Endogenous formation of coalitions. *Econometrica* 51(4): 1047–1064

Kalai, E., Pazner, E.A., Schmeidler, D. (1976) Collective choice correspondences as admissible outcomes of social bargaining processes. *Econometrica* 44(2): 233–240

Kalai, E., Smorodinsky, M. (1975) Other solutions to Nash's bargaining problem. *Econometrica* 43(3): 513–518

Maschler, M. (1978) Playing an n-person game — An experiment. In: Sauermann, H. (ed.) *Beiträge zur experimentellen Wirtschaftsforschung, Vol. VIII: Coalition Forming Behavior*. J. C. B. Mohr, Tübingen

Nash, J. (1950) The bargaining problem. *Econometrica* 18(2): 155–162

Owen, G. (1972) Values of games without side payments. *International Journal of Game Theory* 1: 95–109

Ray, D., Vohra, R. (1997) Equilibrium binding agreements. *Journal of Economic Theory* 73: 30–78

Roth, A.E. (1980) Values for games without sidepayments. Some difficulties with current concepts. *Econometrica* 48(2): 457–465

Roth, A.E. (1986) On the non-transferable utility value: A reply to Aumann. *Econometrica* 54(4): 981–984

Roth, A.E., Vande Vate, J.H. (1990) Random paths to stability in two-sided matching. *Econometrica* 58(6): 1475–1480

Scarf, H.E. (1967) The core of an N-person game. *Econometrica* 35(1): 50–69

Shafer, W.J. (1980) On the existence and interpretation of value allocation. *Econometrica* 48(2): 467–476

Shapley, L.S. (1953) A value for n-person games. In: Kuhn, H.W., Tucker, A.W. (eds.) *Contributions to the Theory of Games II (Annals of Mathematics Studies 28)*. Princeton University Press, Princeton

Shapley, L.S. (1969) Utility comparison and the theory of games. In: Éditions du Centre National de la Recherche Scientifique. *La Décision: Agrégation et Dynamique des Ordres de Préférence*. Paris

Shenoy, P.P. (1979) On coalition formation: A game-theoretical approach. *International Journal of Game Theory* 8(3): 133–164

Shenoy, P.P. (1980) A dynamic solution concept for abstract Games. *Journal of Optimization Theory and Applications* 32(2): 151–169

Zhou, L. (1994) A new bargaining set of an N-person game and endogenous coalition formation. *Games and Economic Behavior* 6(3): 512–526

A strategic analysis of network reliability

Venkatesh Bala[1], **Sanjeev Goyal**[2]

[1] Department of Economics, McGill University, 855 Sherbrooke Street West, Montreal, Canada H3A 1A8 (e-mail: vbala2001@yahoo.com)
[2] Econometric Institute, Erasmus University, 3000 DR, Rotterdam, The Netherlands (e-mail: goyal@few.eur.nl)

Abstract. We consider a non-cooperative model of information networks where communication is costly and not fully reliable. We examine the nature of Nash networks and efficient networks.

We find that if the society is large, and link formation costs are moderate, Nash networks as well as efficient networks will be 'super-connected', i.e. every link is redundant in the sense that the network remains connected even after the link is deleted. This contrasts with the properties of a deterministic model of information decay, where Nash networks typically involve unique paths between agents. We also find that if costs are very low or very high, or if links are highly reliable then there is virtually no conflict between efficiency and stability. However, for intermediate values of costs and link reliability, Nash networks may be underconnected relative to the social optimum.

JEL Classification: D82, D83

Key Words: Networks, coordination games, communication

1 Introduction

Empirical studies have emphasized the role played by social networks in communicating valuable information that is dispersed within the society (see e.g. Granovetter 1974, Rogers and Kincaid 1981, Coleman 1966). The information may concern stock market tips, job openings, the quality of products ranging

We are grateful to the editor, Mathew Jackson, and an anonymous referee for very useful comments. A substantial portion of this research was conducted when the first author was visiting the Economics Department at NYU. He thanks them for the generous use of their resources. Financial support from SSHRC and Tinbergen Institute, Rotterdam is acknowledged.

from cars to computers, and new medical advances, among other things.[1] While agents who participate in communication networks receive various kinds of benefits, they also incur costs in forming and maintaining links with other agents to obtain the benefits. Such costs could be in terms of time, effort and money.

In this paper, we study how social networks are formed by individual decisions which trade off the costs of forming links against the potential benefits of doing so. We suppose that once an agent i forms a link with another agent j they can both share information. One example of this type of link formation is a telephone call. The caller typically pays the telephone company, but both parties can exchange information. We suppose that a link with another agent allows access to the benefits available to the latter via his own links. Thus individual links generate externalities. A distinctive aspect of our model is that the costs of link formation are incurred only by the person who initiates the link. This enables us to study the network formation process as a non-cooperative game.[2]

We model the idea of imperfect reliability in terms of a positive probability that a link fails to transmit information. As a concrete example, consider the network of people who are in contact via telephone. Suppose that agent i incurs a cost and calls agent j. It is quite possible that he may not get through to j because the latter is not available at that time. Hence, from an empirical point of view, imperfect reliability seems to be a reasonable assumption. In this setting, we examine the effect of imperfect reliability of links on the nature of stable and efficient networks. Our notion of stability requires that agents play according to a Nash equilibrium.

As the topic exhibits significant analytic difficulty, we consider a relatively simple two-parameter model which attempts to capture the costs and benefits from link formation. Each agent is assumed to possess some information of value 1 to other agents, and a link between the agents allows this information to be trallsferred. Each link formed by an agent costs an amount $c > 0$. The reliability of a link is measured by a parameter $p \in (0, 1)$. Here, p is the probability that an established link between i and j "succeeds", i.e. allows information to flow between the agents, while $1 - p$ is the probability that it "fails". Moreover, link reliability across different pairs of agents is assumed to be independent. An agent's strategy is to choose the subset of agents with whom he forms links. The choices of all the agents specifies a non-directed network which permits information flows between them. As $p < 1$, the network formed by the agents choices is stochastic, since one or more links may fail. In a realization of the network, agent i obtains the information of all the agents with whom he has a path (i.e. either a link or a sequence of links) in the realization. The agent's payoff is his expected benefit over all realizations less his costs of link formation.

[1] Boorman (1975) provides an early model of information flow in networks in the context of job search. Baker and Iyer (1991) analyze the impact of communication networks for stock market volatility, while Bala and Goyal (1998) study information diffusion in fixed networks.

[2] The model is applicable in cases where links are durable, and must be established at the outset of the game by incurring a fixed cost of c. Once in place, the links provide a stochastic flow of benefits to the agents. This specification allows us to abstract from complex timing issues which would arise in a dynamic game of information sharing.

Our findings on stability may be summarized as follows. If agents find it at all worthwhile to form links in a Nash network, then that network must be *connected* (Proposition 3.1). In other words, the network ensures that every agent obtains information with positive probability from every other agent. Our main result (Proposition 3.3) concerns the distinction between what we call *minimally connected* networks and *super-connected* networks. A minimally connected network is one with a unique path between any two agents. In such a network, every link is 'critical' to communication between the agents. Such a network imposes very strong demands upon reliability, because if even one link fails, there will be at least two agents who will not obtain information possessed by the other agent. In a super-connected network, on the other hand, every link is non-critical.[3] Proposition 3.3 shows that if $c < p$ then for large n, every Nash network will be superconnected.

We also consider efficient networks. We find that if costs are very low or very high, or if links are highly reliable then there is virtually no conflict between efficiency and stability. However, for intermediate ranges of costs and link reliability, Nash networks may be "underconnected" relative to the social optimum. Note that links between agents create positive externalities for other agents. Hence if Nash networks are typically super-connected in large societies, this should be even more true for efficient networks. This is in fact the case: in Proposition 4.3 we prove that efficient networks are super-connected for a much larger range of cost values.

The present paper is part of the recent literature on network formation, see e.g., Bala (1996), Bala and Goyal (2000), Dutta and Mutuswami (1997), Goyal (1993), Jackson and Wolinsky (1996) and Watts (1997). This work studies issues of *stability* and *efficiency* in network formation. Here stability refers to the networks which are consistent with agents' incentives – in terms of costs and benefits – to create links with other agents, while efficiency deals with the kinds of networks which maximize some measure of social surplus.

In this paper, we extend the model of network formation presented in Bala and Goyal (2000). In that paper, links are assumed to be perfectly reliable: in other words, the network formed is deterministic. We also analyzed the case of information decay: agents' payoffs decrease geometrically according to a parameter δ, in the distance to other agents. The main point of the present paper is that imperfect reliability has very different effects on network formation as compared to information decay. Specifically, in our earlier paper we showed that with information decay, minimally connected networks (notably the star) are Nash for a large range of c and δ, *independently* of the size of the society. By contrast, our results in the present paper show that minimally connected networks are increasingly replaced by super-connected networks as n increases. Thus, imperfect reliability creates radically different incentives for agents as

[3] For example, a star network, where every agent communicates through a central agent, is minimally connected. A wheel network, where agents are arranged in a circle, is super-connected since the network remains connected, after any single link is deleted.

compared to information decay. We also find that similar differences arise in the case of efficient networks.

Some of the previous work has also considered network reliability. Jackson and Wolinsky's paper provides a discussion of network formation when links formed by agents can break down with positive probability. In a broader perspective, Chwe (1995) presents a model of strategic reliability in communication. His approach defines communication protocols (which are somewhat related to networks since they allow for transmitting messages between agents) and he studies questions of incentive compatibility and efficiency of protocols in games of incomplete information. Our focus, on the other hand, is upon the properties of networks which arise endogeneously from agents' choices in a normal form game of information sharing.

The rest of the paper is organized as follows. Section 2 presents the model. Section 3 considers Nash networks, while Sect. 4 studies efficiency. Section 5 concludes.

2 The model

Let $N = \{1, \ldots, n\}$ be a set of agents and let i and j be typical members of this set. We shall assume throughout that $n \geq 3$. Each agent is assumed to possess some information of value to other agents. He can augment his information by communicating with other people; this communication takes resources, time and effort and is made possible by setting up of *pair-wise* links. Agents form links simultameously in our model.[4] However, such links are not fully reliable: they can fail to transmit information with positive probability.

A (communication) *strategy* of agent $i \in N$ is a vector $g_i = (g_{i,1}, \ldots, g_{i,i-1}, g_{i,i+1}, \ldots, g_{i,n})$ where $g_{i,j} \in \{0, 1\}$ for each $j \in N \setminus \{i\}$. We say agent i has a *link* with j if $g_{i,j} = 1$. A link between agents i and j potentially allows for two-way (symmetric) flow of information. Throughout the paper we restrict our attention to pure strategies. The set of all strategies of agent i is denoted by $\mathcal{G}_i$. Since agent i has the option of forming or not forming a link with each of the remaining $n - 1$ agents, the number of strategies of agent i is $|\mathcal{G}_i| = 2^{n-1}$. The set $\mathcal{G} = \mathcal{G}_1 \times \cdots \times \mathcal{G}_n$ constitutes the strategy space of all the agents. A strategy profile $g = (g_1, \ldots, g_n)$ can be represented as a network with the agents depicted as vertices and their links depicted as edges connecting the vertices. The link $g_{i,j} = 1$ is represented by a *non-directed edge* between i and j, along with a circular token lying on each edge adjacent to the agent who has initiated the link. Figure 1 gives an example with $n = 3$ agents:

[4] Another possibility would be to allow agents to form links sequentially. In such a game, the precise incentives to form and dissolve links will differ. However, we believe that some of the main properties of Nash and efficient networks that we identify in the simultaneous move setting – e.g., super-connectedness – should still obtain.

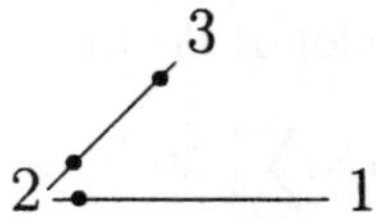

Fig. 1.

Here, agent 2 has formed links with agents 1 and 3, agent 3 has a link with agent 2 while agent 1 does not link up with any other agent.[5] It can be seen that every strategy in $\mathcal{G}$ has a unique representation of the form given in Fig. 1.

For $g \in \mathcal{G}$, define $\mu_i^d(g) = |\{k \in N | g_{i,k} = 1\}|$. Here, $\mu_i^d(g)$ is the number of links formed by agent i. To describe information flows in the network, we introduce the notion of the *closure* of g. This is a non-directed network denoted $h = cl(g)$, and defined by $h_{i,j} = \max\{g_{i,j}, g_{j,i}\}$ for each i and j in N.[6] Each link $h_{i,j} = 1$ *succeeds* (i.e. allows information to flow) with probability p (0, 1) and fails (does not permit information flow) with probability $1 - p$. Furthermore, the success or failure of different links are assumed to be independent. Thus h may be regarded as a random network. Formally, we say that h' is a *realization* of h (denoted $h' \subset h$) if for every $i \in N$ and every $j \in \{i + 1, \ldots, n\}$, we have $h'_{i,j} \leq h_{i,j}$[7]. Given $h' \subset h$, let $L(h')$ be the total number of links in h'. Under the hypothesis of independence, the probability of network h' being realized is

$$\lambda(h'|h) = p^{L(h')}(1 - p)^{L(h)-L(h')} \tag{2.1}$$

For $h' \subset h$ we say there is a path between i and j in h' if either $h'_{i,j} = 1$ or there exists a non-empty set of agents $\{i_1, \ldots, i_m\}$ distinct from each other and from i and j such that $h'_{i,i_1} = \cdots h'_{i_m,j} = 1$. Define $\mathbf{1}_i(j; h')$ to equal 1 if there is a path between i and j in h' and to equal 0 otherwise. We suppose that i observes an agent in h' if and only if there is a path between that agent and i in h'. A network g is said to be connected if there is a path in $h = \text{cl}(g)$, between any two agents i and j. A network is called empty if it has no links. A set $C \subset N$ is called a component of g is there is a path in h for every pair of agents i and j in c, and there is no strict superset C' of C, for which this is true. The geodesic distance $d(i, j; h)$ between two agents i and j is the number of links in the shortest path between them in h.

We can define $\mu_i(h') \equiv \sum_{j \neq i} \mathbf{1}_i(j; h')$ as the total number of people that agent i observes in the realization h'. We assume that each link formed by agent i costs $c > 0$, while each agent that i observes in a realization of the network yields a benefit of $V > 0$. Without loss of generality, we set $V = 1$[8]. Given the

[5] As agents form links independently, it is possible that two agents simultaneously initiate a link with each other, as agents 2 and 3 do in the figure.

[6] Pictorially, the closure of a network simply means removing the circular tokens lying on the edges which show the originator of the links. The network h can be regarded as non-directed because $h_{i,j} = h_{j,i}$ for each i and j.

[7] The network h' should also be regarded as non-directed. Hence, we implicitly assume that $h'_{j,i} = h'_{i,j}$ for all $j \in \{1, \ldots, i - 1\}$.

[8] For simplicity, we assume a lienar specification of payoffs. This implies that the value of additional information is constant. Alternatively, one might expect that the marginal value of information

strategy-tuple g, define the function $B_i(h)$ for agent i as

$$B_i(h) = \sum_{h' \subset h} \lambda(h'|h)\mu_i(h') . \tag{2.2}$$

where $h = \text{cl}(g)$. The probability that the network h' is realized is $\lambda(h'|h)$, in which case agent i accesses the information of $\mu_i(h')$ agents in total. Hence $B_i(h)$ is i's *expected benefit* from the random network h. Using (2.2), we define i's (expected) payoff from the strategy-tuple g as

$$\Pi_i(g) = B_i(h) - \mu_i^d(g)c = \sum_{h' \subset h} \lambda(h'|h)\mu_i(h') - \mu_i^d(g)c . \tag{2.3}$$

The first term in (2.3) is i's expected benefit from the network, while the second is i's cost of forming links, which is incurred at the outset. The expected benefit and the payoff of agent i can be expressed in a different form, which is also useful. Substituting $\mu_i(h') = \sum_{j \neq i} \mathbf{1}_i(j; h')$ in (2.2), we get

$$\begin{aligned} B_i(h) &= \sum_{h' \subset h} \lambda(h'|h) \sum_{j \neq i} \mathbf{1}_i(j; h') \\ &= \sum_{j \neq i} \left[\sum_{h' \subset h} \lambda(h'|h)\mathbf{1}_i(j; h') \right] = \sum_{j \neq i} p_i(j; h) , \end{aligned} \tag{2.4}$$

where $p_i(j; h) = \sum_{h' \subset h} \lambda(h'|h)\mathbf{1}_i(j; h')$ is the *probability* that i observes j in the random network h. From (2.4), using (2.3) we obtain

$$\Pi_i(g) = B_i(h) - \mu_i^d(g)c = \sum_{j \neq i} p_i(j; h) - \mu_i^d(g)c . \tag{2.5}$$

Applying either (2.3) or (2.5) to the network in Fig. 1, we calculate $\Pi_i(g) = p + p^2$, $2p - 2c$, and $p + p^2 - c$ for agents $i = 1, 2$ and 3 respectively. As information is assumed to flow in both directions of a link, agent 1 gets an expected benefit of $p + p^2$ without forming any links. Hence, the payoffs allow for significant "free riding" in link formation.

Given a network $g \in \mathcal{G}$, let g_{-i} denote the network obtained when all of agent i's links are removed. The network g can be written as $g = g_i \oplus g_{-i}$ where the '$\oplus$' indicates that g is formed as the union of the links in g_i and g_{-i}. The strategy g_i is said to be a *best response* of agent i to g_{-i} if

$$\Pi_i(g_i \oplus g_{-i}) \geq \Pi_i(g_i' \oplus g_{-i}) , \quad \text{for all} \quad g_i' \in \mathcal{G}_i . \tag{2.6}$$

The set of all of agent i's best responses to g_{-i} is denoted $BR_i(g_{-i})$. Furthermore, a network $g = (g_1, \ldots, g_n)$ is said to be a *Nash network* if $g_i \in BR_i(g_{-i})$ for each i, i.e. agents are playing a Nash equilibrium. A *strict* Nash network is one where

declines as more information becomes available. However, we believe that our simplification is not crucial for our results. For a study of network formation under a fairly general class of payoff functions and with perfectly reliable links, see Bala and Goyal (2000).

agents get a strictly higher payoff with their current strategy than they would with any other strategy. Our welfare measure is given by a function $W : \mathcal{G} \to \mathcal{R}$, where $W(g) = \sum_{i=1}^{n} \Pi_i(g)$ for $g \in \mathcal{G}$. A network g is *efficient* if $W(g) \geq W(g')$ for all $g' \in \mathcal{G}$. An efficient network is one which maximizes the total expected value of information made available to the agents, less the aggregate cost of communication.

We say that $g \in \mathcal{G}$ is *essential* if $g_{i,j} = 1$ implies $g_{j,i} = 0$. We note that if $g \in \mathcal{G}$ is either a Nash network or an efficient network, then g must be essential. The argument underlying the above observation is as follows. If $g_{i,j} = 1$ then by the definition of Π_j agent j pays an additional cost of c from setting $g_{j,i} = 1$ as well, while neither he nor any one else gets any benefit from it. Hence if g is not essential, it cannot be either Nash or efficient[9]. We denote the set of essential networks as $\mathcal{G}^a$.

We start with the following intuitive property of the benefit function $B_i(\cdot)$ which is useful for our analysis.

Lemma 2.1. *Suppose* $p \in (0, 1)$. *Let* $g^0 \in \mathcal{G}^a$ *be a network with* $h^o = \text{cl}(g^o)$, *and suppose there are two agents* i *and* j *such that* $g^o_{i,j} = g^o_{j,i} = 0$. *Let* g *be the same as* g^o *except for an additional link* $g_{i,j} = 1$ *and let* $h = \text{cl}(g)$. *Then for all agents* m, $B_m(h) \geq B_m(h^o)$. *The inequality is strict if* $m = i$ *or* $m = j$.

The proof of this lemma is omitted. This observation is actually more general than stated, since it also implies that an agent m's benefit is non-decreasing in the addition of any number of links. The following lemma describes some properties of the payoff function.

Lemma 2.2. *The payoff function* $\Pi_i(g)$ *is a polynomial* $\sum_{k=0}^{L(h)} a_i^k p^k$, *where the coefficient* $a_i^o = -\mu_i^d(g)c$ *and* $a_i^1 = |\{j | h_{i,j} = 1\}|$.

The proof is given in the appendix. We shall say that a network is empty if it contains no links, and that is complete if there exists a link between every pair of agents. The empty network is denoted by g^e, while the complete network is denoted by g^c. The star architecture is prominent in this literature: denote a star by g^s, where, for a fixed "central agent" n (say), we have $h^s_{n,j} = 1$ for all $j \neq n$ and $h^s_{j,k} = 0$ for all $j \neq n$, $k \neq n$. In a line network g^l, we have $h^l_{i,i+1} = 1$, for $i = 1, 2, \ldots, n-1$ and $h^l_{i,j} = 0$ otherwise. In a wheel network g^w, we have agents arranged around a circle, i.e., $h^w_{1,n} = 1$, and $h^w_{i,i+1} = 1$ for all $i = 1, 2, \ldots, n-1$ and $h^w_{i,j} = 0$ otherwise. Two networks have the same architecture if one network can be obtained from the other by permuting the strategies of agents in the other network.

---

[9] The payoff function (2.3) assumes that the links $g_{i,j} = 1$ and $g_{j,i} = 1$ are perfectly correlated. The above observation is a consequence of this assumption. An alternative assumption is that $g_{i,j} = 1$ and $g_{j,i} = 1$ are independent: in this case the link $h_{i,j} = 1$ succeeds with probability $q = 1-(1-p)^2 = 2p - p^2$. We briefly discuss the impact of the alternative assumption in Sect. 3.

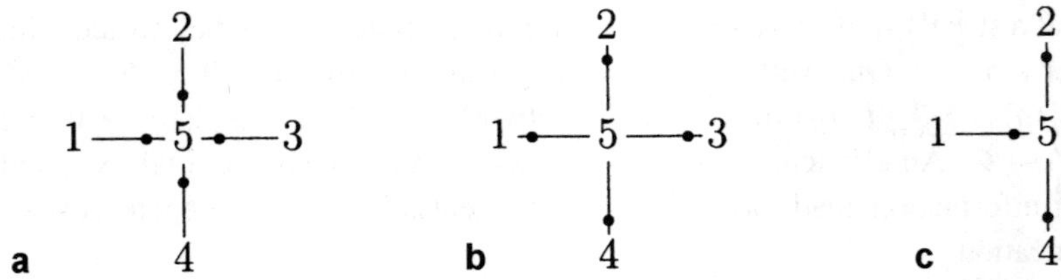

Fig. 2a–c. **a** Center-sponsored; **b** Periphery-sponsored; **c** Mixed-Type

3 Nash networks

We are interested in describing Nash networks for the above model as a function of the link success parameter p and the cost c of link formation.

We start by noting an interesting implication of the assumption that link formation is one-sided. By the definition of payoffs, although a single agent may bear the cost of a link, both agents potentially obtain the benefits associated with it. This asymmetry in payoffs is relevant for defining the architecture of the network. As an illustration, we note that there are now three kinds of 'star' networks, depending upon which agents bear the costs of the links in the network. For a society with $n = 5$ agents, Figs. 2a–c (left to right) illustrate these types.

Figure 2a shows a *center-sponsored* or cs-star, Fig. 2b a *periphery-sponsored* or ps-star and Fig. 2c depicts a *mixed-type* or mt-star. We calculate the payoff obtained by an agent in a star g^s using (2.5). Consider the central agent n. Clearly, $p_n(j;h^s) = p$ for each $j \neq n$, so that $B_n(h^s) = \sum_{i \neq n} p_n(i;h^s) = (n-1)p$. If $i \neq n$ then $p_i(n;h^s) = p$ while $p_i(j;h^s) = p^2$ for each $j \notin \{n,i\}$. Hence $B_i(h^s) = p + (n-2)p^2$. From this, we obtain:

$$\Pi_n(g^s) = (n-1)p - \mu_n^d(g^s)c, \quad \Pi_i(g^s) = p + (n-2)p^2 - \mu_i^d(g^s)c, \; i \neq n. \quad (3.1)$$

To illustrate the range of networks that can be supported as Nash networks, we provide a complete characterization of Nash networks for a society of three agents.

Example 1. Fix $n = 3$. We find that the empty network, the star, and the complete network are the possible Nash architectures. In particular, the empty network is Nash if $c \geq p$. The center-sponsored star and the mixed-type star are Nash if $\{(p,c)|0 < p < 1, p + p^2 - 2p^3 \leq c \leq p\}$, while the periphery-sponsored star is Nash if $\{(p,c)|0 < p < 1, p + p^2 - 2p^3 \leq c \leq p + p^2\}$. The complete network is Nash if $0 < c \leq p + p^2 - 2p^3$. Figure 3 displays the parameter regions where different networks are Nash.

The computations involved in establishing the regions give an idea of some of the general considerations involved in studying equilibrium networks and so we present them here.

First, let g be a center-sponsored star, i.e. $g_{3,1} = g_{3,2} = 1$.[10] Then using (3.1), agent 3's payoff is given by$\Pi_3(g) = 2p - 2c$, while if he does not form a

[10] For brevity in describing a network, we only identify the links $g_{i,j} = 1$.

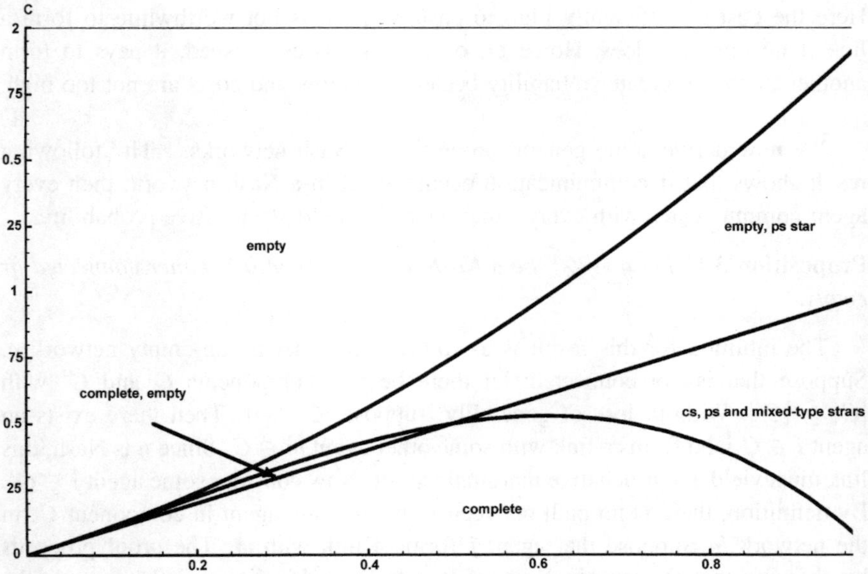

Fig. 3. Nash networks ($n = 3$)

link, his payoff is 0. Thus, he is playing a best response if and only if $p \geq c$. Furthermore, $\Pi_i(g) = p + p^2$ for $i = 1, 2$. We now calculate agent 1's expected payoff if he deviates and forms an additional link with agent 2. In any realization h' where $h'_{1,2} = 1$ (which occurs with probability p) he obtains 2's information. Furthermore, in any realization h' where $h'_{1,2} = 0$ (probability $1 - p$) he still obtains 2's information if the links $h'_{1,3} = h'_{2,3} = 1$. The probability of such a realization is $(1 - p)p^2$. Hence, the probability that 1 obtains 2's information is $p+(1-p)p^2$. By symmetry, the same holds for 1 vis-a-vis 3, so that 1's payoff by deviating is $2(p+(1-p)p^2) - c$. It is, therefore, not worthwhile to deviate if and only if $c \geq 2(p+(1-p)p^2) - (p+p^2) = p+p^2-2p^3$. Thus, the center-sponsored star is Nash in the region $\{(p, c)|0 < p < 1, p+p^2-2p^3 \leq c \leq p\}$. Next, it can be checked that a mixed-type star $g_{3,1} = 1$, $g_{2,3} = 1$ is Nash for the same region identified for the center-sponsored star.

The situation is somewhat different if g is the periphery-sponsored star $g_{1,3} = g_{2,3} = 1$. Here, agent 3 is obviously playing a best response. For agent 1 we have $\Pi_1(g) = p + p^2 - c$. This is better than forming no link at all provided $c < p + p^2$. Furthermore it dominates forming an additional link with agent 2 if $p + p^2 - c \geq 2(p + (1 - p)p^2) - 2c$ or $c \geq p + p^2 - 2p^3$. Hence, the periphery-sponsored star is Nash in the region $\{(p, c)|0 < p < 1, p+p^2-2p^3 \leq c \leq p+p^2\}$.

Furthermore, it is easily seen that the empty network is Nash if and only if $c \geq p$. In the case of complete networks, there are two possibilities. (a) $g_{1,2} = g_{2,3} = g_{3,1} = 1$ and (b) $g_{1,2} = g_{1,3} = g_{2,3} = 1$. In both cases, it can be verified that a complete network is Nash if and only if $0 < c \leq p + p^2 - 2p^3$. Interestingly, the empty network and the complete network are Nash in the non-empty region $\{(p, c)|0 \leq p \leq 1/2, p \leq c \leq p+p^2-2p^3\}$, while no star is Nash.

Here the cost is sufficiently high to ensure that it is not worthwhile to form a link if no one else does. However, once a link is established, it pays to form another to ensure greater reliability because p is low and costs are not too high. □

We now derive some general properties of Nash networks.[11] The following result shows that if communication occurs at all in a Nash network, then every agent communicates with every other agent with strictly positive probability.

Proposition 3.1. *Let $g \in \mathcal{G}^a$ be a Nash network. Then it is either connected or empty.*

The intuition for this result is as follows: consider a non-empty network g. Suppose that is not connected; let there be two components C and C' with $|C| \geq |C'|$. Without loss of generality, suppose $|C| > 1$. Then there exists an agent $i \in C$ who forms a link with some other agent $m \in C$. Since g is Nash, this link must yield a non-negative marginal payoff. Now consider some agent $j \in C'$. By definition, there is no path between j and and any agent in component C, in the network h. Suppose that agent j forms a link with m. The proof proceeds by showing that the marginal payoff to j from such a link strictly exceeds the marginal payoff that i gets from the link with m. Since the latter is non-negative, this implies that g cannot be a Nash equilibrium.[12] We now provide the proof.

Proof. Consider a non-empty Nash network $g^1 \in \mathcal{G}^a$ which is not connected. As g^1 is non-empty, there exist agents i and m such that $g^1_{i,m} = 1$. Let $h^1 = cl(g^1)$ and note that $h^1_{i,m} = 1$. Suppose g^0 denotes the network obtained from g^1 by deleting the link $g^1_{i,m} = 1$, ceteris paribus. Define $h^0 = \text{cl}(g^0)$. We observe that each realization $\hat{h} \subset h^1$ is one of two types: either $\hat{h} = h'$ for some $h' \subset h^0$ (if the link $h^1_{i,m} = 1$ fails) or $\hat{h} = h' \oplus h^1_{i,m}$ for some $h' \subset h^0$ where $h' \oplus h^1_{i,m}$ denotes the network where the link $h^1_{i,m} = 1$ is also present. Moreover, by independence we have $\lambda(\hat{h}|h^1) = (1-p)\lambda(h'|h^0)$ in the former case, and $\lambda(\hat{h}|h^1) = p\lambda(h'|h^0)$ in the latter case. It follows that the marginal payoff $\Pi_i(g^1) - \Pi_i(g^0)$ to agent i from the link $g^1_{i,m} = 1$ is given by

$$\left\{p \sum_{h' \subset h^0} \lambda(h'|h^0)\mu_i(h' \oplus h^1_{i,m}) \ + \ (1-p) \sum_{h' \subset h^0} \lambda(h'|h^0)\mu_i(h')\right\} - c$$

[11] Under the alternative specification, $h_{i,j} = \min\{g_{i,j}, g_{j,i}\}$, the distinction between different types of stars cannot arise in equilibrium, since a link is only formed if both agents involved agree to the link. In this sense, there are fewer networks that can be candidates for Nash equilibrium. However the alternative specification introduces an additional aspect of coordination: it is worthwhile for agent i to form a costly link with agent j only if the latter also wants to form a link. This suggests that, when costs of forming links are small, there will exist a relatively large number of equilibrium architectures – including some partially connected ones – corresponding to varying levels of successful coordination between pairs of agents. For example, in the above example with $n = 3$, if $c < p$ then the partially connected network with $g_{1,2} = 1$, $g_{2,3} = 0$, and $g_{1,3} = 0$, is a Nash network under the alternative specification, while it is not a Nash network under our formulation. See Dutta et al. (1998), for a study of the alternative formulation.

[12] It is worth emphasizing that this argument exploits the fact that link formation is one-sided; hence we only have to check the incentives of individual players to form or delete links.

$$- \sum_{h' \subset h^0} \lambda(h'|h^0)\mu_i(h')$$

$$= p\left[\sum_{h' \subset h^0} \lambda(h'|h^0)(\mu_i(h' \oplus h^1_{i,m}) - \mu_i(h'))\right] - c. \tag{3.2}$$

Since g^1 is Nash, the expression in (3.2) is non-negative. As g^1 is not connected, there exists an agent j such that $\mathbf{1}_i(j; h^1) = 0$. Since $\mathbf{1}_i(m; h^1) = 1$ this also implies $h^1_{j,m} = 0$. We shall show that j can be made strictly better off by forming a link with m, contradicting the supposition that g^1 is Nash.

Using the same logic as with agent i, agent j's benefit from h^1 is given by:

$$B_j(h') = p \sum_{h' \subset h^0} \lambda(h'|h^0)\mu_j(h' \oplus h^1_{i,m}) + (1-p) \sum_{h' \subset h^0} \lambda(h'|h^0)\mu_j(h'). \tag{3.3}$$

We consider the marginal payoff obtained by agent j if, starting from the network g^1, he forms an additional link with agent m, ceteris paribus. Let g^2 denote the new network and let $h^2 = \text{cl}(g^2)$. Clearly, a realization $h^* \subset h^2$ takes one of four forms for some $h' \subset h^0$: (a) $h^* = h' \oplus h^1_{i,m} \oplus h^2_{j,m}$, (b) $h^* = h' \oplus h^1_{i,m}$, (c) $h^* = h' \oplus h^2_{j,m}$ and (d) $h^* = h'$. By independence, it follows that agent j's benefit from h^2 is:

$$\begin{aligned} B_j(h^2) &= p^2 \sum_{h' \subset h^0} \lambda(h'|h^0)\mu_j(h' \oplus h^1_{i,m} \oplus h^2_{j,m}) \\ &+(1-p)p \sum_{h' \subset h^0} \lambda(h'|h^0)\mu_j(h' \oplus h^2_{j,m}) \\ &+p(1-p) \sum_{h' \subset h^0} \lambda(h'|h^0)\mu_j(h' \oplus h^1_{i,m}) \\ &+(1-p)^2 \sum_{h' \subset h^0} \lambda(h'|h^0)\mu_j(h') \end{aligned} \tag{3.4}$$

Using (3.3) and (3.4) and simplifying, we can write agent j's marginal benefit $B_j(h^2) - B_j(h^1)$ from his link with m as:

$$\begin{aligned} B_j(h^2) - B_j(h^1) &= p\left\{p\left[\sum_{h' \subset h^0} \lambda(h'|h^0)(\mu_j(h' \oplus h^1_{i,m} \oplus h^2_{j,m}) - \mu_j(h' \oplus h^1_{i,m}))\right]\right. \\ &+ \left.(1-p)\left[\sum_{h' \subset h^0} \lambda(h'|h^0)(\mu_j(h' \oplus h^2_{j,m}) - \mu_j(h'))\right]\right\} \end{aligned} \tag{3.5}$$

Consider the term $\mu_j((h' \oplus h^1_{i,m} \oplus h^2_{j,m}) - \mu_j(h' \oplus h^1_{i,m})$ in the first set of square brackets in (3.5). Note that $\mu_j(h' \oplus h^1_{i,m}) = \mu_j(h')$ for each $h' \subset h^0$, since agent j cannot access any agent in the component of h^1 containing i and m, when the link $h^2_{j,m} = 1$ fails. Thus, $\mu_j(h' \oplus h^1_{i,m} \oplus h^2_{j,m}) - \mu_j(h' \oplus h^1_{i,m}) = \mu_j(h' \oplus h^1_{i,m} \oplus h^2_{j,m}) - \mu_j(h')$. Suppose now that there is some agent u who is accessed by i in a realization $h' \oplus h^1_{i,m}$ but is not accessed in the realization h'. Then it follows that agent u is certainly accessed by j in $h' \oplus h^1_{i,m} \oplus h^2_{j,m}$. Moreover, since every path between j and u must involve the link $h^2_{j,m} = 1$, agent u cannot be accessed by j in h'. Hence

$$\begin{aligned} \mu_j(h' \oplus h^1_{i,m} \oplus h^2_{j,m}) - \mu_j(h' \oplus h^1_{i,m}) &= \mu_j(h' \oplus h^1_{i,m} \oplus h^2_{j,m}) - \mu_j(h') \\ &\geq \mu_i(h' \oplus h^1_{i,m}) - \mu_i(h') \qquad (3.6) \end{aligned}$$

Note also that if $h' \subset h^0$ is empty then $h' \oplus h^1_{i,m} \oplus h^2_{j,m}$ allows for agent j to access i in addition to accessing m. Thus there exists $h' \subset h^0$ for which the inequality in (3.6) is strict. As $h' \subset h^0$ is arbitrary, it follows from (3.5)–(3.6) that:

$$\begin{aligned} & p\left[\sum_{h' \subset h^0} \lambda(h'|h^0)(\mu_j(h' \oplus h^1_{i,m} \oplus h^2_{j,m}) - \mu_j(h' \oplus h^1_{i,m}))\right] \\ & \quad > p\left[\sum_{h' \subset h^0} \lambda(h'|h^0)(\mu_i(h' \oplus h^1_{i,m}) - \mu_i(h'))\right] \qquad (3.7) \end{aligned}$$

Consider next the term $\mu_j(h' \oplus h^2_{j,m}) - \mu_j(h')$ in the second square brackets of (3.5). If some agent u is contacted by i in $h' \oplus h^1_{i,m}$, *due* to the link $h^1_{i,m} = 1$, then it follows that this same agent u is also accessed by j in the network $h' \oplus h^2_{j,m}$, *due* to the link $h^2_{j,m} = 1$. Hence, for $h' \subset h^0$,

$$\mu_j(h' \oplus h^2_{j,m}) - \mu_j(h') \geq \mu_i(h' \oplus h^1_{i,m}) - \mu_i(h') \qquad (3.8)$$

Since $h' \subset h^0$ is arbitrary, we get

$$\begin{aligned} & (1-p)\left[\sum_{h' \subset h^0} \lambda(h'|h^0)(\mu_j(h' \oplus h^2_{j,m}) - \mu_j(h'))\right] \\ & \quad \geq (1-p)\left[\sum_{h' \subset h^0} \lambda(h'|h^0)(\mu_i(h' \oplus h^1_{i,m}) - \mu_i(h'))\right] \qquad (3.9) \end{aligned}$$

Summing both sides of (3.7) and (3.9) and using the definition $B_j(h^2) - B_j(h^1)$ in (3.5), we see that:

$$B_j(h^2) - B_j(h^1) > p\left[\sum_{h' \subset h^0} \lambda(h'|h^0)(\mu_i(h' \oplus h^1_{i,m}) - \mu_i(h'))\right]. \qquad (3.10)$$

By (3.2) however, the right hand side of (3.10) is at least as large as c. Hence, the marginal benefit to player j from forming a link with m strictly exceeds its marginal cost, which contradicts the supposition that g^1 is Nash. The result follows. □

Our next result provides conditions under which some familiar architectures are Nash.

Proposition 3.2. *Let the payoffs be given by (2.3). (a) Given* $p \in (0,1)$ *there exists* $c(p) > 0$ *such that a complete network* g^c *is (strict) Nash for all* $c \in (0, c(p))$. *(b) Given* $c \in (0,1)$ *there exists* $p(c) \in (c,1)$ *such that* $p \in (p(c),1)$ *implies that all types of stars (center-sponsored, periphery-sponsored and mixed-type) are Nash. If* $n \geq 4$, *they are in fact strict Nash. (c) Given* $c \in (1, n-1)$ *there exists* $p(c) < 1$ *such that* $p \in (p(c),1)$ *implies that the periphery-sponsored star is (strict) Nash. (d) The empty network is (strict) Nash for all* $c > p$.

Proof. We begin with (a). Let $g = g_i \oplus g_{-i}$ be a complete network and suppose that agent i has one or more links in his strategy g_i. Let g^0 be a network where some of these links have been deleted, ceteris paribus. From Lemma 2.1 we get $B_i(h^0) < B_i(h)$ where $h^0 = \text{cl}(g^0)$ and $h = \text{cl}(g)$. It follows that if $c = 0$ then g_i is a strict best response for agent i. By continuity, there exists $c_i(p) > 0$ for which g_i is a strict best response for all $c \in (0, c_i(p))$. Statement (a) follows by setting $c(p) = \min_i c_i(p)$ over all agents i who have one or more links in their strategy g_i.

For (b), choose $p(c) \in (c,1)$ to satisfy $(1-p) + (n-2)(1-p^2) < c$ for all $p \in (p(c),1)$. In what follows, fix $p \in (p(c),1)$. Let g be a mixed-type star and let agent n (say) be the "central agent" of the star. Consider an agent $i \neq n$ for whom $g_{i,n} = 1$. By (3.1) i's payoff is $p + (n-2)p^2 - c$. If he forms no link at all, he obtains a payoff of 0. Since $p > c$ it is worthwhile for him to form at least one link. Next, if i deletes his link with n and forms it with an agent $j \notin \{i,n\}$ instead, ceteris paribus, his payoff can be calculated to be $p + p^2 + (n-3)p^3 - c$ which is dominated by forming one with n. Hence if he forms one link, we can assume he forms it with agent n. Moreover, by forming $k \geq 2$ links, his payoff is bounded above by $(n-1) - kc$. Subtracting the payoff from the star, his maximum incremental payoff from two or more links is no larger than $(1-p)+(n-2)(1-p^2)-c$ which is negative, by choice of p. Hence i's best response is to maintain a single link with agent n. For agent n, if $g_{j,n} = 0$, then $p > c$ implies it is worthwhile for n to form a link with agent j. Thus, the mixed-type star is Nash. Similar arguments apply for the center-sponsored and periphery-sponsored stars.[13]

For part (c), note that $c \in (1,n)$ implies $c > p$. Hence the center-sponsored star and the mixed-type star cannot be Nash. However, the periphery-sponsored star g^s can be supported. From (3.1), the payoff of agent $i \neq n$ is $\Pi_i(g^s) = p + (n-2)p^2 - c$. Given $c \in (0,1)$ there clearly exists $p(c) \in (c,1)$ such that $p \in (p(c),1)$ implies $\Pi_i(g^s) > 0$, so that i will form at least one link in his best response. Arguments analogous to (b) above establish that $p(c)$ may be additionally chosen to ensure that i will not wish to form more than one link for any $p \in (p(c),1)$.

For part (d), if $c > p$ and no other agent forms a link, it will not be worthwhile for an agent to form a link. Hence the empty network is strict Nash. □

---

[13] Note that if $n \geq 4$, then agent $i \neq n$ has a strict incentive to form a link with n rather than $j \notin \{i,n\}$, but only a weak one if $n = 3$. Hence the mixed-type and periphery-sponsored stars are strict Nash for $n \geq 4$ but only Nash when $n = 3$. On the other hand, the center-sponsored star is strict Nash for all $n \geq 3$.

The main result of this section shows that if $c < p$, then for large societies, every link in a Nash network is 'redundant'. By way of motivation, we provide an example concerning the stability of the star in large societies.

Example 2. Let $c \in (0, 1)$ and $p \in (c, 1)$ be fixed. Suppose g^s is an center-sponsored star with agent n as the "central" agent. We set $h^s = \text{cl}(g^s)$. From (3.1), agent 1's (say) payoff is $\Pi_1(g^s) = p + (n-2)p^2$. We consider what happens when agent 1 forms a link with agent 2, ceteris paribus. Denote the resulting network as g with $h = \text{cl}(g)$. Agent 1 can now obtain n's information in two ways: when the link $h^s_{n,1} = h_{n,1} = 1$ succeeds (probability p) and when it fails (probability $1-p$), provided the links $h^s_{1,2} = h^s_{2,n} = 1$ succeed (probability p^2). Hence $p_1(n; h) = p + (1-p)p^2 = p + p^2 - p^3$. Similar arguments show that $p_1(2; h) = p + p^2 - p^3$, and that $p_1(j; h) = (p + p^2 - p^3)p = p^2 + p^3 - p^4$ for $j \in \{3, \ldots, n-1\}$. Hence $B_1(h) = \sum_{j \neq 1} p_1(j; h) = 2(p + p^2 - p^3) + (n-3)(p^2 + p^3 - p^4)$ and $\Pi_1(g) = 2p + (n-1)p^2 + (n-5)p^3 - (n-3)p^4 - c$. Furthermore, $\Pi_1(g) - \Pi_1(g^s) = p + p^2 + (n-5)(p^3 - p^4) - 2p^4 - c$. Since $p < 1$, we have $p^3 - p^4 > 0$. It follows that there exists an integer $\bar{n}$ such that $\Pi_1(g) - \Pi_1(g^s) > 0$ for all $n \geq \bar{n}$, i.e. the center-sponsored star is not Nash. □

Proposition 3.2 shows that for fixed c, if p is sufficiently close to 1 then the center-sponsored star is (strict) Nash. However, Example 2 above reveals that p is not independent of n: no matter how close p is to 1, for sufficiently large societies the center-sponsored star cannot be supported as a Nash network. The intuition for the above finding is as follows: in a star, the solitary link between agent 1 and the central agent is crucial for him to obtain benefits from the rest of society. Furthermore, the loss of benefits due to the failure of the critical link with the central agent becomes unboundedly large as n increases. From some point onwards, it becomes worthwhile for agent 1 to establish another route to the remaining agents in order to recover the foregone benefits, at which point the star is no longer Nash.

Formally, let h be a connected network. We shall say that a link $h_{i,j} = 1$ is critical if it constitutes the unique path between agents i and j in h. The network h is called *minimally connected* if every link $h_{i,j} = 1$ is critical.[14] Correspondingly, a link $h_{i,j} = 1$ is called *redundant* if there is a path between i and j in $\hat{h}$, where $\hat{h}$ is the network where the link $h_{i,j} = 1$ has been deleted, ceteris paribus. We shall say that a network h is *super-connected* if every link in h is redundant.[15] A minimally connected network is very demanding in terms of reliability: even if one link falls, two non-empty subsets of agents will no longer be able to transmit information to each other. On the other hand, super-connected networks have additional built-in protection from communication failure: even when a link happens to fail, all agents will still be able to communicate amongst themselves

[14] This is equivalent to saying that there is a unique path between any two agents i and j in h.

[15] This is equivalent to saying that there are at least two paths between any two agents in the society.

with positive probability. Note that the star and the line are minimally connected, while the wheel and the complete network are super-connected.[16]

The above classification leads to ask: how importart is redundancy in Nash networks? Is it the case that agents rely on single paths for communicating with others, or do they allow for multiple pathways? The following result addresses these questions.

Proposition 3.3. *Suppose $p(1-p^{n/2}) > c$. If $g^1 \in \mathcal{G}^a$ is a Nash network, then it must be super-connected.*

The proof of this result requires the following lemma, whose proof is in the appendix.

Lemma 3.1. *Let $\sum_{k=1}^{t} a^k p^k$ be a polynomial where the coefficients $\{a^k\}$ are integers satisfying (1) a^k is a non-negative integer for each $k \in \{1,\ldots,t\}$, (2) $\sum_{k=1}^{t} a^k = t$ and (3) $a^k \geq 1$ for some $k \in \{2,\ldots,t\}$ implies $a^{\bar{k}} \geq 1$ for all $\bar{k} \in \{1,\ldots,k-1\}$. Then*

$$\sum_{k=1}^{t} a^k p^k \geq \sum_{k=1}^{t} p^k \tag{3.11}$$

for all $p \in (0,1)$.

We now show:

Proof of Proposition 3.3. The proof is by contradiction. Since $p > p(1-p^{n/2}) > c$, and g^1 is Nash, it must be connected. Suppose g^1 is not super-connected, so that there exists a link $h^1_{i,j} = 1$ in h^1 which is critical. Let g^0 be the network where the link $g^1_{i,j} = 1$ has been deleted, ceteris paribus. Then $h^1 = \text{cl}(g^1)$ has two components, C_1 and C_2, with $i \in C_1$ and $j \in C_2$. Let $|C_1| = n_1$ and $|C_2| = n_2$. Suppose, without loss of generality, that $n_1 \geq n_2$. Then it follows that $n_1 \geq n/2$. Let $r \in C_1$ be an agent furthest away from j in h^1. Since j's sole link with agents in C_1 is $h'_{i,j} = 1$, it follows that $d(j,r;h^1) \geq 2$. Also note that since C_1 is a component of h^0 there exists a path in h^0 between r and m, for any $m \in C_1$.

We now suppose that starting from the network g^1, agent j forms an additional link with agent r, ceteris paribus. Denote the new network as g^2. There are now at least two paths in $h^2 = \text{cl}(g^2)$ between j and each agent $m \in C_2$: via the link $h^2_{i,j} = h^1_{i,j} = 1$ and via the link $h^2_{j,r} = 1$, using a path between r and m in h^0 (which does not involve the link $h^1_{i,j} = 1$ by choice of r). Thus even if the link $h^1_{i,j} = 1$ fails (with probability $1-p$) agent j can still obtain the information of m if the link $h^2_{j,r} = 1$ succeeds, as do all the links in the path between r and m. Let $m \in C_1$. By definition, we have

$$p_j(m;h^1) \;=\; p \sum_{h' \subset h^0} \lambda(h'|h^0)\mathbf{1}_j(m;h' \oplus h^1_{i,j}) + (1-p) \sum_{h' \subset h^0} \lambda(h'|h^0)\mathbf{1}_j(m;h')$$

[16] Propositions 3.3 and 4.3 below deal with the notion of super-connectedness. They replace earlier versions of these results using a weaker notion of this concept. We thank Matt Jackson for suggesting the stronger concept and indicating the appropriate modifications to our earlier proofs.

$$= p \sum_{h' \subset h^0} \lambda(h'|h^0) \mathbf{1}_j(m; h' \oplus h^1_{i,j}). \tag{3.12}$$

since $\mathbf{1}_j(m; h') = 0$ when the link $h^1_{i,j} = 1$ fails. Furthermore,

$$\begin{aligned} p_j(m; h^2) &= p^2 \sum_{h' \subset h^0} \lambda(h'|h^0) \mathbf{1}_j(m; h' \oplus h^1_{i,j} \oplus h^2_{j,r}) \\ &\quad +(1-p)p \sum_{h' \subset h^0} \lambda(h'|h^0) \mathbf{1}_j(m; h' \oplus h^1_{i,j}) \\ &\quad +(1-p)p \sum_{h' \subset h^0} \lambda(h'|h^0) \mathbf{1}_j(m; h' \oplus h^2_{j,r}) \end{aligned} \tag{3.13}$$

where we have omitted the term $(1-p)^2 \sum_{h' \subset h^0} \lambda(h'|h^0) \mathbf{1}_j(m; h')$ since $\mathbf{1}_j(m; h') = 0$ for each $h' \subset h^0$. Consider the first term on the right hand side of (3.13). Clearly $\mathbf{1}_j(m; h' \oplus h^1_{i,j} \oplus h^2_{j,r}) \geq \mathbf{1}_j(m; h' \oplus h^1_{i,j})$. Hence the first two terms in (3.13) are at least as large as $(p^2 + (1-p)p) \sum_{h' \subset h^0} \lambda(h'|h^0) \mathbf{1}_j(m; h' \oplus h^1_{i,j}) = p_j(m; h^1)$, where we employ (3.12). We now consider the third term in (3.13). Let H consist of those realizations $h' \subset h^0$ where all the links in the shortest path between r and m succeed. [Since C_1 is a component of h^0, such a path exists]. For each $h' \in H$ we clearly have $\mathbf{1}_j(m; h' \oplus h^2_{j,r}) = 1$. Hence $\sum_{h' \subset h^0} \lambda(h'|h^0) \mathbf{1}_j(m; h' \oplus h^2_{j,r}) \geq \sum_{h' \in H} \lambda(h'|h^0) = p^{d(r,m;h^0)}$. Summarizing these arguments, we obtain, for $m \in C_1$:

$$p_j(m; h^2) \geq p_j(m; h^1) + (1-p)pp^{d(r,m;h^0)} = p_j(m; h^1) + (1-p)p^{d(r,m;h^0)+1}. \tag{3.14}$$

On the other hand it is easy to see that $p_j(m; h^2) = p_j(m; h^1)$ for all $m \in C_2$. Summing over all $m \in N$ and using the facts that $B_j(h^2) = \sum_{m \in N} p_j(m; h^2)$ and $B_j(h^1) = \sum_{m \in N} p_j(m; h^1)$, we get:

$$B_j(h^2) - B_j(h^1) \geq (1-p) \sum_{m \in C_1} p^{d(r,m;h^0)+1} = (1-p) \sum_{k=1}^{n_1} d^k_j(h^0 \oplus h_{j,r}) p^k \tag{3.15}$$

where $d^k_j(h^0 \oplus h_{j,r})$ is the number of agents in C_1 at distance k from j in the network $h^0 \oplus h_{j,r}$. Clearly, $\sum_{k=1}^{n_1} d^k_j(h^0 \oplus h_{j,r}) = n_1$. Also, if there are agents in C_1 at distance $k \geq 2$ from agent j in $h^0 \oplus h_{j,r}$ there must be agents at all lower distances as well, i.e. $d^k_j(h^0 \oplus h_{j,r}) \geq 1$ implies $d^{k'}_r(h^0 \oplus h_{j,r}) \geq 1$ for all $1 \leq k' < k$. Hence conditions (1)–(3) of Lemma 3.1 are satisfied, and we have $\sum_{k=1}^{n_1} d^k_j(h^0 \oplus h_{j,r}) p^k \geq \sum_{k=1}^{n_1} p^k$. Using (3.15) and the above inequality, we obtain

$$\begin{aligned} B_j(h^2) - B_j(h^1) &\geq (1-p) \sum_{k=1}^{n_1} d^k_j(h^0 \oplus h_{j,r}) p^k \\ &\geq (1-p) \sum_{k=1}^{n_1} p^k = p(1-p^{n_1}) \geq p(1-p^{n/2}) \end{aligned} \tag{3.16}$$

where we use the fact that $n_1 \geq n/2$. It follows that when $p(1-p^{n/2}) > c$, agent j's marginal benefit from his additional link with r will exceed the marginal cost, in which case g^1 cannot be Nash. This contradicts our original supposition. □

We now interpret the significance of Proposition 3.3. Note that as n becomes large, the term $p(1-p^{n/2})$ approaches p. Hence, the result states that for fixed parameters (p, c) with $p > c$, all equilibrium networks will be super-connected for sufficiently large societies, i.e. agents will have multiple pathways to communicate with each other. In particular, for large societies, minimally connected networks such as the star will not be observed in equilibrium.

Several additional comments are in order concerning Proposition 3.3. First, we would ideally like it to be complemented by a result which shows that for each n, Nash networks exist in all regions of the parameter space. Due to the formidable computational difficulties, we have been unable to answer this question fully, though our investigations for small values of n lead us to conjecture that this is true.[17]

Next, we also note that the assumption that the links $g_{i,j} = 1$ and $g_{j,i} = 1$ are perfectly correlated plays a role in the above result, by ensuring that if i forms a link with j, then j has no incentive to form a link with i. An alternative assumption is that if $\min\{g_{i,j}, g_{j,i}\} = 1$ then the link $h_{i,j} = 1$ succeeds with probability $q = 1 - (1-p)^2 = 2p - p^2$. While this may alter the parameter regions where specific networks are Nash, the intuition and the main results of the paper should still hold.[18]

Finally, it is interesting to contrast this model with the one developed in our earlier paper (Bala and Goyal 1999). In that paper, the payoff of agent i in a network g is given by

$$\Pi_i(g) = \sum_{\{j;\mathbf{1}_i(j;h)=1\}} \delta^{d(i,j;h)} - \mu_i^d(g)c \tag{3.17}$$

where $\delta \in (0,1)$ is a parameter which measures "information decay" and $h = \text{cl}(g)$. Here, the network g (and h) are deterministic. However, the value of information obtained from another agent decays geometrically based on the geodesic distance to that agent. In general, Nash networks in that model are all minimally connected. For instance, for all $n \geq 3$, star networks (of all types) are Nash in the region $\{(\delta, c) | \delta \in (0,1), \delta - \delta^2 < c < \delta\}$. This contrasts sharply with the finding of Proposition 3.3, if $c < p$, then for large n, the star network (or any other minimally connected network) is not Nash. Thus, the presence of uncertainty creates very different kinds of incentives in network formation as compared to the model specified by (3.17).

[17] In the paper we focus on pure strategies only. We note that the network formation game we examine is a finite game, and so existence of equilibrium in mixed strategies follows directly from standard results in game theory.

[18] The notion of multiple paths between agents has to be suitably extended so that if $\min\{g_{i,j}, g_{j,i}\} = 1$ then i and j are said to have multiple paths with each other.

4 Efficient networks

We now turn to the study of networks which are optimal from a social viewpoint. Our emphasis will be on the relationship between Nash networks and efficient ones as p and c are allowed to vary over the parameter space. Due to the difficulty of the topic, however, our analysis will be fairly limited. The welfare function $W : \mathcal{G} \to \mathcal{R}$ is taken to be the sum of payoffs, i.e. $W(g) = \sum_{i=1}^{n} \Pi_i(g) = \sum_{i=1}^{n}(B_i(h) - \mu_i^d(g)c)$ where $h = \text{cl}(g)$. Recall that a network g is said to be *efficient* if $W(g) \geq W(g')$ for all $g' \in \mathcal{G}$. We restrict ourselves to networks in the set $\mathcal{G}^a$. In the analysis of efficiency, this is without loss of generality.

Let g be a network and let $h = \text{cl}(g)$. From Lemma 2.2, each $B_i(h) = \sum_{k=0}^{n-1} a_i^k p^k$ where $a_i^1 = |\{j|h_{i,j} = 1\}|$. Thus, each link $g_{i,j} = 1$ contributes 1 each to the coefficient a_i^1 and a_j^1. Since the total number of links in h is $L(h)$, we get $\sum_{i=1}^{n} a_i^1 = 2L(h)$. On the other hand, $\sum_{i=1}^{n} \mu_i^d(g) = \sum_{i=1}^{n} |\{j|g_{i,j} = 1\}| = L(h)$. Thus, the welfare function $W(g)$ can be expressed as a polynomial

$$W(g) = 2L(h)p + \sum_{k=2}^{L(h)} a^k p^k - L(h)c \tag{4.1}$$

for some coefficients $\{a^k\}$. In particular, (4.1) indicates that the welfare properties of efficient networks depend only upon their non-directed features. This is a consequence of the linearity of payoffs in the costs of link formation. In what follows, our analysis is in terms of h rather than g.

Example 3. Fix $n = 3$. There are four possible architectures: (a) the empty network h^e. (b) a single link network h^n given by $h_{1,2}^n = 1$. (c) the star network h^s given by $h_{1,2}^s = h_{1,3}^s = 1$. (d) the complete network h^c given by $h_{1,2}^c = h_{2,3}^s = h_{3,1}^s = 1$. Figure 4 depicts the parameter regions where different networks are efficient.

We compute $W(h^e) = 0$, $W(h^n) = 2p - c$, $W(h^s) = 4p + 2p^2 - 2c$ and $W(h^c) = 6(p + p^2 - p^3) - 3c$. If $c > 2p$ then $W(h^e) > W(h^n)$. Likewise, if $c < 2p + 2p^2$ then $W(h^s) > W(h^n)$. Since these two regions cover the entire parameter space, h^n can never be efficient.

For the remaining three networks, straightforward calculations show that the empty network h^e is efficient in the region

$$\{(p,c)|p \in (0,1/2), c > 2p + 2p^2 - 2p^3\} \cup \{(p,c)|p \in [1/2,1), c > 2p + p^2\} \tag{4.2}$$

the star h^s is efficient in the region

$$\{(p,c)|p \in [1/2,1), 2p + 4p^2 - 6p^3 < c < 2p + p^2\} \tag{4.3}$$

while the complete network h^c is efficient in the region

$$\{(p,c)|p \in (0,1/2), c < 2p+2p^2-2p^3\} \cup \{(p,c)|p \in [1/2,1), c < 2p+4p^2-6p^3\}. \tag{4.4}$$

□

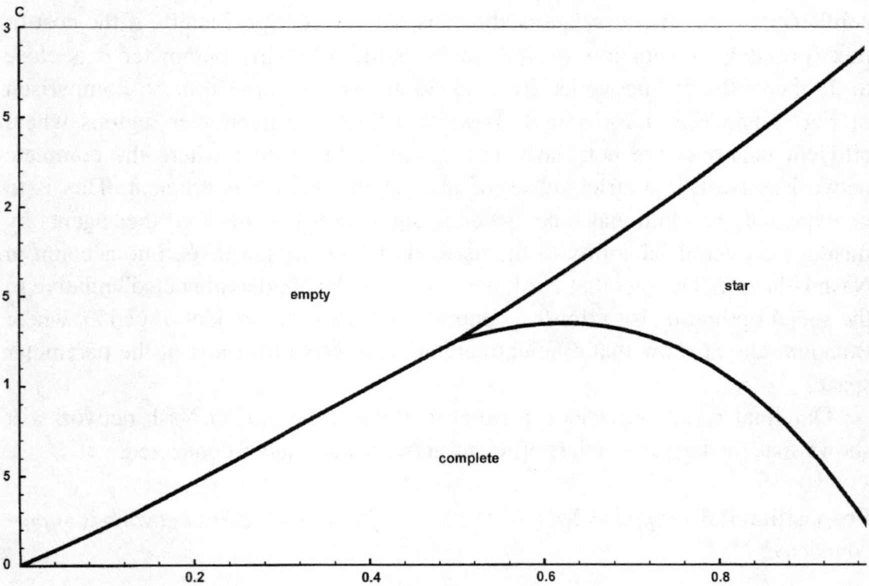

Fig. 4.

We observe that there exist points (p, c) where two or more networks with a different number of links can simultaneously be efficient. (For example, at a point (p, c) where $c = 2p + 4p^2 - 6p^3$ for $p \in [1/2, 1)$, the star and the complete network are both efficient, even though the former has two links while the latter has three). The result below shows for general n that such points are "rare". Specifically, the number of links in an efficient network is generically constant. Second, if we take $p = 0.8$ (say), we see that the number of links in an efficient network is non-increasing in c. This is also true more generally.

Proposition 4.1. *(a) For almost all values of $(p, c) \in (0, 1) \times (0, \infty)$, if h and h^0 are efficient networks, then $L(h) = L(h^0)$. (b) For fixed $p \in (0, 1)$, the number of links in an efficient network is a non-increasing function of $c \in (0, \infty) \setminus V$, where V is a finite set.*

Example 3 also shows that when costs are very low or very high the efficient network is the complete network and the empty network, respectively. Moreover, when links are highly reliable, the star is efficient. These properties hold for general n.

Proposition 4.2. *(a) Given $p \in (0, 1)$, there exists $c_2(p) > c_1(p) > 0$ such that the complete network g^c is efficient for all $c \in (0, c_1(p))$ (b) the empty network g^e is efficient for all $c > c_2(p)$. (c) Given $c \in (0, n)$ there exists $p(c) < 1$ such that the star network is efficient for all $p \in (p(c), 1)$.*

The proof is available in the appendix. A comparison of Proposition 3.2 with the above result shows that there are regions of the parameter space where the conflict between efficiency and stability is not severe. Specifically, if the cost of link formation is very low or very high, or the reliability parameter p is close to 1, then efficient networks are also Nash. At the same time, a comparison of Fig. 3 and Fig. 4 for $n = 3$ shows that there are parameter regions where efficient networks are not Nash. For example, the region where the complete network is Nash is a strict subset of the region where it is efficient. This is to be expected, as additional links generate significant benefits for other agents by raising the overall reliability of the network, which are not taken into account in Nash behavior. The fact that Nash networks may be "underconnected" relative to the social optimum also affords a contrast with the decay model of (3.17), where Bala and Goyal show that efficient networks are Nash for most of the parameter space.

Our final result provides a parallel to Proposition 3.3 on Nash networks. It shows that for large societies, efficient networks are super-connected.

Proposition 4.3. *Suppose* $2p(1 - p^{n/2}) > c$. *Then an efficient network is super-connected.*

Proof. Suppose that h^0 is efficient and not connected. Then there are two agents i and j such that there is no path between them in h^0, so that $p_i(j; h^0) = p_j(i; h^0) = 0$. Let h be the network formed when a link $h_{i,j} = 1$ is added, ceteris paribus. Then $p_i(j; h) = p_j(i; h) = p$. Moreover, from Lemma 2.1 all other agents payoffs either stay the same or increase. Hence, welfare increases by at least $2p - c$, which is strictly positive under the hypothesis that $c < 2p(1 - p^{n/2})$, thus contradicting the supposition that h^0 is efficient. We now show that h^0 must in fact be super-connected. The proof is by contradiction. Suppose this is not true, i.e., there exists a link $h_{i,j} = 1$ in h which is critical. Then h^0 has two components, C_1 and C_2, with $i \in C_1$ and $j \in C_2$. Let $|C| = n_1$ and $|C'| = n_2$. Suppose, without loss of generality, that $n_1 \geq n_2$. Then it follows that $n_1 \geq n/2$. Let $r \in C_1$ be an agent furthest away from j in h. Since agent j has only one link with agents in C_1, it follows that $d(j, r; h) \geq 2$. Moreover, since $\{i, r\} \subset C_1$, it also follows that there is at least one path between i and r, which does not involve $h_{i,j} = 1$.

We now suppose that starting from the network g, agent i forms an additional link with agent r, ceteris paribus. Proceeding as in Proposition 3.3, agent j's expected benefit increases by at least $p(1 - p^{n_1})$. Similarly, the payoff of each of the agents $m \in C_1$ increases. Moreover, by Lemma 2.1, every other agent's payoff is non-decreasing in this link. The lower bound to the total increase is $p(1-p^{n_1})$. Hence, welfare rises by at least $2p(1-p^{n_1})-c \geq 2p(1-p^{n/2})-c$. This expression is strictly positive, by hypothesis. This contradicts the supposition that h is efficient. □

We see that if $c < 2p$, efficiency requires the presence of redundant links as n becomes large. In particular, while star networks are efficient for all n, (as demonstrated in Proposition 4.2) they require larger and larger values of p to maximize social welfare as the society expands in size.

We would like a result which shows that efficient networks are either empty or connected, as is the case with Nash networks. However, this does not seem to be an easy question to settle. Proposition 4.3 goes some distance by showing a significant parameter region where efficient networks must be super-connected. We have been unable to develop a better (and more precise) bound for c than that stated in the proposition. This is also the main difficulty in deriving a general result on connectedness of efficient networks.

It is also possible to show, via a simple continuity argument, that for high levels of reliability, an efficient network is either connected or empty. We briefly sketch the argument here: consider the model with full reliability, i.e., $p = 1$. The welfare of a minimally connected network is given by $(n - c)(n - 1)$, while the welfare from a network with $k \geq 2$ (minimal) components is given by $\sum_{i=1}^{k}(n_i - c)(n_i - 1)$, where n_i is the number of agents in component i. It is easily seen that the former is strictly greater than the latter, so long as $n > c$. Finally, note that the welfare from an empty network is 0. Thus so long as $n > c$, the minimally connected network provides a strictly higher welfare than every other network. Similarly, it can be shown that if $n < c$ then the empty network provides a strictly higher welfare than every other network. From the payoff expression (2.5), and the definition of welfare function in (4.1), it follows that the welfare function is continuous with respect to the reliability parameter p. Thus for values of p close to 1, an efficient network is either connected or empty. We note that unlike the case with Nash networks, existence of an efficient network is not a problem as the domain of the welfare function $W(\cdot)$ is a finite set. Moreover, we see that the super-connectedness of efficient networks has been demonstrated for twice the range of c values that was shown for Nash networks. As we are concerned with total welfare, and the addition of a link provides strictly positive expected benefits to at least two agents, this is to be expected. In a very loose sense, it suggests that having redundant links is even more important for efficiency as compared to stability.

Finally, it is worthwhile to contrast the above result with the result of the information decay model (3.17). Proposition 5.5 in Bala and Goyal (2000) shows that the star is the uniquely efficient network in the region $\delta - \delta^2 < c < 2\delta + (n - 2)\delta^2$. As with Nash networks, it affords a sharp contrast to what we find here.

5 Conclusion

We consider a non-cooperative model of social communication in networks where communication is costly and not fully reliable. We show that Nash networks, provided they are not empty, ensure that every agent communicates with every other agent with positive probability. If the society is large, and link formation costs are moderate, Nash networks for the most part must be 'super-connected', i.e. agents will find it worthwhile to establish multiple pathways to other agents in order to increase the reliability of communication. This contrasts with the properties of a deterministic model of information decay, where Nash networks typically involve unique paths between agents. We also study efficient networks and show that if costs are very low or very high, or if links are highly reliable then there is virtually no conflict between efficiency and stability. However, for intermediate ranges of costs and link reliability, Nash networks may be under-connected relative to the social optimum. As with Nash networks, if the society is large, and link formation costs are moderate, efficient networks will typically have redundant links to increase reliability.

6 Appendix

Proof of Lemma 2.2. Recall from (2.2) that $B_i(h) = \sum_{h' \subset h} \lambda(h'|h)\mu_i(h')$ where $\lambda(h'|h) = p^{L(h')}(1-p)^{L(h)-L(h')}$. Hence $B_i(h)$ potentially involves powers of p upto degree $L(h)$. Moreover, since $\mu_i(h') > 0$ requires $L(h') > 0$, all non-zero terms in $B_i(h)$ involve p^q for some $q \geq 1$. Hence

$$B_i(h) = \sum_{k=1}^{L(h)} a_i^k p^k \tag{A.1}$$

for some coefficients $\{a_i^k\}$. It follows that $\Pi_i(g) = B_i(h) - \mu_i^d(g)c = \sum_{k=0}^{L(h)} a_i^k p^k$ is also a polynomial of degree at most $L(h)$, with $a_i^0 = -\mu_i^d(g)c$. Next, suppose that $h_{i,j} = 1$. We characterize the probability $p_i(i;h)$ that i observes j. From (2.4) this is given by $p_i(j;h) = \sum_{h' \subset h} \lambda(h'|h)\mathbf{1}_i(j;h')$. Consider the event $E = \{h' \subset h | h'_{i,j} = 1\}$. Clearly, the probability of E is p, and if E occurs then i observes j. If E does not occur (with probability $1-p$) then i may still observe j in a realization h' where there is a path between i and j involving two or more links. However, the probability of such a realization is of the form $(1-p)^{k_1}p^{k_2}$ where $k_1 \geq 1$ and $k_2 \geq 2$. Hence, such an event can only contribute terms of degree 2 or higher to $p_i(j;h)$. A similar argument shows that if $h_{i,j} = 0$ then $p_i(j;h)$ can only have terms involving p^2 or higher. Thus each j for which $h_{i,j} = 1$ contributes p (and possibly terms of higher degree) to $p_i(j;h)$ and each j for which $h_{i,j} = 0$ contributes either 0 or terms of degree higher than 1 to $p_i(j;h)$. The claim that $a_i^1 = |\{j | h_{i,j} = 1\}|$ follows from the above observation in conjunction with (2.4). □

Proof of Lemma 3.1. We show $\sum_{k=1}^{t}(a^k - 1)p^k \geq 0$ which is equivalent to (3.11). The proof is by induction. If $t = 1$, condition (1) and (2) imply $a^1 = 1$ so that (3.11) is trivially satisfied. Suppose for some $t \geq 1$, $\sum_{k=1}^{t}(a^k - 1)p^k \geq 0$ for all $\{a^k\}$ satisfying (1)–(3). Consider the case $t+1$, i.e. the polynomial $\sum_{k=1}^{t+1}(a^k - 1)p^k$ where $\{a^k\}$ satisfy (1)–(3). If $a^{t+1} \geq 1$, then (3) implies that $a^k \geq 1$ for all $k < t+1$. Since $\sum_{k=1}^{t+1} a^k = t+1$ from (2), we get $a^k = 1$ for all k and $\sum_{k=1}^{t+1}(a^k - 1)p^k = 0$. Suppose instead that $a^{t+1} = 0$. Then (2) implies $\sum_{k=1}^{t} a^k = t+1$. From (1), this means $a^{k'} \geq 2$ for some $k' \in \{1, \ldots, t\}$. Define $b^k = a^k$ for all $k \neq k'$ and $b^k = a^{k'} - 1$. Clearly, $\{b^k\}$ satisfy (1) and (3), while by definition of $\{b^k\}$ we have $\sum_{k=1}^{t} b^k = t$. Hence we can apply the induction step to get $\sum_{k=1}^{t}(b^k - 1)p^k \geq 0$. Since $k' \leq t$ and $p \in (0, 1)$ we have $p^{k'} - p^{t+1} \geq 0$. Hence $\sum_{k=1}^{t+1}(a^k - 1)p^k = \sum_{k=1}^{t}(b^k - 1)p^k + p^{k'} - p^{t+1} \geq 0$ as required. □

Proof of Proposition 4.1. Let h and h^0 be two networks such that $L(h) \neq L(h^0)$. Using (4.1), the set of points (p, c) where $W(h) = W(h^0)$ satisfies the equation $c = 2p + Q(p)/(L(h) - L(h^0))$ for some polynomial $Q(p)$ which involves only terms of degree 2 or higher. The graph of this polynomial has Lebesgue measure 0 in $\mathscr{R}^2$ (see Halmos 1974, Exercise 4, page 145). Let U be the (finite) union of the graphs of all polynomials generated by pairs h and h^0 satisfying $L(h) \neq L(h^0)$, intersected with $(0, 1) \times (0, \infty)$. The result follows since U has Lebesgue measure 0.

We now consider part (b). Let

$$V = \{c \in (0, \infty) | W(h) = W(h^0) \text{ for } h, h^0 \text{ satisfying } L(h) \neq L(h^0)\}. \tag{A.2}$$

Since the set of all networks is a finite set, V is also finite. It follows that the number of links in an efficient network is a well-defined number on the set V^c. Fix $c \in V^c$ and suppose h is efficient. Then $W(h) > W(h^0)$ for all h^0 such that $L(h) \neq L(h^0)$. In particular this also holds for all h^0 such that $L(h) > L(h^0)$. If $c' \in V^c$ satisfies $c' < c$, then clearly $W(h) > W(h^0)$ continues to hold for all h^0 such that $L(h) > L(h^0)$. □

We now show the following lemma.

Lemma 4.1. *Let g be a connected network and let $h = \mathrm{cl}(g)$. Suppose $h = \mathrm{cl}(g)$ is minimally connected. Then $B_i(h) = \sum_{k=1}^{n-1} d_i^k(h)p^k$, where $d_i^k(h)$ is the number of agents at geodesic distance k from agent i in h.*

Proof. Since h is minimally connected, there is a unique path between any two agents i and j. Hence, for agent i to access agent j it is necessary and sufficient that all $d(i,j;h)$ links on the path between j and i succeed. The probability of this event is $p^{d(i,j;h)}$.

Hence $p_i(j; h) = p^{d(i,j;h)}$ and $B_i(h) = \sum_{j \neq i} p_i(j; h) = \sum_{j \neq i} p^{d(i,j;h)}$. Since h is connected, $1 \leq d(i, j; h) \leq n - 1$ for all j. If there are $d_i^k(h) \leq n - 1$ agents at distance k from i, the coefficient of p^k in $B_i(h)$ will be $d_i^k(h)$ as required. □

Proof of Proposition 4.2. If $c = 0$ then Lemma 2.1 implies that welfare is uniquely maximized at a complete network. Part (a) follows by continuity. Part (b) follows trivially because the welfare of any non-empty network is negative for c sufficiently large. For part (c), choose $p(c)$ to ensure that $n(n - 1)(1 - p^{n-1}) < c$ for all $p \in (p(c), 1)$. Let h be an efficient network and suppose that C is a component of it containing at least two agents. Assume that C is minimal, i.e. that there is a unique path between any two agents in C. Let $q = |C|$. Clearly, $q \leq n$. Using Lemma 4.1 above, $B_i(h) = \sum_{k=1}^{q-1} d_i^k(h)p^k$ where $d_i^k(h)$ is the number of agents in C at distance k from agent i. Since $p^k \geq p^{q-1}$ for each $k \leq q - 1$, we have $B_i(h) \geq (q - 1)p^{q-1}$. Furthermore the contribution to special benefit from the agents in C is at least $q(q - 1)p^{q-1}$. Since the maximum expected benefit of any agent in C is $q(q-1)$, the addition of any links between agents in C can raise total expected benefit by no more that $q(q-1)(1-p^{q-1}) \leq n(n-1)(1-p^{n-1}) < c$ by choice of p. Hence, any component of an efficient network must be minimally connected. Within the class of networks whose components are minimally connected, the welfare function coincides with the one in the model of information decay, where payoffs are specified by (3.17). The result then follows from Proposition 5.5 of Bala and Goyal (2000). □

References

Baker, W., Iyer, A. (1992) Information networks and market behaviour. Journal of Mathematical Sociology 16 (4):305–332

Bala, V. (1996) Dynamics of Network Formation. mimeo, McGill University

Bala, V., Goyal, S. (1998) Learning from neighbours. Review of Economic Studies 65: 595–621

Bala, V., Goyal, S. (2000) A non-cooperative model of network formation. Econometrica 68: 1181–1229

Bollobas, B. (1978) An Introduction to Graph Theory. Springer, Berlin

Boorman, S. (1975) A combinatorial optimization model for transmission of job information through contact networks. Bell Journal of of Economics 6(1): 216–249

Chwe, M. (1995) Strategic reliability of communication networks. mimeo, University of Chicago

Coleman, J. (1966) Medical Innovation: A Diffusion Study. 2nd ed., Bobbs-Merrill, New York

Dutta, B., van den Nouweland, A., Tijs, S. (1998) Link formation in cooperative situations. International Journal of Game Theory 27: 245–256

Dutta, B., Mutuswami, S. (1997) Stable Networks. Journal of Economic Theory 76: 322–344

Goyal, S. (1993) Sustainable Communication Networks. Tinbergen Institute, Erasmus University, Discussion Paper 93-250

Granovetter, M. (1974) Getting a Job: A Study of Contacts and Careers. Harvard University Press, Cambridge, MA

Halmos, P. (1974) Measure Theory. Springer, New York

Jackson, M., Wolinsky, A. (1996) A Strategic Model of Economic and Social Networks. Journal of Economic Theory 71(1): 44–74

Rogers, E., Kincaid, D.L. (1981) Communication Networks: Toward a New Paradigm for Research. Free Press, New York

Rogers, E., Shoemaker, F. (1971) The Communication of Innovations. 2nd ed., Free Press, New York

Watts, A. (1997) A Dynamic Model of Networks. mimeo, Vanderbilt University

A Dynamic Model of Network Formation

Alison Watts

Department of Economics, Box 1819, Sattion B, Vanderbilt University, Nashville, Tennessee 37235, USA

Abstract. Network structure plays a significant role in determining the outcome of many important economic relationships; therefore it is crucial to know which network configurations will arise. We analyze the process of network formation in a dynamic framework, where self-interested individuals can form and sever links. We determine which network structures the formation process will converge to. This information allows us to determine whether or not the formation process will converge to an efficient network structure.

JEL Classification: A14, C7, D20

1 Introduction

Network structure plays a significant role in determining the outcome of many important economic relationships. There is a vast literature which examines how network structure affects economic outcomes. For example, Boorman (1975) and Montgomery (1991) examine the relationship between social network structure and labor market outcomes. Ellison and Fudenberg (1995) show that communication structure can influence a consumer's purchasing decisions. Political party networks can influence election results (see Vazquez-Brage and Garcia-Junado, 1996). The organization of workers within a firm influences the firm's efficiency, see Keren and Levhari (1983), Radner (1993) and Bolton and Dewatripont (1994). Hendricks et al. (1997) show that the structure of airline connections influences competition. Finally, in evolutionary game theory, Ellison (1993), Goyal and Janssen (1997) and Anderlini and Ianni (1996) show that network structure affects whether or not coordination occurs.

I thank an associate editor, an anonymous referee, Matt Jackson, Hervé Moulin, Anne van den Nouweland, John Weymark and Giorgio Fagiolo for valuable comments and criticisms.

Since network structure affects economic outcomes, it is crucial to know which network configurations will arise. We analyze the process of network formation in a dynamic framework, where self-interested individuals can form and sever links. We determine which network structures the formation process will converge to. This information allows us to determine whether or not the formation process will converge to an efficient network structure. Specifically, we show that the formation process is path dependent, and thus the process often converges to an inefficient network structure. This conclusion contrasts with the results of Qin (1996) and Dutta et al. (1998) who find that an efficient network almost always forms.

In our model, there is a group of agents who are initially unconnected to each other. Over time, pairs of agents meet and decide whether or not to form or sever links with each other; a link can be severed unilaterally but agreement by both agents is needed to form a link. Agents are myopic, and thus decide to form or sever links if doing so increases their current payoff. An agent's payoff is determined as in Jackson and Wolinsky's (1996) connections model. (Agents receive a benefit from all direct and indirect connections, where the benefit of an indirect connection is smaller than that of a direct connection. Agents also must pay a cost of maintaining a direct connection, which can be thought of as time spent cultivating the relationship.) We show that if the benefit from maintaining an indirect link is greater than the net benefit from maintaining a direct link, then it is difficult for the efficient network to form. In fact, the efficient network only forms if the order in which the agents meet takes a particular pattern. Proposition 4 shows that as the number of agents increases it becomes less likely that the agents meet in the correct pattern, and thus less likely that the efficient network forms.

There are other papers which also address the idea of network formation. The endogenous formation of coalition structures is examined by Aumann and Myerson (1988), Qin (1996), Dutta, van den Nouweland and Tijs (1998), and Slikker and van den Nouweland (1997). The most important difference between their work and ours is that we assume that network formation is a dynamic process in which agents are free to sever a direct link if it is no longer beneficial. In contrast, Aumann and Myerson (1988) assume that once a link forms it cannot be severed, while Qin (1996), Dutta et al. (1998), and Slikker and van den Nouweland (1997) all consider one-shot games.

The three papers most closely related to the issues considered here are Jackson and Wolinsky (1996), Bala and Goyal (2000), and Jackson and Watts (1999). Jackson and Wolinsky (1996) examine a static model in which self interested individuals can form and sever links. They determine which networks are stable and which networks are efficient.[1] Thus, they leave open the question of which stable networks will form. Here, we extend the Jackson and Wolinsky connections model to a dynamic framework. Bala and Goyal (2000) simultaneously examine network formation in a dynamic setting. However, their approach differs signifi-

[1] Dutta and Mutuswami (1997) also examine the tension between stability and efficiency, using an implementation approach.

cantly from ours both in modeling and results. Bala and Goyal restrict attention to models where links are formed unilaterally (one player does not need another player's permission to form a link with him) in a non-cooperative game and focus on learning as a way to identify equilibria. Jackson and Watts (1999) also analyze the formation of networks in a dynamic framework. Jackson and Watts extend the current network formation model to a general network setting where players occasionally form or delete links by mistake; thus, stochastic stability is used as a way to identify limiting networks.

The remainder of the paper proceeds as follows. The model and static results are presented in Sect. 2, and the dynamic results are presented in Sect. 3. The conclusion and a discussion of what happens if agents are not myopic are presented in Sect. 4.

2 Model

2.1 Static Model and Results[2]

There are n agents, $N = \{1, 2, \ldots, n\}$, who are able to communicate with each other. We represent the communication structure between these agents as a network (graph), where a node represents a player, and a link between two nodes implies that two players are able to directly communicate with each other. Let g^N represent the complete graph, where every player is connected to every other player, and let $\{g \mid g \subseteq g^N\}$ represent the set of all possible graphs. If players i and j are directly linked in graph g, we write $ij \in g$. Henceforth, the phrase "unique network" means unique up to a renaming of the agents.

Each agent $i \in \{1, \ldots, n\}$ receives a payoff, $u_i(g)$, from network g. Specifically, agent i receives a payoff of $1 > \delta > 0$ for each direct link he has with another agent, and agent i pays a cost $c > 0$ of maintaining each direct link he has. Agent i can also be indirectly connected to agent $j \neq i$. Let $t(ij)$ represent the number of direct links in the shortest path between agents i and j. Then $\delta^{t(ij)}$ is the payoff agent i receives from being indirectly connected to agent j, where we adopt the convention that if there is no path between i and j, then $\delta^{t(ij)} = 0$. Since $\delta < 1$, agent i values closer connections more than distant connections. Thus, agent i's payoff, $u_i(g)$, from network g, can be represented by

$$u_i(g) = \sum_{j \neq i} \delta^{t(ij)} - \sum_{j : ij \in g} c \,. \tag{2.1}$$

A network, g, is *stable* if no player i wants to sever a direct link, and no two players, i and j, both want to form link ij and simultaneously sever any of their existing links. Thus, when forming a link agents are allowed to simultaneously

2 The static model (with the exception of the definition of stability) is identical to Jackson and Wolinsky's (1996) connections model.

sever any of their existing links.[3] Formally, g is stable if (a) $u_i(g) \geq u_i(g - ij)$ for all $ij \in g$ and (b) if $u_i(g + ij - ig - jg) > u_i(g)$, then $u_j(g + ij - ig - jg) < u_j(g)$ for all $ij \notin g$, where ig is defined as follows. If agent i is directly linked only to agents $\{k_1, \ldots, k_m\}$ in graph g, then ig is any subset (including the empty set) of $\{ik_1, \ldots, ik_m\}$.

Notice that the formation of a new link requires the approval of two agents. Thus, this definition of network stability differs from the definition of stability of a Nash equilibrium, which requires that no single agent prefers to deviate.

Proposition 1. *For all N, a stable network exists. Further,*

(i) if $c < \delta$ and $(\delta - c) > \delta^2$, then g^N is stable,

(ii) if $c \geq \delta$, then the empty network is stable,

(iii) if $c < \delta$ and $(\delta - c) \leq \delta^2$, then a star[4] network is stable.

Jackson and Wolinsky (1996) prove Proposition 1 for the case in which agents can either form or sever links but cannot simultaneously form and sever links. However, their proof can easily be adapted to our context and is thus omitted. Note that in case (i), g^N is the unique stable network. However in the remaining two cases, the stable networks are not usually unique (see Jackson and Wolinsky, 1996).

A network, g^*, is *efficient* (see Jackson and Wolinsky (1996) and Bala and Goyal, (2000) if it maximizes the summation of each agent's utility, thus $g^* = \arg\max_g \sum_{i-1}^{n} u_i(g)$. The proof of the following proposition (on the existence of an efficient network) may be found in Jackson and Wolinsky (1996).

Proposition 2. (Jackson and Wolinsky, 1996). *For all N, a unique, efficient network exists. Further,*

(i) if $(\delta - c) > \delta^2$, then g^N is the efficient network,

(ii) if $(\delta - c) < \delta^2$ and $c < \delta + \frac{(n-2)}{2}\delta^2$, then a star network is efficient,

(iii) if $(\delta - c) < \delta^2$ and $c > \delta + \frac{(n-2)}{2}\delta^2$, then the empty network is efficient.

2.2 Dynamic Model

Initially the n players are unconnected. The players meet over time and have the opportunity to form links with each other. Time, T, is divided into periods and is modeled as a countable, infinite set, $T = \{1, 2, \ldots, t, \ldots\}$. Let g_t represent the

[3] This notion of stability is an extension of Jackson and Wolinsky's (1996) notion of pairwise stability where agents can either form or sever links but cannot simultaneously form and sever links. The current definition of stability is also used in the matching model section of Jackson and Watts (1999).

[4] A network is called a star if there is a central agent, and all links are between that central person and each other person.

network that exists at the end of period t and let each player i receive payoff $u_i(g_t)$ at the end of period t.

In each period, a link ij is randomly identified to be updated with uniform probability. We represent link ij being identified by $i : j$. If the link ij is already in g_{t-1}, then either player i or j can decide to sever the link. If $ij \notin g_{t-1}$, then players i and j can form link ij and simultaneously sever any of their other links if both players agree. Each player is myopic, and so a player decides whether or not to sever a link or form a link (with corresponding severances), based on whether or not severing or forming a link will increase his period t payoff.

If after some time period t, no additional links are formed or broken, then the network formation process has reached a *stable state*. If the process reaches a stable state, the resulting network, by definition, must be a stable (static) network.

3 Dynamic Network Formation Results

Propositions 3 and 4 tell us what type of networks the formation process converges to. This information allows us to determine whether or not the formation process converges to an efficient network.

Proposition 3. *If $(\delta - c) > \delta^2 > 0$, then every link forms (as soon as possible) and remains (no links are ever broken). If $(\delta - c) < 0$, then no links ever form.*

Proof. Assume $(\delta - c) > \delta^2 > 0$. Since $\delta < 1$, we know that $(\delta - c) > \delta^2 > \delta^3 > \ldots > \delta^{n-1}$. Thus, each agent prefers a direct link to any indirect link. Each period, two agents, say i and j, meet. If players i and j are not directly connected, then they will each gain at least $(\delta - c) - \delta^{t(ij)} > 0$ from forming a direct link, and so the connection will take place. (Agent i's payoff may exceed $(\delta - c) - \delta^{t(ij)}$, since forming a direct connection with agent j may decrease the number of links separating agent i from agent $k \neq j$.) Using the same reasoning as above, if an agent ever breaks a direct link, his payoff will strictly decrease. Therefore, no direct links are ever broken.

Assume $(\delta - c) < 0$ and that initially no agents are linked. In the first time period, two agents, say i and j, meet and have the opportunity to link. If such a link is formed, then each agent will receive a payoff of $(\delta - c) < 0$; since agents are myopic, they will refuse to link. Thus, no links are formed in the first time period. A similar analysis proves that no links are formed in later periods.

Q.E.D.

Proposition 3 says that if $(\delta - c) > \delta^2 > 0$, then the network formation process always converges to g^N, which is the unique efficient network according to Proposition 2. This network is also the unique stable network. Therefore, if the formation process reaches a stable state, the network formed must be g^N.

If $(\delta - c) < 0$, then the empty network is always stable (see Proposition 1). However, the empty network is efficient only if $c > \delta + ((n-2)/2)\delta^2$ (see Proposition 2). Thus, the efficient network does not always form. If $c < \delta + ((n-2)/2)\delta^2$, then the star network is the unique efficient network. However, since

$c > \delta$, this network is not stable (the center agent would like to break all links), and so the network formation process cannot converge to the star in this case.

If $(\delta - c) < 0$, then multiple stable networks may exist. In this case, the empty network is the most inefficient stable network. For example, if $n = 5$ and $(\delta^2 - \delta^3 - \delta^4) > (c - \delta) > 0$, then the circle network is stable. Each agent receives a strictly positive payoff in the circle network; therefore, the circle is more efficient than the empty network.

Proposition 4. *Assume that* $0 < (\delta - c) < \delta^2$. *For* $3 < n < \infty$, *there is a positive probability,* $0 < p(star) < 1$, *that the formation process will converge to a star. However, as n increases,* $p(star)$ *decreases, and as n goes to infinity,* $p(star)$ *goes to 0.*

The following lemma is used in the proof of Proposition 4.

Lemma 1. *Assume* $0 < (\delta - c) < \delta^2$. *If a direct link forms between agents i and j and a direct link forms between agents k and m (where agents i, j, k, and m are all distinct), then the star network will never form.*

Proof of Lemma 1. Assume that $0 < (\delta - c) < \delta^2$ and that the star does form. Order the agents so that agent 1 is the center of the star, agent 2 is the first agent to link with agent 1, agent 3 is the second, ..., and agent n is the last agent to link with agent 1. We show that if the star forms, then every agent $i \neq 1$ meets agent 1 before he meets anyone else.

Assume, at time period t, agent 1 meets agent n and all agents $i \in \{2, \ldots, n-1\}$ are already linked to agent 1. Assume agents 1 and n are so far not directly linked. Thus, in order for the star to form, agent 1 must link to agent n. But agent 1 will link to agent n, only if agent n is not linked to anyone else. Assume, to the contrary, that agent n is linked to agent 2. If agent 1 links to agent n, agent 1's payoff will change by $(\delta - c) - \delta^2 < 0$ (regardless of whether of not agent n simultaneously severs his tie to agent 2). Therefore agent 1 will not link with agent n. In order for agent n to be unlinked in period t, agent n can not have met anyone else previously, since a link between two unlinked agents will always form (recall that $\delta > c$), and such a link is never broken unless the two agents have each met someone else and have an indirect connection they like better.

Next consider time period $(t - 1)$ in which agent $(n - 1)$ joins the star. Again, agent $(n - 1)$ must be unlinked to agents $\{2, \ldots, n-2\}$, otherwise agent 1 will refuse to link with agent $(n - 1)$. Also agent $(n - 1)$ cannot be linked to agent n, since agent n must be unlinked in period t. This process can be repeated for all agents. Hence, all agents must meet agent 1 before they meet anyone else. Contradiction. Q.E.D.

Proof of Proposition 4. Lemma 1 states that if two distinct pairs of players get a chance to form a link, then a star cannot form. We show that the probability of this event happening goes to 1 as n becomes large. Fix any pair of players. The probability that a distinct pair of players will be picked to form a link next is $(n - 2)(n - 3)/n(n - 1)$. This expression goes to 1 as n becomes large.Q.E.D.

Lemma 1 states that the dynamic process often does not converge to the star network. When it does not, the process will converge to either another stable network or to a cycle (a number of networks are repeatedly visited), see Jackson and Watts (1999). For certain values of δ and c, no cycles exist, and thus the dynamic process must converge to another stable network. For example, if c is large or if δ is close to 1, then a player only wants to add a link if it is to a player he is not already directly or indirectly connected to. Thus, the dynamic process will converge to a network which has only one (direct or indirect) path connecting every pair of players. For further discussion of cycles and conditions which eliminate cycles, see Jackson and Watts (1999).

Lemma 1 can be interpreted as follows. First, recall from Propositions 1 and 2 that if $0 < (\delta - c) < \delta^2$, then a star network is stable, but it is not necessarily the only stable network. However, the star is the unique efficient network. Therefore, Lemma 1 says that if $0 < (\delta - c) < \delta^2$, then it is difficult for the unique efficient network to form. In fact, the only way for the star to form is if the agents meet in a particular pattern. There must exist an agent j who acts as the center of the star. Every agent $i \neq j$, must meet agent j before he meets any other agent. If, instead, agent k is the first agent player i meets, then players i and k will form a direct link (since $\delta > c$) and, by Lemma 1, a star will never form. These points are illustrated by the following example.

For $n = 4$, a star will form if the players meet in the order (1:2, 1:3, 1:4, 2:3, 2:4, 3:4), but not if the players meet in the order (1:2, 3:4, 1:3, 1:4, 2:3, 2:4). If the players meet in the order (1:2, 1:3, 1:4, 2:3, 2:4, 3:4), then every agent $i \neq 1$ meets agent 1 before he meets any other agent. Since $\delta > c$, every agent $i \neq 1$ will form a direct link with agent 1. Thus, a star forms in three periods, with agent 1 acting as the center (see Fig. 1).

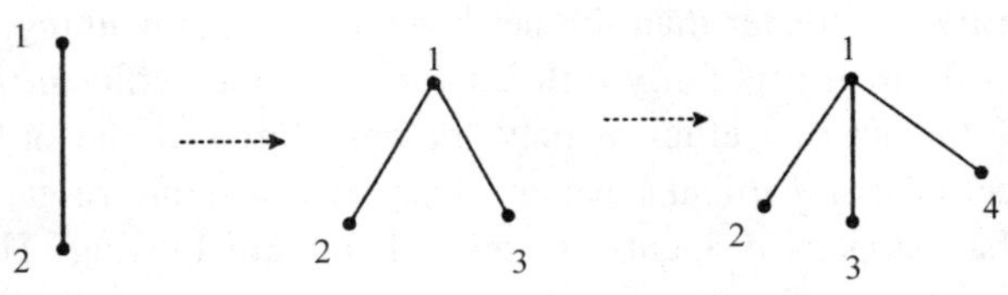

Fig. 1.

If the players meet in the order (1:2, 3:4, 1:3, 1:4, 2:3, 2:4), then the network formation process will converge to a circle if $(\delta - c) > \delta^3$, and the formation process will converge to a line if $(\delta - c) < \delta^3$. Next we briefly outline the formation process. Since $\delta > c$, we know that agents 1 and 2 will form a direct link in period 1, agents 3 and 4 will form a direct link in period 2, and agents 1 and 3 will form a direct link in period 3 (see Fig. 2). In period 4, agent 4 would like to delete his link with agent 3 and simultaneously form a link with agent 1; however, agent 1 will refuse to link with agent 4 since $\delta^2 > (\delta - c)$. Similarly, in period 5, agent 3 will refuse to link with agent 2. In period 6, agents 2 and 4 will agree to link only if $(\delta - c) > \delta^3$.

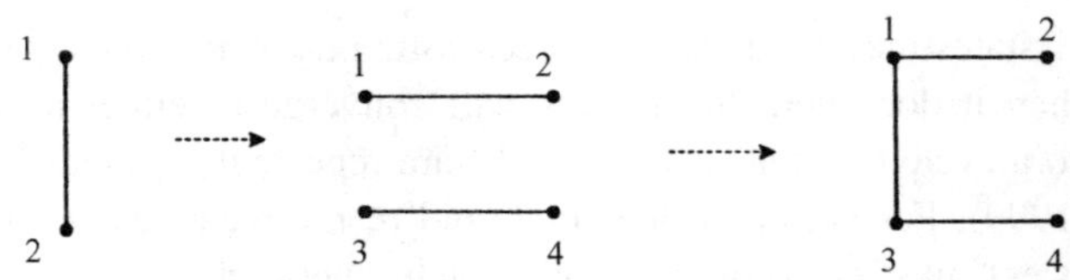

Fig. 2.

Proposition 4 states that as the number of players increases, it becomes less likely that the players will meet in the pattern needed for the star to form. Thus as n increases, the probability of a star forming decreases. For example, if $n = 3$, the probability that the star will form is $p(\text{star}) = 1$. For $n = 4$, the probability that the star forms is $p(\text{star}) = 0.27$, while for $n = 5$, $p(\text{star}) = 0.048$.

The intuition for why it rapidly becomes more difficult for the star to form can be gained by examining the $n = 4$ case. Assume that in period 1, players 1 and 2 meet; players 1 and 2 will form a link since $\delta > c$. Therefore, if a star forms, then we know from Lemma 1, that either agent 1 or agent 2 must act as the center. So, in period 2, the star continues to form as long as agents 3 and 4 do not meet each other. Assume, in period 2, that agents 1 and 3 meet. If the star forms, agent 1 must act as the center. Thus the star can only form if agents 1 and 4 meet before agents 3 and 4 or agents 2 and 4 meet. As the number of agents grows, it becomes less likely that the correct pairs of agents will meet each other early in the game, and thus it becomes less likely that the star forms.

4 Conclusion

We show that if agents are myopic and if the benefit from maintaining an indirect link of length two is greater than the net benefit from maintaining a direct link ($\delta^2 > \delta - c > 0$), then it is fairly difficult for the unique efficient network (the star) to form. In fact, the efficient network only forms if the order in which the agents meet takes a particular pattern. One area of future research would be to explore what happens if agents are instead forward looking. The following example gives intuition for what might happen in such a non-myopic case.

First, consider a myopic four player example where $\delta^2 > \delta - c > 0$. Suppose that agents have already formed the line graph where 1 and 2 are linked, 2 and 3 are linked, and 3 and 4 are linked. If agents 1 and 3 now have a chance to link, then agent 1 would like to simultaneously delete his link with 2 and link with 3. However, agent 3 will refuse such an offer since he prefers being in the middle of the line to being the center agent of the star. This example raises the question: will player 1 delete his link with agent 2 and wait for a chance to link with 3 in a model with foresight?

To answer this question, we first observe that even though the star is the unique efficient network, the payoff from being the center agent is $3\delta - 3c$, which is much smaller than the payoff from being a non-center agent (which equals $(\delta - c) + 2\delta^2$). Thus, in a model with foresight, player 1 may delete his link with agent 2 and wait for a chance to link with agent 3. However, agent 3 would rather that someone else be the center of the star; thus, when 3 is offered a chance to link with 1, he has an incentive to refuse this link in the hope that agent 1 will relink with agent 2 and that agent 2 will then become the center of the star. However, agent 2 will also have incentive not to become the center of the star. Thus, it is unlikely that forward-looking behavior will increase the chances of the star forming.

References

Anderlini, L., Ianni, A. (1996) Path Dependence and Learning from Neighbors. *Games and Economic Behavior* 13: 141–177.

Aumann, R.J., Myerson, R.B. (1988) Endogenous Formation of Links between Players and of Coalitions: An Application of the Shapley Value. In A. Roth (ed.) *The Shapley Value*, New York, Cambridge University Press.

Bala, V., Goyal, S. (2000) A Non-Cooperative Model of Network Formation, forthcoming in *Econometrica*.

Bolton, P., Dewatripont, M. (1994) The Firm as a Communication Network. *The Quarterly Journal of Economics* 109: 809–839.

Boorman, S. (1975) A Combinatorial Optimization Model for Transmission of Job Information through Contact Networks. *Bell Journal of Economics* 6: 216–249.

Dutta, B., Mutuswami, S. (1997) Stable Networks. *Journal of Economic Theory* 76: 322–344.

Dutta, B., van den Nouweland, A., Tijs, S. (1998) Link Formation in Cooperative Situations. *International Journal of Game Theory* 27: 245–256.

Ellison, G., (1993) Learning, Local Interaction and Coordination. *Econometrica* 61: 1047–1072.

Ellison, G., Fudenberg, D. (1995) Word-of-Mouth Communication and Social Learning. *The Quarterly Journal of Economics* 110: 93–126.

Goyal, S., Janssen, M. (1997) Non-Exclusive Conventions and Social Coordination. *Journal of Economic Theory* 77: 34–57.

Hendricks, K., Piccione, M., Tan, G. (1997) Entry and Exit in Hub-Spoke Networks. *The Rand Journal of Economics* 28: 291–303.

Jackson, M.O., Watts, A. (1999) The Evolution of Social and Economic Networks, forthcoming, *Journal of Economic Theory*.

Jackson, M.O., Wolinsky, A. (1996) A Strategic Model of Social and Economic Networks. *Journal of Economic Theory* 71: 44–74.

Keren, M., Levhari, D. (1983) The Internal Organization of the Firm and the Shape of Average Costs. *Bell Journal of Economics* 14: 474–486.

Montgomery, J. (1991) Social Networks and Labor Market Outcomes. *The American Economic Review* 81: 1408–1418.

Qin, C.Z. (1996) Endogenous Formation of Cooperation Structures. *Journal of Economic Theory* 69: 218–226.

Radner, R. (1993) The Organization of Decentralized Information Processing. *Econometrica* 61: 1109–1146.

Slikker, M., van den Nouweland, A. (1997) A One-Stage Model of Link Formation and Payoff Division. CentER Discussion Paper No. 9723.

Vazquez-Brage, M., Garcia-Jurado, I. (1996) The Owen Value Applied to Games with Graph-Restricted Communication. *Games and Economic Behavior* 12: 42–53.

A Theory of Buyer-Seller Networks

Rachel E. Kranton[1], **Deborah F. Minehart**[2]

[1] Department of Economics, University of Maryland, College Park, MD 20742, USA
[2] Department of Economics, Boston University, 270 Bay State Road, Boston, MA 02215, USA

This paper introduces a new model of exchange: networks, rather than markets, of buyers and sellers. It begins with the empirically motivated premise that a buyer and seller must have a relationship, a "link," to exchange goods. Networks – buyers, sellers, and the pattern of links connecting them – are common exchange environments. This paper develops a methodology to study network structures and explains why agents may form networks. In a model that captures characteristics of a variety of industries, the paper shows that buyers and sellers, acting strategically in their own self-interests, can form the network structures that maximize overall welfare.

JEL Classification: D00, L00

This paper develops a new model of economic exchange: networks, rather than markets, of buyers and sellers. In contrast to the assumption that buyers and sellers are anonymous, this paper begins with the empirically motivated premise that a buyer and a seller must have a relationship, or "link," to engage in exchange. Broadly defined, a "link" is anything that makes possible or adds value to a particular bilateral exchange. An extensive literature in sociology, anthropology, as well as economics, records the existence and multifaceted nature of such links. In manufacturing, customized equipment or any specific asset is a link between two firms.[1] Relationships with extended family members, co-ethnics, or "fictive kin" are links that reduce information asymmetries.[2] Personal connections between

---

We thank Larry Ausubel, Abhijit Banerjee, Eddie Dekel, Matthew Jackson, Albert Ma, Michael Manove, Dilip Mookherjee, two anonymous referees, and numerous seminar participants for invaluable comments. Rachel Kranton thanks the Russell Sage Foundation for its hospitality and financial support. Both authors are grateful for support from the National Science Foundation under Grants Nos. SBR9806063 (Kranton) and SBR9806201 (Minehart).

[1] For example, Brian Uzzi's (1996) study reveals the nature of links in New York City's garment industry. Links embody "fine-grained information" about a manufacturer's particular style. Only with this information can a supplier quickly produce a garment to the manufacturer's specifications.

[2] See, for example, Janet Tai Landa (1994), Avner Greif (1993), and Rachel E. Kranton (1996). These links are particularly important in developing countries, e.g. Hernando de Soto (1989). They also facilitate international trade (Alessandra Casella and James E. Rauch, 1997).

managers and bonds of trust are links that facilitate business transactions.[3] There is now a large body of research on how such bilateral relationships facilitate cooperation, investment, and exchange. Some research also considers how an alternative partner or "outside option" affects the relationship.[4] However, there has been virtually no attempt to examine the realistic situation in which both buyers and sellers may have costly links with multiple trading partners.

This paper develops a theory of investment and exchange in a *network*, where a *network* is a group of buyers, sellers, and the pattern of the links that connect them. An economic theory of networks must consider questions not encountered when buyers and sellers are assumed to be anonymous. Because a buyer can obtain a good from a seller only if the two are linked, the pattern of links affects competition for goods and the potential gains from trade. Many new questions arise: Given a pattern of links, how might exchange take place? Who trades with whom and at what "equilibrium" prices? Is the outcome of any competition for goods efficient? The link pattern itself is an object of study. What are the characteristics of efficient link patterns? What incentives do buyers and sellers have to build links, and when are these individual incentives aligned with social welfare?

Networks are interesting, and complex, exchange environments when buyers have links to multiple sellers and sellers have links to multiple buyers. We see multiple links in many settings. The Japanese electronics industry is famous for its interconnected network structure (e.g., Toshihiro Nishiguchi, 1994). Manufacturers work with several subcontractors, transfering know-how and equipment, and "qualify" these subcontracters to assemble specific final products and ship them to customers. Subcontractors, in turn, shift production to fill the orders of different manufacturers. Similarly, in Modena, Italy, the majority of artisans who assemble clothing for garment manufacturers work for at least three clients. These manufacturers in turn spread their work among many artisans (Mark Lazerson, 1993).[5] Annalee Saxenian (1994) attributes the innovative successes of Silicon Valley to its interconnected, rather than vertically integrated, industrial structure, and Allen J. Scott (1993) reaches a similar conclusion in his study of electronics and engineering subcontracting in the Southern Californian defense industry.

In this paper, we explore two reasons why networks emerge, one economic, the other strategic. First, networks can allow buyers and sellers collectively to pool uncertainty in demand, a motive we see in many of the above examples. When sellers have links to more buyers, they are insulated from the difficulties

[3] For a classic description see Stewart Macauley (1963). John McMillan and Christopher Woodruff (1999) show the importance of on-going relations between firms in Vietnam for the extension of trade credit.

[4] The second sourcing literature considers how an alternate source alters the terms of trade between a buyer and supplier. See for example, Joel S. Demski, David E. Sappington, and Pablo T. Spiller (1987), David T. Scheffman and Spiller (1992), Michael H. Riordan (1996) and Joseph Farrell and Nancy T. Gallini (1988). Susan Helper and David Levine (1992) study an environment where the "outside option" is a market.

[5] Elsewhere in the garment industry, we find a similar pattern (Uzzi, 1996, and Pamela M. Cawthorne, 1995).

facing any one buyer. And when buyers purchase from the same set of sellers, there is a saving in overall investment costs. As for the strategic motivation, multiple links can enhance an agent's competitive position. With access to more sources of supply (demand), a buyer (seller) secures better terms of trade.

To capture these motivations we specify a game where buyers form links, then compete to obtain goods from their linked sellers. We implicitly assume that agents do not act cooperatively; they cannot write state-contingent, long-term binding contracts to set links, future prices, or side payments.[6] We consider a stylized general setting: Sellers can each produce one (indivisible) unit of output. Buyers desire one unit each and have private uncertain, valuations for a good.[7] A buyer can purchase from a seller if and only if the two are linked. We then ask what is the relationship between agents' individual self-interests and collective interests? Can buyers and sellers, acting non-cooperatively to maximize their own profits, form a network structure that maximizes overall economic surplus?

To answer these questions, we first explore the relationship between the link pattern and agents' competitive positions in a network. We represent competition for goods by a generalization of an ascending-bid auction, analogous to the fictional Walrasian auctioneer in a market setting.[8] Our first set of results shows that this natural price formation process can lead to an efficient allocation of goods in a network. The buyers that value the goods the most obtain the goods, subject only to the physical constraints of the link patterns. Furthermore, the prices reflect the link pattern. A buyer's revenues are exactly the marginal social value of its participation in the network.[9]

Our main result shows that, when buyers compete in this way, their individual incentives to build links can be aligned with economic welfare. Efficient network structures are always an equilibrium outcome. Indeed, for small link costs, efficient networks are the only equilibria. These results may seem surprising in a setting where buyers build links strategically, and especially surprising in light of our finding that buyers may have very asymmetric positions in efficient networks. Yet, it is the *ex post* competition for goods that yields efficient

---

[6] Such contracts may be difficult to specify and enforce and are even likely to be illegal. An established literature in industrial organization considers how contractual incompleteness shapes economic outcomes (Oliver E. Williamson, 1975; Sanford J. Grossman and Oliver D. Hart, 1986; Hart and John Moore, 1988).

[7] This setting captures the characteristics of at least the following industries particularly well: clothing, electronic components, and engineering services. They share the following features: uncertain demand for inputs because of frequently changing styles and technology, supply-side investment in quality-enhancing assets, specific investments in buyer-seller relationships, and small batches of output made to buyers' specifications. In short, sellers in these industries could be described as "flexible specialists," to use Michael J. Piore and Charles F. Sabel's (1984) term. See above references for studies of apparel industries. Scott (1993), Nishiguchi (1994), and Edward H. Lorenz (1989) study the engineering and electronics industries in southern California, Japan and Britain, and France, respectively.

[8] This auction model can be used whenever there are multiple, interlinked buyers and sellers and has several desirable properties including ease of calculating payoffs.

[9] These revenues are robust to different models of competition. By the payoff equivalence theorem (Roger B. Myerson, 1981), any mechanism that allocates goods efficiently must yield the same marginal revenues. We discuss this point further below.

outcomes. Because of competition, no buyer can capture surplus generated by the links of other buyers. Rather, a buyer's profit is equal to its contribution to overall economic welfare.

By studying competitive buyers and sellers, this paper advances the economic theory of networks.[10] The most closely related work is by Matthew O. Jackson and Asher Wolinsky (1996) who examine strategic link formation in a general setting.[11] They find, using a value function that allocates network surplus to the nodes (players), that efficient networks need not be stable. In our economic environment agents face uncertainty, asymmetric information, and contractual incompleteness. These features constrain the possible allocations of surplus and make efficiency more difficult to achieve. Furthermore, we focus on a specific environment, that of buyers and sellers. The combinatoric methods we develop may be used to examine other bipartite settings, such as supervisory hierarchies in firms and international trading blocs.[12]

More generally, this research adds to our understanding of economic institutions. Following Coase (1937) economists have distinguished between market and non-market institutions. Networks are non-market institutions with important market-like characteristics: exchange is limited to linked pairs, but buyers and sellers may form links strategically and compete. The theory here captures both aspects of networks. We can use this theory to compare networks to other institutions on either side of the spectrum - markets and vertically integrated firms.[13] Furthermore, there are many institutional features of networks that can be built onto the basic structure developed here. We indicate directions for future research in the conclusion.

The rest of the paper is organized as follows. Section 1 constructs the basic model of networks and develops notions of efficiency and competition for the network setting. Sections 2 and 3 consider individual incentives to build links and the efficiency of link patterns. Section 4 concludes.

[10] By theory of "networks," we mean theory that explicitly examines links between individual agents. The word "networks" has been used in the literature to describe many phenomena. "Network externalities" describes an environment where an agent's gain from adopting a technology depends on how many other agents also adopt the technology (see Michael L. Katz and Carl Shapiro, 1994). In this and many other settings, the links between individual agents may be critical to economic outcomes, but have not yet been incorporated in economic modeling.

[11] Much previous research on networks (e.g. Myerson, 1977, and Bhaskar Dutta, Anne van-den-Nouweland, and Stef Tijs, 1998) employs cooperative equilibrium concepts. There is also now a growing body of research on strategic link formation (see e.g. Venkatesh Bala and Sanjeev Goyal, 1999; Jackson and Alison Watts, 1998). Ken Hendricks, Michele Piccione, and Guofo Tan (1997) study strategic formation of airline networks.

[12] In our analysis we use a powerful, yet intuitive, result from the mathematics of combinatorics known as the Marriage Theorem. With this Theorem we can systematically analyze bipartite network structures.

[13] See Kranton and Deborah F. Minehart (1999b).

1 Competition and Exchange in Buyer-Seller Networks

This section develops a theory of competition and exchange in networks. We begin with the basic model of buyers, sellers, and links. We then develop a model of competition in a network.

1.1 The Basic Model of Buyer-Seller Networks

There are $\overline{B}$ buyers, each of whom demands one indivisible unit of a good. We denote the set of buyers as **B**. Each buyer i, or b_i, has a random valuation v_i for a good. The valuations are independently and identically distributed on $[0, \infty)$ with continuous distribution F. We assume the distribution is common knowledge, and the realization of v_i is private information. There are $\overline{S}$ sellers who each have the capacity to produce one indivisible unit of a good at zero cost. We denote the set of sellers by **S**.

A buyer can obtain a good from a seller if and only if the two are *linked.* E.g., a link is a specific asset, and with this asset the buyer has a value $v_i > 0$ for the seller's good. We use the notation $g_{ij} = 1$ to indicate that a buyer i and a seller j are linked and $g_{ij} = 0$ when they are not. These links form a *link pattern*, or graph, **G**.[14] A *network* consists of the set of buyers and sellers and the link pattern.

In a network, the link pattern determines which buyers can obtain goods from which sellers; that is, the link pattern determines the feasible allocations of goods. An allocation **A** is *feasible* only if it respects the pattern of links. That is, a buyer i that is allocated seller j's good must actually be linked to seller j.[15] In addition, no buyer can be allocated more than one seller's good, and no seller's good can be allocated to more than one buyer.[16]

To tell us when an allocation of goods is feasible in a given network, we use the Marriage Theorem – a result from the mathematics of combinatorics and an important tool for our analysis.[17] The theorem asks: Given populations of women and men, when it is possible to pair each woman with a man that she knows, and no man or woman is paired more than once. In our setting, the buyers are "women," the sellers are "men," and the links indicate which women know which men. To use this theorem, it is convenient to define the set of sellers linked to a particular set of buyers, and vice versa. For a subset of buyers $\mathcal{B}$, let $L(\mathcal{B})$ denote the set of sellers linked to any buyer in $\mathcal{B}$. We call $L(\mathcal{B})$, $\mathcal{B}$'s *linked set* of sellers and say the buyers in $\mathcal{B}$ are linked, *collectively,* to these sellers.

[14] It is often convenient to write **G** as a $\bar{B} \times \bar{S}$ matrix where the element g_{ij} indicates whether buyer i and seller j are linked.

[15] An *allocation of goods*, **A**, can also be written as a $\bar{B} \times \bar{S}$ matrix, where $a_{ij} = 1$ when b_i is allocated a good from s_j and $a_{ij} = 0$ otherwise.

[16] Formally, an allocation **A** is feasible given graph **G** if and only if $a_{ij} \leq g_{ij}$ for all i,j and for each buyer i, if there is a seller j such that $a_{ij} = 1$ then $a_{ik} = 0$ for all $k \neq j$ and $a_{lj} = 0$ for all $l \neq i$.

[17] Also known as Hall's Theorem, see R.C. Bose and B. Manvel (1984, pp. 205–209) or other combinatorics/graph theory text for an exposition.

Similarly, for a subset of sellers $\mathcal{S}$, let $L(\mathcal{S})$ denote these sellers' *linked set* of buyers.

The Marriage Theorem. *For a subset of sellers $\mathcal{S}$ containing S sellers and for a subset of buyers $\mathcal{B}$ containing B buyers, there is a feasible allocation of goods such that every buyer in $\mathcal{B}$ obtains a good from a seller in $\mathcal{S}$ if and only if every subset $\mathcal{B}' \subseteq \mathcal{B}$ containing k buyers is linked, collectively, to at least k sellers in $\mathcal{S}$, for each $k, 1 \leq k \leq B$.*[18]

To determine whether an allocation of goods is feasible in a given network, we simply use the counting algorithm provided by the Marriage Theorem. Our first example demonstrates.

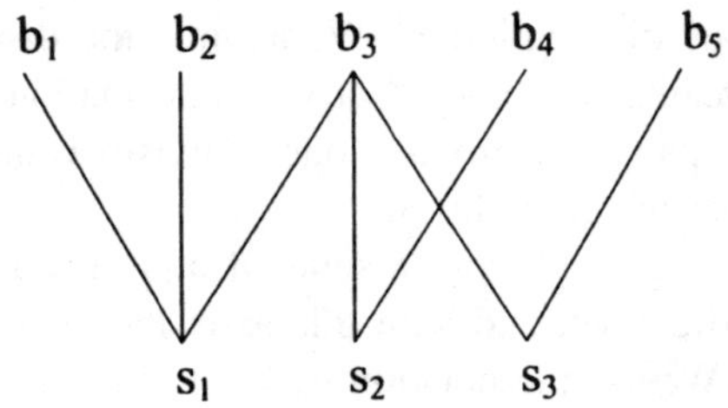

Fig. 1. Example of a buyer-seller network

Example 1. Feasible Allocations of Goods in a Network. In Fig. 1 ask whether there is a feasible allocation in which buyers b_1, b_2, and b_3 all obtain goods. Eyeing the graph, it is clear the answer is no. The Marriage Theorem gives us the following method to prove such an outcome in general. Take the set of buyers $\{b_1, b_2, b_3\}$. Consider all subsets with $k = 1, 2, 3$ buyers. Then count the number of sellers in their linked sets. In this network, there are three sellers in the linked set of $\{b_1, b_2, b_3\}$, that is, $L(\{b_1, b_2, b_3\}) = \{s_1, s_2, s_3\}$. The subset $\{b_1, b_2\}$, however, has only one seller in its linked set: $L(\{b_1, b_2\}) = \{s_1\}$. It must contain at least two sellers to satisfy the theorem. Therefore, there is no feasible allocation in which buyers 1, 2, and 3 all obtain goods. In contrast, there is a feasible allocation in which buyers b_2, b_3, and b_4 all obtain goods (b_2 from s_1, b_3 from s_3, and b_4 from s_2). The condition of the theorem is satisfied; there are three sellers in the linked set of $\{b_2, b_3, b_4\}$. There are two sellers in the linked set of each subset of two buyers, and each single buyer is linked to at least one seller.

1.2 Gains from Exchange and Efficient Allocations of Goods

Economic surplus is generated when buyers procure goods from sellers. The level of surplus will depend on which buyers obtain goods, since buyers' valuations

[18] Note that not all sellers need to be paired to a buyer, and a necessary condition for the proposition to hold is that $S \geq B$.

differ. Let $\mathbf{v} = (v_1, ..., v_{\overline{B}})$ be a vector of buyers' realized valuations. The economic surplus associated with an allocation $\mathbf{A}$ is the sum of the valuations of the buyers that secure goods in $\mathbf{A}$. We denote the surplus as $w(\mathbf{v}, \mathbf{A})$.[19]

We focus on the allocations that yield the highest possible surplus, given the network link pattern. As we saw above, the link pattern constrains the allocation of goods. It may not be feasible for a buyer to obtain a good even though it has a high valuation. Of the feasible allocations, an *efficient allocation* yields the highest surplus from exchange. In this allocation, the buyers with the highest valuations of goods obtain goods whenever possible given the link pattern.[20] We denote the efficient allocation by $\mathbf{A}^*(\mathbf{v}; \mathbf{G})$.

The next example demonstrates the efficient allocation of goods in a network for a particular realization of buyer's valuations. In this allocation, a buyer that has a high valuation does not obtain a good. Yet, the allocation yields the highest possible surplus, given the pattern of links.

Example 2. [Efficient Allocation of Goods in a Network.] Consider again the network in Fig. 1. Suppose buyers' realized valuations have the following order: $v_1 > v_2 > v_3 > v_4 > v_5$. For these valuations, the efficient allocation pairs b_1 with s_1, b_3 with s_3, and b_4 with s_2. The surplus from this allocation is $v_1 + v_3 + v_4$. The only other allocations that could yield higher surplus would allocate goods to buyers $\{b_1, b_2, b_3\}$ or $\{b_1, b_2, b_4\}$. But, using the Marriage Theorem, we see that these allocations are not feasible given this link pattern.

By taking the efficient allocation of goods for each realization of buyers' valuations, we can determine the highest possible expected surplus from exchange in a given network. Let $H(\mathbf{G})$ be the *maximal gross economic surplus* obtainable for a link pattern $\mathbf{G}$.[21] We have

$$H(\mathbf{G}) \equiv \mathbf{E}_{\mathbf{v}} \left[\mathbf{w}(\mathbf{v}, \mathbf{A}^*(\mathbf{v}; \mathbf{G})) \right]$$

where the expectation is taken over all the possible realizations of buyers' valuations.[22]

1.3 Competition in a Network

With the basic model in hand, we now develop a model of competition in a network. We use this model to examine how link patterns affect prices and allocations of goods. The generalization of an English, or ascending-bid, auction that

[19] We can write $w(\mathbf{v}, \mathbf{A}) = \mathbf{v} \cdot \mathbf{A} \cdot \mathbf{1}$, where $\mathbf{1}$ is an $\overline{S} \times 1$ matrix where each element is 1.

[20] Several allocations may yield the same surplus. However, without loss of generality, we can restrict attention to *equivalent allocations* - allocations in which the same subsets of buyers obtain goods and the same subset of sellers produce goods. This is because buyer i earns v_i in any allocation in which it obtains a good and seller j's cost does not depend on which buyer obtains its good.

[21] Later, when we introduce link and investment costs, we will distinguish this value from *net* economic surplus.

[22] This expectation is straightforward to calculate. The efficient allocation depends only on the *ordering* of buyers' valuations, not their absolute levels. The expectation, $H(\mathbf{G})$, can be written as an expectation over the order statistics of the distribution F (see Kranton and Minehart, 2000b).

we construct allows for easy, reasonable, yet exact analysis where we can "see" the competition.[23] For any realization of buyers' valuations, we can compute the equilibrium allocation, prices, and division of surplus.

We view the auction as an abstraction of the way goods are actually allocated and prices negotiated in a network setting, in a sense similar to the fiction of the Walrasian auctioneer.[24] As in a market, the outcome of the competition has several desirable features. First, the allocation of goods is efficient. Despite that buyers' valuations are private information, the buyers with the highest valuations obtain goods whenever possible given the link pattern. Second, the resulting payoffs are "stable;" no buyer and seller can renegotiate the prices or allocation to their mutual benefit.[25] The prices themselves reflect the social opportunity costs of exchange. We will see below that these properties are critical for buyers to have the correct incentives to build links.

We provide an overview of the auction here and refer the reader to Appendix A for a formal analysis.

Recall, first, a standard ascending-bid auction with one seller. The price rises from zero, and each buyer decides at each moment whether to remain in the bidding or drop out. As is well known, it is a weakly dominant strategy for each buyer to remain in the bidding until the price reaches its valuation. The price then rises until it reaches the second highest valuation, and the buyer with the highest valuation secures the good at this price. As long as the number of buyers in the bidding exceeds the supply (of one), the price rises. As soon as the number of bidders equals the supply, the auction "clears."

In our generalization, sellers simultaneously hold ascending-bid auctions, where the going price is the same across all sellers. As this price rises from zero, each buyer decides whether to drop out of the bidding of each of its linked sellers' auctions. The price rises until enough buyers have dropped out so that there is a subset of sellers for whom "demand equals supply." We call such a subset a *clearable set of sellers*. The auctions of these sellers "clear" at the current price. (Appendix A shows the clearing rule is well-defined). If there are remaining sellers, the price continues to rise until all sellers have sold their goods. We prove that it is an equilibrium (following elimination of weakly dominated strategies) for each buyer to remain in the bidding in each of its linked sellers' auctions up to its valuation of a good.

---

[23] Gabrielle Demange, David Gale, and Marilda Sotomayor (1986) develop an ascending-bid auction for multiple buyers and sellers and general preferences. They do not analyze, however, strategic bidding. We solve for a perfect Bayesian equilibrium of the auction game. Independently, Faruk Gul and Ennio Stacchetti (2000) also show that truthful bidding is an equilibrium outcome of such an ascending-bid auction.

[24] There are, however, some instances where auctions are actually used as in defense subcontracting and the shoe industry in Brazil (Hubert Schmitz, 1995, p. 14).

[25] Because only buyer-seller pairs generate surplus, the outcome is also in the core (Lloyd S. Shapley and Martin Shubik, 1972). Kranton and Minehart (2000a) considers general properties of pairwise stable payoffs in networks. The auction yields the lowest payoffs for sellers in the set of all pairwise stable payoffs.

The next example illustrates the auction and demonstrates how a link pattern shapes the competition for goods.

Example 3. [Auction Representation of Competition in a Network.] In Fig. 1, we suppose that buyers' valuations are realized in the order $v_1 > v_2 > v_3 > v_4 > v_5 > 0$. At $p = 0$, the demand exceeds supply for all subsets of sellers. The price rises until it reaches v_5 when b_5 drops out of the auction of s_3. Now $\{s_2, s_3\}$ constitutes a clearable set of sellers; i.e., there are only two buyers (b_3 and b_4) remaining in the bidding for these sellers' goods and there is a feasible allocation in which these buyers obtain goods from these sellers. The buyers each pay a price of $p = v_5$. Buyer 3 purchases from s_3, while buyer 4 purchases from s_2. There are still two buyers, b_1 and b_2, who demand the remaining good of s_1. Since there is excess demand for s_1, the price continues to rise. The price rises until it reaches v_2 at which point b_2 drops out. Buyer 1 purchases from s_1 at $p = v_2$. Note that the auction achieves the efficient allocation of goods. As we discussed in the previous example, the efficient allocation given the link pattern is for b_1, b_3, and b_4 to purchase goods.

For any network, we can easily calculate the payoffs from this equilibrium of the auction; indeed, the ease of calculation is a useful feature of this model of competition. Given a link pattern **G**, let $V_i^b(\mathbf{G})$ denote the expected payoff to buyer i in this equilibrium, and let $V_j^s(\mathbf{G})$ denote the expected payoff to seller j. We will refer to these payoffs as "*V-payoffs.*" We can calculate firms' V-payoffs using the order statistics of the distribution F.[26] Let $X^{n:B}$ be the random variable which is the nth highest valuation of the B buyers, that is $X^{n:B}$ is the nth order statistic. The following example demonstrates the calculation of expected payoffs.

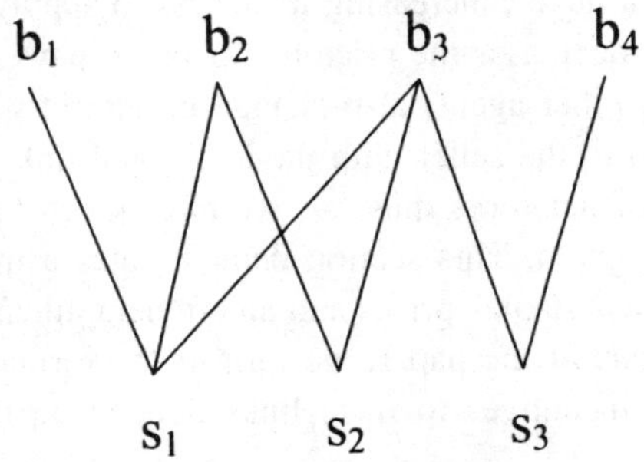

Fig. 2. Network for Example 4

Example 4. [Expected Payoffs from Competition in a Network.] In Fig. 2, for any realization of buyers' valuations, the price will rise until it reaches the lowest valuation of the buyers, $X^{4:4}$. The three buyers with the top three valuations will

[26] See Kranton and Minehart (2000b) for the proof that we can restrict attention to the ordering of buyers' valuations.

then purchase at that price. A buyer expects to have the highest, second highest, third highest, or lowest valuation with equal probability. The V-payoffs for each buyer are therefore $\frac{1}{4}EX^{1:4} + \frac{1}{4}EX^{2:4} + \frac{1}{4}EX^{3:4} - \frac{3}{4}EX^{4:4}$.

This representation of competition in a network has several properties that any frictionless model of network competition should satisfy. First, the equilibrium allocation of goods is efficient. The buyers that value goods the most obtain goods, subject to the constraints of the link pattern.[27] Example 3 demonstrates this feature. Despite that buyers have private information, buyers bid truthfully. They do not hedge their bids or adopt other strategies that would lead to inefficient allocation of goods. Second, the allocation and prices together are *pairwise stable.* That is, the surplus that any linked buyer and seller could generate by exchanging a good does not exceed their joint payoffs. Intuitively, no linked buyer and seller, or indeed any coalition of agents, can renegotiate and strike a better deal.

Proposition 1. *Given the link pattern, for each realization of buyers' valuations, the equilibrium allocation of goods is efficient and the allocation and prices are pairwise stable.*

The intuition behind these two properties is that, in this equilibrium, a buyer pays a price exactly equal to the social opportunity cost of obtaining a good. The price a buyer pays does not depend on its own valuation, so there is no incentive not to bid truthfully. In equilibrium, a buyer i pays the seller the valuation of the "next-best" buyer, that is, the highest valuation buyer that would have obtained a good in i's place. This price is just high enough so that no competing buyer will want to offer a seller a higher price. For example, in the equilibrium described above for Fig. 1, buyer b_3 pays the valuation of b_5.[28]

A third feature of this model of competition is that the payoffs in this equilibrium satisfy desirable comparative statics properties of "supply and demand" in a link pattern. For a buyer, increasing its access to supply by adding a link to another seller weakly decreases the price it expects to pays, and vice versa for a seller. The payoffs of other agents also change in natural ways, e.g. the payoffs of other buyers linked to the seller with the additional link weakly decrease.[29]

In our treatment of networks thus far, we have taken the links that connect buyers and sellers as given. This section demonstrates a model of competition that uniquely associates "stable" prices and an efficient allocation with every link pattern. In the next part of the paper, we turn to the formation of the network. We will see that the incentives to form links depend on the properties of the competitive process.

---

[27] Since all efficient allocations are unique up to equivalence, the auction selects a unique allocation.

[28] These equilibrium properties are well known features of Vickrey-Clark-Grove mechanisms. For discussion in pairwise settings, see Herman B. Leonard (1983) and Alvin E. Roth and Marilda A. Oliveira Sotomayor (1990). Another way to describe this result is that buyers pay the lowest "Walrasian" price for a clearable set of sellers. That is, for any clearable set of sellers, there is a continuum of prices such that demand equals supply. The equilibrium of the auction picks out the lowest price.

[29] Kranton and Minehart (2000a) provides a general analysis of how changes in the link pattern affect different agents' payoffs.

2 Strategic Link Formation and Efficient Link Patterns

In this section we examine buyers' strategic incentives to form links. We consider, in particular, the relationship between buyers' individual incentives and overall economic welfare. We examine the simplest setting: buyers choose links to an exogenously given set of productive sellers.[30] In a two stage game, buyers first simultaneously choose links. Second, buyers learn their valuations and compete in the auction specified above. As mentioned earlier, implicit in this structure is that the buyers cannot use long-term complete contingent contracts to assign individual links, future prices, or allocations. We think of investments in links as long-run investments in anticipation of short-run uncertain demand for goods.

In this analysis, we assume that there are welfare gains when buyers share the productive capacity of sellers. We call these welfare gains *economies of sharing* and assume that there are fewer sellers than buyers, $\bar{S} < \bar{B}$. The gain from fewer sellers arises from the variability of buyers' valuations and the implicit assumption that productive capacity is costly. (In the next section, we explicitly introduce these costs.) To see why, consider the case of three buyers. If sellers' capacity is costly, it may be optimal for the buyers to share the capacity of just two sellers. The capacity could be allocated to the two buyers that ultimately have the highest valuations for goods. While only two buyers obtain goods, there is a savings in the cost of one unit of capacity.[31] The same economies of scale underly the "repairman problem" (William Feller, 1950; Michael Rothschild and Gregory J. Werden, 1979) where agents use repair services only when needed. Similar economies are also exploited by intermediary firms that hold inventories (see Daniel F. Spulber, 1996).[32]

In a network, the link pattern determines the extent to which economies of sharing are realized. For the economies to be fully realized in our three-buyer-two-seller example, there must be enough links so that whichever buyers have the highest two valuations they can always obtain goods. When links are costly, it might not be optimal, from a social welfare point of view, to fully realize the economies of sharing. The gains from adding one more link need not exceed its cost.

We consider *efficient link patterns*, those that balance the link costs with the expected gains from exchange. Let $W(\mathbf{G})$ denote the *net economic surplus* from

[30] In Section III, we endogenize the number of productive sellers. Other questions, such as sellers' incentives to build links to buyers are also of interest. The current analysis can be used to give insight to such network formation games. For instance, if the sellers had uncertain costs and invested in links and if the buyers had fixed common valuations, then our subsequent anaysis could be carried out identically just with buyers' and sellers' places exchanged. We could also easily examine a setting where buyers and sellers invest in individual links.

[31] If there were three units of capacity and each buyer always purchased a good, the expected surplus from exchange would be $3(EX^{1:1})$; that is, three times the mean of buyers' valuations. If there were only two units of capacity and only the buyers with the top two valuations obtained goods, the expected surplus from exchange would be $EX^{1:3} + EX^{2:3}$. While this surplus is smaller than $3(EX^{1:1})$, overall economic welfare may be higher because two units of productive capacity would be less costly than three.

[32] See also Dennis W. Carlton (1978) who assumes that firms must make irreversible decisions before demand uncertainty is resolved.

a link pattern **G**. This net surplus consists of the maximal gross surplus, $H(\mathbf{G})$, minus total link costs. Recall that $H(\mathbf{G})$ is the highest possible surplus from exchange given the link pattern **G**. It is obtained from the efficient allocation of goods for that link pattern. We have

$$W(\mathbf{G}) \equiv \mathbf{H}(\mathbf{G}) - \mathbf{c}\sum_{\mathbf{i=1}}^{\overline{\mathbf{B}}}\sum_{\mathbf{j=1}}^{\overline{\mathbf{S}}} \mathbf{g_{ij}}$$

where recall $g_{ij} = 1$ when buyer i is linked to seller j and $g_{ij} = 0$ otherwise. We say a link pattern **G** is *efficient* if it yields the highest net economic surplus of all graphs.[33]

Our central question is whether buyers, acting noncooperatively, can form efficient link patterns.

2.1 A Network Formation Game with Exogenous Sellers

We consider the following network formation game.
Stage One: Buyers simultaneously choose links to sellers. A buyer incurs a cost $c > 0$ for each link. Restricting attention to pure strategies, we describe the links of buyer i by the vector $\mathbf{g}_i = (g_{i1}, ..., g_{i\overline{S}})$, where $g_{ij} = 1$ when buyer i forms a link to seller j, and $g_{ij} = 0$ otherwise. The buyers' links form a link pattern **G**,[34] and we assume **G** is observable to all players.
Stage Two: Each buyer b_i privately learns its valuation, v_i, of a good. Buyers compete in the auction specified above. We consider the equilibrium in which buyers bid up to their valuations, and summarize the outcome in buyers' V-payoffs. A buyer's final payoff, its *profit*, is its V-payoffs minus its link costs. For b_i, profits are $\Pi_i^b(\mathbf{G}) = \mathbf{V_i^b}(\mathbf{G}) - \mathbf{c}\sum_{\mathbf{j=1}}^{\overline{\mathbf{S}}} \mathbf{g_{ij}}$. For s_j, profits are $V_j^s(\mathbf{G})$.

We solve for a perfect Bayesian equilibrium of this game. In the first stage, a buyer's choice of links must maximize its expected profits, given the strategies of other buyers. A buyer cannot have an incentive to add, break, or rearrange any of its links. We say a link profile $\mathbf{G}^*$ is an equilibrium strategy profile if and only if for each buyer b_i

$$\mathbf{g}_i^* = \arg\max_{\mathbf{g}_i} V_i^b(\mathbf{g}_i, \mathbf{g}_{-i}^*) - c\sum_{j=1}^{\overline{S}} g_{ij}$$

2.2 Efficient Networks and Equilibrium

We show here our main result: Efficient link patterns are always equilibrium outcomes of the game. The second-stage competition for goods aligns the incentives

[33] Below we derive the structure of efficient link patterns using the Marriage Theorem.

[34] Again, we can write this link pattern as the $\overline{B} \times \overline{S}$ matrix $\mathbf{G} = \left[\mathbf{g_{ij}}\right]$.

of buyers to build links with the social goal of welfare maximization. This result is presented below as Proposition 2.

The result follows from our assumption that buyers compete for goods. In our model of competition, the resulting allocation of goods is efficient, given the link pattern. The maximal surplus from exchange for the network is achieved. Furthermore, the price a buyer pays is equal to the social opportunity cost of obtaining a good. With these properties, buyers' competitive payoffs are exactly equal to the contribution of their links to economic welfare. That is, if we remove any number of a buyer's links (holding constant the rest of the link pattern), the loss in a buyer's V-payoffs is the loss in gross economic surplus. The next example illustrates this outcome. The formal result follows.

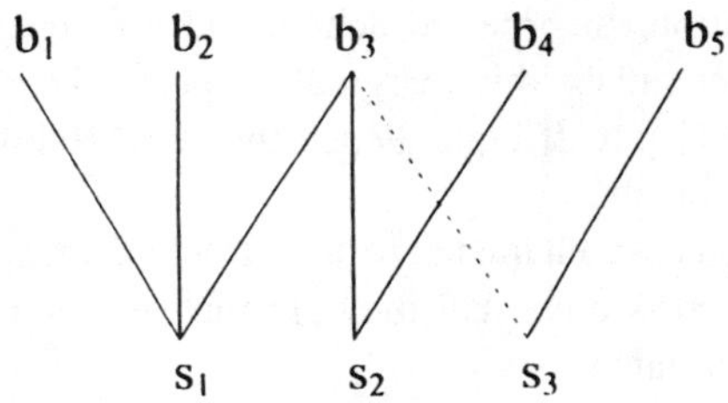

Fig. 3. Removing a link

Example 5. [Competitive Buyers Earn Social Value of Links.]Consider the link pattern in Fig. 3 and suppose buyers' realized valuations are in the order $v_1 > v_2 > v_3 > v_4 > v_5 > 0$. Let us compare the difference in surplus from exchange and the difference in buyer 3's payoffs if we remove buyer 3's link to seller 3. With this link, in the efficient allocation and in the auction, buyers 1, 3, and 4 obtain goods, yielding a total surplus of $v_1 + v_3 + v_4$. Buyer 3 pays a price equal to v_5, and its payoffs are $v_3 - v_5$. When we remove the link, buyer 5 purchases from seller 3, buyer 3 purchases from seller 2, and buyer 4 no longer purchases. The total surplus from exchange is now $v_1 + v_3 + v_5$, and the reduction in the total surplus is $v_4 - v_5$. Buyer 3 pays a higher price to obtain a good - rather than pay a price of v_5, it now pays $p = v_4$ giving it a payoff of $v_3 - v_4$. The reduction in its payoff is $v_4 - v_5$, and this amount is exactly equal to the reduction in total surplus.

Formally, we have:

Lemma 1. *Consider a link pattern* $\mathbf{G}$. *Remove any number of buyer i's links to create a new pattern* $\mathbf{G}'$. *The difference in buyer i's V-payoffs is the same as the difference in expected gross surplus:* $V_i^b(\mathbf{G}) - \mathbf{V_i^b}(\mathbf{G}') = \mathbf{H}(\mathbf{G}) - \mathbf{H}(\mathbf{G}')$. *Therefore,* $\Pi_i^b(\mathbf{G}) - \Pi_\mathbf{i}^\mathbf{b}(\mathbf{G}') = \mathbf{W}(\mathbf{G}) - \mathbf{W}(\mathbf{G}')$.

(Appendix B provides proof of this Lemma and subsequent results.)

That efficient link patterns are equilibrium outcomes follows directly from this result. Consider any efficient link pattern,[35] and ask whether any buyer has an incentive to deviate. The answer is no. By keeping its links in place, the buyer makes the largest contribution possible to surplus from exchange, and the buyer earns all this additional surplus in its V-payoffs. In an efficient link pattern, this additional surplus exceeds the link costs.

Proposition 2. *For any c, each efficient link pattern is an equilibrium outcome of the game.*

Proposition 2 shows that when buyers compete for goods, networks can be formed efficiently. This result would hold for any competitive process that yields an efficient allocation of goods and in which buyers' revenues are the marginal surplus from exchange. Moreover, these revenues are not special. When buyers have private information, in order to achieve an efficient allocation of goods, buyers' revenues must satisfy this marginal property. This requirement follows from Myerson's (1981) payoff equivalence theorem. Below we discuss further the robustness of our results.

In the next sections we characterize the structure of efficient networks and show that when link costs are small they are the only equilibrium outcomes of the network formation game.

2.3 The Structure of Efficient Networks and a Uniqueness Result

Efficient link patterns balance the cost of links with ex post gains from exchange. When link costs are small, a network should have enough links so that the buyers with the highest valuations can all obtain goods. All economies of sharing should be realized. In a network with three buyers and two sellers, for example, any set of two buyers should all be able to obtain goods. We say such as network is *allocatively-complete* (AC) and characterize it formally as follows:

Definition 1. *A network of buyers and sellers* $(\mathbf{B}, \mathbf{S})$ *is allocatively-complete if and only if for every subset of buyers* $\mathcal{B} \subset \mathbf{B}$ *of size* $\overline{S}$ *there is a feasible allocation such that every* b_i *in* $\mathcal{B}$ *obtains a good.*

A network where all the buyers are linked to all the sellers is, obviously, allocatively complete. When $c = 0$, this network is efficient. When $c > 0$, however, this network is not efficient. As we show next, allocative completeness can be achieved with fewer links.

Least-link allocatively complete (LAC) networks achieve all the economies of sharing with the minimal number of links. Using the Marriage Theorem we show that in these networks each seller has exactly $\overline{B} - \overline{S} + 1$ links. We see how these links must be "spread out" so that whichever buyers have the top valuations, there is a feasible allocation in which all these buyers obtain goods.

[35] We will see later that there are generally several efficient link patterns for each specification of the model's primitives.

There are many ways to distribute these links among buyers, and some buyers can have more links than others.

Proposition 3. *In a LAC network of buyers and sellers* (**B**, **S**)*, each seller is linked to exactly* $\overline{B} - \overline{S} + 1$ *buyers. Each buyer has from 1 to* $\overline{S}$ *links.*

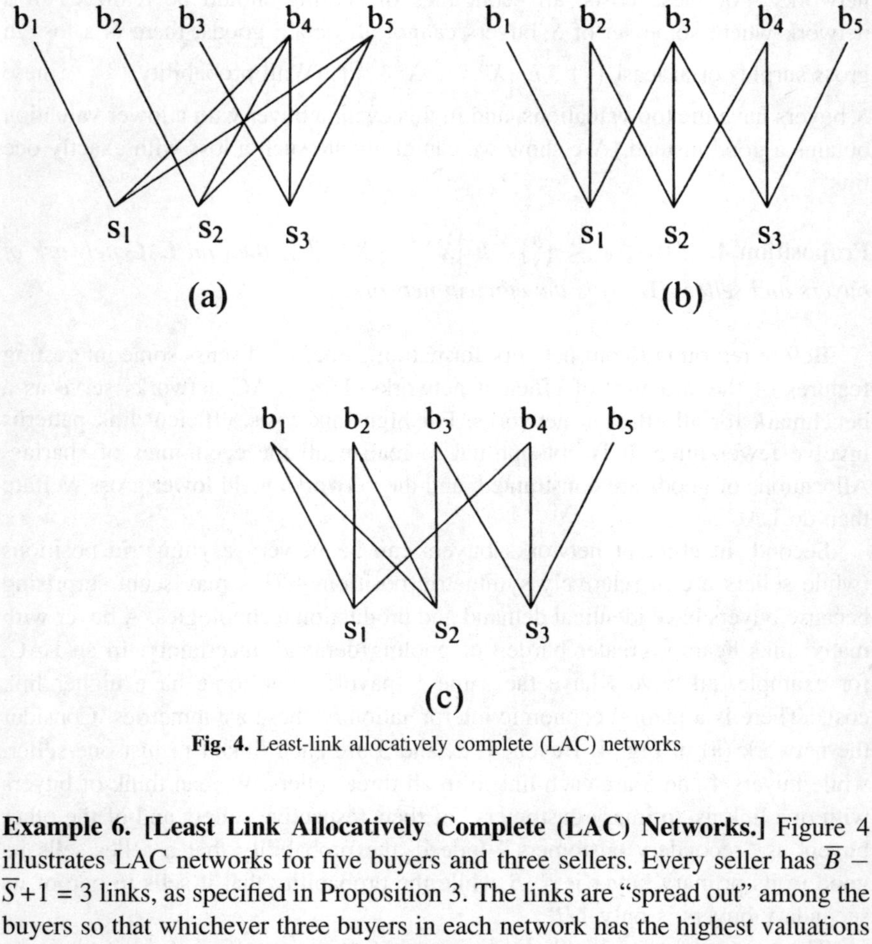

Fig. 4. Least-link allocatively complete (LAC) networks

Example 6. [Least Link Allocatively Complete (LAC) Networks.] Figure 4 illustrates LAC networks for five buyers and three sellers. Every seller has $\overline{B} - \overline{S} + 1 = 3$ links, as specified in Proposition 3. The links are "spread out" among the buyers so that whichever three buyers in each network has the highest valuations ex post, there is a feasible allocation in which these three obtain goods. The Figure shows that 3 links per seller is sufficient for allocative completeness. This result is easiest to see in network (a). The first three buyers are each linked to a distinct seller. This set of three buyers can then always obtain goods. The remaining two buyers are each linked to every seller. With these links the Marriage Theorem is satisfied for any set of three buyers. Therefore, any set of three buyers can always obtain goods. To see that 3 links per seller is necessary for allocative completeness, consider the possibility (in any of the networks in the Figure) that some seller j only has 2 links. This means that seller j is not linked to 3 buyers. When these particular 3 buyers have the top valuations, they cannot all obtain goods, and the network is not allocatively complete. Note that there is more than one way for sellers' links to be placed. Buyers may have different numbers of links. From the point of view of social welfare, however, there is no distinction between these networks.

We identify a range of small link costs where LAC networks are the efficient networks. For these costs, all economies of sharing should be realized. In a network where some set of $\bar{S}$ buyers cannot all obtain goods, there is a loss in gross surplus of at least $\binom{\bar{B}}{\bar{S}}^{-1} E\left[X^{\bar{S}:\bar{B}} - X^{\bar{S}+1:\bar{B}}\right]$. (With probability $\binom{\bar{B}}{\bar{S}}^{-1}$ these $\bar{S}$ buyers have the top valuations, and in this event a buyer with a lower valuation obtains a good instead.) We show we can eliminate such a loss with exactly one link.

Proposition 4. *If* $0 < c \leq \binom{\bar{B}}{\bar{S}}^{-1} E\left[X^{\bar{S}:\bar{B}} - X^{\bar{S}+1:\bar{B}}\right]$, *then an LAC network of buyers and sellers* (**B**, **S**) *is the efficient network.*

Before returning to our network formation game, we discuss some interesting features of the structure of efficient networks. First, LAC networks serve as a benchmark for all efficient networks. For high link costs, efficient link patterns involve fewer links. It is not optimal to realize all the economies of sharing. Allocations of goods are constrained, and the networks yield lower gross welfare than do LAC's.

Second, in efficient networks buyers can be in very asymmetric positions (while sellers are in relatively symmetric positions). This may seem surprising because buyers have identical demand and production technologies. A buyer with many links bears a greater burden of pooling demand uncertainty. In an LAC, for example, all buyers have the same V-payoffs, but some have higher link costs. There is a natural economic interpretation of these asymmetries. Consider the network (a) in Fig. 4. Buyers 1, 2, and 3 are each linked to just one seller, while buyers 4 and 5 are each linked to all three sellers. We can think of buyers with one link as "primary customers" of their respective sellers and of the other buyers as "secondary customers." Indeed, the probability that a seller sells its good to its primary buyer is $\overline{S}/\overline{B}$ while the probability that it sells to one of its secondary buyers is only $1/\overline{B}$.

These asymmetries highlight our result that every efficient network is an equilibrium. No matter how asymmetric are buyers' profits, a buyer with many links is willing to invest in these links because his V- payoff incorporates their value to economic welfare. Given the link holdings of the other buyers, this is the best the buyer can do.[36]

[36] An interesting direction for future research would be to explore how buyers compete for these different positions in a network. Consider sequential link investments by buyers. By investing early, a buyer might be able to establish itself as a primary customer.

We next show that for the range of small link costs discussed above, LAC networks are the unique equilibrium outcomes of the game.[37] Therefore, efficient link patterns are the only equilibria for this range of link costs.

Proposition 5. *For* $0 < c \leq \binom{\bar{B}}{\bar{S}}^{-1} E\left[X^{\bar{S}:\bar{B}} - X^{\bar{S}+1:\bar{B}}\right]$, *only efficient link patterns, that is, LAC link patterns, are equilibrium outcomes of the game.*

For this range of link costs, some buyer has an incentive to add or break a link in any network that is not LAC. There are two types of non-LAC networks to consider. First, the network could be allocatively complete, but with more links than an LAC. In this case, a buyer would have an incentive to cut a link. We show that there is always at least one link that can be removed with no change in the gross surplus from exchange.[38] By Lemma 1, if the buyer cuts this link, its profits increase exactly by c, the gain in welfare. For $c > 0$, then, a buyer would have an incentive to cut this "redundant" link.

Second, the network might not be allocatively complete. In this case, some buyer would have an incentive to add a link. We know that in non-AC networks, there is some set of $\bar{S}$ buyers that cannot all obtain goods. When these buyers have the top valuations, at least one of them will not obtain a good, even though it values a good more than other buyers. We show that it is possible for at least one buyer to add a link and obtain a good in this event, when it would not otherwise. Importantly, the buyer can achieve this greater access to goods without any change in other buyers' links. The buyer earns a gain in revenues of at least $\binom{\bar{B}}{\bar{S}}^{-1} E\left[X^{\bar{S}:\bar{B}} - X^{\bar{S}+1:\bar{B}}\right]$. Since a buyer's gain in revenues is exactly equal to its contribution to economic surplus, it is also efficient for a buyer to add this link.

2.4 Discussion of Efficiency Results

We discuss briefly here the robustness of our equilibrium results. Efficient networks would be equilibrium outcomes for any model of competition that yields an efficient allocation of goods and where buyers earn the marginal surplus from exchange. Models of competition that do not share these features, however, could lead to inefficient networks.

First, in a setting where competition does not yield an efficient allocation of goods, the reduction in surplus would lead to suboptimal investment in links. Depending on the features of competition, more subtle distortions in incentives might also be associated with allocation inefficiencies.[39] However, absent time delays or other frictions, we posit that any reasonable model of competition

[37] Although there may be several LAC link patterns, the equilibrium is unique in the sense that every equilibrium outcome involves an LAC pattern.

[38] That is, we show that for every AC network, there is an LAC network that is a subgraph.

[39] Buyers may build extra links to affect their bargaining position, for example.

should yield an efficient allocation of goods. Otherwise, some buyer and seller could renegotiate the allocation and terms of trade to their mutual benefit.[40]

Second, if buyers receive less than the full marginal value of an exchange, they could have insufficient incentives to invest in links. Setting aside the problem of achieving an efficient allocation, a priori, there could be any split of the marginal surplus from an exchange. In general, when the split of surplus is less than the share of investment, there would be underinvestment in links. This suggests an efficiency argument that the split of surplus should match the investment environment. The division of the surplus in the auction fits our investment environment because buyers bear the entire cost of links.

We next consider a more complex network formation game, where both sellers and buyers make investments in the network.

3 Network Formation with Endogenous Sellers

In this section we study network formation when productive capacity is costly and the set of sellers that invest in capacity is endogenous. We develop a network formation game, define efficient networks, and analyze equilibrium outcomes. We identify two reasons why networks may be formed inefficiently in this more complex environment.

3.1 The Game with Sellers' Investments

Stage One: Buyers simultaneously choose links to sellers and incur a cost $c > 0$ per link. As before, let $\mathbf{g}_i = (g_{i1}, ..., g_{i\overline{S}})$ denote buyer i's strategy, and let $\mathbf{G}$ denote the link pattern. When buyers choose links, sellers simultaneously choose whether to invest in assets that costs $\alpha > 0$. This asset allows it to produce one indivisible unit of a good at zero marginal cost for any linked buyer. A seller that does not invest cannot produce. Let $z_j = 1$ indicate seller j invests in an asset and $z_j = 0$ when seller j does not invest, where $\mathbf{Z} = \{\mathbf{z}_1, \ldots, \mathbf{z}_{\overline{S}}\}$ denotes all sellers' investments. The investments $(\mathbf{G}, \mathbf{Z})$ are observable at the end of the stage.[41]
Stage Two: Each buyer b_i privately learns its valuation v_i. Buyers compete for goods in the auction constructed above. As before, we consider the equilibrium in the auction in which buyers bid up to their valuations. An agent's profits are its V-payoff minus any investment costs. For b_i, profits are $V_i^b(\mathbf{G}, \mathbf{Z}) - \mathbf{c}\sum_{\mathbf{j=1}}^{\overline{\mathbf{S}}} \mathbf{g_{ij}}$.

[40] Another future direction for research would be to characterize network outcomes when sellers also have private information over costs. In this case, no trading mechanism can always yield efficient allocations (Myerson and Mark A. Satterthwaite, 1983).

[41] To derive the link pattern that results from players' investments, it is convenient to write the sellers' investments as $\overline{S} \times \overline{S}$ diagonal matrix $\mathbf{Z}$, where $z_{ii} = 1$ if seller i has invested, $z_{ii} = 0$ otherwise (and $z_{ij} = 0$ for $i \neq j$). The link pattern at the end of stage one will then be $\mathbf{G} \cdot \mathbf{Z}$. In equilibrium, since links are costly, a buyer will not build a link to a seller that does not invest, and we will have $\mathbf{G} \cdot \mathbf{Z} = \mathbf{G}$.

For a seller j, profits are $V_j^s(\mathbf{G}, \mathbf{Z}) - \alpha$ if it has invested in an asset. Profits are zero for all other sellers.

As previously, we solve for a pure-strategy perfect Bayesian equilibria. Given other agents' investments, a buyer invests in its links if and only if no other choice of links generates a higher expected profit. A seller invests in capacity if and only if it earns positive expected profit.

3.2 Efficiency and Equilibrium

Efficient networks allow the highest economic welfare from investment in links, productive assets, and exchange of goods. The *net economic surplus* from a network, $W(\mathbf{G}, \mathbf{Z})$, is the gross economic surplus minus the investment and link costs: $W(\mathbf{G}, \mathbf{Z}) \equiv \mathbf{H}(\mathbf{G}, \mathbf{Z}) - \mathbf{c} \sum_{\mathbf{i=1}}^{\overline{\mathbf{B}}} \sum_{\mathbf{j=1}}^{\overline{\mathbf{S}}} \mathbf{g_{ij}} - \alpha \sum_{\mathbf{j=1}}^{\overline{\mathbf{S}}} \mathbf{z_j}$. A network is an *efficient network* if and only if no other network yields higher net economic surplus.

In contrast to our previous game, here the efficient network is not always an equilibrium outcome. Buyers' incentives are aligned with economic welfare, but sellers sometimes have insufficient incentives to invest in assets. A seller's investment is efficient whenever its cost, α, is less than what it generates in expected surplus from exchange. The price a seller receives, however, is less than the surplus from exchange. As discussed above, the price is not equal to the purchasing buyer's valuation but to the valuation of the "next-best" buyer. Each seller's profit, therefore, is less than its marginal contribution to economic welfare.

The next example illustrates that there is a divergence between efficiency conditions and sellers' investment incentives when sellers' costs are high. When α is sufficiently low, the efficient network is an equilibrium outcome. But when α is high enough, sellers will not invest.

Example 7. [Sellers' Investment Incentives]. Consider the LAC networks in Fig. 4. In these networks the buyers with the top three valuations always obtain goods. They are efficient if the valuation of the third-highest buyer justifies the link and investment costs; that is, if $\alpha + 3c \leq E[X^{3:5}]$ and $c \leq \binom{5}{3}^{-1} E[X^{3:5} - X^{4:5}]$.[42] In these networks, each seller always receives a price of $X^{4:5}$, since at this price the supply of three units equals the demand for three units. Each seller's profit is then $E\left[X^{4:5}\right] - \alpha$, and a seller will invest if and only if $\alpha \leq E\left[X^{4:5}\right]$. It is easy to that efficiency conditions for these networks diverge from the sellers' investment incentives. When $c = 0$, to take an extreme case, the networks are efficient for $\alpha \leq E[X^{3:5}]$. Sellers will invest in assets when $\alpha \leq E[X^{4:5}]$, but not when $E[X^{3:5}] \geq \alpha \geq E[X^{4:5}]$.

[42] A proof available upon request shows that, in general, a sufficient condition for an LAC network with $\overline{B}$ buyers and $\overline{S}$ sellers to be efficient is that (c, α) be such that $\alpha + (\overline{B} - \overline{S} + 1)c \leq E[X^{\overline{S}:\overline{B}}]$ and $c \leq \binom{\overline{B}}{\overline{S}}^{-1} E[X^{\overline{S}:\overline{B}} - X^{\overline{S}+1:\overline{B}}]$.

The problem of covering sellers' costs is a consequence of the private information and contractual incompleteness in our environment.[43] We have assumed that no payments from buyers to sellers are determined until the second stage of the game, after buyers realize their valuations for goods. To cover their costs at this point, sellers could charge a fixed fee or (equivalently) set a reserve price in their auctions. But, since buyers' valuations are private information, any such fee would lead to an inefficient allocation of goods.[44] Buyers with low realized values will not pay the fee; for some realizations of buyers valuations, goods will not be sold. Therefore, in our setting, there is either an inefficiency in the allocation of goods (which would distort buyers' investment incentives) or some underinvestment on the part of the sellers when sellers' investment costs are high.

Coordination failure is a second source of inefficiency in this game. When both buyers and sellers make investment decisions, there are many equilibria where not enough investment takes place. The intuition is simple. Sellers invest in assets only if they expect enough future demand so that they can cover their investment costs. Buyers only establish links to sellers if they expect the sellers to invest. Therefore, the possibility arises that some sellers do not invest because they do not have links to sufficiently many buyers, and buyers do not build links to these sellers because they do not invest. Indeed, the null network is always an equilibrium of this model.

Such coordination failures may be preventable, of course, in an expanded model of network formation. For example, if buyers and sellers can engage in discussions or "cheap talk" prior to investment, they may be able to coordinate on the efficient network without any formal contracting.[45] Indeed, professional associations, chambers of commerce, and other institutions that foster business relations may facilitate this coordination.[46]

---

[43] It would always be possible to cover sellers' costs if long-term contracts were available. Buyers could commit to pay sellers for their investments regardless of which buyers ultimately purchase goods. Such agreements, however, are likely to be difficult to enforce or violate antitrust law. These payments might also be difficult to determine. As we have seen, buyers can be in very asymmetric positions in efficient networks, and the payments may need to reflect this asymmetry. The more complex the fees need to be and the more buyers and sellers need to be involved, the less plausible are long term contracts.

[44] This result again follows from the payoff equivalence theorem (Myerson, 1981). Since buyers have private information, for an efficient allocation of goods buyers must earn the marginal surplus of their exchange, plus or minus a constant ex ante payment. That is, buyers must be bound to make the payment regardless of their realized valuations.

[45] For an overview of the cheap talk literature, see Farrell and Matthew Rabin (1996). Cheap talk can improve coordination, but it can also have no effect at all depending on which equilibrium is selected. Another way to solve coordination failure is for the agents to invest sequentially with buyers choosing links in advance of sellers choosing assets. This specification, however, introduces more subtle coordination problems as discussed in Kranton and Minehart (1997).

[46] See, for example, Lazerson (1993) who describes the voluntary associations and government initiatives that helped establish the knitwear districts in Modena, Italy.

4 Conclusion

This paper addresses two fundamental economic questions. First, what underlying economic environment may lead buyers and sellers to establish links to multiple trading partners? That is, why do networks, which we see in a variety of settings and industries, arise? Second, should we expect such networks to be efficient? Can buyers and sellers, acting non-cooperatively in their own self-interest, build the socially optimal network structure?

Our answer to the first question is that networks can enable agents to pool uncertainty in demand. When sellers' productive capacity is costly and buyers have uncertain valuations of goods, it is socially optimal for buyers to share the capacity of a limited number of sellers. The way in which buyers and sellers are linked, however, plays a critical role in realizing these *economies of sharing*. Because links are costly, there is a tradeoff between building links and pooling risk. Using combinatoric techniques, we show that the links must be "spread out" among the agents and characterize the efficient link patterns which optimize this tradeoff.

We then address the second question: when can buyers and sellers, acting non-cooperatively, form the efficient network structure. A priori there is no reason to expect that buyers will have the "correct" incentives to build links and sellers the correct incentives to invest in productive capacity. We identify properties of the ex post competitive environment that are sufficient to align buyers' incentives with social welfare. First, the allocation of goods is efficient. Second, the buyer earns the marginal surplus from exchange, and thus, the value of its links to economic welfare. However, it is also possible that sellers may not receive sufficient surplus to juistify efficient investment levels. And buyers and sellers may fail to coordinate their link and investment decisions.

We find evidence for our positive results in studies of industrial supply networks. In many accounts, buyers are aware of the potential consequence for their suppliers of uncertainty in their demand. Buyers share suppliers, explicitly to ensure that suppliers have sufficiently high demand to cover investment costs. Buyers "spread out" their orders – reflecting the structure of efficient link patterns. In a study of engineering firms and subcontracting in France, we find a remarkably clear description of this phenomenon. According to Lorenz (1989), buyers keep their orders between 10–15 per cent of a supplier's sales. This "10–15 per cent rule" is explained as follows: "The minimum is set at 10 per cent because anything less would imply a too insignificant a position in the subcontractor's order book to warrant the desired consideration. The maximum is set at 15 per cent to avoid the possibility of uncertainty in the [buyer's] market having a damaging effect on the subcontractor's financial position...." (p. 129).

In another example, Nishiguchi's (1994) study of the electronics industry in Japan reveals that buyers counter the problem of "erratic trading" with their subcontracters by spreading orders among the firms, warning their contracters of shortfalls in demand, and even asking other firms to buy from their subcontracters when they have a drop in orders. In an interview, a buyer explains: "We regard our subcontractors as part of our enterprise group. ...Within the group we try to allocate the work evenly. If a subcontractor's workload is down, we help him find a new job. Even if we have to cut off our subcontractors, we don't do it harshly. Sometimes we even ask other large firms to take care of them." (p.175). These practices are part of long-term economic calculation to maintain a subcontractor's invesment in value-enhancing assets.[47]

There is also evidence of our less optimistic results: firms may fail to coordinate on the efficient network structure, or even in establishing any links at all. In many developing countries, there is hope that local small-scale industries can mimic the success of European vertical supply networks. However, researchers have found that firms do not always coordinate their activities (John Humphrey, 1995). There is then a role for community and industry organizations, such as chambers of commerce, in establishing efficient networks.

By introducing a theory of link patterns, this paper opens the door to much future research on buyer-seller networks. Here we have explored one economic reason for networks: economies of sharing. There are many other reasons why multiple links between buyers and sellers are socially optimal. Buyers may want access to a variety of goods. Sellers may have economies of scope or scale. Sellers could be investing in different technologies, and buyers may want to maintain relationships with many sellers to benefit from these efforts. In many environments, a firm's gain from adopting a technology may depend on the number of other firms adopting the technology. Using the model here, a buyer's adoption of a seller's technology can be represented by a "link," allowing a more precise microeconomic analysis of "network externalities" and "systems competition." Future studies of networks may give other content to the links. Links to sellers or buyers may contain information about product market trends, or even competitors. There may then be a tradeoff between gathering information and revealing information by establishing links.

Future research on networks could build on to the bipartite structure introduced here. For example, in addition to the links between buyers and sellers, there may be links between the sellers themselves (or between the buyers themselves). These links could represent a sellers' cooperative or industry group. There are many settings where sellers, formally and informally, share inventories and otherwise cooperate to increase their collective sales.[48] In another example, a product market could be added to the buyer side of the network. In industrial supply settings, the buyers could be manufacturers that in turn sell output to consumers. We could then ask how the nature of consumers' demand and the final product market affect network structure.

This paper suggests a new, network approach to the study of personalized and group-based exchange. A growing literature shows how long-term, personalized exchange can shape economic transactions. Greif (1993) studies the 11th Century

---

[47] For more evidence of the need for suppliers to serve several buyers, see, for example, Cawthorne (1995, p. 48) and Roberta Rabellotti (1995, p. 37).

[48] For instance, we have seen this phenomenon among jewelry retailers – in San Francisco's Chinatown, Boston's Jewelry Market, and the traditional jewelry district in Rabat, Morocco.

Maghribi traders who successfully engaged in long-distance commerce by hiring agents from within their group. Kranton (1996) shows how exchange between friends and relatives and the use of "connections" supplants anonymous market exchange in many settings. The analysis here suggests a link-based strategy for evaluating such forms of exchange and interlocking groups of buyers and sellers. In our study of efficient link patterns, for example, we saw that all agents need not be linked to all other agents. Sparse links between agents or across groups, then, may not be evidence of trading inefficiencies. The pattern may also reflect the optimal tradeoff between the cost of links and the potential gains from exchange.

Appendix A. Competition for Goods in Buyer-Seller Networks: An Auction Model

In this appendix, we develop our ascending-bid auction model of competition in networks. We first show that it is possible to construct, in a network, a process of "auction clearing" that is well defined. We then construct the auction game and show that it is an equilibrium following iterated elimination of weakly dominated strategies for each buyer to remain in the bidding of each seller's auction up to its valuation of a good. This argument requires a proof beyond that for a single-seller ascending-bid auction. A priori we might think a buyer could gain by dropping out of some auctions at a price below its valuation. The auction could clear at a lower price, and fewer buyers would be bidding in remaining auctions. The buyer could then procure a good at a lower price. The proof shows this reasoning is false. We end the Appendix by proving Proposition 1. In the equilibrium, (i) the allocation is efficient, and (ii) the allocation and prices are pairwise stable.

First make precise what we mean by "demand is weakly less than supply" for a subset of sellers in a network. The auction will specify that whenever this situation occurs, this subset of sellers will "clear" at the going price.

As the auction proceeds, there will be *interim link patterns* that reflect whether buyers are still actively using their links to secure goods. Starting from any link pattern, when a buyer drops out of the bidding of an auction, we can think of it as no longer linked to that seller. Similarly, when a buyer secures a good, it is effectively no longer linked to any remaining seller.

In these interim patterns, we will ask whether any subset of sellers is *clearable*. Formally, consider any link pattern **G**. A subset of sellers $\mathcal{C}$ is *clearable* if and only if there exists a feasible allocation such that all buyers $b_i \in L(\mathcal{C})$ obtain a good from a seller $s_j \in \mathcal{C}$. (Note that, by definition, for a clearable set of sellers $\mathcal{C}$, total demand is weakly less than supply (i.e., $|L(\mathcal{C})| \leq |\mathcal{C}|$).

We use Lemma A.1. to show that there is always a unique *maximal clearable set* of sellers. This set, denoted by **C**, is the union of all clearable sets of sellers in a given "interim" link pattern. If there are no clearable sets, $\mathbf{C} = \varnothing$.

Lemma A.1. *Consider two clearable sets of sellers $\mathcal{C}'$ and $\mathcal{C}''$. The set $\{\mathcal{C}' \cup \mathcal{C}''\}$ is also a clearable set of sellers.*

Proof. If $\mathcal{C}'$ and $\mathcal{C}''$ are disjoint, then clearly the union is a clearable set. For the case when they are not disjoint, the first task is to show that

$$|L(\mathcal{C}' \cup \mathcal{C}'')| \leq |\mathcal{C}'| + |\mathcal{C}''| - |\mathcal{C}' \cap \mathcal{C}''| .$$

That is, the number of buyers linked to the sellers in $\mathcal{C}' \cup \mathcal{C}''$ does not exceed the number of sellers in that set. To show this, we will add up the buyers from linked buyers of each subset and show that the sum cannot exceed $|\mathcal{C}'|+|\mathcal{C}''|-|\mathcal{C}' \cap \mathcal{C}''|$.

Because $\mathcal{C}'$ is a clearable subset, by definition $|L(\mathcal{C}')| \leq |\mathcal{C}'|$. Consider $L(\mathcal{C}' \cup \mathcal{C}'')$. How many buyers are in this set? First of all, we have the buyers in $L(\mathcal{C}')$. Now we add buyers from $L(\mathcal{C}'')$. We add those buyers that are in $L(\mathcal{C}'')$, but not in $L(\mathcal{C}')$. The largest number of buyers that we can add is $|\mathcal{C}''| - |\mathcal{C}' \cap \mathcal{C}''|$. Why? We cannot add any buyers that are linked to the sellers in $\{\mathcal{C}' \cap \mathcal{C}''\}$ because they have already been counted as part of $L(\mathcal{C}')$. So we can only add buyers that are linked exclusively to the remaining sellers in $\mathcal{C}''$, which number $|\mathcal{C}''| - |\mathcal{C}' \cap \mathcal{C}''|$. At most $|\mathcal{C}''| - |\mathcal{C}' \cap \mathcal{C}''|$ buyers are linked exclusively linked to these sellers. If there were more than than this number of buyers exclusively linked to these sellers, the Marriage Theorem would be violated for $\mathcal{C}''$, that is, there would be a subset k of the buyers in $L(\mathcal{C}'')$ that are collectively linked to less than k sellers. So we have

$$|L(\mathcal{C}' \cup \mathcal{C}'')| \leq |L(\mathcal{C}')| + |\mathcal{C}''| - |\mathcal{C}' \cap \mathcal{C}''|$$

which shows that inequality above is satisfied.

Next we show that there exists a feasible allocation in which all the buyers in $L(\{\mathcal{C}' \cup \mathcal{C}''\})$ obtain goods. Assign the buyers in $L(\mathcal{C}')$ to sellers in $\mathcal{C}'$. This is possible because $\mathcal{C}'$ is a clearable set. Assign the additional $|\mathcal{C}''| - |\mathcal{C}' \cap \mathcal{C}''|$ (or less) buyers from $L(\mathcal{C}'')$ to sellers in the set $\{\mathcal{C}'' \setminus (\mathcal{C}' \cap \mathcal{C}'')\}$. This is possible because these buyers are exclusively linked to sellers in $\{\mathcal{C}'' \setminus (\mathcal{C}' \cap \mathcal{C}'')\}$ and $\mathcal{C}''$ is a clearable set – every subset of k buyers must be collectively linked to at least k sellers and thus all are able to secure goods.

We now construct the auction game. First, sellers simultaneously decide whether or not to hold ascending-bid auctions as specified below. This choice is observed by all players. Sellers make no other decisions. (E.g., they cannot set reserve prices. We discuss the implications of reserve prices in the text.) Auctions of participating sellers proceed as follows: there is a common price across all auctions. Buyers can bid only in auctions of sellers to whom they are linked. Initially, all buyers are active at a price $p = 0$. The price rises. At each price, each buyer decides whether to remain in the bidding or drop out of each auction. Once a buyer drops out of the bidding of an auction, it cannot re-enter that auction at a later point in time. Buyers observe the history of the game.

The price rises until a clearable set of sellers occurs in an interim link pattern. The buyers that are linked to the sellers in the maximal clearable set, $L(\mathbf{C})$, secure a good at the current price. If there is more than one feasible allocation where all $b_i \in L(\mathbf{C})$ obtain goods, but where different sellers provide goods, one feasible allocation is chosen at random. Note that this rule implies no buyer is ever allocated more than one good. Removing these sellers and their buyers from the network creates another interim link pattern. If there are remaining sellers, the price continues to rises until another clearable set arises in further "interim" link patterns. This procedure continues until there are no remaining bidders.

In this game, a strategy for a seller is simply a choice whether or not to hold an auction. A strategy for a buyer i specifies the auctions in which it will remain active at any price level p, as a function of v_i, any remaining sellers, any remaining buyers, any interim link pattern, and any prices at which any buyers dropped out of the bidding of any auctions.

We solve for a perfect Bayesian equilibrium following iterated elimination of weakly dominated strategies. It is a weakly dominant strategy for a seller to hold an auction since it earns nothing if it does not. For buyers, we have the following result.

Proposition A.1. *For a buyer, the strategy to remain in the bidding of each of its linked sellers' auctions up to its valuation of a good is an equilibrium following iterated elimination of weakly dominated strategies.*

Proof. We do not consider the possibility that two buyers have the same valuation. This is a probability zero event, and we are interested only in expected payoffs from the auction.

1. First we argue that the proposed strategy constitutes a perfect Bayesian equilibrium. Does any buyer have an incentive to deviate from the above strategy? Clearly, no buyer would have an incentive to stay in the bidding of an auction after the price exceeds its valuation. But suppose that for some history of the game, a buyer i drops out of the auction of some linked seller s_j when the price reaches $\widetilde{p} < v_i$. The buyer's payoff can only increase from the deviation if the buyer obtains a good, so we will assume that this is the case. Let seller h be the seller from whom buyer i obtains a good after it deviates.
We argue that the buyer cannot lower the price it pays for a good by dropping out of an auction early. There are two cases to consider:
(i) Buyer i obtains a good from seller h at the price $\widetilde{p}$. We argue that this outcome can never arise. Consider the maximal clearable set of sellers, $\mathbf{C}$, and the set of buyers that obtain goods from these sellers $\overline{L}(\mathbf{C})$ at price $\widetilde{p}$, given buyer i drops out of seller j's auction a price $\widetilde{p}$. Since buyer i obtains a good, we have $b_i \in \overline{L}(\mathbf{C})$. At some price just below $\widetilde{p}$ (just before buyer i drops out) the set $\overline{L}(\mathbf{C})$ is exactly the same. Hence, if $\mathbf{C}$ is a clearable set at $\widetilde{p}$ it also a clearable set at the lower price.
(ii) Buyer i obtains a good from seller h at a price above $\widetilde{p}$. Consider the buyer that drops out of the bidding so that the auction of s_h clears. Label this buyer b' and its valuation v'. Buyer i pays seller h the price v'. Let $\mathcal{S}$ denote the set of sellers that clear at any price weakly below v'. Seller h is in this set. Consider the set of buyers linked to at least one seller in $\mathcal{S}$ in the original graph; we denote these buyers $L(\mathcal{S})$. We can divide into $L(\mathcal{S})$ into two subsets: those buyers that obtain a good at a price weakly below v', and those that drop out of the bidding at some prices weakly below v'. Every buyer in

the second group drops out of the bidding because it has a valuation below v'. Buyers in the first group obtain their goods from sellers in $\mathcal{S}$, because by definition all sellers whose auctions clear by v' are in $\mathcal{S}$.

Now consider the equilibrium path, where buyer i does not drop out early from seller j's auction. Consider the allocation of goods from sellers in $\mathcal{S}$ to buyers in $L(\mathcal{S})$ from the previous paragraph. Any buyer in $L(\mathcal{S})$ that does not obtain a good has a valuation below v'. Using this allocation, we could "clear" $\mathcal{S}$ at the price v'. It follows that the sellers in $\mathcal{S}$ clear at or before the price v'. Since buyer i is in $L(\mathcal{S})$, buyer i obtains a good at a price weakly below v'. That is, buyer i gets a weakly lower price on the equilibrium path.

To see that a buyer can never decrease the price it pays by dropping out of several auctions, simply order the auctions by the price at which the buyer drops out from lowest to highest and apply the above argument to the last auction. (The argument works unchanged if a buyer drops out of several auctions at once.)

2. We now show that the proposed strategies are an equilibrium following iterated elimination of weakly dominated strategies.

First, suppose that each buyer i chooses a bidding strategy that depends only on its own valuation v_i and not on the history of the game. That is, buyer i's strategy is to stay in the bidding of auction j until the price reaches $b_i(v_i, j)$. The same argument as in part I shows that it is a weak best response for each buyer to stay in the bidding of all auctions until the price reaches its valuation. In the parts of the argument above where a buyer k's valuation v_k determines the price at which an auction clears, replace the buyer's valuation with the price from the bidding strategy $b_k(v_k, j)$.

Second, suppose that buyers choose strategies that depend on the history of the game. These strategies specify that, for some histories, buyers will drop out of some auctions before the price reaches their valuations and/or remain in some auctions after the price exceeds their valuations. There are only two reasons for a buyer i to adopt such a strategy. First, by dropping out of an auction early, a b_i allows another buyer k to purchase a good from s_j and thereby lowers the price b_i ultimately pays for a good. We showed above that this reduction never occurs. Second, dropping out of an auction early or staying in an auction late may lead to a response by other buyers. Consider any play of the game in which, as a consequence of buyer i dropping out, some other buyers no longer remain in all auctions until the price reaches their valuation or stay in an auction after the price exceeds their valuation. Since the population of buyers is finite, there are only a finite number of buyers who would bid in this way. Consider the last such buyer. When it bids in this way, the buyer does not affect the bidding of any other buyers. Therefore, it can only lose by adopting the strategy to drop out of auction before the price reaches its valuation, or remain in an auction after the price exceeds its valuation. This strategy is weakly dominated by staying in each auction exactly until the price reaches its valuation. Eliminating this strategy, the second-to-last buyer's strategy to drop out early or remain late is then also weakly dominated. And so on.

We finish our treatment of the auction by proving Proposition 1: (i) in this equilibrium of the auction, the allocation of goods is efficient for any realization of buyers' valuations **v**, and (ii) the allocation and prices are pairwise stable.

Proof of Proposition 1.

(i) We show that in equilibrium, the highest valuation buyers obtain goods whenever possible given the link pattern. Therefore, the equilibrium allocation of goods is efficient for any realization of buyers' valuations **v**.

At the price $p = 0$ and the original link pattern, consider any maximal clearable set of sellers, **C**, and the buyers in $L(\mathbf{C})$. It is trivially true that these buyers have the highest valuations of buyers linked to sellers in **C** in the original link pattern.

Now consider the remaining buyers $\mathbf{B}\backslash\mathbf{L}(\mathbf{C})$, the interim link pattern that arises when the set **C** is cleared, and the next maximal clearable set of sellers, $\mathbf{C}'$, that arises at some price $p > 0$. We let $L(\mathbf{C}')$ denote the set of buyers linked to any sellers in $\mathbf{C}'$ in the interim link pattern. By definition of a clearable set, $|L(\mathbf{C}')| \leq |\mathbf{C}'|$, but for $p > 0$, it can be shown that $|L(\mathbf{C}')| = |\mathbf{C}'|$.[49] Consider the buyers in $L(\mathbf{C}')$. We argue that these buyers must have the highest valuations of the buyers in $\mathbf{B}\backslash L(\mathbf{C})$ linked to any seller in $\mathbf{C}'$ in the original link pattern. Suppose not. That is, suppose there is a buyer $b_i \in \mathbf{B}\backslash L(\mathbf{C})$ that was linked to a seller in $\mathbf{C}'$ in the original graph and has a higher valuation than some buyer in $L(\mathbf{C}')$. For b_i not to be in $L(\mathbf{C}')$, it either obtained a good from a seller in **C** or it dropped out of the auction at a lower price. The first possibility contradicts the assumption that $b_i \in \mathbf{B}\backslash L(\mathbf{C})$. The second possibility contradicts the equilibrium strategy. So any buyer in $\mathbf{B}\backslash L(\mathbf{C})$ with a higher valuation than some buyer in $L(\mathbf{C}')$ was not linked to any seller in $\mathbf{C}'$ in the original link pattern. Thus, the $|\mathbf{C}'|$ buyers that obtain goods from the sellers in $\mathbf{C}'$ are the buyers with the highest valuations of those linked to the sellers in $\mathbf{C}'$ in the original link pattern who are not already obtaining goods from other sellers. And so on, for the next maximal clearable set of sellers $\mathbf{C}''$.

(ii) We show here that the allocation and prices are pairwise stable. Suppose seller j sells its good to buyer k and in the original graph a seller j is linked to a buyer i that has a higher valuation than buyer k. From part (i), either buyer i purchases from a seller that clears at the same price as seller j or it bought previously at a lower price. Therefore, buyer i would not be willing to pay seller j a higher price that seller j receives from buyer k. The bidding mechanism also ensures that no buyer that does not obtain a good is linked to seller willing to accept a price below the buyer's valuation. (The fact that the buyer is not obtaining a good implies that the price all of its linked sellers are receiving is above the buyer's valuation). There is also no linked seller providing a good at a

---

[49] The proof that $|L(\mathbf{C}')| = |\mathbf{C}'|$ when $p > 0$ is available from the authors on request. Intuitively, if any sellers do not sell goods, they are part of a clearable set at $p = 0$.

lower price than it is paying (otherwise the set of sellers would not be clearable at that price).

Appendix B.

Proof of Lemma 1.
We show below that for any link pattern and for each realization of buyers' valuations, a buyer's payoff in the auction is equal to its contribution to the gross surplus of exchange. That is, a buyer i earns the difference between $w(\mathbf{v}, \mathbf{A}^*(\mathbf{v}, \mathbf{G}))$ and the surplus that would arise if buyer i did not purchase. Taking expectations over all the valuations, then gives us that a buyer's V-payoff is equal to the buyer's contribution to expected gross surplus. The difference in a buyer's $V-$ payoff in any two link patterns is then the difference in the buyer's contribution to expected gross surplus in each link pattern. If two link patterns differ only in the buyer's own link holdings, the difference in the buyer's contribution to expected gross surplus in each link pattern is the same as the difference in total expected gross surplus in each link pattern. This gives the result.

Consider a realization $\mathbf{v}$ of buyers' valuations. Suppose a buyer b_i obtains a good in the equilibrium outcome of the auction given this realization. This buyer obtains a good when there arises a maximal clearable set of sellers $\mathbf{C}$ such that $b_i \in L(\mathbf{C})$. Suppose the price at which this clearable set arises is $p = 0$. The buyer earns its valuation v_i from the exchange, and so if this buyer did not have any links – that is, were not participating in the network – its loss in payoffs would be v_i. This loss is the same as the same as the loss in gross surplus. By the argument in the proof of Proposition 1, in equilibrium the buyers that obtain goods from the sellers in $\mathbf{C}$ are the buyers with the highest valuations of those linked to those sellers in the original graph. When b_i's links are removed, the only change in this set is the removal of b_i. Thus, the loss in gross surplus is also simply v_i.

Suppose the price at which b_i obtains a good is some $p > 0$. Label the set of sellers that have already cleared $\widetilde{\mathbf{C}}$, and the buyers that obtained goods from these sellers $\overline{L}(\widetilde{\mathbf{C}})$. (This set of buyers need not include all the buyers linked to sellers in $\widetilde{\mathbf{C}}$ in the original graph, as some buyers may have dropped out of the bidding.) By the proof of Proposition 1, these buyers are the buyers with the highest valuations of those linked to the sellers in $\widetilde{\mathbf{C}}$ in the original graph. The remaining buyers are $\mathbf{B}\backslash\overline{L}(\widetilde{\mathbf{C}})$. There is some buyer $b_k \in \mathbf{B}\backslash\overline{L}(\widetilde{\mathbf{C}})$ with valuation v_k ($v_k < v_i$) that drops out the bidding and creates the maximal clearable set $\mathbf{C}$. Let $\overline{L}(\mathbf{C})$ be the set of buyers that obtain goods from the sellers in $\mathbf{C}$. (These buyers are all the buyers linked to any sellers in $\mathbf{C}$ in the interim link pattern at p.) Note that b_k must be linked to some seller in $\mathbf{C}$ in the original graph. Otherwise, its bid would not affect whether or not the set is clearable. Of the buyers in $\mathbf{B}\backslash\overline{L}(\widetilde{\mathbf{C}})$ linked to some seller in $\mathbf{C}$ in the original graph, b_k is the buyer with the next highest valuation after the $|\mathbf{C}|$ buyers in $\overline{L}(\mathbf{C})$. Otherwise, b_k would not be the buyer that caused the set to clear. A buyer with higher valuation than b_k but not in $\overline{L}(\mathbf{C})$ still remains linked to some seller in $\mathbf{C}$, and $\mathbf{C}$

would not be clearable. In equilibrium, buyer b_i obtains the good and pays the price $p = v_k$. Its surplus from the exchange is $v_i - v_k$. Now suppose that b_i is not participating in the network. What is the loss in welfare? By the argument in the proof above, the buyers with the highest valuations connected to the sellers in $\mathbf{C} \cup \widetilde{\mathbf{C}}$ in the original graph obtain goods from them in equilibrium. When b_i is not participating in the network, b_i is no longer in this set of buyers. In its place is buyer b_k. This is because we know that buyer b_k is connected to some seller(s) in $\mathbf{C}$ in the original graph. And, of those buyers in $\mathbf{B}\backslash\overline{L}(\widetilde{\mathbf{C}})$, the buyer b_k has the next highest valuation after the $|\mathbf{C}|$ buyers in $\overline{L}(\mathbf{C})$. So in the graph without b_i's links, the $|\mathbf{C}|$ buyers with the highest valuations of those buyers in $\mathbf{B}\backslash\overline{L}(\widetilde{\mathbf{C}})$ includes b_k. Therefore, the loss in welfare is $v_i - v_k$. The same argument holds for any realization $\mathbf{v}$ in which the buyer obtains good.

Proof of Proposition 2.
This result follows immediately from Lemma 1. Let $\mathbf{G^0} = (\mathbf{g_i^0}, \mathbf{g_j^0})$ be an efficient graph. Formally, we can write a buyer's equilibrium conditions as follows:

$$\mathbf{g}_i^* = \arg\max_{\mathbf{g}_i}\{\Pi_i^b(\mathbf{g}_i, \mathbf{g}_{-i}^o) - 0\} = \arg\max_{\mathbf{g}_i}\{W(\mathbf{g}_i, \mathbf{g}_{-i}^o) - W(\mathbf{0}, \mathbf{g}_{-i}^o)\}$$

Since the efficient graph $\mathbf{G^o} = (\mathbf{g_i^o}, \mathbf{g_{-i}^o})$ maximizes $W(\cdot, \mathbf{g}_{-i}^o)$, the solution to the buyer's maximization problem is $\mathbf{g}_i = \mathbf{g}_i^o$.

Proof of Proposition 3.
By the Marriage Theorem, a network of buyers and sellers $(\mathbf{B}, \mathbf{S})$ is allocatively-complete if and only if every subset of k buyers in $\mathbf{B}$ is linked to at least k sellers in $\mathbf{S}$ for each k, $1 \leq k \leq \overline{S}$.

First we show that $\overline{B} - \overline{S} + 1$ links per seller is necessary for allocative completeness. Suppose for some seller $s_j \in \mathbf{S}$, $|L(s_j)| \leq \overline{B} - \overline{S}$. Then there are at least $\overline{B} - (\overline{B} - \overline{S}) = \overline{S}$ buyers in the network that are *not* linked to s_j. No buyer in this set of $\overline{S}$ buyers can obtain a good from s_j. Therefore, there is not a feasible allocation in which this set of buyers obtain goods, and the network is not allocatively complete.

To show sufficiency, we construct a network as follows: $\overline{S}$ of the buyers have exactly one link each to a distinct seller. The remaining buyers are linked to every seller in $\mathbf{S}$. It is easy to check that this network satisfies the Marriage Theorem condition above and involves $\overline{B} - \overline{S} + 1$ links per seller.

Proof of Proposition 4.
Let $\mathbf{G}$ be an LAC link pattern and let $\mathbf{G}'$ be any other link pattern which forms an AC but not LAC network on $(\mathbf{B}, \mathbf{S})$. It is clear that $W(\mathbf{G}) > W(\mathbf{G}')$, since $H(\mathbf{G}) = H(\mathbf{G}')$, and $\mathbf{G}$ involves fewer links.

Next, let $\mathbf{G}'$ be any graph that is not an AC network and yet yields higher welfare than $\mathbf{G}$. In $\mathbf{G}'$ there is at least one set of $\overline{S}$ buyers that cannot all obtain goods when they have the highest $\overline{S}$ valuations. Label one such set of buyers $\widetilde{\mathcal{B}}$. Below we prove that we can add exactly one link between a buyer in $\widetilde{\mathcal{B}}$ and a seller not currently linked to any buyer in $\widetilde{\mathcal{B}}$ so that for any realization $\mathbf{v}$ of buyers' valuations such that the buyers in $\widetilde{\mathcal{B}}$ have the top $\overline{S}$ valuations, one more buyer in $\widetilde{\mathcal{B}}$ obtains goods in the $\mathbf{A}^*(\mathbf{v};\mathbf{G}'')$ than in $\mathbf{A}^*(\mathbf{v};\mathbf{G}')$, where $\mathbf{G}''$ is the new graph formed from adding the link. Therefore, $\mathbf{A}^*(\mathbf{v};\mathbf{G}'')$ yields higher expected surplus than $\mathbf{A}^*(\mathbf{v};\mathbf{G}')$. What is precisely the gain in surplus? The lowest possible valuation of the buyer that obtains the good in $\mathbf{A}^*(\mathbf{v};\mathbf{G}'')$ but not in $\mathbf{A}^*(\mathbf{v};\mathbf{G}')$ is $X^{\bar{S}:\bar{B}}$. The highest possible valuation of the buyer outside of $\widetilde{\mathcal{B}}$ that obtains the good in $\mathbf{A}^*(\mathbf{v};\mathbf{G}')$ but not in $\mathbf{A}^*(\mathbf{v};\mathbf{G}'')$ is $X^{\bar{S}+1:\bar{B}}$. Thus, $\mathbf{A}^*(\mathbf{v};\mathbf{G}'')$ yields an expected increase in surplus of least $E\left[X^{\bar{S}:\bar{B}} - X^{\bar{S}+1:\bar{B}}\right]$. Since adding a link does not decrease the surplus from the efficient allocations for other realizations of $\mathbf{v}$, and since $\binom{\bar{B}}{\bar{S}}^{-1}$ is the probability that the set $\widetilde{\mathcal{B}}$ has the top valuations, the graph $\mathbf{G}''$ yields an expected increase in gross surplus of of at least $\binom{\bar{B}}{\bar{S}}^{-1} E\left[X^{\bar{S}:\bar{B}} - X^{\bar{S}+1:\bar{B}}\right]$ over $\mathbf{G}'$. Hence, for $c < \binom{\bar{B}}{\bar{S}}^{-1} E\left[X^{\bar{S}:\bar{B}} - X^{\bar{S}+1:\bar{B}}\right]$, $\mathbf{G}'$ is not an efficient network. Therefore, there does not exist any graph $\mathbf{G}'$ which yields strictly higher welfare than an LAC link pattern for $c \leq \binom{\bar{B}}{\bar{S}}^{-1} E\left[X^{\bar{S}:\bar{B}} - X^{\bar{S}+1:\bar{B}}\right]$.

To finish the proof, we show that it is possible to add exactly one link between a buyer in $\widetilde{\mathcal{B}}$ and a seller not currently linked to any buyer in $\widetilde{\mathcal{B}}$ so that for any realization $\mathbf{v}$ of buyers' valuations such that the buyers in $\widetilde{\mathcal{B}}$ have the top $\overline{S}$ valuations, one more buyer in $\widetilde{\mathcal{B}}$ obtains goods in the $A^*(\mathbf{v};\mathbf{G}'')$ than in $A^*(\mathbf{v};\mathbf{G}')$, where $\mathbf{G}''$ is the new graph formed from adding the link.

First we need a few definitions. We say that a set of k buyers, for $k \leq \overline{S}$, is *deficient* if and only if it is not collectively linked to k sellers. A set of k buyers, for $k \leq \overline{S}$, is a *minimal deficient set* if and only if it is a deficient set and no proper subset is deficient. For a minimal deficient set of $k \leq \overline{S}$ buyers, the k buyers are collectively linked to exactly $k-1$ sellers. (Otherwise, if they were linked to fewer buyers, the set is not a minimal deficient set.) Hence, adding one link between any buyer in the set and any seller not linked to any buyer in the set removes the deficiency.

In $\mathbf{G}'$, by assumption, there is no feasible allocation in which the set of buyers $\widetilde{\mathcal{B}}$ obtains goods. The Marriage Theorem implies that there is some subset $\widetilde{\widetilde{\mathcal{B}}}$ of k buyers, $\widetilde{\widetilde{\mathcal{B}}} \subset \widetilde{\mathcal{B}}$, that is not collectively linked to k sellers and is thus a deficient set. Label $\mathcal{B}^M$ the minimal deficient set of buyers contained in $\widetilde{\widetilde{\mathcal{B}}}$. Let $NL(\mathcal{B})$ denote the set of sellers that are *not* linked to any buyer in $\mathcal{B}$. Add one link between any buyer in $\mathcal{B}^M$ and any seller in $NL(\widetilde{\mathcal{B}})$. Since adding one link removes the deficiency, $\mathcal{B}^M$ is not deficient in $\mathbf{G}''$, the new graph formed from adding the link. Therefore, for any ordering of valuations in which the buyers in $\widetilde{\mathcal{B}}$ have the top $\overline{S}$ valuations, there is a feasible allocation in which each buyer in $\mathcal{B}^M$ obtains a good in $\mathbf{G}''$ but not in $\mathbf{G}'$.

Proof of Proposition 5

Suppose now that some link pattern $\mathbf{G}$ is an equilibrium outcome, and $\mathbf{G}$ is an AC but not LAC link pattern. A proof available upon request from the authors shows that all LAC link patterns are subgraphs of AC link patterns. Since buyers earn the same V-payoffs in any AC (or LAC) link pattern, in $\mathbf{G}$ some buyer has a link that is redundant in the sense that the buyer can cut the link and not change its V-payoffs. Since $c > 0$, the buyer would want to cut this link to increase its profits. Therefore, **G** can not be an equilibrium outcome.

Suppose some link pattern **G** is an equilibrium outcome and is not AC. Because the graph is not AC, there is at least one minimal deficient set $\mathcal{B}^M$ of buyers. By Proposition 4, there is a link that a buyer $b_i \in \mathcal{B}^M$ can add to some seller s_j, and this link increases gross surplus by $\binom{\bar{B}}{\bar{S}}^{-1} E\left[X^{\bar{S}:\bar{B}} - X^{\bar{S}+1:\bar{B}}\right]$. By Lemma 1, buyer i earns this increase in surplus in its V-payoffs. Hence, b_i has an incentive to add the link for $c < \binom{\bar{B}}{\bar{S}}^{-1} E\left[X^{\bar{S}:\bar{B}} - X^{\bar{S}+1:\bar{B}}\right]$. This contradicts the assumption that **G** is an equilibrium for this range of link costs.

We have shown that the only equilibrium outcomes that are possible for the hypthesized range of link costs are LAC networks . By Proposition 2 the efficient link pattern is always an equilibrium outcome, and by Proposition 4 LAC's are the efficient patterns for this range of costs. Hence, in this range of costs, only efficient networks (i.e., LAC's) are equilibrium outcomes of the game.

References

Bala, V., Goyal, S. (2000) A Non-Cooperative Model of Network Formation. Mimeo, McGill University and the Econometric Institute, Erasmus University. *Econometrica* 68(5): 1181–1229.

Bose, R.C., Manvel, B. (1984) Introduction to Combinatorial Theory. New York: Wiley.

Carlton, D.W. (1978) Market Behavior with Demand Uncertainty and Price Inflexibility. *American Economic Review* 68(4): 571–587.

Casella, A., Rauch, J.E. (1997) Anonymous Market and Group Ties in International Trade. Centre for Economic Policy Research, Discussion Paper: 1748.

Cawthorne, P.M. (1995) Of Networks and Markets: The Rise and Rise of a South Indian Town, the Example of Tiruppur's Cotton Knitwear Industry. *World Development* 23(1): 43–56.

Coase, R.H. (1937) The Nature of the Firm. *Economica* 4(15): 386–405.

Demange, G., Gale, D., Sotomayor, M. (1986) Multi-Item Auctions. *Journal of Political Economy* 94(4): 863–872.

Demski, J.S., Sappington, D.E.M., Spiller, P.T. (1987) Managing Supplier Switching. *RAND Journal of Economics* 18(1): 77–97.

de Soto, H.. (1989) *The Other Path*, New York: Harper & Row.

Dutta, B. Van-den-Nouweland, A., Tijs, S. (1998) Link Formation in Cooperative Situations. *International Journal of Game Theory* 27(2): 245–256.

Farrell, J., Gallini, N.T. (1988) Second-Sourcing as a Commitment: Monopoly Incentives to Attract Competition. *The Quarterly Journal of Economics* 103(4): 673–694.

Farrell, J., Rabin, M. (1996) Cheap Talk. *Journal of Economic Perspectives* 10(3): 103–118.

Feller, W. (1950) *An Introduction to Probability Theory and its Applications.* New York: Wiley.

Greif, A. (1993) Contract Enforceability and Economic Institutions in Early Trade: the Maghribi Traders' Coalition. *American Economic Review* 83(3): 525–458.

Grossman, S.J., Hart, O.D. (1986) The Costs and Benefits of Ownership: A Theory of Vertical and Lateral Integration. *Journal of Political Economy* 94(4): 691–719.

Gul, F., Stacchetti, E. (2000) The English Auction with Differentiated Commodities. *Journal of Economic Theory* 92(1): 66–95.

Hart, Oliver and Moore, John. (1990) Property Rights and the Theory of the Firm. *Journal of Political Economy* 98(6): 1119–1158.

Helper, S., Levine, D. (1992) Long-Term Supplier Relations and Product-Market Structure. *Journal of Law, Economics, and Organization* 8(3): 561–581.

Hendricks, K., Piccione, M., Tan, G. (1999) Equilibria in Networks. *Econometrica* 67(6): 1407–1434.

Humphrey, J. (1995) Industrial Organization and Manufacturing Competitiveness in Developing Countries: Introduction. *World Development* 23(1): 1–7.

Jackson, Matthew O., Watts, A. (1998) The Evolution of Social and Economic Networks. Mimeo, California Institute of Technology and Vanderbilt University, 1998.

Jackson, M.O., Wolinsky, A. (1996) A Strategic Model of Social and Economic Networks. *Journal of Economic Theory* 71(1): 44–74.

Katz, Michael L., Shapiro, Carl. (1994) Systems Competition and Network Effects, *Journal of Economic Perspectives* 8(2): 93–115.

Kranton, R.E. (1996) Reciprocal Exchange: A Self-Sustaining System. *American Economic Review* 86(4): 830–851.

Kranton, R.E. (1997) Link Patterns in Buyer-Seller networks: Incentives and Efficiency in Graphs. Mimeo, University of Maryland and Boston University.

Kranton, R.E., Minehart, D.F. (2000a) Competition for Goods in Buyer-Seller Networks. *Review of Economic Design* 5(3): 301–331.

Kranton, R.E., Minehart, D.F. (2000b) Networks versus Verticial Integration. *RAND Journal of Economics* 31(3): 570–601.

Landa, J.T. (1994) *Trust, Ethnicity, and Identity: Beyond the New Institutional Economics of Ethnic Trading Networks, Contract Law, and Gift Exchange*. Ann Arbor: University of Michigan Press.

Lazerson, M. (1993) Factory or Putting-Out? Knitting Networks in Modena. In: G. Grabher (ed.) *The Embedded Firm: On the Socioeconomics of Industrial Networks*. New York: Routledge, pp. 203–226.

Leonard, H.B. (1983) Elicitation of honest preferences for the assignment of individuals to positions. *Journal of Political Economy* 91(3): 461–79.

Lorenz, Edward H. (1989) The Search for Flexibility: Subcontracting in British and French Engineering. In P. Hirst, J. Zeitlin (eds.) *Reversing Industrial Decline? Industrial Structure and Policy in Britain and Her Competitors*. New York: St. Martin's Press, pp. 122–132.

Macauley, S. (1963) Noncontractual Relations in Business: A Preliminary Study. *American Sociological Review* 28(1): 55–70.

McMillan, J., Woodruff, C. (1999) Interfirm Relationships and Informal Credit in Vietnam. *The Quarterly Journal of Economics* 114(4): 1285–1320.

Myerson, R.B. (1977) Graphs and Cooperation in Games. *Mathematics of Operations Research* 2(3): 225–229.

Myerson, R.B. (1981) Optimal Auction Design. *Mathematics of Operations Research* 6(1): 58–73.

Myerson, R.B., Satterthwaite, M.A. (1983) Efficient Mechanisms for Bilateral Trading. *Journal of Economic Theory* 29(2): 265–281.

Nishiguchi, T. *Strategic Industrial Sourcing*. New York: Oxford University Press.

Piore, M.J., Sabel, C.F. (1984) *The Second Industrial Divide*. New York: Basic Books.

Rabellotti, R. (1995) Is There an 'Industrial District Model'? Footwear Districts in Italy and Mexico Compared. *World Development* 23(1): 29–41.

Riordan, M.H. "Contracting with Qualified Suppliers. *International Economic Review* 37(1): 115–128.

Roth, A.E., Sotomayor, Marilda A. Oliveira. *Two-Sided Matching: A Study in Game-Theoretic Modeling and Analysis*. New York: Cambridge University Press, 1990.

Rothschild, M., Werden, G.J. (1979) Returns to Scale from Random Factor Services: Existence and Scope. *Bell Journal of Economics* 10(1): 329–335.

Saxenian, A. (1994) *Regional Advantage: Culture and Competition in Silicon Valley and Route 128*. Cambridge: Harvard University Press.

Scheffman, D.T., Spiller, P.T. (1992) Buyers' Strategies, Entry Barriers, and Competition. *Economic Inquiry* 30(3): 418–436.

Schmitz, Hubert. (1995) Small Shoemakers and Fordist Giants: Tale of a Supercluster. *World Development* 23(1): 9–28.

Scott, A.J. (1993) *Technopolis*. Berkeley: University of California Press.

Shapley, L.S., Shubik, M. (1972) The Assignment Game I: The Core. *International Journal of Game Theory* 1: 111–130.

Spulber, D.F. (1996) Market Microstructure and Intermediation, *Journal of Economic Perspectives* 10(3): 135–152.

Uzzi, B. (1996) The Sources and Consequences of Embeddness for the Economic Performance of Organizations: The Network Effect. *American Sociological Review* 61: 674–698.

Williamson, O.E. (1975) *Markets and Hierarchies: Analysis and Antitrust Implications*. New York: Free Press.

Competition for Goods in Buyer-Seller Networks

Rachel E. Kranton[1], **Deborah F. Minehart**[2]

[1] Department of Economics, University of Maryland, College Park, MD 20742, USA
[2] Department of Economics, Boston University, 270 Bay State Road, Boston, MA 02215, USA

Abstract. This paper studies competition in a network and how a network structure determines agents' individual payoffs. It constructs a general model of competition that can serve as a reduced form for specific models. The paper shows how agents' outside options, and hence their shares of surplus, derive from "opportunity paths" connecting them to direct and indirect alternative exchanges. Analyzing these paths, results show how third parties' links affect different agents' bargaining power. Even distant links may have large effects on agents' earnings. These payoff results, and the identification of the paths themselves, should prove useful to further analysis of network structure.

Key Words: Bipartite graphs, outside options, link externalities

JEL Classification: D00, D40

1 Introduction

Networks of buyers and sellers are a common exchange environment. Networks are distinguished from markets by specific assets, or "links," between particular buyers and sellers that enhance the value of exchange. In many industries, for example, manufacturers train particular suppliers or otherwise "qualify" suppliers to meet certain criteria. The asset may also be less formal, as when a supplier's

We thank Bhaskar Dutta and an anonymous referee for comments that greatly improved the paper. Both authors are grateful for support from the National Science Foundation under grants SBR-9806063 (Kranton) and SBR-9806201 (Minehart). Deborah Minehart also thanks the Cowles Foundation at Yale University for its hospitality and generous financial support.

understanding of a manufacturer's idiosyncratic needs develops through repeated dealings.[1]

This paper studies competition for goods in a network. The theory we develop explains how third parties may affect the terms of a bilateral exchange. Bargaining theory focuses primarily on bilateral negotiations. Yet strictly bilateral settings seem to be the exception rather than the rule. The network model we develop allows for arbitrarily complex multilateral settings. New links introduce new potential exchange partners. We show how such changes in opportunities affect matchings and divisions of surplus. In particular, we evaluate how third parties can affect an agent's "bargaining power" by changing, perhaps indirectly, its outside options.

The networks we consider consist of buyers, sellers, and the pattern of links that connect them. Each buyer demands a single unit of an indivisible good, and each seller can produce one unit. A buyer can only purchase from a seller to whom it is linked. Competition for goods will then depend on the link pattern. If a linked buyer and seller negotiate terms of trade, each agent's links to other agents determine their respective "outside options." Since alternative sellers or buyers may be linked to yet other agents, the entire pattern of links affects the value of these outside options and each agent's "bargaining power." For example, in Fig. 1 below, the price buyer 3 would pay to either seller 1 or seller 3 depends on its links to buyers 1 and 2 or buyers 4 and 5 respectively, and further depends on these buyers' links to other sellers. The heart of this paper is identifying precisely how such indirect links affect competitive prices and each agent's ability to extract surplus from an exchange.

The paper first develops a theory of competition in a network. There are potentially many ways to model negotiations or competition for goods. This paper takes a general approach: we characterize competitive prices and allocations as those that satisfy a "supply equals demand" condition for the network setting. We show that these prices yield payoffs that are individually rational and pairwise stable. These are the minimal conditions that payoffs resulting from any negotiation process or competition should satisfy. We show that there is range of such competitive prices, as in Demange and Gale (1985). Basic results also show the equivalence between these payoffs and the core of an assignment game (Shapley and Shubik 1997). Competitive prices, then, distribute the surplus generated by an efficient allocation of goods.

Armed with our general results, we turn to our central objective: studying the relationship between network structure and agents' competitive gains from trade.

We first characterize the range of competitive prices in terms of network structure. To do so, we define the notion of an *opportunity path.* An opportunity path is a path of links from a buyer to a seller to another buyer to a seller, and so on. These paths capture direct and indirect competition for a set of sellers' goods.

[1] Kranton and Minehart (2001) introduces a model of buyer-seller networks, and Kranton and Minehart (2000) explores the role of neworks in indurstrial organization and discusses industry examples.

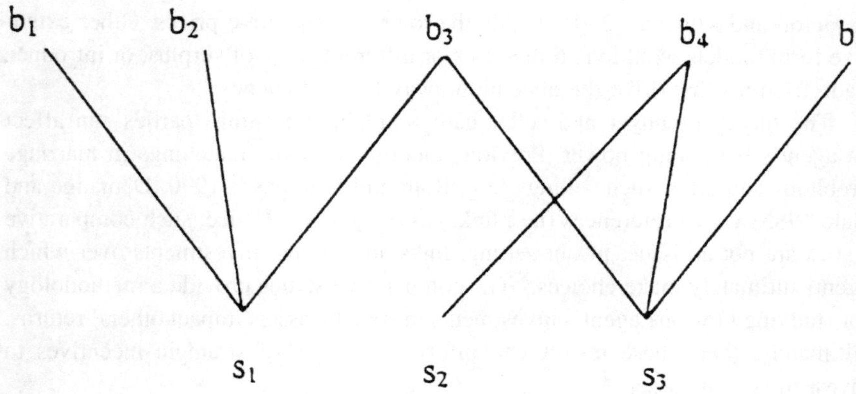

Fig. 1.

The influence of network structure, expressed in these paths, is quite intuitive. For example, for the lower bound of a competitive price, we show that a buyer i that obtains a good must pay at least the valuation of a particular buyer. This buyer is not obtaining a good and is connected by an opportunity path to buyer i. It is therefore buyer i's (perhaps indirect) competitor and can replace buyer i in an allocation of goods. To prevent this replacement, buyer i must pay a sufficiently high price: it must pay at least what this buyer would be willing pay to obtain a good.[2]

The paper then asks how new links affect the prices paid by third parties. Consider a manufacturer that "qualifies" a particular supplier. This investment increases the value of an exchange between the two. It also affects the payoffs of all other manufacturers and suppliers. Since the supplier now has an additional sales option, it could extract greater rents from its other buyers. We call this *a supply stealing effect.* On the other hand, since the manufacturer has access to another source of supply, there is also a *supply freeing effect.* We evaluate these effects by identifying two types of paths in a network, what we call *buyer paths* and *seller paths*. A path between a buyer i and the seller with the new link, a *seller path* is detrimental to buyer i's payoffs. The link confers the *supply stealing effect*. A path between a buyer i and the buyer with the new link, a *buyer path,* is beneficial to buyer i's payoffs. It confers a *supply freeing effect*.

For example, consider the network in Fig. 1 and add a link between b_2 and s_2. The link, for example, frees supply for b_1 which can more often obtain goods from s_1. Buyers with buyer paths benefit from the supply freeing effect. In contrast, b_4 suffers from a supply stealing effect. This effect will extend to b_5, which now must sometimes compete with b_4 for s_3's output. As for sellers, the supply freeing effect hurts s_1 and the supply stealing effect helps s_2 and s_3.

These results provide a general framework to understand competition in a network setting. With this general model, we can place specific models of network competition in context. For example, an ascending-bid auction for a network

[2] This value is also the social opputunity cost of buyer i obtaining a good.

(Kranton and Minehart 2001) yields the lowest competitive prices. Other extensive form models would yield the same or different splits of surplus, or introduce trade frictions that drive the allocation away from efficiency.

The theory of buyer and seller paths explains how third parties can affect an agent's bargaining power. Previous theories of stable matchings in marriage problems and other such settings (e.g. Roth and Sotomayer 1990, Demange and Gale 1985) view preferences (i.e., links) as exogenous. Hence, such comparative statics are not an issue. In our setting, links are specific investments over which agents ultimately make choices.[3] The comparative statics provide a methodology for studying how one agent's investments in specific assets impact others' returns. Ultimately, then, these results can inform the study of strategic incentives to invest in specific assets.[4]

Section 2 builds a model of buyer-seller networks and develops a general notion of competition for goods. Section 3 characterizes the range of competitive prices in terms of the network structure. Section 4 considers how changes in the link pattern impact agents' competitive payoffs. Section 5 concludes.

2 Competition in a Network

2.1 Model of Buyer-Seller Networks[5]

There is a finite set of sellers **S** that number $\overline{S} \equiv |\mathbf{S}|$ who each have the capacity to produce one indivisible unit of a good at zero marginal cost. There is a finite set of buyers **B** that number $\overline{B} \equiv |\mathbf{B}|$ who each demand one indivisible unit of a good. Each buyer i, or b_i, has valuation v_i for a good, where $\mathbf{v} = (v_1, \ldots, v_{\overline{B}})$ is the vector of buyers' valuations. We restrict attention to generic valuations where $v_i > 0$ for all buyers i and $v_i \neq v_k$ for all buyers $i \neq k$.[6]

A buyer and seller can engage in exchange only if they are "linked." A *link pattern*, or graph, $\mathscr{G}$ is a $\overline{B} \times \overline{S}$ matrix, $[g_{ij}]$, where $g_{ij} \in \{0, 1\}$, which indicates linked pairs of buyers and sellers. For buyer i and seller j, $g_{ij} = 1$ when b_i and s_j are linked, and $g_{ij} = 0$ when the pair is not linked. For a given link pattern and a set of buyers $\mathscr{B} \subseteq \mathbf{B}$, let $L(\mathscr{B}) \subseteq \mathbf{S}$ denote the set of sellers linked to any buyer i in $\mathscr{B}$. We call $L(\mathscr{B})$ the buyers' *linked set* of sellers. Similarly, for a set of sellers $\mathscr{S} \subseteq \mathbf{S}$, let $L(\mathscr{S}) \subseteq \mathbf{B}$ denote the sellers' *linked set* of buyers.

Allocations of goods are feasible only when they respect the links between buyers and sellers. An *allocation of goods*, A, is a $\overline{B} \times \overline{S}$ matrix, $[a_{ij}]$, where

[3] In Kranton and Minehart (2001), we develop a model of network formation in which agents invest in links. All the results in this paper apply to that model.

[4] Incentives to invest in specific assets are a major theme in industrial organization and theory of the firm literature. Classic contributions include Grossman and Hart (1990), Hart and Moore (1990), and Williamson (1975). Most studies to date consider specific asset investment in bilateral settings; the "outside option" is assumed but not modeled.

[5] The following model of buyer-seller networks is from Kranton and Minehart (2001).

[6] This assumption is without loss of generality when buyers' valuations are independently and identically distributed with a continuous distribution. In this case, we would be concerned with expected valuations, and non-generic valuations arise with probability zero.

$a_{ij} \in \{0, 1\}$, where $a_{ij} = 1$ indicates that buyer i obtains a good from seller j. For a given link pattern $\mathscr{G}$, an allocation of goods is *feasible* if and only if $a_{ij} \leq g_{ij}$ for all i, j and for each buyer i, if there is a seller j such that $a_{ij} = 1$ then $a_{ik} = 0$ for all $k \neq j$ and $a_{lj} = 0$ for all $l \neq i$. The social surplus associated with an allocation A is the sum of the valuations of the buyers that secure goods in A. We denote the surplus as $w(A; \mathbf{v})$.[7]

2.2 Competitive Prices and Allocations

We next consider competition for goods in this setting. Consider a price vector $\mathbf{p} = (p_1, \ldots, p_{\overline{S}})$ which assigns a price p_j to each seller s_j. Let u_i^b and u_j^s denote payoffs for each buyer i and seller j, respectively. Let $\mathbf{u}^b = (u_1^b, \ldots, u_{\overline{B}}^b)$ and $\mathbf{u}^s = (u_1^s, \ldots, u_{\overline{S}}^s)$ denote payoff vectors. For a price vector and allocation $(\mathbf{p}, A)$, payoffs are as follows: For seller j, $u_j^s = p_j$. For buyer i, $u_i^b = v_i - p_j$ if it obtains a good from seller j in A. Otherwise, $u_i^b = 0$.

We say a price vector and allocation $(\mathbf{p}, A)$ is competitive when it satisfies the following "supply equals demand" conditions for the network setting:

Definition 1. *For a graph $\mathscr{G}$ and valuation $\mathbf{v}$, a price vector and allocation $(\mathbf{p}, A)$ is competitive if and only if (1) if a buyer i and a seller j exchange a good, then $v_i \geq p_j \geq 0$ and $p_j = \min\{p_k | s_k \in L(b_i)\}$,(2) if a buyer i does not buy a good then $v_i \leq \min\{p_k | s_k \in L(b_i)\}$ and (3) if a seller j does not sell a good then $p_j = 0$.*[8]

The first two requirements are that there is no excess demand: given prices $\mathbf{p}$, a buyer would want to buy a seller's good if and only if it is assigned the good in A. That is, no more than one buyer demands any seller's good. The last requirement is that there is no excess supply: given $\mathbf{p}$, a seller would want to supply a good if and only if it provides a good in A. That is, no seller that does not have a buyer would wish to sell a good.

Our first set of results characterizes these competitive price vectors and allocations.

We show that each competitive price vector and allocation $(\mathbf{p}, A)$ yields payoffs $(\mathbf{u}^b, \mathbf{u}^s)$ that are both *individually rational* and *pairwise stable.* Individual rationality simply requires that no agent earn negative payoffs. Pairwise stablity requires no linked buyer and seller can generate more surplus together than they earn in their joint payoffs. Formally,

---

[7] We can write $w(\mathbf{v}, A) = \mathbf{v} \cdot A \cdot 1$, where 1 is an $\overline{S} \times 1$ matrix where each element is 1.

[8] While these conditions may appear asymmetric with respect to buyers and sellers, they are not. More complicated notation would allow us to define competitive prices in an obviously symmetric manner. We have chosen to use the simpler notation in the text. The following definition is payoff equivalent to the one above with appropriate translation of notation: Consider a price vector (p_j^i) for $i = 1, \ldots, \overline{B}$ and $j = 1, \ldots, \overline{S}$. A price vector and allocation A are then competitive if and only if (1) if a buyer i and a seller j exchange a good, then $v_i \geq p_j^i \geq 0$ and $p_j^i = \min\{p_k^i | s_k \in L(b_i)\}$ and $p_j^i = \min\{p_j^k | b_k \in L(s_j)\}$; (2) if a buyer i does not buy a good then $v_i \leq \min\{p_k^i | s_k \in L(b_i)\}$ and (3) if a seller j does not sell a good then $0 = \min\{p_j^k | b_k \in L(s_j)\}$.

Definition 2. *A feasible*[9] *payoff vector* $(\mathbf{u}^b, \mathbf{u}^s)$ *is stable if and only if (i) (individual rationality)* $u_i^b \geq 0$; $u_j^s \geq 0$ *for all* i, j; *and (ii) (pairwise stability)* $u_i^b + u_j^s \geq v_i$ *for all linked pairs* b_i *and* s_j.[10]

We should expect any model of competition or negotiations in networks to yield stable payoffs, absent undue frictions in the negotiation process. It is straightforward to show that our "supply equals demand" conditions for prices and allocations are equivalent to these stability conditions. For example, if at given prices, only one buyer demands any seller's good, then there is no buyer that could offer a seller a different price that would make them both better off.

Proposition 1. *If* $(\mathbf{p}, A)$ *is competitive, then* $\mathbf{u}^s = \mathbf{p}$ *and the associated payoffs for buyers* $\mathbf{u}^b$ *are stable. If* $(\mathbf{u}^b, \mathbf{u}^s)$ *are stable payoffs, then there is an allocation* A *and the price vector* $\mathbf{p} = \mathbf{u}^s$ *such that* $(\mathbf{p}, A)$ *is competitive.*

Proof. Proofs are provided in the Appendix.

Our next result shows that a competitive price and allocation $(\mathbf{p}, A)$ always involves an efficient allocation of goods.[11] For a graph $\mathcal{G}$ and valuation $\mathbf{v}$, an *efficient allocation* of goods yields the greatest possible social surplus and is defined as follows:

Definition 3. *For a given* $\mathbf{v}$, *a feasible allocation* A *is efficient if and only if, given* $\mathcal{G}$, *there does not exist any other feasible allocation* A' *such that* $w(A'; \mathbf{v}) > w(A; \mathbf{v})$.

It is easy to understand why competitive prices are always associated with efficient allocations. If it were not the case, then there would be excess demand for some seller's good. A buyer that is not purchasing but has a higher valuation than a purchasing buyer would also be willing to pay the sales price.

Proposition 2. *For a graph* $\mathcal{G}$ *and valuation* $\mathbf{v}$, *if a price vector and allocation* $(\mathbf{p}, A)$ *is competitive, then* A *is an efficient allocation.*

We next present a result that greatly simplifies the analysis of competitive prices and allocations. The first part of the proposition shows the "equivalence" of efficient allocations: in any efficient allocation, the same set of buyers obtains goods.[12] The second part of the proposition shows that the set of competitive price

[9] Feasiblility requires that payoffs can derive from a feasible allocation of goods. The payoffs $(\mathbf{u}^b, \mathbf{u}^s)$ are *feasible* if there is a feasible allocation A such that (i) $u_i^b = 0$ for any buyer i who does not obtain a good, (ii) $u_j^s = 0$ for any seller j who does not sell a good, and (iii) $\sum_i u_i^b + \sum_j u_j^s = w(A; \mathbf{v})$.

[10] We do not write the stability condition for buyers and sellers that are not linked because it is always trivially satisfied.

[11] This and the remainder of the results in this section derive from basic results on assignment games. Assignment games consider stable pairwise matching of agents in settings such as marriage "markets." In our setting, the value of matches would be given by $\mathbf{v}$ and the graph $\mathcal{G}$. Shapley and Shubik (1972) develop the basic results we use in this section. Roth and Sotomayor (1990, Chap. 8) provide an excellent exposition. We refer the reader to their work and our (1998) working paper for proofs and details.

[12] Proposition 3 below requires generic valuations. Otherwise, efficient allocations could involve different sets of buyers. For example, in a network with one seller and two linked buyers, if the two

vectors is the same for all efficient allocations. With this result we can ignore the particular efficient allocation and refer simply to the set of competitive price vectors for a graph $\mathcal{G}$ and valuation $\mathbf{v}$. The result implies that the set of agents' competitive payoffs is uniquely defined; it is the same for all efficient allocations of goods.

Proposition 3. *For a network $\mathcal{G}$ and valuation $\mathbf{v}$:*

(a) If A and A' are both efficient allocations, then a buyer obtains a good in A if and only if it obtains a good in A'.

(b) If for some efficient allocation A, $(\mathbf{p}, A)$ is competitive, then for any efficient allocation A', $(\mathbf{p}, A')$ is also competitive.

Our final result of this section shows that the set of competitive price vectors for a graph $\mathcal{G}$ and valuation $\mathbf{v}$ has a well-defined structure. Competitive price vectors exist, and convex combinations of competitive price vectors are also competitive. There is a maximal and a minimal competitive price vector. The maximal price vector gives the best outcome for sellers, and the minimal price vector gives the best outcome for buyers. We will later examine how changes in the network structure affect these bounds.

Proposition 4. *The set of competitive price vectors is nonempty and convex. It has the structure of a lattice. In particular, there exist extremal competitive prices $\mathbf{p}^{\max}$ and $\mathbf{p}^{\min}$ such that $\mathbf{p}^{\min} \leq \mathbf{p} \leq \mathbf{p}^{\max}$ for all competitive prices $\mathbf{p}$. The price $\mathbf{p}^{\min}$ gives the worst possible outcome for each seller and the best possible outcome for each buyer. $\mathbf{p}^{\min}$ gives the opposite outcomes.*

2.3 Competitive Prices, Opportunity Cost and Network Structure

In this section, we determine the relationship between network structure and the set of competitive prices. To do so, we use the notion of "outside options" to characterize the extremal competitive prices; that is, we relate $\mathbf{p}^{\max}$ and $\mathbf{p}^{\min}$ to agents' next-best exchange opportunities. We will see that the private value of these opportunities can be determined by quite distant indirect links. The relationships we derive below are a basis for our comparative static results on changes in the link pattern.

We first formalize the physical connection between a buyer, its exchange opportunities, and its direct and indirect competitors. A buyer's exchange opportunities and competitors in a network are determined by its links to sellers, these sellers' links to other buyers, and so on. In a graph $\mathcal{G}$, we denote a *path* between two agents as follows: a path between a buyer i and a buyer m is written as $b_i - s_j - b_k - s_l - b_m$, meaning that b_i and s_j are linked, s_j and b_k are

buyers have the same valuation v, then either one could obtain the good. The assumption of generic valuations simplifies our proofs, but the results obtain for all valuations. If two efficient allocations A and A' involve different sets of buyers, and if $(\mathbf{p}, A)$ is competitive then there is a $(\mathbf{p}', A')$ that gives the same payoffs and is also competitive. That is, each allocation is associated with the same set of stable payoffs. In the two buyer example, for instance, the buyer obtaining the good always pays $p = v$ and both buyers earn $u^b = 0$.

linked, b_k and s_l are linked, and finally s_l and b_m are linked. For a given feasible allocation A, we use an arrow to indicate that a seller j's good is allocated to a buyer $k : s_j \rightarrow b_k$.

For a feasible allocation A, we define a particular kind of path, an *opportunity path,* that connects an agent to its alternative opportunities and the competitors for those exchanges. Consider some buyer which we label b_1. We write an opportunity path connecting buyer 1 to another buyer n as follows:

$$b_1 - s_2 \rightarrow b_2 - s_3 \rightarrow \ldots b_{n-1} - s_n \rightarrow b_n \ .$$

That is, buyer 1 is linked to seller 2 but not purchasing from seller 2. Seller 2 is selling to buyer 2, buyer 2 is linked to seller 3, and so on until we reach b_n. An opportunity path begins with an "inactive" link, which gives buyer 1's alternative exchange. The path then alternates between "active" links and "inactive" links, which connect the direct and indirect competitors for that exchange. Since the path must be consistent both with the graph and the allocation, we refer to a path as being "in $(A, \mathcal{G})$. " We say a buyer has a "trivial" opportunity path to itself.

Opportunity paths determine the set of competitive prices. We next show that $\mathbf{p}^{\max}$ and $\mathbf{p}^{\min}$ derive from opportunity paths in $(A^*, \mathcal{G})$, where A^* is an efficient allocation of goods for a given valuation v. The results show how prices relate to third party exchanges along an opportunity path and build on the following reasoning. Suppose for given competitive prices, some buyer 1 obtains a good from a seller 1 at price p_1. Suppose further that buyer 1 is also linked to a seller 2, through which it has an opportunity path to a buyer n, as specified above. Because buyer 1 does not buy from seller 2 and prices are competitive, it must be that $p_2 \geq p_1$. That is, seller 2's price is an upper bound for p_1. Furthermore, since buyer 2 buys from seller 2 but not seller 3, it must be that $p_3 \geq p_2$. That is, seller 3's price provides an upper bound on p_2 and hence on p_1. Repeating this argument tells us that buyer 1's price is bounded by the prices of all the sellers on the path. That is, *if a buyer buys a good, the price it pays can be no higher than the prices paid by buyers along its opportunity paths.*[13]

Building on this argument, let us characterize $\mathbf{p}^{\max}$. No price paid by any buyer is higher than its valuation. Therefore, $p_1^{\max}$ is no higher than the lowest valuation of any buyer linked to buyer 1 by an opportunity path. We label this valuation $v^L(b_1)$.[14] Our next result shows that when $p_1^{\max} \neq 0$, it exactly equals $v^L(b_1)$. To prove this, we argue that we can raise $p_1^{\max}$ up to $v^L(b_1)$ without violating any stability conditions. When the price of exchange between a buyer and a seller changes, the stability conditions of all linked sellers and buyers change as well. The proof shows that we can raise the prices simultaneously for a particular group of buyers in such a way as to maintain stability for all buyer-seller pairs. For $p_1^{\max} = 0$, we show that buyer 1 has an opportunity path to a buyer that is linked to a seller that does not sell its good. This buyer obtains a price of 0, which then forms an upper bound for buyer 1's price. We have:

[13] This observation is central to the proofs of most of our subsequent results. We present is as a formal lemma in the appendix.

[14] Since buyer 1 has an opportunity path to itself, $p_1^{\max} \leq v_l$.

Proposition 5. *Suppose that in $(A^*, \mathscr{G})$, a buyer 1 obtains a good from a seller 1. If $p_1^{\max} > 0$, then $p_1^{\max} = v^L(b_1)$ where $v^L(b_1)$ is the lowest valuation of any buyer linked to buyer 1 by an opportunity path. If $p_1^{\max} = 0$, then buyer 1 has an opportunity path to a buyer that is linked to a seller that does not sell its good.*

We can understand the value $v^L(b_1)$ as buyer 1's "outside option" when purchasing from seller 1. If buyer 1 does not purchase from seller 1, the worst it can possibly do is pay a price of $v^L(b_1)$ to obtain a good. This is the valuation of the buyer that buyer 1 would displace by changing sellers. This displaced buyer could be arbitrarily distant from buyer 1. Buyer 1 can purchase from a new seller and, in the process, displace a buyer n on an opportunity path pictured above as, follows: buyer 1 obtains a good from seller 2 whose former buyer 2 now purchases from seller 3 whose former buyer 3 now purchases from seller 4 and so on, until we reach buyer n who no longer obtains a good. In order to accomplish this displacement, buyer 1 must pay its new seller a price of at least v_n. This price becomes a lower bound for the prices paid along the opportunity path, and is just high enough so that buyer n is no longer interested in purchasing a good. As indicated by the above Proposition, the easiest such buyer to displace is the one with the lowest valuation on opportunity paths from buyer 1.

We next characterize the minimum competitive price $\mathbf{p}^{\min}$ in terms of opportunity paths. The opportunity paths from a seller also determine a seller's "outside option." Consider a seller 1 that is selling to buyer 1. We write an opportunity path connecting seller 1 to another buyer n as follows:

$$s_1 - b_2 \rightarrow s_2 - b_3 \rightarrow \ldots s_{n-1} - b_n \ .$$

The path begins with an "inactive" link, then alternates between active and inactive links and ends with a buyer. If s_1 has opportunity path(s) to buyers that do not obtain goods, s_1 will receive $p_1 > 0$. The non-purchasing buyers at the end of the paths set the lower bound of p_1. If p_1 were lower than these buyers' valuations, there would be excess demand for goods. Therefore, $p_1^{\min}$ must be no lower than the highest valuation of these non-purchasing buyers. We label this valuation $v^H(s_1)$; it is the highest valuation of any buyer that does not obtain a good and is linked to seller 1 by an opportunity path. The proof of the next result shows that if $p_1^{\min} > 0$, then $p_1^{\min}$ is exactly equal to $v^H(s_1)$. As in the previous proposition, we show this by supposing $p_1^{\min} > v^H(s_1)$ and showing it is possible to decrease the price in such a way as to maintain all stability conditions. If and only if $p_1^{\min} = 0$, then s_1 has no opportunity paths to buyers that do not obtain goods. We have

Proposition 6. *Suppose that in $(A^*, \mathscr{G})$, a buyer 1 obtains a good from a seller 1. If $p_1^{\min} > 0$ then $p_1^{\min} = v^H(s_1)$, where $v^H(s_1)$ is the highest valuation of any buyer that does not obtain a good and is linked to seller 1 by an opportunity path. If and only if $p_1^{\min} = 0$, then all buyers linked to seller 1 by an opportunity path obtain a good in A.*

We can understand the value $v^H(s_1)$ as seller 1's "outside option" when selling to buyer 1. The worst seller 1 can do if it does not sell to buyer 1 is earn a price $v^H(s_1)$ from another buyer. This price is the valuation of the buyer that would replace buyer 1 in the allocation of goods. The replacement occurs along an opportunity path from seller 1 to a buyer n as follows: seller 1 no longer sells to buyer 1, but sells instead to buyer 2, whose former seller 2 now sells to buyer 3, and so on until seller $n-1$ now sells to buyer n. To accomplish this replacement, seller 1 can charge its new buyer a price no more than v_n. This price forms a new lower bound on the opportunity path, and is just low so that the new buyer b_n is willing to buy. Out of all the buyers n that could replace buyer 1 in this way, the best for seller 1 is the buyer with highest valuation.[15]

We conclude the section with a summary of our results on the set of competitive prices and opportunity paths.

Proposition 7. *A price vector* $\mathbf{p}$ *is a competitive price vector if and only if for an efficient allocation A,* $\mathbf{p}$ *satisfies the following conditions: if a buyer i and a seller i exchange a good, then* $v^L(b_i) \geq p_i \geq v^H(s_i)$ *and* $p_i = \min\{p_k | s_k \in L(b_i)\}$*, (ii) if a seller i does not sell a good then* $p_i = 0$

We illustrate these results in the example below. We show the efficient allocation of goods and derive the buyer-optimal and the seller-optimal competitive prices, $\mathbf{p}^{\min}$ and $\mathbf{p}^{\max}$, from opportunity paths.

Example 1. For the network in Fig. 2 below, suppose buyers' valuations have the following order: $v_2 > v_3 > v_4 > v_5 > v_6 > v_1$. The efficient allocation of goods involves b_2 purchasing from s_1, b_3 from s_2, b_4 from s_3, and b_6 from s_4, as indicated by the arrows. In a competitive price vector $\mathbf{p}, p_1$ is in the range $v_3 \geq p_1 \geq v_1$: To find $p_1^{\min}$, we look for opportunity paths from s_1. Seller 1 has only one opportunity path to a buyer that does not obtain a good – to b_1. Therefore, $v^H(s_1) = v_1$. For $p_1^{\max}$ we look for opportunity paths from b_2. Buyer 2 has only one opportunity path – to b_3.[16] Therefore, $v^L(b_2) = v_3$. The price p_2 for seller 2 is in the range $v_3 \geq p_2 \geq v_5$: Seller 2 has two opportunity paths to buyers who do not obtain a good – to b_1 and b_5. Since $v_5 > v_1$, $v^H(s_2) = v_5$. Buyer 3 has only a "trivial" opportunity path to itself. Therefore, $v^L(b_3) = v_3$. We can, similarly, identify the maximum and minimum prices for s_3 and s_4, giving us $\mathbf{p}^{\min} = (v_1, v_5, v_5, 0)$ and $\mathbf{p}^{\max} = (v_3, v_3, v_4, v_6)$. Any convex combination of these upper and lower bounds, $(\beta v_1 + (1-\beta)v_3, \beta v_5 + (1-\beta)v_3; \beta v_5 + (1-\beta)v_4, (1-\beta)v_6)$ where $\beta \in [0, 1]$, are also competitive prices.

3 Network Comparative Statics

In this section we explore how changes in a network impact agents' competitive payoffs.

[15] Note that $v^H(s_1)$ is exactly the social opportunity cost of allocating the good to buyer 1. If buyer 1 did not purchase, the buyer that would replace it in an efficient allocation of goods, the "next-best" buyer, has valuation of $v^H(s_1)$.

[16] The path from b_2 to b_4, for example, is not an opportunity path, because it does not alternative between inactive and active links.

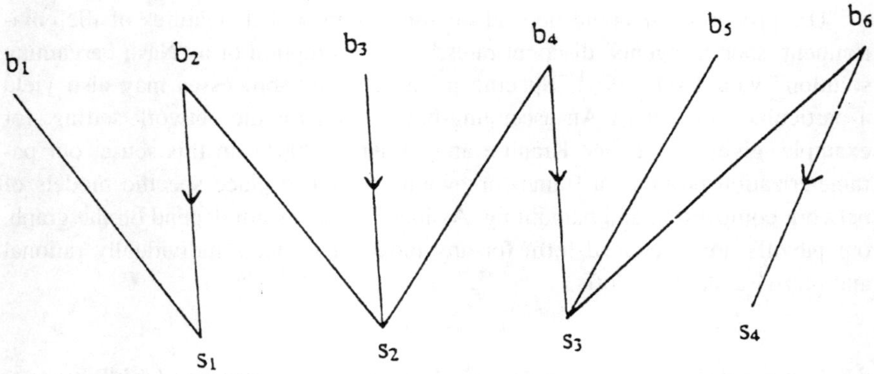

Fig. 2.

3.1 Payoffs as Functions of the Graph

To compare payoffs between graphs, we first make a unique selection from set of competitive payoffs for each graph. For a graph $\mathscr{G}$ and valuation $\mathbf{v}$, we define the price vector $\mathbf{p}(\mathscr{G}) \equiv q\mathbf{p}^{\min}(\mathscr{G})+(1-q)\mathbf{p}^{\max}(\mathscr{G})$, where $q \in [0,1]$ and $\mathbf{p}^{\min}(\mathscr{G})$ and $\mathbf{p}^{\max}(\mathscr{G})$ are the lowest and highest competitive prices for $\mathscr{G}$ given $\mathbf{v}$. We assume that q is the same for all graphs and valuations. By Proposition 4, the set of competitive prices is convex, so the price vector $\mathbf{p}(\mathscr{G})$ is competitive. For a given q and given valuation $\mathbf{v}$, let $u_i^b(\mathscr{G})$ and $u_j^s(\mathscr{G})$ denote the competitive payoffs of buyer i and seller j as a function of $\mathscr{G}$. Taking an efficient allocation for $(\mathscr{G}, \mathbf{v})$, for a buyer i that purchases from seller j, we have $u_j^s(\mathscr{G}) = p_j(\mathscr{G})$ and $u_i^b(\mathscr{G}) = v_i - p_j(\mathscr{G})$. Buyers who do not obtain a good receive a payoff of zero, as do sellers who do not sell a good.

This parameterization allows us to focus on how changes in a network affect an agent's "bargaining power." With q fixed across graphs, the difference in an agent's ability to extract surplus depends on the changes in the outside options, as determined by the graphs. We can see this as follows: The total surplus of an exchange between a buyer i and a seller j is v_i. Of this surplus, in graph $\mathscr{G}$ a buyer i earns at least its outside option $v_i - p_j^{\max}(\mathscr{G}) = v_i - v^L(b_i)$, where $v^L(b_i)$ is derived from the opportunity paths in $\mathscr{G}$. Similarly, seller j earns at least its outside option $p_j^{\max}(\mathscr{G}) = v^H(s_i)$. The buyer then earns a proportion q of the remaining surplus, and the seller earns a proportion $(1-q)$. We have

$$u_i^b(\mathscr{G}) = v_i - v^L(b_i) + q\left[v_i - \left(v_i - v^L(b_i) + v^H(s_j)\right)\right] = v_i - p_j(\mathscr{G}),$$
$$u_j^s(\mathscr{G}) = v^H(s_j) + (1-q)\left[v_i - \left(v_i - v^L(b_i) + v^H(s_j)\right)\right] = p_j(\mathscr{G}).$$

A change in the graph would impact $v^L(b_i)$ and $v^H(s_j)$ through a change in an agent's set of opportunity paths, and thereby affect agents' shares of the total surplus from exchange.[17]

[17] This approach to "bargaining power" is often used in the literature on specific assets. For instance, in a bilateral settting Grossman and Hart (1986) fix a 50/50 split ($q = 1/2$) of the surplus net agents' outside options. They then analyze how different property rights change agents' outside options.

The proportion q could depend on some (unmodeled) features of the environment, such as agents' discount rates.[18] An assumption of a "Nash bargaining solution" would set $q = \frac{1}{2}$. Specific price formation processes may also yield a particular value of q. An ascending-bid auction for the network setting, for example, gives $q = 1$ (see Kranton and Minehart 2001). In this sense, our parameterization provides a framework within which to place specific models of network competition and bargaining. As long as q does not depend on the graph, our payoffs are a reduced form for any model that yields individually rational and pairwise stable payoffs.

3.2 Comparative Statics on an Agent's Network: Population and Link Pattern

We now study how changes in a network affect agents' competitive payoffs. We show changes in payoffs for any valuation $\mathbf{v}$. We first consider the payoff implications of adding a link, holding fixed the number of buyers and sellers. We then consider adding new sets of buyers or sellers to a network. A priori, the impact of these changes is not obvious. As mentioned in the introduction, there are possibly many externalities from changing the link pattern.

Adding Links. We begin with preliminary results to help identify the source of price changes when a link is added to a network. Consider a link pattern $\mathcal{G}$ and add a link between a buyer and a seller that are not already linked. Denote the buyer b_a, the seller s_a, and the augmented graph $\mathcal{G}'$. The first result shows that an efficient allocation A' for $\mathcal{G}'$ involves at most one new buyer with respect to an efficient allocation A for $\mathcal{G}$. We can trace all price changes to this buyer. This buyer either replaces a buyer that purchased in A or is simply added to this set of buyers. It is also possible that no new buyer obtains a good. In this case, the second result says we can simply restrict attention to an allocation that is efficient in both graphs.

If a new buyer does obtain a good, the efficient allocation changes along what we call a *replacement path,* a form of opportunity path. The new buyer n, which we call the *replacement buyer,* obtains a good from a seller, whose previous buyer obtains a good from a new seller, and so on, along an opportunity path from buyer n to some buyer 1. Buyer 1 either no longer obtains a good or obtains a good from a previously inactive seller. Critically, we show that this replacement involves the new link, and no other changes can strictly improve economic welfare. (If any such improvement were possible, it could not involve the new link and so would have been possible in the original graph $\mathcal{G}$, and, hence, the original allocation A could not have been efficient.)

Lemma 1. *For a given* $\mathbf{v}$*, if an efficient allocation* A *for* $\mathcal{G}$ *and an effcient allocation* A' *for* $\mathcal{G}'$ *involve different sets of buyers, then* A *and* A' *are identical except on a set of* n *or* $n+1$ *distinct agents* $\mathcal{H} = \{(s_1), b_1, s_2, b_2, \ldots, s_n, b_n\}$ *that*

[18] In bilateral bargaining with alternating offers, Rubinstein (1982) and others derives q from agents' relative rates of discount.

may or may not include the seller s_1. *These agents are connected by an opportunity path* $(s_1) \to b_1 - s_2 \to b_2 - s_2 \to \ldots b_{n-1} - s_n \to b_n$ *in* $(A', \mathcal{G}')$. *The path includes (relabeled) the agents with the additional link* $s_a \to b_a$. *In* $(A, \mathcal{G})$, *this path is in two pieces* $b_n - s_n \to b_{n-1} - \ldots s_{a+1} \to b_a$ *and* $s_a \to b_{a-1} - \ldots s_2 \to b_1 - (s_1)$, *with the new link between* s_a *and* b_a *the "missing" link. Buyer n obtains a good in* A' *but not in A. Buyer 1 obtains a good in A. Buyer 1 obtains a good in* A' *if and only if* $s_1 \in \mathcal{H}$.

Lemma 2. *For a given* **v**, *if an efficient allocation A for* $\mathcal{G}$ *and an efficient allocation* A' *for* $\mathcal{G}'$ *involve the same set of buyers, then A is efficient for both graphs.*

With these preliminary results, we can evaluate the impact of an additional link on payoffs for different buyers and sellers in a network. Our first result considers the direct effects of the link. We show that the buyer and seller with the additional link (b_a and s_a) enjoy an increase in their competitive payoffs. Intuitively, the buyer (seller) is better off with more direct sources of supply (demand).

Proposition 8. *For the buyer and seller with the additional link* (b_a *and* s_a), $u_a^b(\mathcal{G}') \geq u_a^b(\mathcal{G})$ *and* $u_a^s(\mathcal{G}') \geq u_a^s(\mathcal{G})$.

The result is proved by examining opportunity paths. Suppose that when the link is added, a new buyer (the replacement buyer) obtains a good. By Lemma 1, this buyer, b_n, has an opportunity path to b_a in $(A, \mathcal{G})$. Because b_n does not obtain a good in A, it must be facing a prohibitive "best" price of at least v_n. This can only happen if b_a's price, which is an upper bound of the prices of sellers along the opportunity path, is at least v_n. In $(A', \mathcal{G}')$, the direction of the opportunity path is reversed. That is, b_a now has an opportunity path to b_n. The price that b_n pays is now an upper bound on b_a's price. Since b_n pays at most its valuation, b_a's price is at most v_n. We have, thus, shown that b_a pays a (weakly) lower price and receives a higher payoff in $(A', \mathcal{G}')$.

Our next results consider the indirect effects of a link. One effect, as mentioned in the introduction, is the *supply stealing effect*. When a link is added between b_a and s_a, b_a can now directly compete for s_a's good. Buyers with direct or indirect links to s_a, then, should be hurt by the additional competition. Sellers should be helped. On the other hand, there is a *supply freeing effect*. When a link is added between b_a and s_a, b_a depends less on its other sellers for supply. Some buyer n that is not obtaining a good may now obtain a good from a seller $s_k \in L(b_a)$. With less competition for these sellers' goods, sellers should be hurt and buyers helped.

We identify the two types of paths in a network that confer these payoff externalities. If in $\mathcal{G}$, there is a path connecting an agent and b_a, we say the agent has a *buyer path* in $\mathcal{G}$. If in $\mathcal{G}$ there is a path connecting an agent and s_a, we say the agent has a *seller path* in $\mathcal{G}$. Buyer paths confer the supply freeing effect: A buyer i with a buyer path is indirectly linked to buyer a and is, thus, in competition for some of the same sellers' output. When b_a establishes a link with another seller, it frees supply for b_i. Sellers along the buyer path face lower demand and receive weakly lower prices. Seller paths confer the supply stealing effect: If b_i has a seller path, b_i faces more competition for s_a's good; that is, b_a steals supply from b_i. Competition for goods increases, hurting b_i and helping sellers along the seller path.

We use these paths to show how new links affect the payoffs of third parties. It might seem natural that the size of the externality depends on the length of the path. The more distant the new links, the weaker the effect. The next example shows that this is not the case. It is not the length of a path that matters, but how it is used in the allocation of goods.

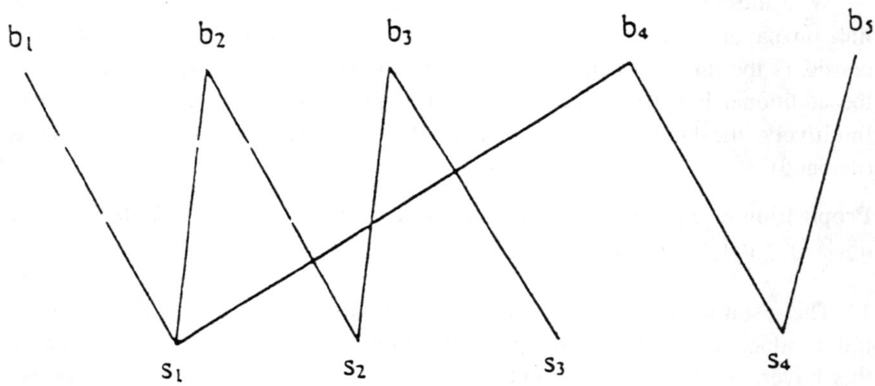

Fig. 3.

Example 2. In Figure 3 above, consider the impact on b_2 of a link between b_4 and s_3. b_2 has a short buyer path and a long seller path. However, the supply stealing effiect (through the seller path) dominates. Without the link between b_4 and s_3, b_2 always obtains a good (b_2 always buys from s_2, and b_3 always buys from s_3). With the link, b_2 is sometimes replaced by b_4 and no longer obtains a good. This occurs for particular valuations $\mathbf{v}$. For other $\mathbf{v}, b_2$ is not replaced, but the price it pays is weakly higher. Therefore, b_2's competitive payoffs fall for any $\mathbf{v}$.

We next show how the impact of buyer and seller paths depend on the network structure. Our first result demonstrates the payoff effects when an agent only has one type of path. Following results indicate payoff effects when agents have both buyer and seller paths.

If an agent has only a buyer path or only a seller path in $\mathscr{G}$, the effect of the new link on its payoffs is clear. A buyer that has only a buyer path (seller path) is helped (hurt) by the additional link. A seller that has only a buyer path (seller path) is hurt (helped) by the additional link. We have the following proposition, which we illustrate below.

Proposition 9. *For a buyer i that has only buyer paths in $\mathscr{G}$, $u_i^b(\mathscr{G}') \geq u_i^b(\mathscr{G})$. For a buyer i that has only seller paths in $\mathscr{G}$, $u_i^b(\mathscr{G}') \leq u_i^b(\mathscr{G})$. For a seller j that has only buyer paths in $\mathscr{G}$, $u_j^s(\mathscr{G}') \leq u_j^s(\mathscr{G})$. For a seller j that has only seller paths in $\mathscr{G}$, $u_j^s(\mathscr{G}') \geq u_j^s(\mathscr{G})$.*

Example 3. In the following graph, consider adding a link between buyer 4 and seller 3. Sellers 1 and 2 have only seller paths and are better off. Seller 4 with only a buyer path is worse off. Buyer 5 is better off because it has only a buyer path. Buyers 1, 2, and 3 with only seller paths are worse off.

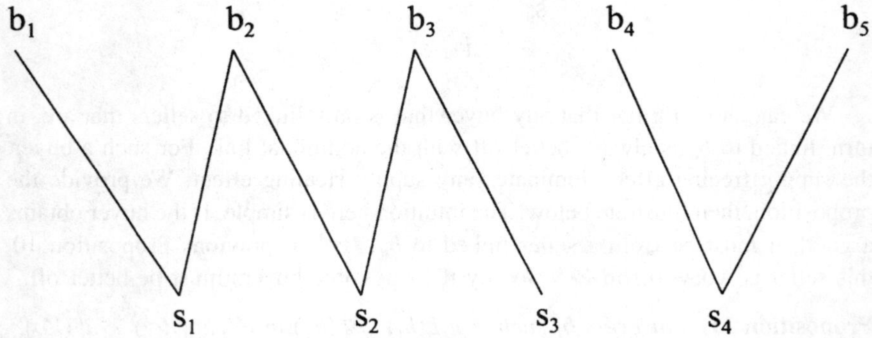

Fig. 4.

When agents have both buyer and seller paths, the overall impact on payoffs is less straightfoward. Supply freeing and supply stealing effects go in opposite directions. In many cases, however, we can determine the overall impact of a new link. We begin with the agents that have links to b_a or s_a. We show that the buyers (sellers) linked to the seller (buyer) with the additional link are always weakly worse off. For buyers (sellers), the supply stealing (freeing) effect dominates.

Proposition 10. *For every $b_i \in L(s_a)$ in $\mathscr{G}$, $u_i^b(\mathscr{G}') \leq u_i^b(\mathscr{G})$. For $s_j \in L(b_a)$ in $\mathscr{G}$, $u_j^s(\mathscr{G}') \leq u_j^s(\mathscr{G})$.*

The proof argues that for these buyers (sellers), there is in fact no supply freeing (stealing) effect associated with the new link. To see this, consider a $b_i \in L(s_a)$. Potentially, b_i could benefit from the fact that b_a's new link frees the supply of b_a's *other* sellers. We show that this hypothesis contradicts the efficiency of the allocation A in $\mathscr{G}$. Suppose, for example, that b_i benefits from the new link because it is the replacement buyer. That is, b_i obtains a good in $(A', \mathscr{G}')$, but not in $(A, \mathscr{G})$. By Lemma 1, b_i replaces a buyer 1 along an opportunity path such as $b_1 - s_a \rightarrow b_a - s_i \rightarrow b_i$ in $(A', \mathscr{G}')$, as pictured below in Fig. 5 where the new link is dashed. That is, in this example, b_i obtained a good directly from s_a in A and does not obtain a good at all in A'. Then, since b_i is linked to s_a by hypothesis, b_i could have replaced buyer 1 along the path

$b_1 - s_a \to b_i$ in $\mathscr{G}$. If the replacement is efficient in $\mathscr{G}'$ then it is also efficient in $\mathscr{G}$. Hence, A could not have been an efficient allocation.[19]

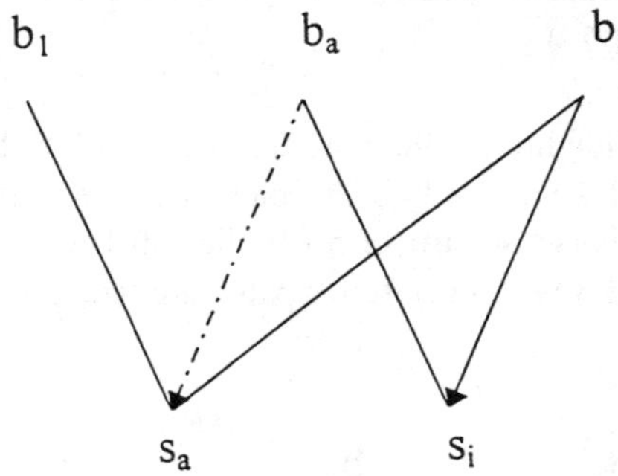

Fig. 5.

We can show further that any buyer that is only linked to sellers that are, in turn, linked to b_a is always better off with the additional link. For such a buyer, the supply freeing effect dominates any supply stealing effect. We provide the proposition, then illustrate below. The intuition here is simple. If the buyer obtains a good, it must be from a seller linked to b_a. By our previous Proposition 10, this seller is worse off in $\mathscr{G}'$. So any if its possible buyers must be better off.

Proposition 11. *For every b_i such that $L(b_i) \subseteq L(b_a)$ in $\mathscr{G}$, $u_i^b(\mathscr{G}') \geq u_i^b(\mathscr{G})$.*

The next example shows how to apply this and previous results to evaluate the impact of a link in a given graph.

Example 4. In Fig. 6 below, consider the impact of a link between b_3 and s_2. By Proposition 8, b_3 and s_2 both enjoy an increase in their competitive payoffs. By Proposition 10, b_2, s_1, b_4, and s_3 all have lower payoffs. By Proposition 11, b_1 and b_5 have higher payoffs. We can further show that b_6 has higher payoffs and s_4 has lower payoffs, since their only paths to the agents with the additional link is through b_5.

Changing network by adding buyers of sellers. We conclude our analysis by placing the above propositions in the context of earlier results on assignment games. The literature on assignment games considered adding agents on one side of a matching "market." In our framework, this would be equivalent to adding new buyers, or sellers, to a network. In this case, what we call the supply stealing/freeing effects are easier to analyze because the new buyers or sellers do not have any existing links; buyers and sellers are added along with all their links. Intuitively, adding a new seller can only free supply and adding a new buyer can only steal it.

[19] The proof of Proposition 10 involves some subtlety. For example, the result does not generalize to buyers linked to sellers linked to buyers linked to s_a.

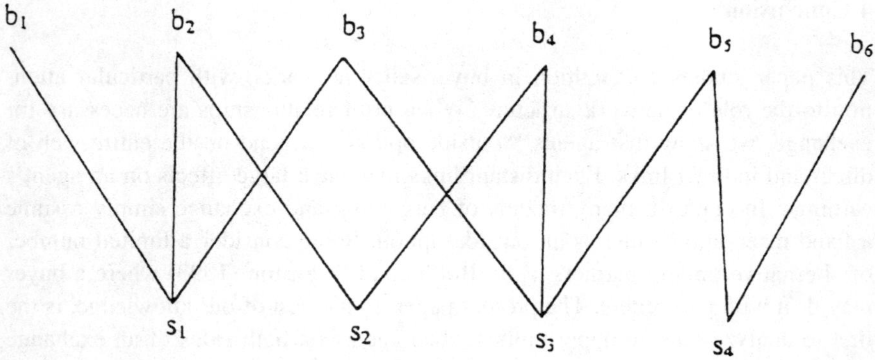

Fig. 6.

The results below show that indeed adding a seller (buyer) along with all its links must cause a net supply freeing (stealing) effect. The above intuition aside, these results are interesting because, given the necessity of links for exchange in a network, adding buyers (sellers) does not necessarily increase (decrease) the *effective* buyer/seller ratio of an agent. For example, in the network in Fig. 1, suppose s_1 is subtracted from the network. Because s_1 provides the links to the rest of the network for buyers 1 and 2, these buyers are also effectively removed, and the buyer-seller ratio would decrease for the remaining agents. At first glance, it would seem that a lower buyer-seller ratio should help some buyers and hurt some sellers. The next result, however, shows that this is not the case.

A buyer is always better off when sellers are added to its network, regardless of the number of new buyers that compete for the albeit increased supply. Sellers are always worse off. The proposition is proved by an application of an earlier result due to Demange and Gale (1985, Corollary 3).

Proposition 12. *Consider a graph $\mathscr{G}$ linking a collection of buyers and sellers. Add any set of sellers S together with arbitrary links to the graph, and let $\mathscr{G}'$ denote the new graph. For all buyers i we have $u_i^b(\mathscr{G}) \leq u_i^b(\mathscr{G}')$. For every seller $j \notin S$, we have $u_j^s(\mathscr{G}) \geq u_j^s(\mathscr{G}')$.*

We have an analogous result for adding buyers to a network. A seller is always better off, regardless of the number of competing sellers that are effectively added to the seller's network. A buyer is always worse off.

Proposition 13. *Consider a graph $\mathscr{G}$ linking a collection of buyers and sellers. Add any set of buyers B together with arbitrary links to the graph, and let $\mathscr{G}'$ denote the new graph. For all sellers j, we have $u_j^s(\mathscr{G}) \leq u_j^s(\mathscr{G}')$. For every buyer $i \notin B$, we have $u_i^b(\mathscr{G}) \geq u_i^b(\mathscr{G}')$.*

4 Conclusion

This paper studies competition in buyer-seller networks, with particular attention to the role of network structure. When prior relationships are necessary for exchange, we show that agents' "outside options" depend on the entire web of direct and indirect links. Even distant links may have large effects on an agent's earnings. In contrast, many models of bargaining and exchange simply assume a fixed reservation value as an outside option. Some consider a limited number of alternative trading partners, as in Bolton and Whinston (1993) where a buyer may deal with two sellers. The present paper, to the best of our knowledge, is the first to analyze outside opportunities when agents on both sides of an exchange can have multiple alternative partners.[20]

We first develop a general model of network competition. This model characterizes prices that satisfy a natural "supply equals demand" condition for the network setting. Resulting payoffs are both individually rational and pairwise stable. No individual agent or pair of agents can do better. Any specific model of competition that yields individually rational, pairwise stable payoffs can be represented by our payoff functions. A parameter $q \in [0, 1]$ allows for different splits of the surplus of exchange net agents' "outside options." It is these outside options, then, that determine an agent's bargaining power.

We show how these outside options derive from a given network structure. We define a particular type of path in a network called an *opportunity path*. *Opportunity paths* connect agents to their alternative exchange opportunities and to their direct and indirect competitors. These paths determine agents' outside options. For example, the (perhaps indirect) demand from a (perhaps distant) buyer along a path ensures that a seller receive at least a certain price elsewhere, if it does not sell to its current buyer. This distant demand gives the seller its outside option, which guarantees at least a certain share of the surplus from exchange.

Finally, we consider how changes in third parties' links impact agents' payoffs. That is, we conduct comparative statics on the link pattern. Again, we parse a network into paths. *Seller (buyer) paths* connect an agent to a particular seller (buyer) and tell us whether it will be helped or hurt by that seller's (buyer's) additional links. Seller paths generate what we call a *supply stealing effect*, since the new link establishes an additional source of demand. Buyer paths generate a *supply freeing effect*, since the new link establishes another source of supply. Using these paths, we prove several results about differently connected buyers and sellers.

In conclusion, the paper provides a general model of "outside options," and hence bargaining power, when exchange is limited by pre-existing relationships. The model of network competition can serve as a reduced form for specific models of competition and bargaining. The network structure, through opportunity

[20] A series of papers considers price formation in a market where anonymous buyers and sellers meet pairwise (e.g. Gale, 1987, Rubinstein and Wolinsky, 1985). This effort differs from ours, because we require buyers and sellers to be linked in order to engage in exchange.

paths, buyer paths, and seller paths should affect the outcomes of these games in the same way as they affect our competitive payoffs. These payoff results, and the identification of the paths themselves, should also prove useful to further analysis of network structure.

Appendix

Proof of Proposition 1.
Part 1: We show that if a price vector and allocation $(\mathbf{p}, A)$ satisfies Conditions (1), (2), and (3) in the definition of a competitive price vector, then the associated payoff vector is stable.

Individual rationality: For each buyer i and seller j that exchange a good, individual rationality is satisfied since $0 \leq p_j \leq v_i$. Buyers and sellers that do not exchange goods all earn a payoff of 0 which is also individually rational.

Pairwise stability: First consider linked buyers and sellers that exchange goods in A: For a buyer i and seller j, $u_i^b + u_j^s = v_i - p_j + p_j = v_i$ satisfying pairwise stability. Next consider linked buyers and sellers that do not exchange goods in A. For each buyer i linked to seller k but obtains a good from seller j, the joint payoffs of buyer i and seller k are $u_i^b + u_k^s = v_i - p_j + p_k$. By condition (1), $p_j \leq p_k$, which implies that $u_i^b + u_k^s \geq v_i$, satisfying pairwise stability. For each buyer i linked to seller k who does not obtain a good from any seller, the joint payoffs of buyer i and seller k are $u_i^b + u_k^s = p_k$. Condition (2) implies that $p_k \geq v_i$, so $u_i^b + u_k^s \geq v_i$, satisfying pairwise stability.

Part 2: A stable payoff vector $(\mathbf{u}^b, \mathbf{u}^s)$ is defined for a feasible allocation of goods A. We show that $(\mathbf{p}, A)$ is competitive, where $\mathbf{p} = \mathbf{u}^s$. That is, we show that $(\mathbf{p}, A)$ derived from $(\mathbf{u}^b, \mathbf{u}^s)$ satisfies Conditions (1), (2), and (3).

Condition (1): For a buyer i purchasing from a seller j: Individual rationality implies that $0 \leq p_j \leq v_i$. Pairwise stability implies that $u_i^b + u_k^s = v_i - p_j + p_k \geq v_i$ for all $s_k \in L(b_i)$. This implies $p_k \geq p_j$ for all $s_k \in L(b_i)$, or, in other words, $p_j = \min\{p_k | s_k \in L(b_i)\}$.

Condition (2): For a buyer i that is not purchasing a good, by the definition of feasible stable payoffs, $u_i^b = 0$. Pairwise stability then implies that $0 + p_k \geq v_i$ for all $s_k \in L(b_i)$. That is, buyer i's valuation is lower than the price charged by any of its linked sellers: $v_i \leq \min\{p_k | s_k \in L(b_i)\}$.

Condition (3): For a seller j that is not selling a good, by the definition of feasible stable payoffs, $u_j^s = 0$, which implies $p_j = 0$. □

Lemma A1. *Suppose that in* $(A, \mathcal{G})$*, a buyer* 1 *has an opportunity path to a buyer* n*. Let* $\mathbf{p}$ *be a competitive price vector. If buyer* 1 *obtains a good from a seller* 1*, then* $p_1 \leq p_n$*. If buyer* 1 *does not obtain a good, then* $v_1 \leq p_n$*.*

Proof. Since $b_{n-1} \in L(s_n)$ but $s_{n-1} \to b_{n-1}$, we have $p_{n-1} \leq p_n$. Since $b_{n-2} \in L(s_{n-1})$, but $s_{n-2} \to b_{n-2}$, we have $p_{n-2} \leq p_{n-1}$. Repeating this reasoning, we obtain $p_2 \leq \ldots \leq p_n$. If buyer 1 obtains a good from a seller 1, then since $b_1 \in L(s_2)$, we have $p_1 \leq \ldots \leq p_n$ as desired. If buyer 1 does not obtain a good from seller 1, then since $b_1 \in L(s_2)$, we must have $v_1 \leq p_2 \leq \ldots \leq p_n$.

Lemma A2. *Suppose that in* $(A, \mathcal{G})$, *a buyer* 1 *has an opportunity path to a buyer* n. *Let* $\mathbf{p}$ *be a competitive price vector. If buyer* 1 *obtains a good from a seller* 1, *then* $p_1 \leq v_n$. *If buyer* 1 *does not obtain a good, then* $v_1 \leq v_n$.

Proof. By individual rationality, a buyer never pays a price higher than its valuation. Therefore $p_n \leq v_n$. If buyer 1 obtains a good from a seller 1, then Lemma A1 implies that $p_1 \leq v_n$. If buyer 1 does not obtain a good, then Lemma A1 implies that $v_1 \leq p_n$. So we have that $v_1 \leq p_n \leq v_n$ as desired.

Proof of Proposition 5. We show that $p_1^{\max}$ is exactly equal to $v^L(b_1)$, the lowest valuation of any buyer on an opportunity path from buyer 1. The logic is that we can raise p_1 to $v^L(b_1)$ without any violation of pairwise stability, but any higher price would violate pairwise stability.

Let $\hat{p}^{\max}(b_1) = \min\{p_k^{\max} | s_k \in L(b_1), k \neq 1\}$. By individual rationality and pairwise stability for buyer 1, we must have $p_1^{\max} \leq \min\{v_1, \hat{p}^{\max}(b_1)\}$. If $p_1^{\max} < \min\{v_1, \hat{p}^{\max}(b_1)\}$, then we can raise $p_1^{\max}$ up to $\min\{v_1, \hat{p}^{\max}(b_1)\}$ without violating pairwise stability for any buyer-seller pair containing b_1. Raising $p_1^{\max}$ also does not violate pairwise stability for other buyer-seller pairs, since other buyers in $L(s_1)$ already find seller 1's price to be prohibitively high. This contradicts the maximality of $p_1^{\max}$. So we must have $p_1^{\max} = \min\{v_1, \hat{p}^{\max}(b_1)\}$.

Let $B^{\hat{p}^{\max}(b_1)} = \{b_k | b_1$ has an o.p.("opportunity path") to b_k and b_k pays a price $p_k^{\max} = \hat{p}^{\max}(b_1)\}$. Fix any $b_k \in B^{\hat{p}^{\max}(b_1)}$. By pairwise stability, we have that for all $s_m \in L(b_k), p_m^{\max} \geq \hat{p}^{\max}(b_1)$. The inequality is strict ($p^{\max} > \hat{p}^{\max}(b_1)$) if and only if s_m sells its good to a buyer $b_m \notin B^{\hat{p}^{\max}(b_1)}$.

If $b_k \in B^{\hat{p}^{\max}(b_1)}$ then $v_k \geq \hat{p}^{\max}(b_1)$.

Case I: $p_1^{\max} > 0$.
We argue that if $p_1^{\max} > 0$, then there is a $b_k \in B^{\hat{p}^{\max}(b_1)}$ with $v_k = \hat{p}^{\max}(b_1)$. If $b_k \in B^{\hat{p}^{\max}(b_1)}$ then $v_k \geq \hat{p}^{\max}(b_1)$. Suppose that for all $b_k \in B^{\hat{p}^{\max}(b_1)}$, we have $v_k > \hat{p}^{\max}(b_1)$. If there is a $b_k \in B^{\hat{p}^{\max}(b_1)}$ linked to an inactive seller, then it must be that $\hat{p}^{\max}(b_1) = 0$ and hence that $p_1^{\max} = 0$ contradicting our assumption.

Otherwise, it is possible to raise the price $\hat{p}^{\max}(b_1)$ paid by all the buyers in $B^{\hat{p}^{\max}(b_1)}$ without violating stability: that is, individual rationality for any agent or pairwise stability for any pair of agents. The pairs affected are all those (b_i, s_j) for which either u_i^b or u_j^s changes. These are of two types: (i) (b_i, s_j) where $s_j \in L(b_i)$ and $b_i \in B^{\hat{p}^{\max}(b_1)}$ and (ii) (b_i, s_j) where $s_j \in L(b_i)$, $b_i \notin B^{\hat{p}^{\max}(b_1)}$, and s_j sells a good to some $b_k \in B^{\hat{p}^{\max}(b_1)}$.

To preserve stability, we raise $\hat{p}^{\max}(b_1)$ by a small enough amount that (1) for each $b_k \in B^{\hat{p}^{\max}(b_1)}$, the inequality $v_k > \hat{p}^{\max}(b_1)$ is still satisfied; and (2) for each $s_j \not{i}nL(b_k)$ selling to a buyer $b_j \notin B^{\hat{p}^{\max}(b_1)}$, the inequality $p_j^{\max} > \hat{p}^{\max}(b_1)$ is still satisfied.

Requirement (1) insures that individual rationality still obtains for all buyers whose payoffs have changed. Since sellers who sell goods to these buyers get higher prices, their payoffs are also individually rational.

Consider pairwise stability. First consider pairs (b_i, s_j) of type (i) above. We have argued that s_j must sell a good to some b_j. If $b_j \in B^{\hat{p}^{\max}(b_1)}$, then $p_j^{\max} = \hat{p}^{\max}(b_1)$. So s_j receives the same price that b_i pays and pairwise stability for (b_i, s_j) is trivial. If $b_j \notin B^{\hat{p}^{\max}(b_1)}$, then Requirement (2) insures pairwise stability for (b_i, s_j). Next consider pairs (b_i, s_j) of type (ii) above. The seller j receives a higher payoff $\hat{p}^{\max}(b_1)$ than before. Buyer i's payoff is unchanged, so pairwise stability still holds.

We have shown that we can raise $\hat{p}^{\max}(b_1)$ without violating the stability conditions for any agents. This is a contradiction to the assumption that the prices were maximal. Therefore, we must have $v_k = \hat{p}^{\max}(b_1)$ for some $b_k \in B^{\hat{p}^{\max}(1)}$. We can write $p_1^{\max} = \min\{v_1, v_k\}$.

We next argue that $p_1^{\max} = \min\{v_1, v_k\}$ is the lowest valuation out of all buyers linked to buyer 1 by an o.p. (including itself). If b_1 has an o.p. to any other buyer n then by Lemma A2, $p_1^{\max} \le v_n$ so that $\min\{v_1, v_k\} \le v_n$ as desired.

We have argued that if $p_1^{\max} > 0$, then $p_1^{\max} = v^L(b_1)$ where $v^L(b_1)$ is the lowest valuation out of all buyers linked to buyer 1 by an o.p. (including buyer 1 itself).

Case II: $p_1^{\max} = 0$.
Finally, suppose that $p_1^{\max} = 0$. By our genericity assumption, $v_k \neq 0$ for all buyers k. So if $p_1^{\max} = 0$, it must be that $\hat{p}^{\max}(b_1) = 0$. It follows from the proof above that there is a $b_k \in B^{\hat{p}^{\max}(b_1)}$ linked to an inactive seller. (Otherwise, we could raise $\hat{p}^{\max}(b_1)$ to be above 0 without violating pairwise stability.) We have thus shown that b_1 has an opportunity path to a buyer who is linked to a seller who does not sell its good.

Proof of Proposition 6. The proof is similar to the proof of Proposition 5 and is available from the authors on request.

Proof of Proposition 7. We will show the equivalence of conditions (i) and (ii) to the definition of a competitive price vector.

Neccessity: If $(\mathbf{p}, A)$ is a competitive price vector, then conditions (i) and (ii) are an immediate implication of Propositions 5 and 6.

Sufficiency: We show that a price vector satisfying conditions (i) and (ii) satisfies Conditions (1), (2), and (3) in the definition of a competitive price vector.

We first show that if a buyer i and seller j exchange a good, then $0 \le p_j \le v_i$. Since, by condition (i), $p_j \ge v^H(s_i) > 0$ (for generic $\mathbf{v}$), we have $p_j > 0$. Since buyer i has a trivial opportunity path to itself, $v^L(b_i) \le v_i$. Condition (i) that $p_j \le v^L(b_i)$ then implies $p_j \le v_i$.

We next show that if a buyer i does not obtain a good, then $v_i \le \min\{p_k | s_k \in L(b_i)\}$. Consider an $s_k \in L(b_i)$. If s_k is selling its good to some other buyer l, then $p_k \ge v^H(s_k)$. Since buyer i is on an opportunity path to seller k, it must be that $v^H(s_k) \ge v_i$. So $p_k \ge v_i$ as desired. If s_k is not selling its good, then since buyer i is not obtaining a good, there is a violation of efficiency (since $v_i > 0$ by our genericity assumption). □

Proof of Lemma 1. We restate the lemma, because the notation is important.

Lemma 1. *If efficient allocations in $\mathcal{G}$ and $\mathcal{G}'$ involve different sets of buyers, then there are efficient allocations A for $\mathcal{G}$ and A′ for $\mathcal{G}'$ such that A and A′ are identical except on a set of firms $\mathcal{H} = \{(s_1), b_1, s_2, b_2, \ldots, s_n, b_n\}$ that may or may not include the seller s_1. These firms are connected by a path $(s_1) \to b_1 - s_2 \to b_2 - s_2 \to \ldots b_{n-1} - s_n \to b_n$ in $(A', \mathcal{G}')$. The path includes (relabeled) $s_a \to b_a$. In $(A, \mathcal{G})$, the same firms are connected by two paths $b_n - s_n \to b_{n-1} - \ldots s_{a+1} \to b_a$ and $s_a \to b_{a-1} - \ldots s_2 \to b_1 - (s_1)$. Buyer n obtains a good in A′ but not in A. Buyer 1 obtains a good in A. Buyer 1 obtains a good in A′ if and only if $s_1 \in \mathcal{H}$.*

Proof. For each new buyer and any efficient allocations, we first construct paths that have the structure of the paths in the Lemma. We then argue that there can be at most one new buyer in A'. We then argue that we can choose the efficient allocations to have the desired structure.

Choose any efficient allocations A and A'. Let buyer n be a buyer that obtains a good in A' but not in A. Buyer n buys a good in A', say from seller n. If s_n did not sell a good in A then b_n should have obtained s_n's good in $(A, \mathcal{G})$ unless b_n and s_n were not linked. That is, unless $b_n = b_a$ and $s_n = s_a$, we have contradicted the efficiency of A. If $b_n = b_a$ and $s_n = s_a$, the hypothesis in the Lemma about opportunity paths is trivially satisfied.

Otherwise, it must be that s_n did sell a good in A, say to b_{n-1} where $b_{n-1} \neq b_n$. If b_{n-1} does not obtain a good in A', then the efficiency of A' implies that $v_{n-1} \leq v_n$. If $v_{n-1} < v_n$, this contradicts the efficiency of A because b_n could have replaced b_{n-1} in A. We rule out the case $v_{n-1} = v_n$ as non-generic.

So it must be that b_{n-1} does obtain a good in A', say from s_{n-1} where $s_{n-1} \neq s_n$. Repeating the above argument shows that s_{n-1} must have sold its good in A to a b_{n-2} who also obtains a good in A', and so on. Eventually, this process ends with $b_{n-k} = b_a$ and $s_{n-k} = s_a$ and $s_a \to b_a$ in $(A', \mathcal{G}')$. (By construction, the process always picks out agents not already in the path. Also by construction, the process does not end unless we reach $b_{n-k} = b_a$ and $s_{n-k} = s_a$, but it must end because the population of buyers is finite.)

We have constructed two paths. In A', we have constructed an opportunity path from b_a to b_n. In A, we have constructed an o.p. (opportunity path) from b_n to b_a.

If s_a is inactive in A, we now have paths that have the structure of the paths in the lemma. Otherwise, s_a sells its good to a buyer, say b_{n-k-1} in A. If b_{n-k-1} does not obtain a good in A' then we again now have paths that have the structure of the paths in the lemma. Otherwise b_{n-k-1} obtains a good in A' from a seller, say s_{n-k-1}. If this seller is inactive in A, we now have paths that have the structure of the paths in the lemma. And so on. Eventually, this process must end because it always picks out new agents from the finite population of agents. This constructs the paths in the lemma.

We next argue that there can be at most one new buyer in A'. Suppose there are two new buyers n and n'. For each one, we can construct a path to b_a and s_a as above. But this is a contradiction: since each seller has only one unit of capacity, it is impossible for the two paths from buyers n and n' to overlap.

Finally, we show that we can choose A and A' as in the hypothesis of the Lemma. Fix any efficient allocations A and A' and construct the paths as above. Suppose that the path construction process above ends with an inactive seller s_1. In $\mathscr{G}'$ at the allocation A, buyer n has a path to s_1 : $b_n - s_n \to b_{n-1} - \dots s_{a+1} \to b_a - s_a \to b_{a-1} - \dots s_2 \to b_1 - s_1$. We replace this with the path: $s_1 \to b_1 - s_2 \to b_2 - s_2 \to \dots b_{n-1} - s_n \to b_n$. This gives us an allocation $\tilde{A}'$ that is necessarily efficient in $\mathscr{G}'$ (the efficient set of buyers obtains goods) and is related to A as in the hypothesis of the lemma.

Suppose that the path construction process above ends with a buyer b_1 who does not obtain a good in $(A', \mathscr{G}')$. In $\mathscr{G}'$ at the allocation A, buyer n has an opportunity path to b_1 : $b_n - s_n \to b_{n-1} - \dots s_{a+1} \to b_a - s_a \to b_{a-1} - \dots s_2 \to b_1$. We replace this with the path: $b_1 - s_2 \to b_2 - s_2 \to \dots b_{n-1} - s_n \to b_n$. This gives us an allocation $\tilde{A}'$ that is necessarily efficient in $\mathscr{G}'$ (the efficient set of buyers obtains goods) and is related to A as in the hypothesis of the lemma. □

Proof of Lemma 2. Since by hypothesis the same set of buyers obtains goods in A as in A', A yields the same welfare as any efficient allocation in $\mathscr{G}'$. Since $\mathscr{G} \subset \mathscr{G}'$, A is also feasible in $\mathscr{G}'$ and hence efficient. □

We call the set $\mathscr{H}$ from Lemma 1 the *replacement set*. We also refer to the paths in the lemma as the *replacement paths*. Buyer n is the *replacement buyer*, and we say that buyer 1 is replaced by buyer n.

The next four lemmas will be used in proofs below. They use the notation and set up of Lemma 1. The first two characterize the maximal prices for buyers in the replacement set. There are corresponding results for the minimal prices. These second two results (which we state without proof) pin down the minimal prices quite strongly.

Lemma A3. *Let A and A' be efficient allocations in $\mathscr{G}$ and $\mathscr{G}'$ involving different sets of buyers as in Lemma 1. We assume the notation from Lemma 1. In $\mathscr{G}$, $v_n \le p_n^{\max} \le p_{n-1}^{\max} \dots \le p_{a+1}^{\max}$ and $p_a^{\max} = \dots = p_2^{\max}$. If buyer 1 is replaced, then $p_2^{\max} = v_1$. If buyer 1 is not replaced, then $p_2^{\max} = 0$.*

Proof. The inequalities $p_n^{\max} \le p_{n-1}^{\max} \dots \le p_{a+1}^{\max}$ follow from the fact that there is an opportunity path from buyer n to buyer a in $(A, \mathscr{G})$. Since b_n is linked to s_n but does not obtain a good in A, it must be that $v_n \le p_n^{\max}$.

First suppose that buyer 1 is not replaced by buyer n in A'. Then in A, buyer 1 is linked to an inactive seller and so pays a price $p_2^{\max} = 0$ to seller 2. There is an opportunity path from any buyer in $\{b_2, \dots, b_{a-1}\}$ to b_1 so by Proposition 5 we have $0 = p_2^{\max} = p_3^{\max} = \dots p_a^{\max}$.

Now suppose that buyer 1 is replaced by buyer n in $(A', \mathscr{G}')$. Let b_i be one of the set $\{b_1, b_2, \dots, b_{a-1}\}$. If b_i pays a price of $p_{i+1}^{\max} = 0$ *in* $\mathscr{G}$, then by Proposition 5, b_i has an opportunity path to a buyer l who is linked to an inactive seller. In $\mathcal{G}'$ *with the allocation* A, buyer n has an opportunity path to b_i and hence to b_l. But then buyer n could be added to the set of buyers who obtain a good without replacing buyer 1. This contradicts the efficiency of A'.

So b_i pays a positive price $p_{i+1}^{\max}$. Let buyer L be the "price setting" buyer – that is, the buyer with valuation $v^L(b_i) = p_{i+1}^{\max}$. (We will say $v^L(b_i) = v_L$ for short.) By Proposition 5, b_i has an opportunity path to buyer L . In $\mathcal{G}'$ *with the allocation* A, buyer n has an opportunity path to b_i and hence to b_L. If $v_L < v_1$, then it is more efficient for buyer n to replace b_L than to replace b_1. This contradicts the efficiency of A'. So it must be that $p_{i+1}^{\max} \geq v_1$. Buyer i also has an opportunity path to buyer 1. This implies that $p_{i+1}^{\max} \leq v_1$. So it must be that $p_{i+1}^{\max} = v_1$ as desired. □

Lemma A4. *Let A and A′ be efficient allocations in* $\mathcal{G}$ *and* $\mathcal{G}'$ *involving different sets of buyers as in Lemma 1. We assume the notation from Lemma 1. In* $\mathcal{G}'$, $v_n = p_n^{\max\prime} = \ldots p_{a+1}^{\max\prime} = p_a^{\max\prime} \geq \ldots \geq p_2^{\max\prime}$. *If buyer* 1 *is replaced, then* $p_2^{\max\prime} \geq v_1$. *If buyer* 1 *is not replaced, then it buys from* s_1 *and* $p_2^{\max\prime} \geq p_1^{\max\prime}$.

Proof. There is an opportunity path in $(A', \mathcal{G}')$ from b_2 to b_n. This implies that $p_2^{\max\prime} \leq p_3^{\max\prime} \leq p_n^{\max\prime}$. Since b_n buys a good from s_n, $p_n^{\max\prime} \leq v_n$.

We argue that $p_a^{\max\prime} = v_n$. This implies that $v_n = p_n^{\max\prime} = \ldots = p_{a+1}^{\max\prime} = p_a^{\max\prime}$.

By Proposition 5 the price $p_a^{\max\prime}$ is determined by buyer a's opportunity paths in $(A', \mathcal{G}')$. All of these opportunity paths are also opportunity paths in $(A, \mathcal{G})$ except the one from buyer a to buyer n: $b_a - s_{a+1} \rightarrow b_{a+1} - \ldots s_n \rightarrow b_n$. By Lemma A3, buyer a pays a strictly positive price $p_{a+1}^{\max}$ in $(A, \mathcal{G})$ and so has a price setting buyer L. Buyer L has the lowest valuation $v^L(b_a)$ (or v_L for short) of all buyers to which buyer a has an opportunity path in $(A, \mathcal{G})$. There is an opportunity path from buyer n to buyer L in $(A, \mathcal{G})$. (Join the o.p. from buyer n to buyer a $[b_n - s_n \rightarrow b_{n-1} - \ldots s_{a+1} \rightarrow b_a]$ to the o.p. from buyer a to buyer L.) Therefore $v_L \leq v_n$. If $v_L < v_n$, we have a contradiction to the efficiency of A in $\mathcal{G}$ because we could have replaced buyer L with buyer n in $(A, \mathcal{G})$. Therefore $v_L = v_n$ or equivalently $p_a^{\max\prime} = v_n$.

To finish the proof, suppose that buyer 1 is replaced, so that it does not obtain a good in A'. Since $b_1 \in L(s_2)$, it must be that $v_1 \leq p_2^{\max\prime}$. If buyer 1 is not replaced, then it buys from s_1. Since $b_1 \in L(s_2)$ it must be that $p_2^{\max\prime} \geq p_1^{\max\prime}$. □

Lemma A5. *Let A and A′ be efficient allocations in*$\mathcal{G}$ *and* $\mathcal{G}'$ *involving different sets of buyers as in Lemma 1. We assume the notation from Lemma 1. In* $\mathcal{G}$, $v_n \leq p_n^{\min} \leq p_{n-1}^{\min} \cdots \leq p_{a+1}^{\min}$ *and* $p_a^{\min} = \ldots = p_2^{\min}$. *If buyer* 1 *is replaced, then* $p_a^{\min} = v^H(s_a) \leq \min\{v_1, \ldots, v_{a-1}\}$. *If buyer* 1 *is not replaced, then* $p_a^{\min} = 0$.

Proof. The proof is available on request from the authors. It is similar to the proofs of Lemmas A3 and A4.

Lemma A6. *Let A and A′ be efficient allocations in* $\mathcal{G}$ *and* $\mathcal{G}'$ *involvinq different sets of buyers as in Lemma 1. We assume the notation from Lemma 1. In* $\mathcal{G}'$, $p_n^{\min\prime} = \ldots = p_{a+1}^{\min\prime} = p_a^{\min\prime} = \ldots = p_2^{\min\prime}$. *If buyer* 1 *is replaced, then* $p_2^{\min\prime} = v_1$. *If buyer* 1 *is not replaced, then* $p_2^{\min\prime} = p_1^{\min\prime} = 0$.

Proof. The proof is available on request from the authors. It is similar to the proofs of Lemmas A3 and A4.

Proof of Proposition 8. We will prove the result for $q = 0$ ($\mathbf{p} = \mathbf{p}^{\max}$). For $q = 1$ ($\mathbf{p} = \mathbf{p}^{\min}$) we proved this result in Kranton and Minehart (2001) using the fact that revenues are realized by an ascending bid auction. For other q the revenue functions are given by a convex combination of the extremal revenue functions.

Fix $q = 0$. For a valuation $\mathbf{v}$, we will choose efficient allocations A in $\mathcal{G}$ and A' in $\mathcal{G}'$ as in Lemmas 2 and 1. That is either $A = A'$ or A and A' differ only on the replacement set of agents.

I. Buyers: $u_a^b(\mathcal{G}') \geq u_a^b(\mathcal{G})$.
Consider a valuation $\mathbf{v}$. If $A = A'$, then every opportunity path for buyer a in $\mathcal{G}$ is also an opportunity path for buyer a in $\mathcal{G}'$. If buyer a does not obtain a good in A, then its payoff is 0 in both graphs. Otherwise, let buyer a obtain a good from seller j. By Proposition 5, buyer a's price is the lowest valuation of any buyer along an opportunity path. Since buyer a has a larger set of opportunity paths in $\mathcal{G}'$ we have $p_j^{\max} \geq p_j^{\max\prime}$. That is, buyer a earns a higher maximal payoff in $\mathcal{G}'$ than in $\mathcal{G}$.

If $A \neq A'$, we use the notation from Lemma 1. The replacement buyer n has an opportunity path to buyer a in $(A, \mathcal{G})$:

$$b_n - s_n \to b_{n-1} - \ldots s_{a+1} \to b_a \ .$$

Because $b_n \in L(s_n)$ and b_n does not obtain a good, it must be that $p_n^{\max} \geq v_n$. Therefore buyer a's price satisfies $p_{a+1}^{\max} \geq v_n$.

In $(A', \mathcal{G}')$, buyer a has an opportunity path to buyer n:

$$b_a - s_{a+1} \to b_{a+1} - \ldots s_n \to b_n \ .$$

Buyer a obtains a good from seller a. By Lemma A2, $p_a^{\max\prime} \leq v_n$.

We have shown that buyer a pays a lower price in $(A', \mathcal{G}')$ than in $(A, \mathcal{G})$. Therefore, buyer a earns a higher maximal payoff in $\mathcal{G}'$ than in $\mathcal{G}$ for all generic $\mathbf{v}$.

II. Sellers: $u_a^s(\mathcal{G}') \geq u_a^s(\mathcal{G})$.

Consider a valuation $\mathbf{v}$. If $A = A'$, then every opportunity path for buyer a in $\mathcal{G}$ is also an opportunity path for buyer a in $\mathcal{G}'$. If seller a does not sell a good under A, then its payoff is 0 in both graphs. Otherwise, let some buyer b_1 obtain a good from seller a in $(A, \mathcal{G})$. (This buyer cannot be b_a.) Consider an opportunity path for buyer 1 in $\mathcal{G}'$. The path has the form

$$b_1 - s_2 \to b_2 - s_3 \to \ldots b_{n-1} - s_n \to b_n \ .$$

If the path is not an opportunity path in $\mathcal{G}$ then it must contain the link $b_a - s_a$. But then the path has the form

$$b_1 - s_2 \to b_2 - s_3 \to \ldots b_a - s_a \to b_1 - \ldots$$

That is, the path takes us from s_a back to buyer 1. The path therefore does not link buyer 1 to any buyers that it was not already linked to by an opportunity path in $\mathcal{G}$. By Proposition 5, seller a has the same price and hence the same payoff in both graphs.

If $A \neq A'$, we use the notation from Lemma 1. There is a replacement buyer n and a buyer 1 that buyer n may or may not replace. If buyer 1 is not replaced, then by Lemma A3, $p_a^{\max} = 0$. That is, seller a earns a maximal payoff of 0 in $(A, \mathcal{G})$ and so is weakly better off in $(A', \mathcal{G}')$. If buyer 1 is replaced, then by Lemma A3, $p_a^{\max} = v_1$. Efficiency of A' in $\mathcal{G}'$ implies that $v_n \geq v_1$. So $p_a^{\max} \leq v_n$.

By Lemma A4, $p_a^{\max\,\prime} = v_n$. Since seller a earns a weakly higher price in $\mathcal{G}'$, it is weakly better off in $\mathcal{G}'$ than in $\mathcal{G}$.

We have shown that seller a earns a weakly higher maximal payoff in $\mathcal{G}'$ than in $\mathcal{G}$ for all generic $\mathbf{v}$. □

Proof of Proposition 9. We omit a proof of this as it is very similar to the proofs of Propositions 10 and 11. It is available from the authors on request.

Proof of Proposition 10. We will prove these results for $q = 0$ ($\mathbf{p} = \mathbf{p}^{\max}$). For $q = 1$ ($\mathbf{p} = \mathbf{p}^{\min}$), we proved the result in Kranton and Minehart (2001). For other q, the results follow from the fact that the payoffs are a convex combination of the payoffs for $q = 0$ and $q = 1$.

For a valuation $\mathbf{v}$, we will choose efficient allocations A in $\mathcal{G}$ and A' in $\mathcal{G}'$ as in Lemmas 2 and 1. That is either $A = A'$ or A and A' differ only on the replacement set of firms.

I. For every $b_i \in L(s_a)$, $u_i^b(\mathcal{G}') \leq u_i^b(\mathcal{G})$.
Fix a valuation $\mathbf{v}$. Suppose $A = A'$. If b_i does not obtain a good, its payoff is 0 in both graphs, so we are done. Let $\mathcal{O}_i$ denote the set of buyers connected to b_i by an opportunity path in $(A, \mathcal{G})$ and let $\mathcal{O}_i'$ denote the set of buyers connected to b_i by an opportunity path in $(A', \mathcal{G}')$. We argue that these two sets are the same. Clearly $\mathcal{O}_i' \supseteq \mathcal{O}_i$. Suppose there is some $b_k \in \mathcal{O}_i'$ that is not in $\mathcal{O}_i$. The o.p. from b_i to b_k in $(A', \mathcal{G}')$ must contain the link $b_a - s_a$, and since no good is exchanged b_a must precede s_a in the o.p. as follows:

$$b_i - s_{i+1} \to b_{i+1} - \ldots \to b_a - s_a \to b_{a+1} - \ldots \to b_k \ .$$

But then since $b_i \in L(s_a)$ in $\mathcal{G}$, there is a more direct o.p. from b_i to b_k that does not contain the link $b_a - s_a$ given by:

$$b_i - s_a \to b_{a+1} - \ldots \to b_k \ .$$

Since this o.p. does not contain $b_a - s_a$, it is also an o.p. in $(A, \mathcal{G})$, so $b_k \in \mathcal{O}_i$ which contradicts our assumption.

We have shown that $\mathcal{O}_i = \mathcal{O}_i'$ By Proposition 5, b_i pays the same price in both graphs.

If $A \neq A'$, we use the notation from Lemma 1. Suppose that b_i is the replacement buyer b_n. There is an o.p. $b_i - s_a \to b_{a-1} - \ldots s_2 \to b_1$ in $(A, \mathcal{G})$. So b_i could have replaced b_1 in $\mathcal{G}$. This contradicts the efficiency of A.

Suppose that b_i is replaced by b_n, then it is worse off in $\mathscr{G}'$, and we are done.

Otherwise b_i obtains a good in A and A'. Suppose that $b_i \in \mathscr{H}$. If $b_i \in \{b_1, b_2, \ldots, b_{a-1}\}$ then by Lemmas A3 and A4, we are done because b_i pays a higher price in $\mathscr{G}'$ and so is worse off. Because b_i obtains a good in both A and A', we have $b_i \neq b_n$. If $b_i \in \{b_a, \ldots, b_{n-1}\}$, then in $(A, \mathscr{G})$, b_i obtains its good from a seller $s_* \in \{s_{a+1}, s_{a+2}, \ldots, s_n\}$ and there is an o.p. from b_n to b_1 in $(A, \mathscr{G})$ using the link $b_i - s_a$ as follows:

$$b_n - s_n \to b_{n-1} - s_{n-1} \to \ldots s_* \to b_i - s_a \to b_{a-1} - s_{a-1} \to \ldots \to b_1 .$$

Therefore, b_n could have replaced b_1 (or bumped b_1 to an inactive seller) in $\mathscr{G}$ contradicting the efficiency of A.

Suppose that $b_i \notin \mathscr{H}$. First note that b_i does not obtain a good from s_a in either graph. Instead b_i has an o.p. to b_1 in $(A, \mathscr{G})$. So by Lemma A2, b_i has $p_i^{\max} \leq v_1$.

In $\mathscr{G}'$, suppose that b_i pays a positive price. Then by Proposition 5, there is an o.p. from b_i to a price setting buyer l, (that is, $v_l = v^L(b_i)$) so that $p_i^{\max\prime} = v_l$. If this path is also an o.p. in $(A, \mathscr{G})$ we are done because by Lemma A2, $p_i^{\max} \leq v_l$ and so b_i is worse off in $(A', \mathscr{G}')$. Otherwise the o.p. intersects $\mathscr{H}$. The path must come in to a seller and leave at a buyer. Let b_k be the last buyer in the intersection. The portion of the o.p. from b_k to b_l is also an o.p. in $(A, \mathscr{G})$. If $k \leq a - 1$, then b_i has an o.p. to b_k in $(A, \mathscr{G})$ (namely: $b_i - s_a \to b_{a-1} - s_{a-1} \ldots \to b_k$) and hence to b_l in $(A, \mathscr{G})$. (Remark: This last step could not be generalized to $b_i \in L(L(L(s_a)))$.) Then we are done because by Lemma A2, $p_i^{\max} \leq v_l$ and b_i is worse off in $(A', \mathscr{G}')$. If $k \geq a$, then b_n has an o.p. to b_k and hence to b_l in $(A, \mathscr{G})$. This implies that $p_i^{\max\prime} = v_l \geq v_n$, because otherwise b_n should replace b_l contradicting the efficiency of A. Since $v_n \geq v_1$, we then have that $p_i^{\max\prime} \geq v_1$.We have already argued that $p_i^{\max} \leq v_1$ so b_i is weakly better off in $\mathscr{G}$.

In $\mathscr{G}'$, suppose that b_i pays a price of $p_i^{\max\prime} = 0$. Then b_i has an o.p. to a buyer that is linked to an inactive seller. By Lemma 1 this seller was also inactive in A. If the o.p. is also an o.p. in $(A, \mathscr{G})$, we are done because b_i also pays a price of $p_i^{\max} = 0$ in $\mathscr{G}$ and so is weakly better off in $\mathscr{G}$. Otherwise the o.p. intersects $\mathscr{H}$. The path must come in to a seller and leave at a buyer. Let b_k be the last buyer in the intersection. The portion of the o.p. from b_k to the buyer linked to the inactive seller is also an o.p. in $(A, \mathscr{G})$. If $k \geq a$, then there is an o.p. in $(A, \mathscr{G})$ from b_n to b_k. Joining the two paths gives an o.p. from b_n to the inactive seller. But this contradicts the efficiency of A because b_n could obtain a good without replacing any buyer. If $k \leq a - 1$, there is an o.p. in $(A, \mathscr{G})$ from b_i to b_k and hence from b_i to the buyer linked to the inactive seller. (Remark: This last step could not be generalized to $b_i \in L(L(L(s_a)))$.) We are done because b_i also pays a price of $p_i^{\max} = 0$ in $\mathscr{G}$ and so is weakly better off in $\mathscr{G}$.

We have shown that b_i is weakly better off in $\mathscr{G}$ for generic $\mathbf{v}$.

II. For $s_j \in L(b_a)$, $u_j^s(\mathscr{G}') \leq u_j^s(\mathscr{G})$.

Fix a valuation $\mathbf{v}$. Suppose $A = A'$. If s_j is inactive, it gets 0 in both graphs and we are done. If s_j sells a good to b_j, then by Proposition 5, $p_j^{\max}$ is the lowest valuation of a buyer linked to b_j by an o.p. Since every o.p. in $(A, \mathcal{G})$ is also an o.p. in $(A', \mathcal{G}')$, s_j's price must be weakly lower in $\mathcal{G}'$ and we are done.

If $A \neq A'$, we use the notation from Lemma 1. There is an o.p. from b_n to b_a in $(A, \mathcal{G})$. If s_j were inactive in A, then b_n could have been added to the set of buyers who obtain goods by using this o.p. as a replacement path: $s_j \rightarrow b_a - s_{a+1} \rightarrow b_{a+1} \ldots \rightarrow b_n$. This contradicts the efficiency of A.

Suppose that s_j is active in A. If $s_j \in \mathcal{H}$, and $j \geq a + 1$, then by Lemmas A1 and A2, $p_j^{\max} \geq v_n$ and $p_j^{\max\,\prime} = v_n$. So s_j is weakly worse off in $\mathcal{G}'$ and we are done. If $s_j \in \mathcal{H}$, and $j \leq a$, then because s_j is active, $j \neq 1$. In A, s_j sells its good to b_{j-1}. There is an o.p. from b_n to b_a in $(A, \mathcal{G})$ and also one from b_{j-1} to b_1. These paths can be joined by the linkage $b_a - s_j \rightarrow b_{j-1}$ to give an o.p. from b_n to b_1 in $(A, \mathcal{G})$. But this contradicts the efficiency of A, because b_n could be brought in to the set of buyers who obtain goods in $\mathcal{G}$ using this o.p. as the replacement path. (Remark: This last step could not be generalized to $s_j \in L(L(L(b_a)))$ because the buyer that s_j is linked to in $L(L(b_a))$ need not be in the replacement path.)

If $s_j \notin \mathcal{H}$, then s_j sells its good to the same b_j in both graphs where $b_j \notin \mathcal{H}$. If $p_j^{\max} = 0$, then the fact that $b_a \in L(s_j)$ together with Lemma A1 implies that $p_{a+1}^{\max} = p_{a+2}^{\max} = \ldots = p_n^{\max} = 0$. (Buyer n has an o.p. to every buyer in $\{b_a, \ldots, b_{n-1}\}$.) But this contradicts pairwise stability for these prices in $(A, \mathcal{G})$, because b_n would want to buy the good from s_n.

If $p_j^{\max} > 0$, then $p_j^{\max} = v_l$ for the price setting buyer l (that is, $v^L(b_j) = v_l$). There is an o.p. from b_j to b_l and from b_n to b_a. These paths can be joined by the linkage $b_a - s_j \rightarrow b_j$ to give an o.p. from b_n to b_l. Since b_n could replace b_l along this o.p., but does not, it must be that $p_j^{\max} = v_l \geq v_n$.

If the o.p. from b_j to b_l is still an o.p. in $(A', \mathcal{G}')$, we are done because $p_j^{\max\,\prime} \leq v_l$ and so s_j is weakly worse off in $\mathcal{G}'$. Otherwise the o.p. intersects $\mathcal{H}$. The o.p. enters $\mathcal{H}$ for the first time at a seller s_k. Up to that point the path (from b_j to s_k) is the same in $(A', \mathcal{G}')$ as in $(A, \mathcal{G})$. There is an o.p. in $(A', \mathcal{G}')$ from b_k to b_n. So we can join these by the link $s_k \rightarrow b_k$ to form an o.p. in $(A', \mathcal{G}')$ from b_j to b_n. By Lemma A2, we have $p_j^{\max\,\prime} \leq v_n$. Therefore we have $p_j^{\max\,\prime} \leq v_n \leq p_j^{\max}$ and s_j is weakly worse off in $\mathcal{G}'$.

We have shown that s_j is weakly better off in $\mathcal{G}$ for generic $\mathbf{v}$. □

Proof of Proposition 11. We prove this result for $q = 0$ ($\mathbf{p} = \mathbf{p}^{\max}$). Similar techniques prove the result for $q = 1$, and hence for other q between 0 and 1.

For a valuation $\mathbf{v}$, we will choose efficient allocations A in $\mathcal{G}$ and A' in $\mathcal{G}'$ as in Lemmas 2 and 1. That is either $A = A'$ or A and A' differ only on the replacement set of agents.

Fix a valuation $\mathbf{v}$. Suppose $A = A'$. If b_i does not obtain a good, its payoff is 0 in both graphs. Otherwise, it obtains a good from $s_j \in L(b_a)$. By the proof of Proposition 10, s_j receives a weakly lower payoff in $\mathcal{G}'$ than in $\mathcal{G}$. So its price must be weakly lower, which means that b_i receives a weakly higher payoff in $\mathcal{G}'$ and we are done.

If $A \neq A'$, we use the notation from Lemma 1. If b_i does not obtain a good in either graph, its payoff is 0 in both graphs and we are done. If b_i receives a good only in $\mathcal{G}'$ (b_i is the replacement buyer), then it must be weakly better off in $\mathcal{G}'$ and we are done. If b_i receives a good only in $\mathcal{G}$ ($i = 1$ where b_1 is the replaced buyer), then by Lemma A1, b_i paid a price in $\mathcal{G}$ exactly equal to its valuation. So it earns a payoff of 0 in both graphs and we are done.

Suppose that b_i obtains a good in both graphs. Let s_i denote the seller that sells its good to b_i in A. If b_i pays a strictly positive price to s_i, let b_l be the price setting buyer (that is, $v^L(b_i) = v_l$). Then b_i has an o.p. to b_l. If this path is also an o.p. to b_l in $(A', \mathcal{G}')$ then b_i pays a price in $(A', \mathcal{G}')$ that is weakly lower than v_l. So b_i is weakly better off in $\mathcal{G}'$ and we are done.

Otherwise, the path intersects the replacement set $\mathcal{H}$. Suppose that neither b_i nor b_l is in the replacement set. The intersection must begin with a seller and end with a buyer. Let b_k be the last buyer in the intersection. The portion of the o.p. from b_k to b_l is also an o.p. in $(A', \mathcal{G}')$. If $k \leq a - 1$, consider the part of the o.p. from b_i to b_k. Join the o.p. $b_n - s_n \rightarrow \ldots b_{a+1} - s_{a+1} \rightarrow b_a - s_i \rightarrow b_i$ to the beginning (this uses the assumption that $s_i \in L(b_a)$) and the o.p. from b_k to b_1 to the end. This forms an o.p. from b_n to b_1 in $(A, \mathcal{G})$. But this implies that the allocation A' is feasible in $\mathcal{G}$ which contradicts efficiency. If $k \geq a$, then joining the o.p. from b_n to b_k to the o.p. from b_k to b_l, forms an o.p. from b_n to b_l in $(A, \mathcal{G})$. This means that b_n could replace b_l in $(A, \mathcal{G})$. Since it does not, it must be that $v_l \geq v_n$. That is, b_i pays $p_i^{\max} \geq v_n$. In $(A', \mathcal{G}')$, there is an o.p. from b_i to b_n. (Because $k \geq a$, the o.p. from b_i to b_l in $(A, \mathcal{G})$ must intersect the replacement set for the first time at a seller m with $m \geq a + 1$. The part of the o.p. from b_i to s_m is an o.p. in $(A', \mathcal{G}')$.) Join this to the path $s_m \rightarrow b_m - s_{m+1} \rightarrow \ldots b_n$ to form an o.p. from b_i to b_n in $(A', \mathcal{G}')$. But then by Lemma A2 it must be that $p_i^{\max\,\prime} \leq v_n$. That is, b_i pays a weakly lower price in $\mathcal{G}'$ and so is weakly better off. The cases that b_i or b_l are in the replacement set are similar (these are essentially special cases of what we have just proved.)

If b_i pays a price of 0 in $(A, \mathcal{G})$, then it has an o.p. to a buyer who is linked to an inactive seller. A similar argument to the previous paragraph (see also the proof of this case in Proposition 10, I.) implies that b_i pays a price of 0 in $\mathcal{G}'$ and so is equally well off in both graphs.

Therefore, b_i's payoff is at least as high in $(A', \mathcal{G}')$ as in $(A, \mathcal{G})$ for all valuations $\mathbf{v}$. □

Proof of Proposition 12. This result follows from Demange and Gale (1985) Corollary 3 (and the proof of Property 3 of which the Corollary is a special case). Demange and Gale identify two sides of the market, P and Q. We may identify P with buyers and Q with sellers. (This identification could also be reversed.) The way Demange and Gale add agents is to assume that they are already in the initial population, but they have prohibitively high reservation values so that they do not engage in exchange. "Adding" an agent is accomplished by lowering its reservation value. (In our framework, we reduce the reservation value from a very large number to zero.) They show that the minimum payoff for each seller is increasing in the reservation value of *any* other seller (including itself).

That is, when sellers are added, the minimum payoff of each original seller weakly decreases. They also show that the maximum payoff for each buyer is decreasing in the reservation value of any seller. That is, when sellers are added, the maximum payoff of each buyer weakly increases.

To complete the proof, we interchange the role of buyers and sellers. (In Demange and Gale 1985, the game is presented in terms of payoffs, and there is no interpretational issue involved in switching the roles of buyers and sellers. In our framework, there is an interpretational issue in switching the roles, but, technically, in terms of payoffs there is no issue.) Corollary 3 states that when buyers are added, the maximum payoff for buyers is weakly decreasing and the minimum payoff for sellers is weakly increasing. Interchanging the roles of buyers and sellers gives us that when sellers are added to a population the maximum payoff for each original seller weakly decreases and the minimum payoff for each buyer weakly increases.

We have shown that when sellers are added to a network, both the minimum and maximum payoff for each original seller weakly decreases. And both the minimum and maximum payoff for each buyer weakly increases. Convex combinations of these payoffs therefore share the same property. That is, the buyers are better off and the original sellers are worse off. □

Proof of Proposition 13. The proof is analogous to the one above and is available from the authors on request. □

References

Bolton, P., Whinston, M.D. (1993) Incomplete contracts, vertical integration, and supply assurance. *Review of Economic Studies* 60: 121–148

Demange, G., Gale, D. (1985) The strategy structure of two-sided matching markets. *Econometrica* 53: 873–883

Demange, G., Gale, D., Sotomayor, M. (1986) Multi-item auctions. *Journal of Political Economy* 94: 863–872

Gale, D. (1987) Limit theorems for markets with sequential bargaining. *Journal of Economic Theory* 43: 20–54

Grossman, S., Hart, O. (1986) The costs and benefits of ownership. *Journal of Political Economy* 94: 691–719

Hart, O., Moore, J. (1990) Property rights and the nature of the firm. *Journal of Political Economy* 98: 1119–1158

Kranton, R., Minehart, D. (1999) *Competition for Goods in Buyer-Seller Networks.* Cowles Foundation Discussion Paper, Number 1232, Cowles Foundation, Yale University

Kranton, R., Minehart, D. (2001) Theory of buyer-seller networks. *American Economic Review* 91: 485–508

Kranton, R., Minehart, D. (2000) Networks versus vertical integration. *RAND Journal of Economics* 31: 570–601

Roth, A., Sotomayor, M. (1990) *Two-Sided Matching.* Econometric Society Monograph, Vol. 18. Cambridge University Press, Cambridge

Rubinstein, A. (1982) Perfect equilibrium in an bargaining model. *Econometrica* 50: 97–110

Rubinstein, A., Wolinsky, A. (1985) Equilibrium in a mkarket with sequential bargaining. *Econometric* 53: 1133–1150

Shapley, L. Shubik, M. (1972) The assignment game I: The core. *International Jounal of Game Theory* 1: 111–130

Williamson, O. (1975) *Markets and Hierarchies.* Free Press, New York

Buyers' and Sellers' Cartels on Markets With Indivisible Goods

Francis Bloch[1], **Sayantan Ghosal**[2]

[1] IRES, Department of Economics, Université Catholique de Louvain, Belgium
[2] Department of Economics, University of Warwick, Coventry, UK

Abstract. This paper analyzes the formation of cartels of buyers and sellers in a simple model of trade inspired by Rubinstein and Wolinsky's (1990) bargaining model. When cartels are formed only on one side of the market, there is at most one stable cartel size. When cartels are formed sequentially on the two sides of the market, there is also at most one stable cartel configuration. Under bilateral collusion, buyers and sellers form cartels of equal sizes, and the cartels formed are smaller than under unilateral collusion. Both the buyers' and sellers' cartels choose to exclude only one trader from the market. This result suggests that there are limits to bilateral collusion, and that the threat of collusion on one side of the market does not lead to increased collusion on the other side.

JEL classification: C78, D43

Key Words: Bilateral collusion, buyers' and sellers' cartels, collusion in bargaining, countervailing power

1 Introduction

Recent theoretical models of collusion only consider the formation of cartels on one side of the market. Studies of bidding rings in auctions focus on collusion on the part of buyers in models with a unique seller. On oligopolistic markets, the

This research was started while both authors were at CORE. The first author gratefully acknowledges the financial assistance of the European Commission (Human Capital Mobility Fellowship) which made his visit to CORE possible, and the second author a CORE doctoral fellowship. Discussions with Ulrich Hege, Heracles Polemarchakis and Asher Wolinsky helped us formalize the ideas in this paper. We also benefitted from comments by participants at seminars in Namur and Bilbao, at the Hakone Conference in Social Choice, the Valencia Workshop on Game Theory, the Bangalore and Saint Petersburg Conferences in Game Theory and Applications and the Econometric Society European Meeting in Istanbul. This paper is an extended version of a paper formerly entitled "Bilateral Bargaining and Stable Cartels".

formation of cartels of producers is analyzed under the assumption that demand is atomistic, so that buyers react in a competitive fashion to the choices of sellers.

While these models are well-suited to analyze collusion on some markets (simple auctions, markets for consumer goods), their conclusions can hardly be extended to other markets, such as markets for primary commodities, where a small number of buyers and sellers interact repeatedly. However, some of the best known examples of cartels are actually found on thin markets with a small number of traders. The commodity cartels grouping producer countries (OPEC, the Uranium, Coffee, Copper and Bauxite cartels) face a very small number of buyers of primary commodities. Similarly, the famous shipping conferences, legal cartels grouping all shipping companies operating on the same route, interact repeatedly with the same shippers.

On markets with a small number of strategic buyers and sellers, the study of collusion on one side of the market must take into account the reaction of traders on the other side. The formation of a cartel by traders on one side of the market may induce collusion on the other side. In fact, it is often argued that commodity cartels were formed in the 1970's as a response to the increasing concentration of buyers on the market (see the case studies by Sampson (1975) on the oil market, by Holloway (1988) on the aluminium market and by Taylor and Yokell (1979) on the uranium market). In oceanliner shipping, the monopoly power of shipping companies has led to the emergence of cartels of buyers, the shippers' councils which negotiate directly with the shipping conferences (Sletmo and Williams 1980).

Our purpose in this paper is to analyze the formation of buyers' and sellers' cartels on markets with a small number of strategic buyers and sellers. In particular, we study how collusion on one side of the market may induce collusion on the other side. When do cartels emerge on the two sides of the market? What are the sizes of those cartels? Does the formation of cartels on the two sides of the market lead to a higher restriction in trade than in the case of unilateral collusion? Does bilateral collusion induce a "balance" in the market power of buyers and sellers, as suggested by Galbraith (1952)?

In order to answer these questions, we study a sequential model of interaction between an equal number of buyers and sellers of an indivisible good. In the first stage, buyers decide to form a cartel and restrict the number of traders they put on the market. In the second stage, sellers form a cartel and restrict trade in the same way. In the third stage, once the number of buyers and sellers excluded from the market is determined, the remaining traders trade on the market.

The model of trade for an indivisible commodity that we use is inspired by Rubinstein and Wolinsky (1990)'s model of matching and bargaining among a small number of traders. Buyers and sellers bargain over the surplus generated by the indivisible good, which is normalized to one. At each point in time, buyers and sellers are randomly matched and make decentralized offers. However, to guarantee the existence of a unique price at which trade occurs, we add to the model a centralization mechanism: trade only occurs when all offers are accepted in the same round. It is easy to see that, as the discount factor converges to 1, the outcome of this model of trade converges to the competitive outcome, giving all of the surplus to traders on the short side of the market. On the other hand, when the discount factor converges to 0, the trading mechanism approaches a simple bargaining model with take-it-or-leave-it offers.

As the price of the good traded depends on the numbers of buyers and sellers on the markets, traders have an incentive to restrict the quantities of the good they buy or sell on the market. However, given the indivisibility of the good traded, the only way to restrict offer or demand on the market is to exclude some agents from trade. Hence, we assume that cartels are formed in order to exclude some traders from the market and to compensate them for withdrawal.[1]

We model the formation of the cartel as a simple, noncooperative game, where traders simultaneously decide on their participation to the cartel. This participation game implies that a cartel is stable when (i) no trader has an incentive to join the cartel and (ii) no trader has an incentive to leave the cartel.

In a first step, we analyze the formation of a stable cartel on one side of the market and show that there exists at most one stable cartel size. This is the unique cartel size for which, upon departure of a member, the cartel collapses entirely.[2] If there are originally more sellers than buyers on the market, sellers form a cartel in order to equalize the number of active buyers and sellers. If there are more buyers than sellers, there does not exist any stable cartel of sellers. Finally, if originally buyers and sellers are on equal number on the market, sellers form a cartel and exclude one trader from the market.

Next, we analyze the response of one side of the market to collusion on the other side – when buyers form a cartel, they anticipate that sellers will respond by colluding themselves. We suppose that buyers and sellers are originally in equal number on the market. Using our earlier characterization of stable cartels on one side of the market, we show that in the sequential game of bilateral cartel formation, there exists a unique stable cartel configuration, where both buyers and sellers form cartels, the cartels are of equal size, and both cartels exclude one trader from the market. It thus appears that the formation of cartels on the two sides of the market leads to the *same* restriction in trade as in the case of unilateral collusion. Furthermore, the size of the cartels formed under bilateral collusion is *smaller* than the size of the cartel formed under unilateral collusion. We interpret these results by noting that there exist *limits to bilateral collusion*. The threat of collusion on one side of the market does not lead to a higher level of collusion among traders on the other side.

In order to gain some insights about these results, it is instructive to consider the limiting case of a competitive market, where traders on the short side of the market almost obtain the entire surplus.[3] Suppose that originally buyers form a

[1] Alternatively, we could assume that cartels are formed for traders to coordinate their actions at the bargaining stage. This is a much more complex issue that we prefer to leave for further research.

[2] This characterization of the stable cartel emphasizes the role played by the indivisibility of the good traded. On markets with divisible goods, the formation of a cartel is prevented by the traders' incentives to leave the cartel and free ride on the cartel's trading restriction.

[3] It has long been noted that traders have an incentive to collude on these competitive markets for indivisible commodities. See Shapley and Shubik (1969), fn. 10 p. 344.

cartel and exclude one buyer from the market. What will the seller's response be? Clearly, by forming a cartel which excludes two sellers from the market they could capture a surplus of $1-\epsilon$ per unit traded, whereas by excluding one trader they obtain a surplus of $\frac{1}{2}$. Hence, at first glance, it seems that sellers should form a cartel which excludes two traders. However, we argue that this cartel cannot be stable, and that the only stable cartel is a cartel of size three which excludes one seller from the market. To see this, note that the minimal cartel size for which two sellers are excluded is four. Each cartel member then receives a payoff of $\frac{2-2\epsilon}{4} = \frac{1}{2} - \frac{\epsilon}{2}$. By leaving the cartel, a member would obtain a higher payoff of $\frac{1}{2}$, so that the cartel is unstable. By the same free-riding argument, no cartel of size greater than four can be stable. Hence, the only stable cartel is the cartel of size three, showing that free-riding prevents the formation of a cartel in which sellers could capture the entire trading surplus.

While our analysis departs from recent studies of collusion in auctions and competitve markets, its roots can be traced back to the debate surrounding Galbraith's (1952) book on "countervailing power". In this famous book, Galbraith (1952) argues that the concentration of market power on the side of buyers is the only check to the exercise of market power on the part of sellers. (see Scherer and Ross (1990), Ch. 14, for a survey of recent contributions to the theory of "countervailing power"). As was already noted by Stigler (1954) in his discussion of the book, Galbraith's (1952) assertions are not easily supported by formal economic arguments. In fact, we show that the existence of countervailing power may balance the market power of buyers and sellers, but does not help to reduce the inefficiencies linked to the existence of market power.

In the 1970's, the formation of stable cartels has been studied in general equilibrium models, using various solution concepts (see, for example the survey by Gabszewicz and Shitovitz 1992 for the core, Legros 1987 for the nucleolus, Hart 1974 for the stable sets and Okuno et al. 1980 for a strategic market game). While these models provide some general existence and characterization results on stable cartels, these results cannot be easily compared to the results we obtain here.

Finally, our analysis relies strongly on the study of stable cartels on oligopolistic markets initiated by d'Aspremont et al. (1983), Donsimoni (1985) and Donsimoni et al. (1986). The stability concept we use is due to d'Aspremont et al. (1983). In spite of differences in the models of trade, our results bear some resemblance to the characterization of stable cartels in Donsimoni et al. (1986). As in their analysis, we find that free-riding greatly limits the size of stable cartels, thereby reducing collusion on the market.

The rest of the paper is organized as follows. We present and analyze the model of trade and describe the cartel's optimal choice in Sect. 2. In Sect. 3, we define the game of cartel formation and characterize stable cartels, both when cartels are formed only on one side of the market and when cartels are formed sequentially by buyers and sellers. Finally, Sect. 4 contains our conclusions and directions for future research.

2 Trade and Collusion on the Market

In this section, we present the basic model of trade and analyze the behavior of cartels formed on the market. We consider a market for an indivisible good with a finite set B of identical buyers and a finite set S of identical sellers and let b and s denote the cardinality of the sets B and S respectively. Each buyer i in B wants to purchase one unit of the indivisible good traded on the market, and each seller j in S owns one unit of the good. Without loss of generality, we normalize the gains from trade to 1.

The interaction between participants on the market is modeled as a three-stage process. In the first stage, a cartel is formed; in the second stage, members of the cartel choose the number of active traders they put on the market. Finally, in the third stage of the game, buyers and sellers trade on the market. Since the model is solved by backward induction, we start our formal description of the game by the final stage of the game and proceed backwards to the first stage.

2.1 A Model of Matching and Trade

After cartels are formed, and the number of active traders on the market is determined, agents engage in trade. The trading mechanism we analyze combines elements of bilateral bargaining as in Rubinstein and Wolinsky (1990) and a centralization mechanism. We suppose that, at each period of time, traders are matched randomly and engage in a bilateral bargaining process. However, in order for trade to be concluded, we require that all agents unanimously agree on the offer they receive.

Formally, we let $t = 1, 2, \ldots$ denote discrete time periods. At each period t, the traders remaining on the market are matched randomly. If s_t and b_t denote the numbers of sellers remaining on the market in period t, and if $b_t < s_t$, this implies that each buyer i is matched with a seller j, whereas a seller j is matched with a buyer i only with probability $\frac{b_t}{s_t}$. Each match (i, j) is equally likely.

Once a match (i, j) is formed, one of the traders is chosen with probability $\frac{1}{2}$ to make an offer. The other trader then responds to the offer. If, at some period t, *all* offers are accepted, the transactions are concluded and traders leave the market. If, on the other hand, one offer is rejected, all traders remain on the market and enter the next matching stage. If a transaction is concluded at period T for a price of p, the seller obtains a utility equal to $\delta^T p$ whereas the buyer obtains $\delta^T(1 - p)$.

While our model shares some formal resemblance to models of bilateral bargaining, it differs sharply from traditional models of decentralized trade since transactions are concluded only when all traders unanimously accept the offers. While this coordination device is clearly unnatural in a model of decentralized trade, we need it to guarantee that the bargaining environment is stationary, so that classical methods of characterization of stationary perfect equilibria can be used (see for example, Rubinstein and Wolinsky 1985). Without this coordination device, utilities obtained in any subgame following the rejection of an offer would depend on the number of pairs of buyers and sellers who have concluded trade at this stage. Since bargaining occurs simultaneously in all matched pairs of buyers and sellers, traders cannot observe the outcome of bargaining among other traders. Hence, this is a game of incomplete information, and in order to compute utilities obtained after the rejection of offer, we need to specify the players' beliefs about the behavior of the other traders in the game. As in any extensive form game with incomplete information, there is some leeway in defining those beliefs off the equilibrium path, and there is no natural way to select a sequential equilibrium in this game.[4]

It is well known that the sequential game we consider may have many equilibria. In order to restrict the set of equilibria, we assume that traders use *stationary* strategies, which only depend on the number of active traders in the game, and on the current proposal.

Formally, a *stationary perfect equilibrium* of the trading game is a strategy profile σ such that

- For any player i, σ_i only depends on the number of active traders and the current proposal.
- At any period t at which player i is active, the strategy σ_i is a best response to the strategy choices σ_{-i} of the other traders.

Proposition 1. *In the model of trade, there exists a unique stationary perfect equilibrium. The equilibrium is symmetric and all offers are accepted immediately. If* $b \leq s$*, sellers propose a price* $p_s = \frac{(1-\delta)(2s-\delta b)}{(2-\delta)s-\delta b}$ *and buyers propose a price* $p_b = \frac{\delta(1-\delta)b}{(2-\delta)s-\delta b}$*. If* $s \leq b$*, sellers propose a price* $p_s = \frac{(b-\delta s)(2-\delta)}{b(2-\delta)-\delta s}$ *and buyers propose a price* $p_b = \frac{\delta(b-\delta s)}{b(2-\delta)-\delta s}$*.*

Proof. Without loss of generality, suppose $b \leq s$. First observe that, since unanimous agreement is needed for the conclusion of trade, in a stationary perfect equilibrium, all players must make acceptable offers. If, on the other hand, one trader makes an offer which is rejected, the game moves to the next stage and, by stationarity, the bargaining process continues indefinitely. Hence, given a fixed set of buyers indexed by $i = 1, 2, \ldots b$ and a fixed set of sellers indexed by $j = 1, 2, \ldots, s$, the strategy profile σ must be characterized by price offers $(p_b(i,j), p_s(i,j))$ where $p_b(i,j)$ represents the price offered by buyer i in the match (i,j) and $p_s(i,j)$ the price offered by seller j in the match (i,j). In a subgame perfect equilibrium, the price $p_b(i,j)$ is the minimum price accepted by seller j and must satisfy

$$p_b(i,j) = \delta \frac{1}{s} \sum_i \frac{p_b(i,j) + p_s(i,j)}{2}.$$

[4] Rubinstein and Wolinsky (1990) suggest to specify the beliefs in such a way that, after observing an offer off the equilibrium path, each agent believes that the other agents stick to their equilibrium behavior. Under this restriction on beliefs, we are able to characterize a symmetric stationary sequential equilibrium in the game without the coordination device. Unfortunately, we have been unable to characterize the formation of cartels of buyers and sellers in that setting.

Similarly, $p_s(i,j)$ is the maximum price accepted by buyer j and must satisfy

$$1 - p_s(i,j) = \delta \frac{1}{s} \sum_j \frac{1 - p_b(i,j) + 1 - p_s(i,j)}{2}.$$

By the two preceding equations, the minimum price accepted by seller j is independent of the identity of the buyer i and the maximum price accepted by buyer i is independent of the identity of the seller j. Hence $p_b(i,j) = p_b(j)$ and $p_s(i,j) = p_s(i)$. Replacing in the two equations, we get

$$\begin{aligned} p_b(j) &= \delta \frac{1}{s} \sum_t \frac{p_b(j) + p_s(i)}{2} \\ 1 - p_s(i) &= \delta \frac{1}{s} \sum_j \frac{1 - p_b(j) + 1 - p_s(i)}{2}. \end{aligned}$$

Hence, the strategies (p_b, p_s) are independent of the identity of the buyers and sellers, and can be found as the unique solution to the system of equations

$$\begin{aligned} p_b &= \delta \frac{b}{2s} (p_s + p_b) \\ (1 - p_s) &= \delta \frac{1}{2} (1 - p_s + 1 - p_b). \end{aligned}$$

It remains to check that the strategies (p_b, p_s) form indeed a subgame perfect equilibrium. First note that no seller has an incentive to lower the price it offers and no buyer can benefit from increasing the price. Suppose next that a seller deviates and proposes a price $p' > p_s$. If the buyer rejects the offer, no trade is concluded at this period, and the buyer obtains her continuation value $\delta \frac{1}{2}(1 - p_s + 1 - p_b) = 1 - p_s > 1 - p'$. Hence, the offer p' will be rejected. Similarly, if a buyer offers a price $p' < p_b$, her offer will be rejected by the seller. Hence the strategy (p_b, p_s) forms a subgame perfect equilibrium of the game. To complete the proof, it suffices to use the same arguments to obtain the unique stationary perfect equilibrium in the case $s \leq b$. □

Proposition 1 allows us to compute the expected payoff of a seller, $U(b, s)$ as a function of the number of buyers and sellers on the market.[5]

$$\begin{aligned} U(b,s) &= \frac{b(1-\delta)}{s(2-\delta) - \delta b} \quad \text{if} \quad b \leq s \\ U(b,s) &= \frac{b - \delta s}{b(2-\delta) - \delta s} \quad \text{if} \quad s \leq b. \end{aligned}$$

Figure 1 graphs the utility of a seller as a function of the number of sellers s on the market for various values of δ when $b = 50$. Notice that for all values of δ, this function has an inverted-S shape. The marginal effect of a reduction in the number of sellers on the price of the good is highest around the point

[5] The expected payoff of a buyer is simply given by $V(b, s) = 1 - U(b, s)$.

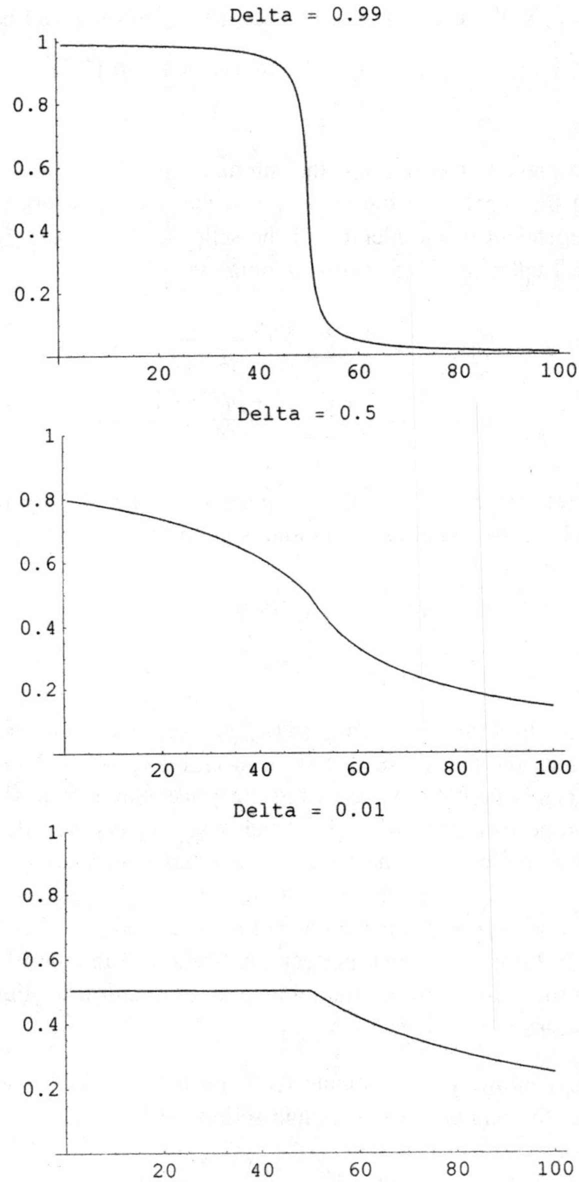

Fig. 1.

where the market is balanced ($b = s$) and lowest when the numbers of buyers and sellers are very different. As δ converges to 1, the outcome of the bargaining game approaches the symmetric competitive solution, with a price of 1 if $s < b$, 0 if $s > b$ and $\frac{1}{2}$ if $s = b$. On the other hand, as δ converges to 0, the trading mechanism converges to a model where each agent makes a take-it-or-leave-it-offer, and the seller's expected payoff converges to $\frac{1}{2}$ if $b \geq s$ and $\frac{b}{2s}$ if $b \leq s$.

2.2 Collusion and Cartel Behavior

In the model of trade we consider, agents benefit from the exclusion of traders on the same side of the market. We assume that cartels are formed precisely to withdraw some traders from the market and compensate them for their exclusion. These collusive arrangements, whereby some traders agree not to participate on the market, have commonly been observed among bidders in auctions. The well-known "phase of the moon" mechanism, used by builders of electrical equipment in the 50's, specified exactly which of the companies was supposed to participate in an auction at any period of time (see Mac Afee and Mac Millan 1992 and the references therein). Clearly, agreements to exclude traders from the market face two types of enforcement issues. First, the excluded traders could decide to renege on the agreement and reenter the market after receiving compensation. We assume that the market for the indivisible good opens repeatedly, so that this deviation can be countered by an appropriate dynamic punishment strategy. Second, members of the cartel could find it in their interest to organize different rounds of trade, and sell (or buy) the goods of the excluded traders in a second trading round. However, in equilibrium, this behavior will be perfectly anticipated by traders on the other side of the market. As the time between trading rounds goes to zero, the Coase conjecture indicates that trade should then occur as if no good was ever excluded from the market (see Gul et al. 1986). In order to avoid this problem, we assume that cartel members have access to a technology which allows them to credibly commit not to sell (or buy) the goods of excluded traders. For example, we could assume that buyers and sellers can ostensibly destroy their endowments in money and goods.

Formally, we analyze in this section the formation of a cartel of sellers on the market. If a cartel K of size k forms on the market, and decides to withdraw r sellers, the total number of active sellers is given by $s - r$ since independent sellers always participate in trade. Hence the total surplus obtained by cartel members is given by

$$U_K(b, s) = (k - r)U(b, s - r)\ .$$

Members of the cartel K thus select the number r of traders excluded from the market, $0 \leq r \leq k$, in order to maximize

$$\begin{aligned} U_K(b, s) &= (k - r)\frac{b(1-\delta)}{(s-r)(2-\delta) - \delta b} && \text{if} \quad r \leq s - b \\ &= (k - r)\frac{b - \delta(s-r)}{b(2-\delta) - \delta(s-r)} && \text{if} \quad r \geq s - b\ . \end{aligned}$$

This optimization problem is a simple integer programming problem, and it admits generically a unique solution. For a fixed value b and s of the total number of traders on the market, we let $\rho(k)$ denote the optimal choice of traders excluded from the market by the cartel K.

Proposition 2. *The optimal choice of a cartel K of sellers, $\rho(k)$ is given by the following expressions.*

$$\begin{aligned}
\rho(k) &= 0 && \text{if } s - \frac{\delta b}{2-\delta} \geq k, \\
\rho(k) &= \max\{0, s-b\} && \text{if } s + 2 - \frac{(3\delta - 2)b}{\delta} \geq k \geq s - \frac{\delta b}{2-\delta}, \\
\rho(k) &= r^* && \text{if } k \geq s + 2 - \frac{(3\delta-2)b}{\delta},
\end{aligned}$$

where r^* is the first integer following the root of the equation

$$-(b - \delta(s - r + 1))(b(2-\delta) - \delta(s - r)) + b(k - r)\delta(1-\delta) = 0. \qquad (1)$$

Proof. In order to characterize the optimal choice of a cartel, we consider the incremental value of excluding one additional seller from the market

$$f(r) = (k - (r+1))U(b, s - (r+1)) - (k - r)U(b, s - r).$$

Assume first that $r \leq s - b - 1$, then

$$\operatorname{sign} f(r) = \operatorname{sign} \delta b - (s - k)(2 - \delta).$$

It thus appears that the sign of the incremental value of an additional exclusion is independent of r. If $k \leq s - \frac{\delta b}{2-\delta}, f(r) \leq 0$ and, if $k \geq s - \frac{\delta b}{2-\delta}, f(r) \geq 0$.

Next suppose that $r \geq s - b$. Then

$$\operatorname{sign} f(r) = \operatorname{sign} \delta(1-\delta)b(k-r) - (b - \delta(s - r - 1))(b(2-\delta) - \delta(s-r)).$$

Notice that the function $f(\cdot)$ is a quadratic function in r and has at most two roots. We show that it has at most one root on the domain $[s - b, k]$. First, notice that $f(k) < 0$. Next, note that, at $\overline{r} = s - \frac{b(2-\delta)}{\delta} < s - b, f(\overline{r}) > 0$. This implies that $f(\cdot)$ has at most one root in the relevant domain. Furthermore, note that $f(s-b) < 0$ if and only if

$$k < s + 2 - \frac{(3\delta - 2)b}{\delta}.$$

Otherwise, as long as $k \geq s + 2 - \frac{(3\delta-2)b}{\delta}$, there is a unique root to the equation $f(r) = f(r+1)$ in the relevant domain.

Finally, note that $\frac{3\delta - 2}{\delta} < \frac{\delta}{2-\delta}$ so that we can summarize our findings as follows. If $k \leq s - \frac{\delta b}{2-\delta}, f(r) < 0$ so that $\rho(k) = 0$. If $s + 2 - \frac{(3\delta-2)b}{\delta} > k \geq s - \frac{\delta b}{2-\delta}$, $f(r) > 0$ for $r \leq s - b$ and negative afterwards, so that $\rho(k) = \max\{0, s - b\}$. Finally, if $k \geq s + 2 - \frac{(3\delta-2)b}{\delta}$, there exists a unique integer r^* such that $f(r) > 0$ for all integers $r < r^*$ and $f(r) < 0$ for all integers $r \geq r^*$. □

Proposition 2 shows that three different situations may arise depending on the size k of the cartel and the numbers b and s of traders on the market. If k is small, the cartel has no incentive to restrict the number of traders on the market. If, on the other hand, k is large, the optimal choice of the cartel is to exclude enough sellers from the market so that the number of active sellers becomes smaller than the number of active buyers. Finally, there exists an intermediate situation where the cartel chooses to restrict the number of sellers in order to match the number of buyers on the market. Notice that, if the discount parameter δ is too low, ($\delta < \frac{2}{3}$), the cartel never chooses to restrict the number of sellers below the number of buyers. In fact, when δ is low, the increase in per unit surplus obtained by excluding sellers is too small to outweigh the decrease in total surplus due to the restriction in trade.

We now derive some properties of the function $\rho(k)$ assigning to each cartel of size k the number of traders withdrawn from the cartel.

Lemma 1. *For any two cartels k and k' with $k' > k$, $\rho(k') \geq \rho(k)$.*

Proof. To prove that the function $\rho(\cdot)$ is weakly increasing, it suffices to check that it is monotonic for $k \geq s + 2 - \frac{(3\delta-2)b}{\delta}$. In that case, $\rho(k)$ is defined by the unique root of Eq. (1). It is easy to see that this root is strictly increasing in k, so that the function $\rho(k)$ is weakly increasing. □

Lemma 1 shows that the number of sellers withdrawn from the market is a *weakly increasing* function of the size of the cartel. Hence, larger cartels choose to exclude more sellers from the market.

Lemma 2. *For any k such that $\rho(k) \geq s - b$, $\rho(k+1) \leq \rho(k) + 1$.*

Proof. Suppose by contradiction $\rho(k+1) \geq \rho(k) + 2$. By definition of $\rho(k+1)$, we have

$$k + 1 - \rho(k+1) \geq \frac{(b - \delta(s - \rho(k+1) - 1))(b(2-\delta) - \delta(s - \rho(k+1))}{b\delta(1-\delta)}.$$

Observe that the right hand side of the inequality is increasing in r. Since $\rho(k+1) \geq \rho(k) + 1$, we thus have

$$k + 1 - \rho(k+1) \geq \frac{(b - \delta(s - \rho(k)))(b(2-\delta) - \delta(s - \rho(k) - 1))}{b\delta(1-\delta)}.$$

Next note that, since $\rho(k+1) \geq \rho(k) + 2$, $k + 1 - \rho(k+1) \leq k - (\rho(k) + 1)$. Hence we obtain

$$k - (\rho(k) + 1) \geq \frac{(b - \delta(s - (\rho(k) + 1) - 1))(b(2-\delta) - \delta(s - \rho(k) + 1))}{b\delta(1-\delta)},$$

contradicting the fact that $\rho(k)$ is the first integer following the root of Eq. (1). □

Lemma 2 shows that the total number of active sellers in the cartel $(k - \rho(k))$ is an increasing function of the size of the cartel, as long as $\rho(k) > s - b$. Hence, larger cartels put more active traders on the market. Alternatively, Lemma 2 shows that, for any value r of excluded members, where $s - b \leq r \leq \rho(n)$, there exists a unique cartel size $\kappa(r)$ such that for any $k < \kappa(r)$, $\rho(k) < r$, $\rho(\kappa(r)) = r$ and, for any $k > \kappa(r)$, $\rho(k) \geq r$. The function $\kappa(r)$ thus assigns to each number r of excluded traders, the minimum size of a cartel excluding r traders. We conclude this section by establishing the following Lemma.

Lemma 3. *For any* r, $\rho(n) \geq r \geq \max\{0, s-b\}$, $\kappa(r+1) > \kappa(r)+1$.

Proof. Notice first, that, by Lemma 1, the function $\kappa(r)$ is weakly increasing. Hence, to prove the Lemma, it suffices to show $\kappa(r+1) \neq \kappa(r)+1$. Suppose by contradiction that $\kappa(r+1) = \kappa(r)+1$. Let $k = \kappa(r)$. We then have $\rho(k-1) < r$ and $\rho(k+1) = k+1$. Hence we obtain

$$\begin{aligned} k-1 &< r + \frac{(b-\delta(s-r+1))(b(2-\delta)-\delta(s-r))}{b\delta(1-\delta)} \\ k+1 &\geq \frac{(b-\delta(s-r))(b(2-\delta)-\delta(s-r-1))}{b\delta(1-\delta)}. \end{aligned}$$

After some manipulations, this system can be rewritten as

$$\begin{aligned} k &< r + \frac{b(2-\delta)}{1-\delta} - \frac{(3-\delta)(s-r)}{1-\delta} - \frac{1}{1-\delta} + \frac{\delta(s-r)(s-r+1)}{b(1-\delta)} \\ k &\geq r + \frac{b(2-\delta)}{1-\delta} - \frac{(3-\delta)(s-r)}{1-\delta} + \frac{1}{1-\delta} + \frac{\delta(s-r)(s-r-1)}{b(1-\delta)} \end{aligned}$$

implying

$$\delta(s-r) > b,$$

which contradicts the fact that $r \geq s-b$. □

Lemma 3 shows that, for any fixed value r, there are at least two cartel sizes k and k' such that $\rho(k) = \rho(k') = r$. Summarizing the findings of the three preceding lemmas, Fig. 2 depicts a typical function $\rho(k)$ in the case where $s > b$.

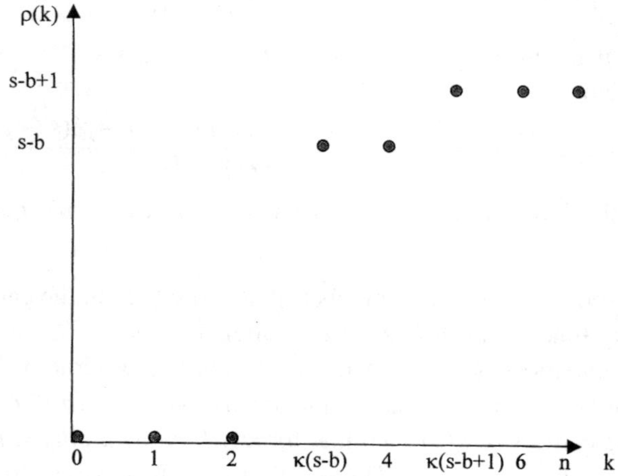

Fig. 2.

3 Stable Cartels

In this section, we analyze the formation of cartels by buyers and sellers on the market. We first describe the noncooperative game of coalition formation, and then consider both the situation where a cartel is formed only on one side of the market and where cartels are formed on the two sides of the market.

3.1 Cartel Formation

The formation of the cartel is modeled as a noncooperative participation game. All traders on one side of the market simultaneously announce whether they want to participate in the cartel (C) or not (N). The cartel is formed of all the traders who have announced (C). A cartel K of size $k > 1$ is *stable* if it is a Nash equilibrium outcome of the game of cartel formation.

This simple game of cartel formation embodies the notions of internal and external stability proposed by d'Aspremont et al. (1983). In a stable cartel, no insider wants to leave the cartel, and no outsider wants to join. It should be noted that this model imposes several crucial restrictions on the formation of cartels. First, it is assumed that only one cartel is formed. The issue of the formation of several cartels and the possible competition between cartels is ignored in the analysis. Second, our focus on the Nash equilibria of the game, excluding coordination between cartel members at the participation stage, implies that whenever a member leaves the cartel, all other cartel members remain in the cartel. Given that outsiders benefit from the formation of a cartel, this assumption makes deviations *easy* and thus leads to a more stringent concept of cartel stability than if we allowed cartel members to respond to a defection. On the other hand, by focussing on Nash equilibria, we ignore the possibility that traders deviate together. Hence, we make deviations *harder* than in games where coalitions of traders can cooperate[6].

Once a cartel is formed, we assume that sellers share *equally* the profits of the cartel. Since all sellers are identical, this assumption can be made without loss of generality in the following way. If a stable cartel with unequal sharing exists, it must be that the cartel is also stable under equal sharing. Hence, the stable cartels under equal sharing are the only candidates for stable cartels under arbitrary sharing rules. On the other hand, note that a cartel may be stable under equal sharing but unstable under an alternative sharing rule.[7]

---

[6] In games of cartel formation, where the formation of a coalition induce externalities on the other traders, there is no simple way to define a stable cartel structure. The concept of stability depends on the assumptions made on the reaction of external players to a deviation. See Bloch (1997) for a more complete discussion of the alternative concepts of stability in games with externalities across coalitions.

[7] In a different model of coalition formation, where traders make sequential offers consisting both of a coalition and the distribution of gains within the coalition, Ray and Vohra (1999) show that, when traders are symmetric, the assumption of equal sharing can also be made without loss of generality.

3.2 Cartel Formation on one Side of the Market

We first analyze the formation of a cartel on one side of the market. As before, we assume that only sellers can organize on the market and characterize the stable cartels of sellers.

Proposition 3. *There exists at most one stable cartel sixe. If $s < b$, no cartel is stable. If $s = b \leq \frac{2\delta}{3\delta-2}$, no cartel is stable. If $s = b \geq \frac{2\delta}{3\delta-2}$, the unique stable cartel size is the first integer k^* following $2 + \frac{2b(1-\delta)}{\delta}$ and $\rho(k^*) = 1$. If $s > b$, the unique stable cartel size is the first integer k^* following $s - \frac{\delta b}{2-\delta}$ and $\rho(k^*) = s - b$.*

Proof. Pick a cartel K of size k. First observe that, if there exists a number r such that $\kappa(r) < k < \kappa(r+1)$, the cartel K cannot be stable, since, following the departure of a member, the cartel of size $k - 1$ still selects to exclude the same number r of sellers. Hence, the only candidates for stable cartels are the cartels of size $k = \kappa(r)$ for some $r \geq s - b$.

Suppose now that the cartel K of size $k = \kappa(r)$ is indeed stable. Since no member wants to leave the cartel,

$$\frac{k-r}{k} U(b, s-r) > U(b, s-r+1). \tag{2}$$

Furthermore, by Lemma 2, any cartel of size $k - 1$ chooses $\rho(k-1) = r - 1$,

$$\frac{k-r-1}{k-1} U(b, s-r) < \frac{k-r}{k-1} U(b, s-r+1). \tag{3}$$

Inequalities 2 and 3 imply

$$\frac{k-r}{k} > \frac{k-r-1}{k-r}.$$

Rearranging, we obtain

$$\frac{r^2}{r-1} > k\ .$$

Next observe that, using Eq. (1), we derive the following inequality

$$k - r > 2(r - (s - b)).$$

By the two previous inequalities, we obtain

$$O > -r + 2(r - (s - b))(r - 1)),$$

a condition which can only be satisfied by $r = 1$ if $s = b$ and $r = s - b$ is $s > b$. Hence, the only candidates for stable cartels are $k = \kappa(1)$ if $s = b$ and $k = \kappa(s - b)$ if $s > b$. It is clear that no insider wants to leave the cartel. Furthermore, by Lemma 3, no outsider wants to join the cartel, since the addition of a new member does not change the cartel's optimal choice. Hence, the two cartels we have found are indeed stable. □

Proposition 3 characterizes the unique stable cartel formed by sellers on the market. It appears that the formation of cartels cannot lead to a large imbalance in the number of buyers and sellers on the market. If sellers are initially on the long side of the market, the cartel they form leads to an equal number of active buyers and sellers on the market. If initially, buyers and sellers are present in equal numbers on the market, the cartel formed only leads to the exclusion of one seller.[8]

As opposed to the classical model of Cournot competition with divisible goods studied by Salant et al. (1983), we show that, when the good traded is indivisible, there exists a stable cartel size. To understand the differences between the two models, it is useful to recall the free-rider problem associated to the formation of a cartel. Since outsiders obtain a higher payoff than insiders, most cartels are not sustainable, because members have an incentive to leave the cartel. As a result, the only sustainable cartels are those which break down entirely upon the departure of a member. These cartels can be characterized when the good traded is indivisible, but do not exist in models with divisible commodities.

The characterization of Proposition 3 allows us to compute the stable cartel sizes when the market outcome converges to the competitive outcome. As δ approaches 1, the stable cartel sizes converge to 3 if $s = b$ and to $s - b + 1$ when $s > b$. It is easy to see that these are the minimal cartel sizes for which it is profitable to exclude some traders from the market.

While the arguments underlying Proposition 3 rely on computations using the specific trading mechanism based on Rubinstein and Wolinsky's (1990) bargaining model, we believe that some of the conclusions are robust to changes in the model of trade. In fact, the characterization of the unique stable cartel relies on the fact that gains from excluding a trader are highest when the market is balanced. Hence, the trader's incentives to join a cartel are largest when the cartel forms to match the number of buyers and sellers or to exclude one trader when the market is originally balanced. The characterization of the unique stable cartel thus seems to rely primarily on the inverted S-shape of the function relating the utility of a trader to the number of traders on his side of the market. The extension of our results to other trading mechanisms is thus related to the shape of this function.

The Shapley value of this market, analyzed by Shapley and Shubik (1969), also yields an inverted S-shape relation. Similarly, small perturbations around the competitive equilibrium solution also result in an inverted S shape relation. On the other hand, the distribution of the trading surplus obtained by Rubinstein and Wolinsky (1985) at the steady-state of a matching and bargaining market with entry yields a completely different picture. In that case, the utility of a seller is given by $\frac{\delta b}{2(1-\delta)b+\delta s+\delta b}$ if $s \leq b$ and $\frac{\delta b}{2(1-\delta)s+\delta s+\delta b}$ if $s \geq b$. These functions are not inverted S-shape functions of the number s of sellers, and our conclusions do not extend to this trading mechanism. In fact, with this specification of gains

[8] Note that, for a stable cartel to exist when $s = b$, the following condition must be satisfied: $s = b \geq \frac{2\delta}{3\delta - 2}$. If this condition is violated, the stable cartel size k^* is larger than s.

from trade, it turns out that no cartel is stable since it is never optimal to exclude any trader from the market.

3.3 Cartel Formation on the Two Sides of the Market

In this section, we characterize the stable cartels formed *on the two sides of the market*. In order to analyze the response of agents on one side of the market to collusion on the other side, we adopt a *sequential* framework where buyers form a cartel first, and sellers organize in response to the formation of the buyers' cartel. Hence, the sequence of stages in the game is as follows. First, buyers engage in the game of cartel formation; second, the buyers' cartel selects the number of active traders; third, sellers engage in the game of cartel formation; fourth, the sellers' cartel chooses the number of active traders, and finally buyers and sellers meet and trade on the market.

In order to focus on the endogenous response of sellers to collusion on the part of buyers, we assume that the market is originally balanced and let n denote the initial number of buyers and sellers on the market. In order to analyze the behavior of the buyers, we note that, for any choice b of active buyers, there exists a unique stable cartel size on the side of sellers as given by Proposition 3. The expected utility obtained by a cartel of buyers K of size k can then be computed as

$$\begin{aligned} U_K &= \frac{k}{2} \quad \text{if } r = 0 \text{ and } n \leq \frac{2\delta}{3\delta - 2} \\ U_K &= kU(n-1, n) \quad \text{if } r = 0 \text{ and } n \geq \frac{2\delta}{3\delta - 2} \\ U_K &= \frac{k-r}{2} \quad \text{for all } k \geq r > 0. \end{aligned}$$

This definition of the cartel's surplus takes into account the reaction of sellers to the formation of the cartel of buyers. If buyers do not collude on the market, sellers either respond by forming a cartel which withdraws one trader from the market (if $n \geq \frac{2\delta}{3\delta-2}$) or else do not collude. If buyers withdraw some traders from the market ($r > 0$), a cartel of sellers is formed in order to match the number of active buyers on the market.

The definition of the cartel's surplus can be used to determine the optimal behavior of the cartel. First observe that, as long as $n \leq \frac{2\delta}{3\delta-2}$, for any size k of the cartel, the optimal restriction is $\rho(k) = 0$. When $n \geq \frac{2\delta}{3\delta-2}$, the cartel must choose between a restriction $r = 0$, yielding a surplus of $kU(n-1, n) = \frac{k(n-1)(1-\delta)}{2n(1-\delta)+\delta}$ and a restriction $r = 1$ yielding a payoff of $\frac{k-1}{2}$. Simple computations show that the optimal choice is given by

$$\begin{aligned} \rho(k) &= 0 \text{ if } k \leq \frac{2n(1-\delta)+\delta}{2-\delta} \\ \rho(k) &= 1 \text{ if } k \geq \frac{2n(1-\delta)+\delta}{2-\delta}. \end{aligned}$$

Using the same arguments as in the proof of Proposition 3, one can easily show that the only stable cartel size is the first integer k^* following $\frac{2n(1-\delta)+\delta}{2-\delta}$. We then obtain the following Proposition.

Proposition 4. *When cartels are formed sequentially on the two sides of the market, there is at most one stable cartel configuration. If* $n \leq \frac{2\delta}{3\delta-2}$, *no cartel is stable. If* $n \geq \frac{2\delta}{3\delta-2}$, *buyers and sellers form cartels of size* k^* *where* k^* *is the first integer following* $\frac{2n(1-\delta)+\delta}{2-\delta}$. *Both the buyers' and sellers' cartels choose to withdraw one trader from the market.*

Proof. The determination of the buyers' stable cartel follows from the preceding arguments. The determination of the sellers' stable cartel is obtained by applying Proposition 3 to the case $b = n - 1$, $s = n$. □

Proposition 4 shows that, when the numbers of buyers and sellers are initially equal, the stable cartels formed on the two sides of the market have the same size. Furthermore, both cartels select to withdraw one trader from the market so that bilateral collusion leads to a "balanced market" where the number of active buyers and sellers are identical. The requirement that the cartels formed be stable greatly limits the scope of collusion and the sizes of the cartels. For example, as δ converges to 1, the stable cartel sizes converge to 2: in equilibrium, the cartels formed by buyers and sellers only group two traders on both sides.

In order to understand how the threat of collusion on one side of the market affects the incentives to form cartels on the other side, it is instructive to compare the sizes of the cartels obtained under bilateral collusion with the size of the cartel formed under unilateral collusion with $b = s = n$. It appears that the cartel formed under unilateral collusion is always *larger* than the cartels formed under bilateral collusion. Furthermore, the total numbers of trades obtained under unilateral and bilateral collusion are equal. We interpret this result by noting that there are *limits to bilateral collusion*. The threat of collusion on one side of the market does not lead to a higher level of collusion on the other side. In fact, there is no "escalation" process by which buyers would choose to form large cartels and to restrict trade by a large amount, anticipating the reaction of sellers on the market.

4 Conclusion

This paper analyzes the formation of cartels of buyers and sellers in a simple model of trade inspired by Rubinstein and Wolinsky's (1990) bargaining model. We show that, when cartels are formed only on one side of the market, there is at most one stable cartel size. When cartels are formed sequentially on the two sides of the market, there is also at most one stable cartel configuration. It appears that, under bilateral collusion, buyers and sellers form cartels of equal sizes, and that the cartels formed are smaller than under unilateral collusion. Both cartels choose to exclude only one trader from the market. This result suggests that there are limits to bilateral collusion, and that the threat of collusion on one side of the market does not lead to increased collusion on the other side.

Our results thus show that the formation of a cartel of buyers induces the formation of a cartel of sellers yielding a "balance" in market power on the two sides of the market. Clearly, our model is much too schematic to account for the emergence of cartels of producers of primary commodities. However, we believe that our model gives credence to the view that these cartels were formed partly as a response to increasing concentration on the part of sellers. Furthermore, our results indicate that the cartels formed would only group a fraction of the active traders on the market, in accordance with the actual evidence at the time of the formation of OPEC, the Copper and Uranium cartels.

While our analysis provides a first step into the study of bilateral collusion on markets with a small number of buyers and sellers, we are well aware of the limitations of our model. The process of trade we postulate, the specific model of cartel formation we analyze, allow us to derive some sharp characterization results, but clearly restrict the scope of our analysis. Furthermore, we assume in this paper that collusion results in the exclusion of traders from the market. In reality, the cartel can choose to enforce different collusive mechanisms. It could for example specify a common strategy to be played by its members at the trading stage or delegate one of the cartel members to trade on behalf of the other members. The analysis of these forms of collusion is a difficult new area of investigation in bargaining theory, and we plan to tackle this issue in future research.

References

d'Aspremont, C., Jacquemin, A., Gabszewicz, J.J., Weymark, J. (1983) The stability of collusive price leadership. Canadian Journal of Economics 16: 17–25

Bloch, F. (1997) Noncooperative models of coalition formation in games with spillovers. In: Carraro, C., Siniscalco, D. (eds.) New Directions in the Economic Theory of the Environment. Cambridge University Press, Cambridge

Donsimoni, M.P. (1985) Stable heterogeneous cartels. International Journal of Industrial Organization 3: 451–467

Donsimoni, M.P., Economides, N.S., Polemarchakis, H.M. (1986) Stable cartels. International Economic Review 27: 317–336

Gabszewicz, J.J., Shitovitz, B. (1992) The core in imperfectly competitive economies. In: Aumann, R.J., Hart, S. (eds) *Handbook of Game Theory with Economic Applications.* Chap. 15, Elsevier Science, Amsterdam

Galbraith, J.K. (1952) American Capitalism: The Concept of Countervailing Power. Houghton Mifflin, Boston

Gul, F., Sonnenschein, H., Wilson, R. (1986) Foundations of dynamic monopoly and the coase conjecture. Journal of Economic Theory 39: 155–190

Halloway, S.K. (1988) The Aluminium Multinationals and the Bauxite Cartel. Saint Martin's Press, New York

Hart, S. (1974) Formation of cartels in large markets. Journal of Economic Theory 7: 453–466

Legros, P. (1987) Disadavantageous syndicates and stable cartels. Journal of Economic Theory 42: 30–49

Mac Afee, P., Mac Millan, J. (1992) Bidding rings. American Economic Review 82: 579–599

Okuno, M., Postlewaite, A., Roberts, J. (1980) Oligopoly and Competition in Large Markets. American Economic Review 70: 22–31

Ray, D., Vohra, R. (1999) A theory of endogenous coalition structures. Games and Economic Behavior 26: 286–336

Rubinstein, A., Wolinsky, A. (1985) Equilibrium in a market with sequential bargaining. Econometrica 53: 1133–1150

Rubinstein, A., Wolinsky, A. (1990) Decentralized trading, strategic behavior and the Walrasian outcome. Review of Economic Studies 57: 63–78

Salant, S., Switzer, S., Reynolds, R. (1983) Losses from horizontal mergers: The effects of an exogenous change in industry structure on Cournot-Nash equilibrium. Quart. J. Econ. 98: 185–199

Sampson, A. (1975) The Seven Sisters: The Great Oil Companies and the World they Made. Viking Press, New York

Scherer, F., Ross, D. (1990) Industrial Market Structure and Economic Performance, 3rd ed. Houghton Mifflin, Boston, MA

Shapley, L.S., Shubik, M. (1969) Pure competition, coalitional power and fair division. International Economic Review 10: 337–362

Sletmo, G.K., Williams, E.W. (1980) Liner Conferences in the Container Age: U.S. Policy at Sea. Mac Millan, New York

Stigler, G.J. (1954) The economist plays with blocs. American Economic Review 44: 7–14

Taylor, J.H., Yokell, M.D. (1979) Yellowcake: The International Uranium Cartel. Pergamon Press, New York

Network Exchange as a Cooperative Game

Elisa Jayne Bienenstock, Phillip Bonacich

Department of Sociology, Stanford University. SU/120, 160, Stanford, CA 94305, USA.
(e-mail: ejb@leland.stanford.edu)

Abstract. This paper presents parallels between network exchange experiments and N-person cooperative games with transferable utility, to show how game theory can assist network exchange researchers, not only in predicting outcomes, but in properly specifying the scope of their models. It illustrates how utility, strategy and *c-games*, concepts found in game theory, could be used by exchange theorists to help them reflect on their models and improve their research design. One game theoretic solution concept, the kernel, is compared to recent network exchange algorithms as an illustration of how easy it is to apply game theory to the exchange network situation. It also illustrates some advantages of using a game theory solution concept to model network exchange.

Key Words. social networks, game theory, coalitions, exchange, kernel

1 Introduction

The topic of power and resource distribution in exchange networks has generated much work and discussion. Not only are there several algorithms that attempt to predict which positions have power, there have also recently been articles comparing these algorithms (Skvoretz and Fararo 1992; Skvoretz and Willer 1993). In previous work Bienenstock (1992) and Bienenstock and Bonacich (1993) have shown that the current agenda of exchange network theorists overlaps with the literature of N-person cooperative game theory with transferable utility. Their first step was to show how easily and effectively solution concepts, already available in N-person cooperative game theory, can be applied to the exchange network situation. The objective was to merge the two fields, so that the wealth of insight and discussion about bargaining, coalition formation and the effect of diadic negotiation on larger groups, that game theorists have accumulated over the past 70 years, could be utilized by researchers studying exchange. Unfortunately, most of the response to this work has focused on the usefulness of one of four game theoretic solution concepts introduced as an algorithm to predict ordinal power in networks, the core. This article will show how game theory can assist network exchange researchers, not only in predicting outcomes, but in properly specifying the scope of their models.

This article reviews prominent approaches in the literature. The intent is to focus on underlying assumptions common to all theories, not the predictive power of algorithms. Several assumptions are essential and implicit in all exchange network theories. All theories acknowledge that structure matters: structures provide some positions with advantages that emerge over time. All theories also accept that subjects act. Subjects are sentient beings and structural advantages emerge, in part, as a result of the strategies or actions of the actor. Finally, there is consensus that experimental outcomes can test both assumptions and resulting predictions, regarding differential power and resource distribution in networks.

For the most part, these assumptions have been universally adopted and so have gone unchallenged. Research instead has focused on comparisons of different algorithms and their predictive capabilities. The result is better prediction about resource distribution, with little theoretical reflection on which aspects of these algorithms lead to the better predictive capabilities. While many current theories are the results of several revisions of older theories, most revisions came about in an attempt to match empirical findings and were not theoretically inspired. In fact, questions about the relationship between the behavioral assumptions of these theories and the structural outcomes have largely been ignored. This article questions some of these assumptions and points to some implications for the predictive capacity of these algorithms. Concepts borrowed from game theory will anchor some of these concerns.

Work in network exchange attempts to answer questions about both the behavior of subjects (actors) and the structural distribution of resources in groups (networks) without developing a formal theory about that relationship. Network exchange theorists will eventually have to address this issue to clarify the scope of their experiments. The next section is a statement of the problem. We will make explicit what issues need to be addressed and explain why it is important these issues are addressed.

The section that follows is a review of two exchange network theories: Cook et al. (1983), and Markovsky et al. (1988). These early works were selected because it is in these articles that the assumptions of rational choice and structure were introduced. Later works built on the assumptions adopted here. This article's focus on the function of behavioral assumptions in a structural theory is designed to frame a dialogue among network exchange researchers about this important topic. A secondary related focus is on the formulation of 'rational actors', adopted by network exchange theory. Finally, the design of exchange experiments are examined, focusing on what they measure, their scope, and what conclusions can be drawn from their findings.

Game theory approaches are then introduced as a mechanism for evaluating some of these implicit assumptions.[1] The behavioral assumption of game theory is based on rational choice: actors are expected to maximize utility. This is not a simple assertion. Utility theory, an area closely associated with game theory, is an axiomatic theory that attempts to create a cardinal measure of preferences. It is within utility theory that game theorists define utility and determine how to measure it empirically. Implicitly, exchange theorists designed their experiments to measure power in accordance with utility theory, yet Skvoretz and Willer (1993) distinguish between behavioral (social psychological) assumptions and rational choice assumptions. A discussion of utility theory combined with the concept of the 'form' of a game illustrates that what Skvoretz and Willer (1993) refer to as 'social psychology' defines a 'strategy' in game theory, and that the underlying behavioral assumptions of all exchange algorithms is rational choice.

Although network exchange theory is concerned with behavior, its primary concern is structural outcome. Game theorists have also been concerned with both behavioral assumptions and structural outcomes. Network exchange theorists have thus far not clearly defined the connection between their behavioral assumptions and their structural theory of distribution of resources and power in networks. The concept of the form of games will prove useful in clarifying the association between the behavioral assumptions of network theories and the structural outcomes that are measured experimentally. In game theory, situations similar to that of the network exchange experiments are modeled using the characteristic function form. This paper will demonstrate the advantage to characterizing the exchange network situation as a game in characteristic function form.

Following the discussion of game theory we describe how exchange network theories have evolved. Exchange resistance theory, an amalgam of Markovsky's GPl approach and Willer's resistance theory, will be reviewed. Also Cook and Yamagishi (1992) have revised their theory and have a new approach known as equidependence. We will compare these two approaches as they stand today and compare them with a game theoretic solution concept: the kernel.

1.1 Rationality and Structure

All theories of power in exchange networks have behavioral and structural components. The behavioral component refers to the theory's conception of how the individual in a network makes choices. The structural component concerns the identification of positions of power within the network. The structural component of a theory must consider not only individual choice but how the complete pattern of choice and network constraints create power differences. The psychological

[1] What follows is a *general* introduction to how game theory approaches topics of interest to network exchange theorists. The particulars of how solution concepts could be applied to specific issues, such as positive versus negative exchange or weak versus strong power, are beyond the scope of this discussion. These are topics that we find interesting and intend on pursuing in our future work.

components of all the major theories are, either formally or informally, theories of rational choice and maximizing behavior.

Skvoretz and Willer (1993) focused on the differences in the rational choice assumptions of different algorithms. They compared four theories in their investigation and deemed three theories to be more social psychological and less rational than the one game theoretic solution: the core. The core (Bienenstock 1992; Bienenstock and Bonacich 1993; Bonacich and Bienenstock 1993) was judged to be not truly 'social psychological' while the other three theories were thought to be more social psychological and less rational than the core.

This focus on the social psychology inherent in the algorithms is important to a discussion of the interplay between the underlying behavioral assumptions of these theories and the structural implications. If these are truly structural theories and structure determines differential outcomes for different players then what place do behavioral assumptions have in the theory at all? Would actors who behave randomly with no strategy defy the structural outcomes? Could very strategic actors defy structural determinism? These important questions have so far not been addressed in the literature on exchange networks.

On the other hand if the theories are social psychological theories why are the tests of these theories measured on a structural level? None of the experiments published to date has tested behavioral assumptions, they have just asserted them.[2] The dependent variable measured to test theories are structural outcome variables. Where is the test of individual cognition, motivation and intention?

Initially exchange network theories addressed these issues. Unfortunately, a preoccupation with the predictive accuracy of algorithms distracted researchers and little theoretical work emerged addressing this issue. It is time that network exchange researchers revisit these issues.

2 Looking at the Past

This section is a review of the two articles that spurred subsequent research in network exchange: Cook et al. (1983) and Markovsky et al. (1988). Some of the underlying assumptions of these two theories differ. Since most of the literature has focused on the predictive power the algorithms generated, there has been little discussion of the subtle differences in the underlying assumptions of these two works. For the most part these works have been grouped together and evaluated as a unit. The next section highlights important theoretical differences in these perspectives.

[2] At Sunbelt XVI: The International Sunbelt Social Network Meetings, February 1996, three of six papers in a session on 'Exchange' either presented results or proposed research testing the social psychological assumptions of these theories: 'Is there any Agency in Exchange Networks?' by Elisa Jayne Bienenstock; 'The Process of Network Exchange: Assumptions and Investigations,' by Shane R. Thye, Michael Lovaglia and Barry Markovsky; and 'A Psychological Basis for a Structural Theory of Power in Exchange Networks', by Phillip Bonacich. This may indicate a new trend.

2.1 Cook, Emerson, Gilmore and Yamagishi, 1983

The article by Cook et al. (1983) was written first to empirically demonstrate that some structural positions had advantages over others in networks of exchange. Their main contribution was to demonstrate the need for an algorithm specific to *exchange* networks. Although standard centrality measures, successful in determining which positions have power in information and influence networks, were not able to predict power in exchange networks, nonetheless structure did matter.

Cook et al. proposed a first attempt to develop an algorithm (point vulnerability) to predict which position had power in networks. This addressed the need for a structural measure that systematically considered all positions within an entire network. Vulnerability was successful in making predictions for the network Cook et al. investigated. The details of their measure are not relevant to this discussion. The actual approach has been superseded by a better model (Cook and Yamagishi 1992). What is relevant was that their algorithm for predicting power demonstrates that power differences can emerge from structural differences.

Cook et al. (1983: 286) assume that subjects behave in a rational manner. Rational behavior in this situation means that, 'each actor maximizes benefits by (a) accepting the better of any two offers, (b) lowering offers when offers go unaccepted, and (c) holding out for better offers when it is possible to do so.'[3] These principles were especially necessary for designing the computer simulations used to support experimental results. Cook et al. wanted to show that even with very simple behavioral assumptions structural outcomes could be predicted. Rationality was assumed only insofar as the actors were expected to use a power advantage if they had one. 'This assumption [rationality] is necessary theoretically since it allows us to derive testable predictions concerning manifest power from principles dealing with potential power.'[4]

Cook et al. did not connect their rationality assumptions to their structural algorithm. The structural algorithm was designed to predict the outcomes of experiments. Since the predicted outcomes were measured as resource differences resulting from the utilization of potential power, Cook et al. required that the subjects exercise power, if they had any. As the following quote shows they did not assume that they were actually modeling behavior, nor, did they believe, necessarily, that their subjects were rational. They recognized that their rationality assumption was just that, an assumption, that needed to be tested independently of the structural component of their model. They said:

> This [rationality] is clearly a testable assumption, but all one could conclude from evidence to the contrary is that sometimes subjects in our laboratory act irrationally. We have examined empirically some of the conditions under which these conditions do not hold (e.g. when equity concerns are operative).

Cook and Emerson (1978), in a separate study, focus on the behavioral component of this question. The 1983 article focused on the structural component.

[3] Cook et al. (1983, 286).

[4] Cook et al. (1983, 286, note 12).

The rational assumptions were not needed for the structural argument. This indicates an awareness of the disjuncture between the rational choice principles and the structural theory. Cook et al. never imply that the assumptions about behavior used in this article were necessary for the model. Their point was that even these simple assumptions, principly maximization, produced the predicted results.[5]

Unlike later models, Cook et al. did not require that actors be aware of their position in a network. Actors were intentionally not informed about the value of their potential exchanges. Subjects could not compare their benefits with that of others so equity concerns could not affect their evaluations. At each point in the negotiation subjects were able to evaluate the utility of each choice presented to them. They did not have knowledge of the network structure or the rewards of other subjects. The only information that was made available to them was (1) with whom they may exchange, (2) what their current offers are, and (3) their prior history.

Vulnerability was unambiguously defined and predicted the results of their experiments and simulations. How it does so is unclear. Subjects with the limited information provided, clearly could not assess vulnerability. Even if subjects were aware of their positions within a network, there is no reason to assume that the hypothetical possibility of their removal could lead them to demand more from exchanges and for other subjects to accede to their requests. There is a disjuncture between individual and structural principles. Vulnerability addresses only structural questions. The measure allows an observer to make predictions about outcomes based on the structure. It does not tell us how subjects arrive at those outcomes.

To make their simulations *go* Cook et al. needed to impart some behavior. They chose rational behavior. Would other behaviors have led to the same structural outcomes? Would any behavior have led to the same outcome? If the answer to both these questions is yes, there is no need for an individual component to the theory, structure accounts for everything. If the answer is no, then the structural theory is not robust. This experimental design only tests the structural theory. The rationality of the subjects is only assumed. This leaves open a big question: what is the relationship between the individual and structural assumptions of this theory?

2.2 Markovsky, Willer and Patton, 1988

Markovsky et al. (1988) introduced another algorithm to address the same question, which they showed was a better predictor of which position would amass resources. Since then, much of the focus in the literature has been on fine tuning algorithms to better match the results of experiments rather than on testing the validity of some of the assumptions implicit in the Cook et al. research design.

[5] This type of reasoning is identical to 'as if' reasoning of economists whose assumption of rational choice, despite the fact that it might not accurately model the actual behavior of actors, has 'not prevented the rational choice model from generating reasonably accurate prediction about aggregate market tendencies' (Macy 1990, 825).

At the individual level, it is clear that Markovsky et al. also assume that individuals behave in a more or less rational fashion. Rational behavior is maximizing behavior. According to condition 4 of their model,[6] people will not exchange with those in more favorable structural positions than themselves because they expect to earn less in such exchanges.

To assess structural power, Markovsky et al. developed a ‘Graph-theoretic Power Index’, or GPI.[7] It is based on this measure that individuals are supposed to evaluate what offers to make.

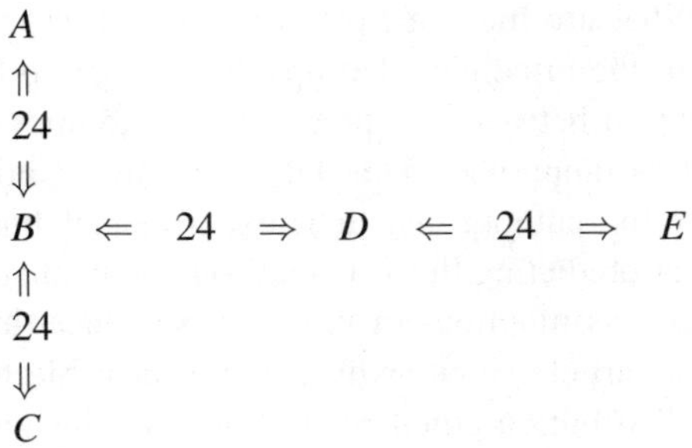

Fig. 1. The five-person ‘T’ network

The GPI is a sum of the number of non-intersecting paths of varying length, where paths of odd length are weighted +1 and paths of even length are weighted −1. For example, in Fig. 1, from position A emanates: one path of length 1 ($A - B$); one non-intersecting path of length 2 ($A - B - C$ or $A - B - D$, which share the $A - B$ edge); and one path of length 3 ($A - B - D - E$). Therefore A's (and C's) power is $1 - 1 + 1 = 1$. B's power is $3 - 1 = 2$. D has two paths of length 1 ($D - E$ and $D - C$), and one path of length 2 ($D - B - A$ or $D - B - C$, which share the $D - B$ edge), so its power is $2 - 1 = 1$. Finally E's GPI index of $1 - 1 + 1 = 1$. Therefore, by the GPI measure, B is the most powerful position.[8]

The experiment designed to test this algorithm was different than that of Cook et al. For this experiment complete information about the network structure was provided to all subjects. They were given complete information, not only about their exchanges and network position, but about every exchange in the network.

---

6 Markovsky et al. accept the best offer they receive, and choose randomly in deciding among the tied best offers (1988, 223).

7 Although the focus of this article is not the algorithm it is included here for two reasons. (1) Even the most current algorithm still uses the GPI. (2) The GPI is a part of the behavioral assumptions of the theory. Calculation of the GPI (Axiom 1) is necessary to determine what positions have power. Markovsky et al.'s Axiom 2 demands that actors seek exchange with partners only if they are more powerful than the partner as determined by the GPI.

8 Actors or positions in figures are represented as capital letters, exchange opportunities exits where arrows connect pairs. In some figures numbers appear between modes. These numbers indicate the value of the exchange. If no values appear all exchanges in the network have the same value.

With this information subjects were expected to devise strategies that allowed them to maximize their resource accumulation.[9]

This design assumes a forward looking rational actor rather than a backward looking or responsive actor.[10] The assumption appears to be that given complete information people will behave in a way that will ensure the predicted structural outcomes. What is striking about this measure is that it is inconceivable that any subject in an experiment would engage in this calculation. The GPI index describes only the behavior of Markovsky et al., rather than subject behavior. The index works for the networks that they examine, but how it works is unclear. The psychology in the article is a kind of rational choice, but the authors do not address the reason that the strategy of rational actors produce results predicted by the GPI index. What are the principles that operate at the individual level? How do they relate to the principles that operate at a group level?[11]

While the connection between the predictive capabilities and the underlying social psychological assumptions of the GPI were never articulated, there has been an assertion by the authors that their experimental findings, which show the GPI to be better at predicting the structural outcomes than vulnerability, also 'challenged some basic assumptions of power-dependence theory'.[12]

Section 3.3 of this article shows why it is not possible to use experimental results measured as distribution outcomes to draw conclusions about underlying assumptions. Even if this could be done, showing that the GPI is a better predictor of outcomes than vulnerability is not a refutation of power dependence theory at all.[13]

2.3 Discussion

The work of Cook et al. and Markovsky et al. have been lumped together despite fundamental differences. The root of the differences can be traced to the fact that

[9] Markovsky et al. appear to believe that actors will use all the information provided. 'Having information on negotiations other than one's own is expected to accelerate the use of power but not affect relative power' (Markovsky et al. 1988, 226, note 12).

[10] Skvoretz and Willer (1993) use Macy's (1990, 811) distinction between backward and forward looking actors to point out that Cook et al.'s actors could be thought of as 'backward looking' in contrast to Markovsky et al.'s 'forward looking' actors.

[11] The specific nature of the prescription for behavior Markovsky et al. define make it unlikely that they are using an 'as if' argument. Their behavioral axioms are explicit and appear to be prescriptive if not descriptive. For that reason the disjuncture between behavior and outcome is more problematic than for Cook et al.

[12] Lovaglia et al. (1995, 124).

[13] In fact, the GPI can be interpreted as a power-dependence measure. Consider Fig. 1. B has power because it is connected to many other nodes. For every direct tie value is added. That is because the more connections B has the more options for exchanges. That makes B more *powerful*, because B is not *dependent* on any one exchange partner. However if the nodes that B is connected to are also connected to others, for example D is connected to E, then D is not as dependent on B, so B's power is reduced. That is captured by the GPI by subtracting 1. If however E had options (which is not the case in this network), then B's power would increase. This is because E is not as dependent on D which makes D more dependent on B, which gives B more power. That is captured by the GPI, which adds 1.

unlike Cook et al., Markovsky et al. distinguished network actors (decision-making entities) from positions (network locations occupied by actors). 'The reason for distinguishing actors and positions is that actor properties (e.g. decision strategies) and position properties (e.g. number of relations) may affect power independently'[14] (Markovsky et al. p. 223). This subtle difference explains many differences previously discussed between the two perspectives. Cook et al. were interested primarily in looking at structure. Markovsky et al. seemed to be more interested in describing behavior that leads to structural outcomes. Distinguishing actors from position in network experiments is not a minor issue. Unfortunately, although Markovsky et al. define the two differently, operationally they do not distinguish them.

One way this difference was expressed was in the detail and importance of behavioral assumptions in the theories. Cook et al. did not expect their assumptions about behavior to actually model how subjects act. Markovsky et al. appear to place more importance on their 'actor conditions'. Cook et al. wanted to show that the structure determined the outcome, independent of the intentions, actions and strategies of subjects. Their only demand on actors was that actors did not try to lose points. It appears that Markovsky et al. were more invested in their detailed account of the behavior of subjects to produce outcomes.

Despite this difference in perspective, both experiments had the same dependent variable: resource distribution after several rounds of negotiation and exchange. This might be a historical remnant since Cook et al. designed their experiment first. This level of analysis is a good test of a structural theory.

Another difference is the amount of information that the experimenters provided subjects. Cook et al. provided no information to subjects about their network position or the value of their exchanges (let alone the value of the exchanges of others). No more information was required for subjects; Cook et al. wanted to show that the structure would determine outcome. Markovsky et al. provided subjects with complete information, because they felt that lack of information might impede rational decisions. They felt that to capitalize on their structural advantage subjects would need to be aware of their position.

There is a myth that for actors to behave in a 'rational' manner, it is necessary that they have complete information and that they must be aware of everything that is going on in their environments.[15] For this reason the lack of complete information provided subjects by Cook et al. may seem like a detour from a rational choice perspective. This is not the case. Game theory which is explicitly based on rational choice, was invented to deal with uncertainty and risk. The limits of information for the subjects in the Cook et al. design does not necessarily indicate a limitation of rationality.

---

[14] Markovsky et al. (1988, 223, note 4).

[15] In his discussion of public goods, Macy (1990) articulates this requirement for rational actors: 'The rational choice formulation requires "forward looking" actors, who are able to compute the expected rate of return on investments ... These demanding calculations seem unlikely to inform the typical volunteer' (p. 811).

In fact subjects in the Cook et al. design are provided with all the information that game theory would require. Even though they were not aware of the exchange opportunities of others, or even the values of exchanges, subjects had enough information to design a plan of action for every contingency. In game theory this type of plan *defines* a strategy.

Despite fundamental differences, all work on network exchange is grouped together. This has caused theoretical confusion. It is important that assumptions about the relationship between behavior and structure be explicitly addressed. Game theorists study a related topic: the relationship between the rules of games and the behavior of actors. The next section will introduce concepts from game theory that can address network exchange concerns.

3 A Game Theoretic Interpretation

The objective of this section is to convince the reader that the exchange networks previously described can and should be analyzed with tools provided by game theory. The first task is to show that the exchange networks are N-person cooperative games with transferable utility. The second task is to show that there is no loss in using game theories definition of rationality rather than those formulated by exchange theorists and that there are some advantages in formulating the condition as a game. For instance, using the game perspective encourages analysis of these networks at an appropriate level for the data collected. The final task is to demonstrate how to convert exchange networks into games.

3.1 Network Exchange as N-person Cooperative Games

Even though there are differences between the Cook et al. and Markovsky et al. theories, and the experiments that they designed to test them, there is no question that both are studying the same thing. What is striking is the similarity between these experiments and many of the experiments designed by game theorists to study bargaining and coalition formation. For a detailed review of these games read Kahan and Rapoport (1984) Chapters 11–14. There follows an example of one game that Kahan and Rapoport review that illustrates the similarity between situations game theorists have modeled and the exchange network experiments previously described.

> *Odd Man Out* Three players bargain in pairs to form a deal. The deal is simply to agree on how to divide money provided by the experimenter. The amount of money the experiment provided depends on which pair concludes the deal. If players A and B combine, excluding C, then they split \$4.00. If players A and C coalesce to the exclusion of B, then they get \$5.00. And if B and C combine, they split \$6.00. Any player alone gets nothing, and all three are not allowed to negotiate together. (Kahan and Rapoport p. 30)

Following this description, Kahan and Rapoport explain how to convert this situation into the characteristic function form of the game. An alternative representation is to display the network representation as we have done in Fig. 2. This should make the parallel to the exchange experiment clear.

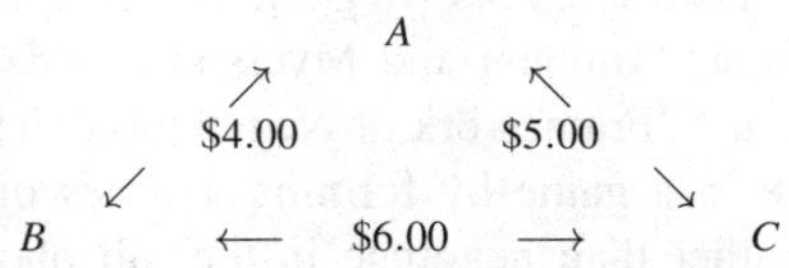

Fig. 2. Odd man out, network representation

The theoretical inspiration of many of the games in N-person game theory is the same as for much of network theory. There are two concurrent themes studied in N-person cooperative game theory. 'Although early interest centered on the question of how members of a coalition would apportion among themselves the fruits of their coalition, recent interest has been directed to the other main question in coalition formation, mainly, which of the possible coalitions will form.'[16] Economists have traditionally been concerned about reward allocations, while other social scientists (political scientists, sociologists and social psychologists) have concerned themselves with the latter.[17] The two main goals of the exchange network researchers is to determine who has power (the reward allocation) and which nodes form an agreement (which coalitions form).

One reason for the reluctance to use game theory is that sociology is supposed to deal with social exchange, while economics deals with economic exchange. In fact, the exchange network experiments studied economic exchange. Blau (1967) identified distinguishing characteristics of social exchange. Exchange is social when the kind of return for a favor is not determined by a contract, and when there is no guarantee that the favor will ever be returned. Social exchange requires and builds trust. None of this is true of exchange network experiments. Experimenters, in fact, went through great pains to ensure these factors were eliminated. Subjects were restricted from face to face negotiation to limit externalities from influencing exchange. Subjects in these experiments were not trading smiles, they were negotiating over points that were later to converted into money.

There is one important feature of the exchange network experiment that does distinguish it from what is studied by most game theorists and does in fact make what is being studied a social exchange: the network structure itself.[18] The economic history of game theory has made the assumption of opened markets and free trade a feature in most games. In many games there is no incentive for certain pairs of players to engage. Even so, they are never restricted from communicating. Although it might not be rational for certain exchanges to occur, the opportunity is always there. The introduction of networks into the literature of games is an important contribution in that it will allow game theorists to begin considering social factors such as lack of access, that have previously not been considered.

16 Kahan and Rapoport (1984, 9).
17 Kahan and Rapoport (1984).
18 Networks represent historical or social forces that limit markets.

There have already been attempts by game theorists to use graph theory to model social phenomenon. Aumann and Myerson (1988) and Myerson (1977) introduce graphs to discuss 'Framework of Negotiation'. The basic idea was that 'players may cooperate in a game by forming a series of bilateral agreements among themselves,[19] rather than negotiate in the 'all player' framework traditional in game theory.[20] They model which links should be expected to form, based on the values of coalitions, using the Myerson value, which is represented by a graph whose vertices are players and edges are links between players. The situation they model is related to exchange experiment. Myerson and Aumann address the question of the emergence of networks. What links can be, added to eliminate power? Despite the similarity to network research, these authors were not aware of the literature on networks.

This illustrates two things. First that there is no compelling reason that network ideas cannot be incorporated into the literature of games. Second that there is a need for communication between the two areas. This relationship would be reciprocal, both perspectives could benefit from opening a dialogue.

If it is clear that game theory could benefit by considering the network exchange situation, it still may not be clear how incorporating game theory into the network literature can enhance that field. One way is by providing solution concepts as algorithms to determine which positions have power. This was the topic of Bienenstock (1992) and Bienenstock and Bonacich (1993). While important, it is only a secondary benefit. An even more important benefit is the distinction that game theory makes between the choice principles that are postulated (usually maximization) and the game outcome. As we have seen, there has been some confusion about this in the exchange theory literature. Choice principles are postulated without any explicit connection to the predicted exchange outcomes.

Two game theory topics will be introduced and applied to the issues discussed previously: the importance of the assumption of rationality and the disjuncture between social psychological and the structural assumptions of theory. Utility theory formally defines rationality for game theorists. The rational assumptions of exchange theorists do not differ from the conceptualization of rational actors defined by game theory. Game theories' use of the term is more explicit and more general. If exchange researchers do not find utility theory adequate, it could be used as a starting point from which they can diverge. The second theme is a discussion of the form of games. Differentiating games based on these guidelines has helped game theorists make clear the scope of their work. The exchange network experiments are similar enough in structure and intent that researchers might also benefit from thinking about their theories and experimental designs with these ideas in mind.

[19] Myerson (1977, 225).

[20] Aumann and Myerson (1988, 175).

3.2 Utility Theory

Markovsky et al. and Cook et al. both assume, underlying the complicated strategies that both theories ascribe to actors, that all actors maximize and that they prefer more money (or points) to less. Cook et al. went out of their way to ensure that equity or other concerns would not confound this. The fact that power differences can be measured as differences in resource attainment was adopted by network exchange theorists with little reflection. It was simply assumed. After much debate and discussion, game theorists agreed that under certain conditions (which are met by the exchange experiments) money can represent utility and that all players prefer more money to less money. Related to this Luce and Raiffa (1957, 50) propose this postulate of rational behavior:

> Of two alternatives which give rise to outcomes, a player will choose the one which yields the more preferred outcome, or, more precisely, in terms of the utility function he will attempt to maximize expected utility.

So there is agreement between game theory and network exchange theory on how to evaluate utility. However, this assumption about money being used as a yardstick of utility or value is not all exchange theorists were concerned with when defining their actors as rational. Rationality was defined as one specific prescription of behavior for subjects.[21] Why this is the case is unclear. If another strategy would ensure more points, would that strategy not be rational? It seems that there are many ways that subjects might go about trying to maximize their outcomes.

Another useful distinction in game theory is between rationality and strategy. Rationality involves maximization of some sort. A strategy is simply a rule that tells the player how to act under every possible circumstance. A particular strategy may or may not be rational according to some criterion. A strategy that works to maximize under some conditions may or may not be rational under other conditions. Thus, when Cook et al. (1983) and Markovsky et al. (1988) define rationality in terms of a particular strategy (raising offers to others if excluded and lowering offers if included), there is conceptual confusion. Although they intend for their actors to maximize, the prescribed strategy may not under all conditions.[22]

Game theory is actually very social psychological. The social psychology of decision making in game theory is housed under the section that deals with utility. Implications for the distribution of resources to the entire group are dealt with separately. When looking at N-person situations, game theory is removed from the details of how individuals attempt to maximize, unlike exchange theory.

Kahan and Rapoport (1984, 4–5) point out that game theory is not one theory but a multiplicity of solutions that allow various aspects of rationality to be studied. The variation of the rationality assumption that distinguishes solution

[21] Scope conditions from Markovsky et al. (1988, 223).

[22] There is some evidence that players in these games raise offers to others when they are excluded but do not lower offers to others when they are included.

concepts allow game theorists to make predictions about behavior, that reflect different underlying social psychological strategies.[23]

3.3 Games in Extensive Form

Both Cook et al. and Markovsky et al. assumed that their subjects were rational actors who wished to maximize the amount of points they accumulated. Both prescribed detailed strategies that their subjects (or simulated actors) were expected to follow. In game theory the details involved informing the actual moves of actors under all possible conditions and would suggest that Cook et al. and Markovsky et al. were both defining games in extensive form. There is a broad literature on games in extensive form, yet the preponderance of work in N-person cooperative game theory distills games further, in order to look at games in their strategic (otherwise known as normal) or further distilled characteristic function forms.[24]

Martin Shubik (1987) defines a strategy as follows:

> A *strategy*, in the technical sense, means a complete description of how a player intends to play a game, from beginning to end. The test of completeness of a strategy is whether it provides for all contingencies that can arise, so that a secretary or agent or programmed computer could play the game on behalf of the original player without ever having to return for further instructions.

The social psychological assumptions of Cook et al. and Markovsky et al. were clearly strategies. Given the finite set of possibilities each subject might encounter, each theory prescribes an action.

> In [extensive form] we set forth each possible move and information state in detail throughout the course of the play. In [strategic form] we content ourselves with a tabulation of overall strategies, together with the outcomes or pay-offs that they generate. (Shubik 1987, 34)

Getting into the minds of the subjects involved in these experiments in order to determine how they make choices is a worthwhile pursuit. The analysis of exchange experimental data has focused on resource distribution as a measure of power. Outcomes have been studied, not the strategies or the preferences of subjects. Looking at outcomes might help us determine what paths were avoided

---

[23] Skvoretz and Willer (1993) attribute the inability of the core to make point predictions to the core's basis on a game theoretic definition of rationality. They say, 'Because no specific social psychological principle is assumed, rationality considerations alone cannot always single out a particular outcome from this set.' The core is not indeterminent because it is based on rational choice, other solution concepts generate point predictions. The core is based on three different conceptions of rationality, that combined, can produce no prediction, a range of predictions, or one point. The core was constructed in this way intentionally. Other solution concepts, also based on rationality, can easily provide the point solutions Skvoretz and Willer seek.

[24] Most researchers are familiar with strategic form. The bimatrix game known as the prisoners' dilemma is represented in strategic form. All possible options are presented for each player in a matrix, and the players have to select the row or column that is best for him/her, considering his/her assessment of the action of the other player.

by subjects, but provide us with little information about which paths were taken. Many different strategies can spawn identical outcomes.

One model for describing games in extensive form is known as the 'Kuhn Tree'. The sketching out of a simple game of fingers, using the Kuhn tree, illustrates the point:

> *Fingers.* The first player holds up one or two fingers, and the second player holds up one, two or three fingers. If the total paid displayed is odd then P_1 pays \$5 to P_2; if it is even, then P_2 pays \$5 to P_1.

If we assume that P_1 moves first, the game tree in Fig. 3 describes the game.

> Each node in the tree represents a position or state in which the game might be found by an observer. A node labeled P_1 is a decision point for player 1: he is called upon to select one of the branches of the tree landing out of that node, that is away from the root. In our example P_1 has two alternatives, one finger or two fingers; accordingly we have labeled 1 and 2 edges leading away from the initial node. After P_1's move, the play progresses to one of the two nodes marked P_2; at either of these P_2 has three alternatives, which we have labeled 1, 2, 3. Finally a terminal position is reached, and an *outcome* O_j is designated. Thus any *path* through the tree, from the initial node to one of the terminals, corresponds to a possible play of the game. (Shubik 1987, 40)

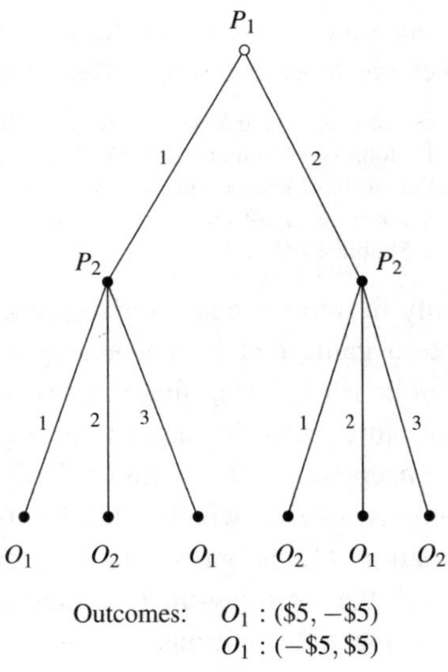

Fig. 3. Kuhn tree for fingers

Imagine that P_2 has the social psychological strategy that follows: 'If P_1 displays a 1 I will display a 2; if P_1 displays a 2 I will display a 1.' That strategy would ensure outcome 2 for each first move of player 1. The outcome is \$5 for player 2. Knowing that outcome, however, does not allow us to retrace the actions or thinking of player 2. An alternative strategy may have been, 'I will show one finger *more* than P_1 shows.' This strategy results in the same pay-off distribution and outcome but from a different strategy and different path.

The experiments being conducted in the area of network exchange are designed to measure structural outcomes, not individual decision. Looking at outcomes allows researchers to rule out strategies that are not used, but does not prove what strategies are used. Furthermore, no conclusion can be drawn about *why* one strategy is successful and another is not.

There are *two* implications to this. First, it indicates that it is beyond the scope of the work of exchange theorists to speculate about the motivation of actors or the strategies they use based on experimental results on outcome. Second, it is not necessary for the theories about structural outcomes to be addressed at the level of games in extensive form. If the goal of these experiments were to provide a mechanism for examining the relationship between the individual assumptions and structural predictions of these theories, the extensive form of the game would be appropriate. If what is being measured is outcome, however, the extensive form of the game provides a great deal of unimportant information and the strategic or characteristic function form of the game may be more appropriate.

N-person cooperative games are usually expressed in the characteristic function form. This form assigns to each coalition of actors, that might possibly form, the value it would earn regardless of the actions of other players. When games are represented in the characteristic function form there is less temptation to interpret structural level results at the micro level. It is easier to view rationality as a preference over outcomes, than as one limited strategy. Different social psychological perspectives are represented by different solution concepts.

> The cornerstone of the theory of cooperative N-person games is the *characteristic function*, a concept first formulated by John von Neumann in 1928. The idea is to capture in a single numerical index the potential worth of each coalition of players.
> With the characteristic function in hand, all questions of tactics, information, and physical transaction are left behind. (Shubik 1987, 128)

Not all situations easily fit into the characteristic function. Shubik coined the term ‘c-game’ to indicate a game that is ‘adequately represented by the characteristic functions’. (Shubik 1987, 131). Shubik does not provide a categorical definition of a *c-game*, because ‘what is adequate in a given instance may well depend on the solution concept we wish to employ.’ (Shubik 1987, 131). There are, however, two conditions, one of which must be met, and are met by the exchange experiment situation: (1) the games must be expressible as a constant sum game, a game in which the total pay-off is a fixed quantity; and (2) it must be a game of consent or orthogonal coalitions; a game where nothing can happen to a player without his/her consent. ‘Either you can cooperate with someone or you can ignore him; you cannot actively hurt him.’ (Shubik 1987, 131).[25]

4 The Present

In the 1990s Markovsky and Willer and their collaborators, and Cook and Yamagishi and their collaborators, have improved and expanded on their theories. In

[25] Bienenstock (1992), Bienenstock and Bonacich (1992) and Bienenstock and Bonacich (1993) interpreted the exchange game in its characteristic function form.

addition, several other authors have attempted to discover algorithms to predict power differences in exchange networks. While predictions are becoming better, the connection between the theory and algorithm has become less distinct. A discussion of some of the ongoing work in this area follows.

4.1 Lovaglia et al. 1995

This article summarizes several advances made recently on the GPI approach. First it recounts the method introduced in Markovsky et al. (1993) for differentiating different types of networks: weak power networks and strong power networks. The GPI works well for predicting power in strong power networks. When strong power is not present additional calculations must be made. Weak power networks are networks in which power differences are more tenuous because no position is assured of inclusion in an exchange or, if there are positions certain of inclusion, no position can be excluded without some cost to the network as a whole (p. 202). For example (Fig. 4), the five-person hourglass network, where, in every completed game one player is left without a trading partner. There are five patterns of exclusion and no position is assured of not being the excluded party. Therefore the five-person hour glass network exhibits weak power differences.

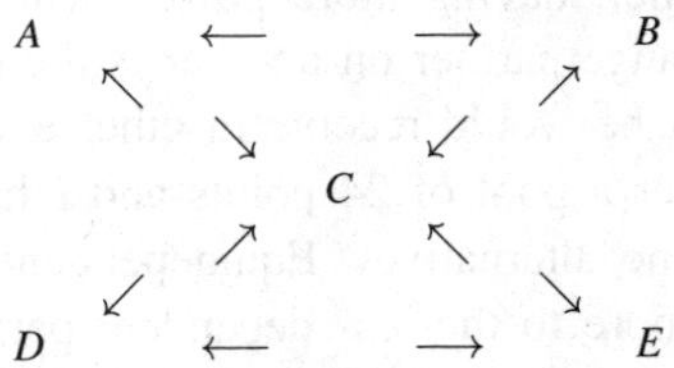

Fig. 4. The five-person hourglass

The predicted power differences in weak power networks are based on a combination of two ideas. First is the calculation of the likelihood of inclusion: the higher the probability of inclusion, the greater the power: The probabilities of inclusion are based on the assumption that positions choose each other randomly, and that an exchange is completed if two connected positions randomly choose one another. For example, in the five-person hourglass network, one can calculate that the probability that the middle C position has a probability of 0.8205 of being included, while the other positions have a probability of 0.7949. Therefore, when A and C exchange, C should have power.

As was the case with previous attempts, there is a disjuncture between the behavioral and structural models. It is inconceivable that network members respond to these values. The values can be quite difficult to compute. Moreover, the process that generates the values, random choice among positions, is not hypothesized to occur in the experimental groups. Why should a position with a value of 0.8205 have power over a position with the nearly equal values of

0.7949? No explanation is offered of the process by which these values affect power in networks.

To perfect the model, resistance (Heckathom 1983, Willer 1981) was also included in this model. In weak networks actors try to balance their 'best hope' against their 'worst fear' in an attempt to simultaneously maximize profit (get the most they can) and minimize loss (avoid exclusion). Actors determine how much they should offer by figuring out how much they reasonably get without asking for so much that they are excluded from negotiation.

This amalgam of likelihood and resistance may result in better prediction, but it makes it even less likely that the model describes the thought process of subjects. Both of these formulations imply an underlying social psychology. Positions that are less certain of inclusion must be especially eager to be accommodating to their exchange partners; otherwise they will be excluded. This, again, is clearly a kind of rational behavior. And yet the authors are resistant to employing game theory.[26]

4.2 Cook and Yamagishi 1992

Working from a power-dependence perspective, Cook and Yamagishi assume that the dependence of two exchanging partners on each other will equalize. The less dependent partner, having more power, will raise his demands. The dependence of one exchange partner on another is the difference between what he is receiving and what he would receive in other exchanges. For example, i and j are negotiating over a pool of 24 points and i has an alternative partner offering 10 while j has no alternatives. Equidependence exists when the more dependent person pays more to the less dependent person, so that both parties receive the same excess points as they would have otherwise. In this example i would get 17 points, j would get seven points, both i and j receiving seven points more than their guaranteed alternative. When this comparison of values occurs simultaneously in all dyads there is network wide equidependence. Each individual is also trying to maximize their share and trade with the partner that will provide the most points.

This elegant theory does not depend on strange structural postulates. It is firmly based on Emerson's power-dependence psychology which is, as we have seen, an informal rational choice model. At the same time, in spirit, if not in procedure, it is similar to the resistance likelihood model presented in Lovaglia et al. (1995). Actors make and accept offers that indicate an attempt to maximize profit and still appear reasonable so they are not excluded. The following section

[26] In footnote 14 (p. 152) Lovaglia et al. argue that despite similarities these network exchange situations that can not be readily applied to N-person non-cooperative game theory. The similarity between the Nash solution and Resistance theory is obvious and has been acknowledged. No convincing argument to not apply game theory solutions has been expressed. The footnote expressed that Bienenstock and Bonacich (1992) have made the most successful use of game theory to study these networks. That may be because they employ cooperative and not non-cooperative game theory.

will examine the similarities between these solutions and a solution concept that exists in game theory: the kernel (see Kahan and Rapoport 1984, 127–36).

4.3 Bienenstock and Bonacich

Several solution concepts borrowed from game theory have been applied to the exchange network situation. Because they are game theory solution concepts they are explicitly based on rational choice. These solution concepts are applicable to all cooperative games with transferable utility. Cooperative games are those in which binding agreements are possible between partners. Transferable utility are goods, like money, that can be transferred freely between members of a coalition. The network exchange experiments are cooperative games with transferable utility. A subject is supposed to form a binding agreement with another subject agreeing on a way of dividing a set number of points between them. The points, which are later convened into money are a transferable utility. Any solution concept developed to study cooperative games with transferable utility can be applied to these network experiments. The core is a solution to the game in characteristic function form.

$$A \leftarrow 24 \rightarrow B \leftarrow 24 \rightarrow C$$

Fig. 5. The three-person chain

The major contribution of Bienenstock (1992), was to show that it is possible to map any experimental exchange network situation into the characteristic function formulation of a game. The fact that this has been overlooked in favor of a preoccupation with the core is disappointing. The illustration of how to convert the network exchange situation into a form readily accessible to all game theorists and that can convert N-person cooperative games into exchange networks opens up both fields for mutual communication and collaboration. This bridge is essential to unite the two fields. Without the translation or mapping it might be difficult to see the parallels between games and networks. Now the isomorphism should be apparent. Consider the three-person chain in Fig. 5.

The characteristic function form of the game (Table 1) assigns values to exchanges. If no agreements are made there is no value. A coalition of one can also only guarantee itself zero points. Similarly, if A and C form a coalition they can only guarantee themselves zero points. The value of either the AB or BC coalition is 24, as is the grand coalition of ABC. Adding to the coalition does not add value to the pay-off. Once a network is converted to the characteristic function *any* solution concept for N-person cooperative game theory with transferable utility can be applied to the exchange network situation.

The solution concept that has received the most attention from network exchange theorists has been the core. Skvoretz and Willer (1993) have criticized

Table 1. The characteristic function representation of the three-person chain

3-person chain	$v(\phi) = 0$	
	$v(A) = v(B) = v(C) = 0$	$v(AC) = 0$
	$v(AB) = v(BC) = 24$	$v(ABC) = 24$

the core for two reasons. First, other exchange algorithms were better at predicting exact cardinal distributions. Second, the core was not as social psychological as the other theories. Bienenstock (1992) and Bienenstock and Bonacich (1993) included the core in their analysis because of its importance to game theory. It happens to also be the solution concept that receives the most attention from game theorists, because of its value to the field.

> The core, or lack of core, is an undeniably important feature of any cooperative game. Its existence, size, shape, location within the space of imputations, and other characteristics are crucial to the analysis under almost any solution concept. The core is usually the first thing we look for after we have completed the descriptive work. (Shubik 1982)

Bienenstock and Bonacich (1992) also introduced three other solution concepts: the kernel, the Shapley value and the semi-value.[27] Each solution concept was designed by game theorists to focus on particular aspects of exchange and specific social psychological assumptions. Although all assume rational actors, the core assumes an actor motivated to minimize loss. The Shapley value and semi-value are considered equity solutions. The kernel, the last solution discussed by Bienenstock (1992) and Bienenstock and Bonacich (1993) is described as an excess solution. It is one of several solutions specifically termed bargaining solutions.

The kernel makes no predictions about which coalitions will form and does not assume group rationality. The kernel predicts only the distribution of rewards given some assumption about the memberships of all coalitions (Kahan and Rapoport 1984, 128–134). To calculate the kernel, assume a complete coalition structure and a hypothetical distribution of rewards within each coalition. Then ask whether this distribution is in the kernel. Consider two players k and l in the same coalition. In the context of this proposal, it means that the two players have agreed to trade with one another. Both k and l consider alternative trading partners. s_{kl}, the maximum surplus of k over l, is the maximum increase in reward to k and to any alternative trading partner j with respect to the present distribution if k and j agree to trade. Similarly, s_{lk} is the maximum increase in reward to l and some alternative trading partner j with respect to the present distribution of rewards if l were to agree to trade with j. A reward distribution is in the kernel if $s_{kl} = s_{lk}$ for every pair of players who are trading.

The appeal of the kernel is that it might model the way players in these networks actually determine how much they are willing to ask. In the three-person chain network, for example, B will trade with A or C and will try to take

[27] Additional solution concepts are available for application to this situation: the $e-core$, nucleolus or bargaining set may be even better predictors of outcomes.

the entire 24 points. This is calculated as follows. Assume B is considering an exchange with A. The coalition is worth 24 points. If B chooses to exclude A, A has no alternative trading partners. B can chose to trade with C and form a coalition worth 24 points. The surplus to B is 0 points. The same logic holds for the BC coalition. The AC coalition is worth no points. If either chose B the surplus is 24 points. This is the logic of the kernel; it is also likely to be the way players might determine how much they are entitled to receive when engaged in exchanges with others.

The kernel for this network is $x(A) = 0, x(B) = 24, x(C) = 0$. For this pay-off configuration $S_{ab} = S_{ba} = S_{ac} = S_{ca} = S_{bc} = S_{cb} = 0$. This is the only pay-off configuration that exist for this game that allows for all *excesses* to be equal. As a counter example consider the pay-off configuration $x(A) = 2$, $x(B) = 22$, $x(C) = 0$. For this distribution of resources the maximum surplus for A with reference to B is: $S_{ab} = -2$. Similarly, $s_{ba} = 2$. Since, for a pay-off configuration to be in the kernel all 'surpluses' must be equal, this pay-off configuration is not in the kernel.

This solution is similar, if not identical, to the equidependence solution.[28] It therefore is also very close to the most recent incarnation of the exchange resistance solution. The fact that there appears to be a convergence is interesting. Game theory might be able to provide some insight into how this relates to other available solutions and why this solution seems to best fit the situation at hand.

The reason the kernel works so nicely is also the reason it is an appropriate measure for the exchange situation described by Cook et al. (1983). Subjects in the Cook et al. experiment are not provided with complete information so that they can not make complicated calculations regarding their value compared to the values of others. When the kernel is the solution concept used, it is not necessary for subjects to have complete information. The perceived violation of rational choice principles for complete information has an effect only for solution concepts that demand that subjects base their worth on global network properties (group rationality). The kernel, however, does not assume group rationality; only coalition and individual rationality. This makes it possible for players to assess their value based only on information about their local environments: their own excess and the excess of those connected to them.

This is the perfect example of the contribution of game theory. Cook et al. (1983) and Lovaglia et al. (1995) hit upon a good solution, but are not able to explain the connection between the underlying social psychological assumptions and the ultimate resource distribution. Game theory makes this possible. The kernel is the solution that simultaneously allows an outside observer, aware of the characteristic function, to make predictions about global distributions, and yet it is still appropriate for almost all incarnations of the exchange experiment, because

[28] If the kernel were calculated for the example used to illustrate equidependence the 'excesses' for each player would be seven points, just as they were for equidependence theory.

it does not demand that the subjects have any more than a local awareness.[29] Furthermore, game theorists have investigated the kernel and are aware of some properties of the kernel that may prove useful to the theoretical development of exchange theory algorithms. For example, it has been proven that the kernel always exists. Yamagishi et al.'s search for a set of pay-offs in which there is equal dependence in every exchanging dyad is not quixotic. Moreover, the kernel is not always unique, Yamagishi et al. can benefit by being aware of this possibility in testing power-dependence.

The advantage of the kernel is that it is part of a set of solutions that have been derived to get at different perspectives of coalition formation and resource distribution. Game theorists are conformable with using different solution concepts for different games. Each solution concept is based on different rational choice assumptions. The kernel is a good solution for this game of network exchange. There may be another solution conception game theory that would work better.

5 Conclusions

This article was written in the hope of weakening the resistance of exchange theorists to the notion of using the arsenal of solution concepts available in game theory to attack their questions. It attempted to show the parallels between the network exchange experiment and what game theorists refer to as N-person cooperative games with transferable utility. The secondary goal was to show how using game theory could help exchange theorists reflect on their models and research design.

There were three related themes interwoven through the text. The first point advocated using utility rather than very specific, ad hoc, yet rational assumptions to express behavioral assumptions. Related to this was a focus on the disjuncture between the social psychological and structural components of these theories.

While the need to have actors behave is important, the social psychological assumptions that were used to derive the structural outcomes were, clearly, not also meant to be descriptions of how subjects actually think or act. Even if these axioms are constructive for theory building they are certainly too complex to be prescriptive. This takes us back to the relevance of utility theory, and game theories' use of rational choice. In game theory rational choice is more general. It implies simple maximization. This includes the option to use solutions that prescribe strategies, but also allows subjects the recourse to use alternate strategies. The assumption is that rational actors may employ different strategies under different circumstances.

To continue this theme, the concept of the extensive form of the game was used to show that although an exchange theory, based on specific prescribed behaviors, may predict outcomes, it does not follow that these outcomes could not

[29] Subjects in Cook et al. (1983) did not have enough information because they were not even aware of the value of the coalition. In later experiments subjects were better able to access the value of different coalitions they could join.

have resulted from different behavior. Since network exchange theorists measure outcome, not strategy, the details of the underlying social psychological assumptions of the theories were not important. Finally, since the details of how subjects behaved to achieve the outcome is not important, games in characteristic function form, not extensive form, are appropriate as models.

Once it was established that actors are rational and that the characteristic function form of the game could be used, a solution concept, the kernel, was elaborated on. This solution is similar to both exchange resistance and equidependence. While exchange theory provided the same result as game theory, game theory also provided a means for reflection on why the algorithm should work. Game theory highlights the differences between solutions concepts based on different assumptions of rationality. Not only are many different solution concepts formally derived to represent different social psychological assumptions, game theorists also provide formal mechanism for comparing the varied implications of the solutions. It is from these comparisons that game theory derives its strength.

The kernel also shed light on why experimental results based on two different experimental paradigms, the full- and restricted information setting, produced similar results (Lovaglia et al. 1995). If subjects are using a strategy like the kernel extra information provided in the full information setting might be superfluous. Subjects might not need or use all the information provided. Of course, while that may be the case, until an experiment is designed specifically to test the social psychological assumptions of the theory, this is only speculation. It might also be the case that the 'remarkable convergence of experimental results in different settings demonstrate'[30] (Lovaglia et al. 1995) that the structural properties of these networks are robust.

All this said, the main point of the article is simply that game theory has much to contribute to the study of exchange networks. Exchange networks fit nicely into the general class of *c-games*. Even so, the network exchange situation is not redundant with any existing game. This article's ultimate goal, then, is to set the stage to open dialogue between these two coexisting fields in the behavioral sciences.

Notes

We thank Michael Macy for comments an earlier drafts that helped us focus our thinking about many issues discussed in this article.

References

Aumann, R.J., Myerson, R.B. (1988) Endogenous Formation of Links Between Players and of Coalitions: An Application of the Shapley Value. In: A.E. Roth (ed.) *The Shapley Value: Essays in honor of Lloyd S. Shapley*. Cambridge, Cambridge University Press.

Bienenstock, E.J. (1992) Game Theory Models for Exchange Networks: An Experimental Study. Doctoral Dissertation, Department of Sociology. University of California, Los Angeles. Ann Arbor, MI, UMI.

[30] Lovaglia et al. (1995, 148) remark on the convergence of results from settings other than the two compared in their paper.

Bienenstock, E.J., Bonacich, P. (1992) The Core as a Solution to Negatively Connected Exchange Networks. *Social Networks* 14: 231–43.

Bienenstock, E.J., Bonacich, P. (1993) Game Theory Models for Social Exchange Networks: Experimental Results. *Sociological Perspectives* 36: 117–36.

Blau, P. (1967) Exchange and Power in Social Life. New York, Wiley.

Bonacich, P., Bienenstock, E.J. (1993) Assignment Games, Chromatic Number and Exchange Theory. *Journal of Mathematical Sociology* 14(4): 249–59.

Coole, K.S., Emerson, R.M. (1978) Power, Equity and Commitment in Exchange Networks. *American Sociological Review* 43: 721–39.

Cook, K.S., Yamagishi, T. (1992) Power in Exchange Networks: A Power Dependence Formulation. *Social Networks* 14: 245–66.

Cook, K.S., Emerson, R.M., Gillmore, M.R., Yamagishi, T. (1983) The Distribution of Power in Exchange Networks: Theory and Experimental Results. *American Journal of Sociology* 89: 275–305.

Heckathorn, D. (1983) Extensions of Power-dependence Theory: The Concept of Resistance. *Social Forces* 61: 1206–1231.

Kahan, J., Rapoport, A. (1984) Theories of Coalition Formation. Hillsdale, N.J., L. Erlbaum.

Lovaglia, M.J., Skvoretz, J., Willer, D., Markovsky, B. (1995) Negotiated Exchange Networks. *Social Forces* 74(1): 123–55.

Luce, R., Raiffa, D.H. (1957) Games and Decisions. New York, John Wiley.

Machina, M.J. (1990) Choice Under Uncertainty: Problems Solved and Unsolved. In: Cook, K.S., Levi, M. (eds.) *The Limits of Rationality*, pp. 90–131. Chicago, University of Chicago Press.

Macy, M.W. (1990) Learning Theory and The Logic of Critical Mass. *American Sociological Review* 55: 809–26.

Markovsky, B., Willer, D., Patton, T. (1988). Power Relations in Exchange: Networks. *American Sociological Review* 53: 220–236.

Markovsky, B., Skvoretz, J., Willer, D., Lovaglia, M., Ergo, J. (1993) The Seeds of Weak Power: an Extension of Network Exchange Theory. *American Sociological Review* 58: 197–209.

Myerson, R.B. (1977) Graphs and Cooperation in Games. *Mathematics of Operations Research* 2: 225–229.

Shubik, M. (1987) Game Theory in The Social Sciences: Concepts and Solutions. Cambridge, MIT Press.

Skvoretz, J., Fararo, T.J. (1992) Power and Network Exchange: An Essay Toward Theoretical Unification. *Social Networks* 14: 325–344.

Skvoretz, J., Willer, D. (1993) Exclusion and Power. A Test of Four Theories of Power in Exchange Networks. *American Sociological Review* 58: 801–818.

Willer, D.E. (1981) Quantity and Network Structure. In: D. Willer and B. Anderson (eds.) *Networks, Exchange, and Coercion: The Elementary Theory and its Application*, pp. 108–127. Oxford, Elsevier.

Incentive Compatible Reward Schemes for Labour-managed Firms

Salvador Barberà[1], **Bhaskar Dutta**[2]

[1] Universitat Autònoma de Barcelona, 08193 Bellaterra, Barcelona, Spain
(e-mail: salvador.barbera@uab.es)
[2] Indian Statistical Institute, 7 SJS Sansanwal Maarg, New Delhi 110016, India
(e-mail: dutta@isid.ac.in)

Abstract. We consider a simple case of team production, where a set of workers have to contribute a single input (say labour) and then share the joint output amongst themselves. Different incentive issues arise when the skills as well as the levels of effort expended by workers are not publicly observable. We study one of these issues in terms of a very simple model in which two types of workers, skilled and unskilled, supply effort inelastically. Thus, we assume away the problem of moral hazard in order to focus on that of adverse selection. We also consider a hierarchical structure of production in which the workers need to be organised in two tiers. We look for reward schemes which specify higher payments to workers who have been assigned to the top-level jobs when the principal detects no lies, distribute the entire output in all circumstances, and induce workers to revel their true abilities. We contemplate two scenarios. In the first one, each individual worker knows only her own type, while in the second scenario each worker also knows the abilities of all other workers. Our general conclusion is that the adverse selection problem can be solved in our context. However, the range of satisfactory reward schemes depends on the informational framework.

Key Words: Incentives, adverse selection, strategy-proofness, reward schemes, labour-managed firms

JEL Classification: D82, J54, D20

We are most grateful to an anonymous referee, J. Crémer, M. Jackson, I. Macho, D. Pérez-Castrillo, and specially A. Postlewaite for very helpful discussions and suggestions.

1 Introduction

In the simplest cases of team production, there is a set of workers who each have to contribute a single input (say labour) and then share the joint output amongst themselves. Different incentive issues arise when the skills as well as the levels of effort expended by workers are not publicly observable. The issue of *moral hazard*, which appears whenever the supply of the input involves some cost, is well recognised in the literature.[1] In contrast, the problem of *adverse selection* which is caused by the presence of workers of differential abilities, seems to have been relatively neglected. The purpose of this paper is to study the possibility of designing suitable incentive schemes which will induce workers to reveal their true abilities.

We study this problem in terms of a very simple model in which *two* types of workers, skilled and unskilled, supply effort *inelastically*.[2] Thus, we assume away the problem of moral hazard in order to focus on the issues raised by adverse selection. We also consider a hierarchical structure of production in which the workers need to be organised in two tiers. The first-best outcome requires that only skilled workers be assigned to the *top* level jobs since these require special skills. Indeed, we specify that unskilled workers are more productive at the low level jobs. The adverse selection problem arises because skilled workers need to be paid more than unskilled workers when the principal[3] can verify that all workers have told the truth.

Since types are not observable, there is a need to design a system of payments which will induce workers to reveal their types correctly. Since the principal can observe the realized output, the payment schedule can be made contingent on realized output as well as on the assignment of tasks. A trivial way to solve the adverse selection problem is to distribute the realized output equally under all circumstances. It will then be in the interests of all workers to maximise total product, and hence to volunteer the true information about abilities so as to achieve an optimal assignment of tasks. However, this extreme egalitarianism may be inappropriate. For example, skilled workers may have better outside options and hence higher reservation prices than the unskilled workers.

Another trivial way to solve the adverse selection problem is to levy very harsh punishment on *all* workers whenever lies are detected. Observe that since the principal observes the realized output, she can detect lies whenever unskilled workers claiming to be skilled have been assigned to the top level jobs. However, such punishments imply that some output has to be destroyed. This will typically not be *renegotiationproof*. Therefore, we look for reward schemes which

[1] See for instance Sen (1966), Israelson (1980) or Thomson (1982) for related work on labour-managed firms. Groves (1973) and Holmstrom (1982) are a couple of papers which deal with the more general framework of teams.

[2] In the last section, we describe a more general model containing more than 2 types in which almost all our results remain valid.

[3] Notice that there is no actual principal as in the standard principal-agent models. Following standard practice in implementation theory, we use the term "principal" to represent the set of agreements or rules used by the workers to run the cooperative.

specify higher payments to workers who have been assigned to the top-level jobs when the principal detects no lies, and which distribute the entire output in all circumstances.

Our general conclusion is that the adverse selection problem can be solved in our context. However, the range of possible reward schemes depends on the informational framework. We contemplate two scenarios. In the first one, where each individual worker knows only her own type, there exist strategyproof (in fact even group strategyproof) reward schemes. But these schemes can only accomodate limited pay differentials between workers of different types. As we shall see, this implies the incompatibility of strategyproofness with some reasonable distributional principles. In the second scenario, each worker also knows the abilities of all other workers.[4] In this case, the class of reward schemes solving the adverse selection problem is much wider.

2 The Formal Framework

Let N be the set of n members of a cooperative enterprise. We assume that workers are of *two* types - *skilled* (or more able) and *unskilled* (or less able). T_1 will denote the set of skilled workers, who will also be called the Type 1 workers. T_2 will denote the set of unskilled workers, who will be labelled Type 2 workers. We assume that both sets are *nonempty* since an adverse selection cannot arise if one of the sets is empty. Note that the type of each worker is *private information*- there are no external characteristics which can be used to identify workers' types.

Two kinds of jobs need to be performed in order to produce output. One type of job is essentially a routine or mechanical activity, and does not require any special skills. So, both types of workers are equally proficient at performing this job, which will henceforth be labelled as J_2 or Type 2 job. In contrast, the *Type 1* job, to be denoted J_1, involves "managerial" responsibilities requiring some skill. Hence, these should ideally be performed by the Type 1 workers. However, if Type 2 workers are assigned to J_1, then they perform their job *inefficiently*, and are responsible for some loss of output. We model this by stipulating that output increases strictly when a Type 2 worker is shifted from the Type 1 job to the Type 2 job. We also assume that the *maximum* cardinality of J_1 is given by some number K, where $K \leq n$.[5] However, it turns out that except in Sect. 4, the possible restriction on the number of Type 1 positions does not affect any of our results.

Let t_{ij} denote the number of workers of type $i (i = 1, 2)$ employed in job $j (j = 1, 2)$. Hence, the "organizational structure" of the enterprise can be described by a vector $t = (t_{11}, t_{12}, t_{21}, t_{22})$. Let T denote the set of such vectors t with (i) $t_{11} + t_{21} \leq K$, and (ii) $t_{11} + t_{12} + t_{21} + t_{22} = n$. So, T represents the set of *feasible*

[4] Notice that an adverse selection problem arises even in this case since the information about other workers' types is not verifiable.

[5] Given our interpretation of jobs, this seems a natural restriction.

structures, with (i) expressing the requirement that no more than K workers can be in J_1, while (ii) states that all the n workers have to be employed.

We also assume that all workers supply one unit of effort *inelastically*. We are therefore assuming away the problem of *moral hazard*. We do this in order to focus on some of the issues raised by *adverse selection*.

Let $f(t)$ represent the function describing output produced by any particular structure. The following assumptions are made on the production function f.

Assumption 1: For all $t, t' \in T$,

(i) $f(t) = f(t')$ if $t_{11} = t'_{11}$ and $t_{21} = t'_{21}$.
(ii) $f(t) > f(t')$ if $t_{11} > t'_{11}$ and $t_{21} = t'_{21}$.
(iii) $f(t) > f(t')$ if $t_{11} = t'_{11}$ and $t_{21} < t'_{21}$.

Condition (i) in the Assumption says that if two structures differ only in the composition of workers performing Type 2 jobs, then the output produced must be the same. This expresses the notion that both skilled and unskilled workers are equally adept at performing the Type 2 job. Condition (ii) essentially captures the idea that skilled workers are more productive doing Type 1 jobs than Type 2 jobs provided no more than K workers are employed at Type 1 jobs. Conversely, Condition (iii) states that the unskilled workers are unsuitable for Type 1 jobs.

Notice that given Assumption 1, the total output produced by the enterprise is determined completely by the composition of workers performing Type 1 jobs. We will sometimes find it convenient to represent the output of the enterprise by $f(k, l)$, where k and l are respectively the numbers of workers in T_1 and T_2 doing Type 1 jobs.

An interesting special case of the general model, which will be used in the next section, is described below. Choose a vector $p = (p_1, p_2, p_3)$ with $p_1 > p_2 > p_3 \geq 0$, and a number $C > 0$. Then, in the p-model, the output produced is given by

$$f(k, l) = kp_1 + (n - k - l)p_2 + lp_3 - C \tag{1}$$

Equation (1) has the following interpretation. C represents the *fixed cost* of running the enterprise. Moreover, each worker in a Type 2 job has a productivity of p_2. In Type 1 jobs, the skilled workers have a productivity of p_1, while the unskilled workers have a productivity of p_3. Since $p_1 > p_2 > p_3$, it is easy to check that the p-model satisfies Assumption 1 above.

If workers' types were publicly observable, then upto K skilled workers would be assigned to Type 1 jobs, while the rest would be assigned to Type 2 jobs. However, since types are private information, the *principal* cannot adopt this naive procedure. So, she has to design a *reward scheme* or payment schedule which will induce workers to reveal their *true* types. Notice that since the principal can observe the organizational structure and the total output realized, the reward to each worker can be made contingent on output as well as the structure $t \in T$. In fact, the principal can, after observing output, actually infer the *number* of workers in T_2 who have actually lied and been assigned to J_1. Of course, the principal cannot infer *who* have lied. Nor can the prinicipal deduce anything about workers in T_1 who have falsely claimed to be in T_2 and hence been assigned to J_2. Nevertheless, it is apparent that the principal in this setting has more information than in the traditional implementation framework.

This suggests the following scenario. First, the principal announces the *assignment rule* which she will use to determine the production structure as a function of the information revealed by the individuals. Second, she also announces the reward scheme which make payments a function of (i) realized output (ii) the structure $t \in T$ which she will choose after hearing the vector of announcements by the workers.

Given the reward scheme, each worker announces his private information. As far as a worker's private information is concerned, we describe two alternative possibilities. In the first case, an individual only knows his or her *own* type. Naturally, in this case, an individual's announcement consists of a declaration of one's own type. The second case corresponds to that of complete information, where each individual knows every other worker's type. In the latter case, an announcement consists of a *profile* of types, one for each worker.

The announcements made by the workers together with the assignment rule chosen by the principal determines the organizational structure. The workers perform their assigned job, output is realized, and subsequently distributed according to the reward scheme announced by the principal. Notice that the organizational structure may be inoptimal if workers have lied about their types. For instance, if worker i *falsely* claims to be skilled, then he may be assigned to J_1, although he would be more productive in a Type 2 job.

The formal framework is as follows. The principal announces an *assignment rule* A which assigns each worker i to either J_1 or J_2 as a function of the information vector announced by the workers. She also announces a reward scheme, which is a pair of functions $\mathbf{r} = (r_1, r_2)$, where

$$r_i : \mathbb{R}_+ \times T \to \mathbb{R}_+. \tag{2}$$

Here, $r_1(y,t)$ is the reward to workers assigned to J_1, contingent on output being y, while $r_2(y,t)$ is the corresponding payment promised to workers assigned to Type 2 jobs. Remembering our earlier remark that output is completely specified by the composition of workers assigned to J_1, we will sometimes represent a reward scheme as $\{r_1(k,l), r_2(k,l)\}$, where k and l are the numbers of skilled and unskilled workers assigned to Type 1 jobs. This formulation assumes that the principal can infer how many unskilled workers have been assigned to Type 1 jobs. Note that knowledge of the production function is enough for this purpose.

Equation (2) also assumes that the principal has to employ *anonymous* schemes - the reward to workers i and j cannot differ if they are assigned to the same job. In particular, workers i and j may both have been assigned to J_2 even though i may have announced that she is *skilled* and j may have announced that she is *unskilled*.[6] In other words, agents' announcements about types matter only in so far as this influences the assignment to jobs. A more general approach[7] would have been to consider schemes in which worker i is paid more than worker j. Notice, however, that if workers announce only their own types, then the principal has no way of verfying whether i has announced the truth if she has been assigned to J_2. Hence, if i is paid more than j, then that would give j an incentive t o declare that she is skilled!

Of course, if worker j wrongly claims to be skilled, then she would also have to take into account the possibility that she is assigned to J_1. If she is indeed assigned to J_1, then the principal would detect that *someone* has lied, and then j (along with others assigned to J_1) would have to pay a penalty. The probability that j is assigned to J_1 depends on the number of other workers who have announced that they are skilled, the number of positions in J_1, and the tie-breaking ru le used by the principal. Clearly, non-anonymous schemes would have to satisfy very complicated schemes in order to be induce truthtelling as a dominant strategy. That is why we have chosen the simpler (but somewhat less general) approach of restricting attention to anonymous schemes.

We also consider the complete information case when workers announce entire type profiles. In this case, *other* workers' announcements could in principle be used to distinguish between two workers assigned to J_2. Here, non-anonymous schemes ca give rise to a differet problem. Suppose skilled worker i is assigned to J_2, and paid more than the unskilled workers. Then, the unskilled workers may have an incentive to declare i to be unskilled. This, by decreasing the amount paid to i will leave more to be distributed to the others. Notice again that there is no way in which the principal can verify that the others have told the truth about i.

In what follows, we will refer to an assignment rule and reward scheme as a *mechanism*.

Clearly, each specification of a mechanism gives rise to a normal form game in which the workers' strategies are to announce either their own types or an entire vector of types, depending upon the structure of information. We assume that the principal's primary objective is to choose mechanisms which will induce workers to reveal their private information truthfully in equilibrium. Of course, this involves the appropriate choice or specification of an equilibrium, depending upon the informational framework. In this paper, we focus on *strategyproof* mechanisms, that is mechanisms under which truthtelling is a *dominant strategy*, in the case when workers know only their own types. In the complete information framework, we restrict attention to Nash equilibria and undominated strategy equilibria. In other words, we are interested in the issue of designing mechanisms under which the sets of these equilibria will coincide with truthtelling or strategies which are equivalent to truthtelling.

[6] Notice that this issue matters only for workers assigned to J_2 since all workers assigned to J_1 must have announced that they are skilled.

[7] We are grateful to the anonymous referee for pointing out the need to clarify this issue.

While these concepts are defined rigorously in subsequent sections, we specify below some restrictions which will be imposed on all reward schemes. These restrictions essentially ensure that the problems we are studying are nontrivial.[8]

Definition 1. *A reward scheme* $\mathbf{r}$ *is admissible if*

(i) $(k+l)r_1(k,l)+(n-k-l)r_2(k,l)=f(k,l)\ \forall k,l$ *such that* $k+l \leq K$
(ii) $r_1(k,0) > r_2(k,0)\ \forall k \leq K$.

Remark 1. In this paper, we are going to restrict attention to admissible reward schemes. Henceforth, reward schemes are to be interpreted as admissible reward schemes.

Feasibility requires that the sum of the payments made to the workers never exceeds realized output. Condition (i) goes a step further, and insists that the principal can never destroy output. As we have mentioned earlier, a feasible reward scheme which leaves some surplus is open to renegotiation.

Condition (ii) states that if the principal observes a level of output which confirms that all workers assigned to J_1 are skilled, then these workers must be paid more than the rest. Notice that unless skilled workers are paid *at least as much* as unskilled workers, the former will not have any incentive to reveal their true types. It is also obvious that under the reward scheme which *always* distributes output equally amongst all workers, the adverse selection problem disappears. The imposition of Condition (ii) can be thought of as a search for "non-trivial" incentive compatible reward schemes. Also, such differentials may be necessary because of superior outside options for the skilled workers.

3 Strategyproof Reward Schemes

In this section, we first define the conditions of strategyproofness and group strategyproofness. We go on to derive a necessary and sufficient condition for strategyproof reward schemes. We then show that the class of such schemes is nonempty - indeed, we prove a stronger result by constructing a reward scheme which is *group strategyproof*. Finally, we explore the possibility of constructing strategyproof schemes which are also "nice" from an ethical point of view.

When workers only know their own types, an *announcement vector* $a = (a_1, \ldots, a_n)$ is an n-tuple of messages sent by the workers, each a_i representing worker i's claim about his type. We will use $a_i = 1$ to denote the claim that i is skilled, while $a_i = 2$ will denote the claim that i is unskilled. Given the assignment rule A employed by the principal, an announcement vector a generates a structure $t = A(a)$. The reward scheme $\mathbf{r}$ applied to t and the realized output then gives the payoff vector $R(a, \mathbf{r})$ associated with a. This is given by

[8] Also, notice that our formulation rules out the use of various ad hoc features such as *tailchasing* which are often incorporated in game forms employed in the traditional literature on implementation. For a review of the criticism against the use of these features, see Dutta(1997), Jackson(1992), Moore(1992).

$$R^i(a, \mathbf{r}) = \begin{cases} r_1(k(a), l(a)) & \text{if } i \text{ is assigned to Type 1 job} \\ r_2(y, A(a)) & \text{otherwise} \end{cases} \tag{3}$$

where $k(a), l(a)$ are the number of skilled and uskilled workers assigned to J_1 according to the anouncement a.[9] Notice that when workers announce only their own types, the principal has essentially no freedom in so far as the assignment rule is concerned. If some workers declare that they are skilled, the principal must treat these claims as if they are true since she cannot detect lies *before* the output is realized. Hence, the "best" chance of achieving efficiency is to assign upto K workers to J_1 from amongst those workers who claim to be in T_1. [10] So, the principal has to use only the *reward scheme* to induce workers to tell the truth. In view of this, we will define strategyproofness to be a property of reward schemes, although strictly speaking it is the combination of the assignment rule and the reward scheme which defines the appropriate game.

Let a^* denote the vector of *true* types of workers.

For any coalition S, a vector a will sometimes be denoted as (a_S, a_{-S}).

Definition 2. *For any coalition S, a_S is a coalitionally dominant strategy profile under reward scheme* $\mathbf{r}$ *iff*

$$\sum_{i \in S} R^i(a_S, a_{-S}, \mathbf{r}) \geq \sum_{i \in S} R^i(\hat{a}_S, a_{-S}, \mathbf{r}) \; \forall \hat{a}_S, \forall a_{-S}.$$

So, a_S is a coalitionally dominant strategy profile for coalition S if it is a best reply to *any* vector of strategies chosen by workers outside the coalition. When the coalition S consists of a single individual, we will use the terminology *dominant strategy*.

Definition 3. *A reward scheme* $\mathbf{r}$ *is group-strategyproof if for all coalitions S, a_S^* is a coalitionally dominant strategy profile under* $\mathbf{r}$.

This definition assumes the possibility of side payments within any coalition. If side payments are not possible, then the corresponding definition of group strategyproofness would be *weaker*. Since our result on group strategyproofness (Proposition 2) demonstrates the existence of group strategyproof schemes, we use the definition which leads to a stronger concept.

Definition 4. *A reward scheme* $\mathbf{r}$ *is strategyproof if for all individuals i, a_i^* is a dominant strategy under* $\mathbf{r}$.

The following notation will be used repeatedly. Call a pair of integers (k, l) *permissible* if $k + l \leq K$ and $k \geq 1, l \geq 1$.

---

[9] Whenever there is no confusion about the anouncement vector a, we will simply write $r_i(k, l)$ instead of $r_i(k(a), l(a))$.

[10] If more than K workers claim to be in T_1, then the principal has to use some rule to select a set of K workers. We omit any discussion of these selection rules since the results of this section are not affected by the choice of the selection rule.

Proposition 1. *An admissible reward scheme* $\mathbf{r}$ *is strategyproof iff* $\mathbf{r}$ *satisfies the following conditions for all permissible pairs* (k, l)*:*

$$r_2(k-1, l) \leq r_1(k, l) \leq r_2(k, l-1) \tag{4}$$

Proof. Consider any $\mathbf{r}$, and suppose for some permissible pair (k, l), $r_2(k-1, l) > r_1(k, l)$. Consider a^* such that $|T_1| = k$, and let $i \in T_1$. Consider a such that $|\{j \in T_2 | a_j = 1\}| = l$ and $a_m = a_m^* \ \forall m \in T_1$. That is, all skilled workers declare the truth about their types, but exactly l unskilled workers claim to be skilled. Then, $R^i(a, \mathbf{r}) = r_1(k, l)$. Suppose i deviates and announces $\bar{a}_i = 2$. Then, $R^i(\bar{a}_i, a_{-i}, \mathbf{r}) = r_2(k-1, l) > R^i(a, \mathbf{r})$. But, then $\mathbf{r}$ is not strategyproof.

Suppose now that $r_1(k, l) > r_2(k, l-1)$. Let a^* be such that T_1 contains k workers. Consider a such that $(l-1)$ unskilled workers declare themselves to be skilled, all other workers telling the truth. Let $j \in T_2, a_j = a_j^*$. Then $R^j(a_j^*, a_{-j}, \mathbf{r}) = r_2(k, l-1) < R^j(\bar{a}_j, a_{-j}, \mathbf{r}) = r_1(k, l)$ when $\bar{a}_j = 1$. Then, $\mathbf{r}$ is not strategyproof. These establish the necessity of (4).

We now want to show that if $\mathbf{r}$ satisfies (4), then it is strategyproof.

Suppose $\mathbf{r}$ satisfies (4). If for some i, a_i^* is *not* a dominant strategy, then there are two possible cases.

Case(i): $i \in T_1$. Let $\bar{a}_i = 2$. Then, there is a_{-i} such that

$$R^i(\bar{a}_i, a_{-i}, \mathbf{r}) > R^i(a_i^*, a_{-i}, \mathbf{r}) \tag{5}$$

But, (5) is not possible if $r_2(k-1, l) \leq r_1(k, l)$ for each permissible pair (k, l).

Case (ii): $i \in T_2$. Let $\bar{a}_i = 1$. Suppose there is a_{-i} such that

$$R^i(\bar{a}_i, a_{-i}, \mathbf{r}) > R^i(a_i^*, a_{-i}, \mathbf{r}) \tag{6}$$

But, (6) is not possible in view of $r_1(k, l) \leq r_2(k, l-1)$ from (4). So, a_i^* must be a dominant strategy for all i. □

In the next Proposition, we construct a group strategyproof reward scheme. The reward scheme has the following features. The payment made to an individual in J_1 exceeds the payment made to an individual in J_2 by a "small" amount when no lies are detected. If the principal detects any lie, then the output is distributed equally. The proof essentially consists in showing that provided the difference in payments to individuals in J_1 and J_2 are small enough, no group can gain by misrepresenting their types.

Proposition 2. *There exists a group-strategyproof reward scheme.*

Proof. Let f be the production function. Define the following:

$$\begin{aligned} a(k,l) &= \frac{f(k,l)}{n} \; \forall k,l \text{ such that } k+l \leq K \\ \gamma &= \min_k \{k(k+1)[a(k+1,0) - a(k,0)]\} \\ \epsilon &= \min_k \{f(k,0) - f(k,1)\} \\ \delta &= \frac{1}{n} \min(\epsilon, \gamma) \end{aligned}$$

Consider the following reward scheme **r**.

$$\begin{aligned} r_1(k,0) &= a(k,0) + \frac{n-k}{k}\delta \\ r_2(k,0) &= a(k,0) - \delta \\ \forall i = 1,2, r_i(k,l) &= a(k,l) \; \forall \text{ permissible pairs } k,l \text{ such that } l \geq 1 \end{aligned}$$

Claim 1. $r_1(k,l)$ is monotonically increasing in k.

The claim is obviously true for all $l \geq 1$ since $f(k,l)$ is increasing in k, and since $r_1(k,l) = a(k,l)$. So, it is sufficient to prove that $r_1(k+1,0) \geq r_1(k,0) \; \forall k < K$. To see this, note that

$$\begin{aligned} r_1(k+1,0) - r_1(k,0) &= a(k+1,0) - a(k,0) - \frac{\delta n}{k(k+1)} \\ &\geq 0 \text{ since } n\delta \leq \gamma. \end{aligned}$$

Claim 2. **r** is group-strategyproof.

Take any coalition S. We need to show that no matter what announcements are made by $(N \setminus S)$, a_S^* is a best reply of S.

Suppose not. Then, there is a_S, a_{-S} such that

$$\sum_{i \in S} R^i(a_S, a_{-S}, \mathbf{r}) > \sum_{i \in S} R^i(a_S^*, a_{-S}, \mathbf{r}) \tag{7}$$

This cannot hold if there is $i \notin S$ such that $i \in T_2 \cap J_1$. For, then the "average rule" applies, and any deviation from the truth by S can only reduce aggregate output, and hence their own share.

So, without loss of generality, let $a_{-S} = a_{-S}^*$. First, suppose there is $i \in S$ such that $a_i^* = 2$, but $a_i = 1$. Then, a lie is detected, and the average rule is applied. However, the choice of δ guarantees that $r_2(k,0) \geq a(k,1) \geq a(k',l)$ $\forall l \geq 1, \forall k' \leq k$. Since $r_1(k,0) > r_2(k,0)$, *no* individual in S can be better-off.

So, the only remaining case is when $\forall i \in S$, $a_i \neq a_i^*$ implies $a_i^* = 1$ and $a_i = 2$. However, given Claim 1, $r_1(k,0) \geq r_1(k',0) \; \forall k' \leq k$. Also, $r_2(k,0) \geq r_2(k',0)$. So, again this deviation from a_S^* cannot benefit anyone in S.

So, **r** is group-strategyproof. □

Since strategyproof reward schemes exist, a natural question to ask is whether it is possible to construct such schemes which are also satisfactory from other perspectives. This is what we pursue in the rest of this section.

First, one ethical principle which is appealing in this context is that workers whose contributions to production are proven to be in accordance with their declared types should not be punished for any loss of output. That is, consider $f(k,0)$ and $f(k,l)$. Although $f(k,0) > f(k,l)$, workers who have been assigned to Type 2 jobs are not responsible for the loss of output. Hence, they should not be punished. We incorporate this principle in the following Axiom.

Axiom 1. $r_2(k,0) \leq r_2(k,l)$ for all permissible pairs k,l.

Unfortunately, it is not possible to construct strategyproof reward schemes which always satisfy Axiom 1. This is the content of the next proposition.

Proposition 3. *There exist production functions such that no strategyproof reward scheme satisfies Axiom 1.*

Proof. Consider the p-model defined in the previous section with $p_3 = 0$. To simplify notation, also assume that $C = 0$.

Let $\mathbf{r}$ be a strategyproof scheme satisfying Axiom 1. Denote $r_2(1,0) = \mu$.

Since $\mathbf{r}$ is strategyproof, we must have $\mu \geq r_1(1,1) \geq r_2(0,1)$. From Axiom 1, $r_2(0,1) \geq r_2(0,0)$. Since $r_2(0,0) = p_2$, we must have

$$\mu \geq p_2 \tag{8}$$

Choose any $i \leq K - 1$. Then,

$$\begin{aligned} (i+1)r_1(1,i) + (n-i-1)r_2(1,i) &= p_1 + (n-i-1)p_2 \\ \text{or } (1+i)r_1(1,i) &= p_1 - (n-i-1)[r_2(1,i) - p_2] \end{aligned}$$

Also, $r_2(1,i) \geq \mu \geq p_2$ from Axiom 1 and (8). Hence,

$$(1+i)r_1(1,i) \leq p_1 - (n-1-i)(\mu - p_2) \tag{9}$$

Since $\mathbf{r}$ is strategyproof, $r_1(1,i) \geq r_2(0,i)$. Also, from Axiom 1, $r_2(0,i) \geq r_2(0,0) = p_2$. Using $r_1(1,i) \geq p_2$ and (9), we get

$$p_1 - (n-1-i)(\mu - p_2) \geq (1+i)p_2 \tag{10}$$

Since $\mu \geq p_2$, this yields

$$p_1 \geq (1+i)p_2 \tag{11}$$

Obviously, a p-model can be specified for which this is not true.

This shows that strategyproofness and Axiom 1 are not always compatible. □

Axiom 1 imposed a restriction on the nature of possible punishments incorporated in reward schemes. Another restriction which one may want to impose on reward schemes is the principle of workers being paid "according to contribution" when the principal detects no lies. Of course, this principle is not always enforceable for the simple reason that the production function may be such that workers' marginal contributions do not add up to the gross output. However, one case in which this principle is a priori feasible is when the production function is described by the p-model. Here, the principle of "payment according to contribution" takes a simple form. For each value of $k \leq K$, one should have $r_1(k,0) = p_1 - \frac{C}{n}$ and $r_2(k,0) = p_2 - \frac{C}{n}$. In other words, all workers are paid their marginal product minus an equal share of the fixed cost. Unfortunately, we show below that the requirement of strategyproofness is not always compatible with this principle of payment.

Proposition 4. *There exists a p-model and a size of society such that the principle of "payment acording to contribution" is not strategyproof.*

Proof. Define for $i = 1,2,3$, $\bar{p}_i = p_i - \frac{C}{n}$. Clearly, $\bar{p}_1 > \bar{p}_2$.

Suppose $\mathbf{r}$ is strategyproof and satisfies the principle of payment according to contribution. So, for all $k \leq K$ and $i = 1,2$, we must have $r_i(k,0) = \bar{p}_i$. From (4), $r_1(k,1) \leq r_2(k,0) = \bar{p}_2$. Since $(k+1)r_1(k,1) + (n-k-1)r_2(k,1) = k\bar{p}_1 + \bar{p}_3 + (n-k-1)\bar{p}_2$, we have $r_2(k,1) = \bar{p}_2 + \frac{\Delta(k)}{n-k-1}$, where $\Delta(k) = k(\bar{p}_1 - r_1(k,1)) + \bar{p}_3 - r_1(k,1)$. Since $(\bar{p}_1 - r_1(k,1)) > 0$, there exists a value of k, say k^*, such that $\Delta(k^*) > 0$. Hence, $r_2(k^*,1) > \bar{p}_2$.

But this contradicts the requirement that

$$r_2(k^*,1) \leq r_1(k^*+1,1) \leq r_2(k^*+1,0) = \bar{p}_2 \, . \qquad \square$$

4 The Complete Information Framework

In the last section, we showed that there are non-trivial strategyproof schemes. Unfortunately, Propositions 3 and 4 show that such schemes may fail to satisfy additional attractive properties. This provides us with the motivation to examine whether an incentive requirement weaker than strategyproofness widens the class of permissible schemes. This is the avenue we pursue here by examining the scope of constructing reward schemes which induce workers to reveal their true information as equilibria in games of complete information.[11]

When each worker knows other workers' types, the principal can ask each worker to report a *type profile*, although of course she may not always utilise all the information. Let $a^i = (a^i_1, \ldots, a^i_n)$ be a typical report of worker i, with $a^i_j = 1$ denoting that i declares j to be in T_1. Similarly, $a^i_j = 2$ represents the statement that i declares j to be in T_2. Let $\mathbf{a} = (a^1, \ldots, a^n)$ denote a typical vector of announced type profiles. Let $m = (A, \mathbf{r})$ be any mechanism where A is the assignment rule specifying whether worker i is in J_1 or J_2 given workers' announcements $\mathbf{a}$. Letting $A(\mathbf{a})$ denote the structure produced when workers an-

[11] Actually, we are interested in a stronger requirement. In line with traditional implementation theory, we also want to ensure that truthtelling and strategies equivalent to truthtelling are the *only* equilibria.

nounce **a** and the principal uses the mechanism **m**, the payoff function of the corresponding game is given by[12]

$$R^i(a, \mathbf{m}) = \begin{cases} r_1(k(\mathbf{a}), \mathbf{l}(\mathbf{a})) & \text{if } i \text{ is assigned to} J_1 \\ r_2(k(\mathbf{a}), \mathbf{l}(\mathbf{a})) & \text{otherwise} \end{cases} \tag{12}$$

where $k(\mathbf{a}), \mathbf{l}(\mathbf{a})$ are the number of skilled ad uskilled workers assigned to J_1 respectively corresponding to the anouncementy vector **a**.[13]

Definition 5. *Given a mechanism* **m**, *an announcement* a^i *is undominated for worker i if there is no announcement* $\bar{a}^i$ *such that for all* a^{-i}, $R^i((\bar{a}^i, a^{-i}), \mathbf{m}) \geq R^i((a^i, a^{-i}), \mathbf{m})$ *with strict inequality for some* $\tilde{a}^{-i}$.

Definition 6. *Given a mechanism* **m**, *two announcement vectors* **a** *and* $\hat{\mathbf{a}}$ *are equivalent if* $R^i(\mathbf{a}, \mathbf{m}) = R^i(\hat{\mathbf{a}}, \mathbf{m})$ *for all* i.

Notice that *all* announcement vectors will be equivalent if the principal uses an assignment rule which is completely insensitive to workers' announcements. Hence, in order to ensure a satisfactory or non-trivial solution to the incentive problem, we need to ensure that only "sensible" assignment rules are used. This provides the motivation for the following definition.

Definition 7. *An assignment rule is seemingly efficient if corresponding to any announcement vector* **a** *satisfying* $a^i = a^j$ *for all* $i, j \in N$, *up to* K *workers declared to be in* T_1 *by all workers are assigned to* J_1 *and all the rest are assigned to* J_2.

The principal of course has no way of verifying whether workers have told the truth or not until the output has actually been realized. However, if all workers unanimously announce the same type profile, then the principal has no basis for disbelieving this announcement. The assignment in this case should assign only workers declared to be in T_1 to J_1. Of course, at most K such workers can be assigned to J_1. Notice that the definition places no restriction on how assignments are made when workers do not make the same anouncement. So, it is a very weak restriction.

In this section, we are interested in the Nash equilibria and undominated strategy equilibria[14] of mechanisms which use seemingly efficient assignment rules. Let $NE(\mathbf{m})$ and $UD(\mathbf{m})$ denote the set of Nash equilibria and undominated strategy equilibria of the mechanism **m**.

Definition 8. *A reward scheme* **r** *is implemented in Nash equilibrium (respectively undominated strategies) with a seemingly efficient assignment rule A if there is a mechanism* **m** *such that for* $\mathbf{m} = (A, \mathbf{r})$, $NE(\mathbf{m})$ *(respectively* $UD(\mathbf{m})$*) consist of truthtelling and strategies which are equivalent to truthtelling.*

---

[12] Note that in contrast to the incomplete information framework, the principal does have some freedom about the assignment rule. That is why we have explicitly introduced the mechanism **m** in the notation.

[13] To simplify notation, we will omit the dependence of k, l on the announcement vector **a**.

[14] An undominated strategy equilibrium is one in which no worker is using a dominated strategy.

Let $\mathbf{r}$ be implemented in Nash equilibrium with a seemingly efficient assignment rule according to the definition given above. Then, at any equilibrium announcement, the "correct" or "efficient" assignment will be made. Furthermore, workers in J_1 will be paid $r_1(k,0)$ while workers in J_2 will be paid $r_2(k,0)$ where $|J_1| = k$. An exactly similar interpretation is valid if $\mathbf{r}$ is implemented in undominated strategies. Thus, if the class of implementable reward schemes is rich enough, then the principal can ensure payments according to various desirable principles, apart from achieving the maximum possible output given workers' true types and the production function.

In our first proposition in this section, we identify sufficient conditions on the production function which ensure that a rich class of anonymous reward schemes are Nash implementable with a seemingly efficient rule. [15]

Proposition 5. *Suppose either (i) $K < n$ or (ii) $\frac{f(k,n-k)}{n} < f(k-1,n-k)$ for all k. Let $\mathbf{r}$ satisfy the following:*

(i) $r_1(k,0) > r_2(k-1,0)$ *for all* $k \leq K$
(ii) $r_1(k,l) = 0$ *and* $r_2(k,l) = \frac{f(k,l)}{n-k-l}$ $\forall l \geq 1$.

Then, $\mathbf{r}$ is implementable in Nash equilibrium with seemingly efficient assignment rule.

Proof. Let $\mathbf{r}$ be any reward scheme satisfying (i) and (ii). Consider the following assignment rule A. For all $\mathbf{a}$, let $T_1(\mathbf{a}) = \{i \in N | a_i^i = 1\}$. Without loss of generality, let $T_1(\mathbf{a}) = \{1, 2, \ldots, L\}$. If $L \leq K$, then all $i \in T_1(\mathbf{a})$ are assigned to J_1. If $L > K$, then $\{1, 2, \ldots, K\}$ are assigned to J_1. So, the assignment rule only depends on what each individual reports about herself. If no more than K workers claim to be in T_1, then they are all assigned to J_1. If more than K workers claim to be skilled, then the first K workers are assigned to J_1.

It is easy to check that this assignment rule is *seemingly efficient*.

Let $a^* = (a_1^*, \ldots, a_n^*)$ be the vector of true types. We first show that any $\mathbf{a}$ such that $a_i^i = a_i^*$ is a Nash equilibrium.

Suppose $i \in T_1$. Then, either (i) i is assigned to J_1 or (ii) $T_1(\mathbf{a})$ contains more than K workers and i is assigned to J_2. Now consider any deviation $\hat{a}^i$ such that $\hat{a}_i^i \neq a_i^*$. If (i) holds, then i' payoff is $r_1(k,0)$ before the deviation, and either $r_2(k,0)$[16] or $r_2(k-1,0)$ after the deviation. In either case, i's deviation is not profitable. If (ii) holds, then i's deviation does not change the outcome.

Suppose now that $i \in T_2$. Then, i's payoff when all workers tell the truth is $r_2(k,0)$. Consider any deviation $\hat{a}^i$ such that $\hat{a}_i^i = 1$. Either this does not change the assignment (if i is not amongst the first K workers who declare they are in T_1) or i is assigned to J_1. But, then since $r_1(k,1) = 0$ $\forall k$, i will not deviate.

Now, we show that any $\mathbf{a} \in NE(\mathbf{m})$ must produce the same payoff vector as the truth.

[15] We are most grateful to A.Postlewaite for suggesting the mechanism used in the proof of the proposition.

[16] i's payoff could be $r_2(k,0)$ if more than K workers had originally declared themselves to be skilled. Of course, in this case $k = K$.

Assume first that $K < n$. Let $\mathbf{a} \in NE(\mathbf{m})$, and suppose that there is $i \in T_2$ such that $a_i^i = 1$. If i is not assigned to J_1, then i's announcement of a_i^i instead of the truth does not change the outcome. If i is not assigned to J_1, then $R^i(\mathbf{a}, \mathbf{m}) = r_1(k, l) = 0$. But, i can deviate by announcing $\hat{a}_i^i = 2$. Then, i's payoff would be strictly positive.

So, if $\mathbf{a} \in NE(\mathbf{m})$, then T_2 must be assigned to J_2. Consider now $i \in T_1$, and suppose $a_i^i = 2$. If i deviates and announces $\hat{a}_i^i = 1$, then either (i) i is assigned to J_1 or (ii) i is not amongst the first K workers in T_1. If i is not assigned to J_1 after the deviation, then she must be better off, so that in case (i), $\mathbf{a}$ cannot be a Nash equilibrium. In case (ii), $\mathbf{a}$ gives the same outcome as the truth.

So, this shows that when $K < n$, any $\mathbf{a} \in NE(\mathbf{m})$ is equivalent to the truth. Suppose now that $K = n$, but $\frac{f(k,n-k)}{n} < f(k-1, n-k)$ for all k.

The only remaining case we have to consider is if $a^i = 1$ for all $i \in N$. Then, for all $i \in N$, $R^i(\mathbf{a}, \mathbf{m}) = \frac{f(k,n-k)}{n}$ for some k. But, then some $i \in T_1$ can deviate and announce $\hat{a}^i = 2$. Then, i's payoff will be $f(k-1, n-k)$. This is a profitable deviation for i. □

Remark 2. An anonymous referee has pointed out that the reward schemes incorporate very heavy punishment since $r_1(k, l) = 0$ for all $l \geq 1$. However, note that this provision will apply only out of equilibrium. Thus, the only stipulation on the reward scheme applying to equilibrium messages is that $r_1(k, 0) > r_2(k-1, 0)$ for all $k \leq K$. Since this is a very weak requirement, Proposition 5 shows that the planner can implement a large class of anonymous reward schemes.

Notice that the smaller is n, the more restrictive is the condition that $\frac{f(k,n-k)}{n} < f(k-1, n-k)$. In our next proposition, we show that practically all reward schemes can be implemented in undominated strategies without this restriction on the production function, provided $K = n$.

Proposition 6. *Let $K = n$. Let* $\mathbf{r}$ *satisfy the following.*

(i) $r_1(k, 0)$ *and* $r_2(k, 0)$ *are increasing in* k.
(ii) $r_1(k, l) = r_2(k, l) = \frac{f(k,l)}{n}$ *for all* $l \geq 1$.

Then, $\mathbf{r}$ *is implementable in undominated strategies.*

Proof. Consider the following assignment rule A. For any $\mathbf{a}$, define $U_1(\mathbf{a}) = \{j \in N | a_j^i = 1 \ \forall i \in N\}$. So, the set $U_1(\mathbf{a})$ is the set of workers who are unanimously declared to be in T_1. Then, $A(\mathbf{a})$ assigns all workers in $U_1(\mathbf{a})$ to J_1, all other workers being assigned to J_2.

Let a^* be the vector of true types.

Step 1. Let $i \in T_1$. Then, the only undominated strategy of i is to announce a^*.

To see this, suppose $a^i \neq a^*$. There are two possible cases. Either (i) there is j such that $a_j^* = 1$ and $a_j^i = 2$ or (ii) there is j such that $a_j^* = 2$ and $a_j^i = 1$.

In all cases, we need only consider announcement vectors in which all other workers have declared j to be in T_1. Otherwise, i cannot unilaterally change j's assignment.

In case (i), consider first $j = i$, that is i lies about herself. Then, i is assigned to J_2. If some unskilled worker is assigned to J_1, then the "average rule" applies. Then, i does strictly better by announcing the truth about oneself since this increases aggregate output and hence the average.

If no unskilled worker is assigned to J_1, then the same conclusion emerges from the fact that $r_1(k,0) > r_2(k,0) \geq r_2(k-1,0)$.

Suppose now that $j \neq i$. Then, i's deviation to the truth about j is strictly beneficial when some unskilled worker is assigned to J_1. For then the average rule applies and aggregate output increases when j is assigned to J_1. To complete this case, note that i never loses by declaring the truth about j since $r_1(k,0)$ is increasing in k.

Consider now Case (ii). Suppose some unskilled worker other than j is assigned to J_1. Then, i's truthful declaration about j increases aggregate output, and hence i's share through the average rule. If no unskilled worker other than j is assigned to J_1, then again i gains strictly since $r_1(k,0) > \frac{f(k,0)}{n} > \frac{f(k,1)}{n}$.

This completes the proof of Step 1.

Step 2. If $i \in T_2$, and if a^i is undominated, then $a^i_j = 1$ for all $j \in T_1$.

Suppose $a^i_j = 2$ for some $j \in T_1$. Again, we need only consider announcement vectors in which all other workers declare j to be in T_1. If some unskilled worker is assigned to J_1, then i gains by declaring the truth about j since $f(k,l) > f(k-1,l)$ and the average rule applies. If only skilled workers are assigned to J_1, then i cannot lose by telling the truth since $r_2(k,0)$ is increasing in k.

This completes the proof of Step 2.

From Steps 1 and 2, $U_1(\mathbf{a}) = T_1$ whenever workers use undominated strategies. Hence, all workers in T_1 will be assigned to J_1 ad all workers in T_2 will be assigned to J_2. □

Remark 3. Notice that while truthtelling is the *only* undominated strategy for individuals in T_1, individuals in T_2 may falsely declare an *unskilled* worker i to be skilled at an undominated strategy. However, this lie or deception does not matter since some $j \in T_1$ will reveal the truth about i. Hence, Proposition 6 shows that for a very rich class of anonymous reward schemes, the outcome when individuals use undominated strategies is equivalent to truthtelling. Of course, this remarkably permissive conclusion is obtained at the cost of a strong restriction on the class of production functions since the proposition assumes that $K = n$. If $K < n$, then workers in T_1 may have to "compete" for the positions in J_1. This implies that declaring another Type 1 worker to be in T_2 is no longer a dominated strategy for some worker in T_1.

5 Conclusion

In this paper, we have used a very simple model in which incentive issues raised by adverse selection can be discussed. The main features of the model are the presence of two types of workers as well as two types of jobs. We conclude by pointing out that our results do not really depend on there being *two* types of workers and jobs. The model can easily be extended to the case of k types of workers and jobs, provided an assumption analogous to Assumption 1 is made. What we need to assume is that workers of Type i are most productive in jobs of type i. They are as productive as workers of Type $(i+j)$ in jobs of type $(i+j)$, and *less* productive in type $(i-j)$ jobs than in type $(i+j)$ jobs. With this specification and the assumption that despite possible capacity restrictions on jobs of a particular type , the first best assignment never places a worker of type i in a job of type $(i-j)$, the principal can still detect whether workers of a particular type have claimed to be of a higher type. Notice that except in Proposition 1, the specification of the reward schemes did not need knowledge of *how many* workers had lied. It was sufficient for the principal to know that realized output was lower than the expected output. Hence, obvious modifications of the reward schemes and assignment rules will ensure that Propositions 2, 5 and 6 can be extended to the k type case. Of course, Propositions 3 and 4 are true since they are in the nature of counterexamples. It is only in the case of Proposition 1 that the reward scheme needs to use detailed information on the *number* of people who have lied. This came for free in the two-type framework, given assumption 1. In the general k type model, we would need to assume that the principal can on the basis of the realized output, "invert" the production function and find out how many workers of each type have lied and claimed to be of a higher type. Note that this will be generically true for the class of production functions satisfying the extension of Assumption 1 outlined above.

References

Dutta, B. (1997) Reasonable mechanisms and Nash implementation. In: Arrow, K.J., Sen, A.K., Suzumura, K. (eds.) *Social Choice Theory Reexamined*. Macmillan, London

Jackson, M. (1992) Implementation in undominated strategies: A look at bounded mechanisms. *Review of Economic Studies* 59: 757-75

Groves, T. (1973) Incentives in teams. *Econometrica* 41: 617-31

Holmstrom, B. (1982) Moral hazard in teams. *The Bell Journal of Economics* 13: 324-340

Moore, J. (1992) Implementation, contracts and renegotiation in environments with complete information. In: Laffont, J.J. (ed.) *Advances in Economic Theory*. Cambridge University Press, Cambridge

Israelson, D.L. (1980) Collectives, communes and incentives. *Journal of Comparative Economics* 4: 99-121

Thomson, W. (1982) Information and incentives in labour-managed economies. *Journal of Comparative Economics* 6: 248-268

Project Evaluation and Organizational Form

Thomas Gehrig[1], **Pierre Regibeau**[2], **Kate Rockett**[2]

[1] Institut zur Erforschung der wirtschaftlichen Entwicklung, Universität Freiburg, D-79085 Freiburg, GERMANY (email: gehrigt@vwl.uni-freiburg.de)
[2] University of Essex, Wivenhoe Park, Colchester CO4 3SQ, UK (email: pregib@essex.ac.uk)

Abstract. In situations of imperfect testing and communication, as suggested by Sah and Stiglitz (AER, 1986), organizational forms can be identified with different rules of aggregating evaluations of individual screening units. In this paper, we discuss the relative merits of polyarchical organizations versus hierarchical organizations in evaluating cost-reducing R&D projects when individual units' decision thresholds are fully endogenous. Contrary to the results of Sah and Stiglitz, we find that the relative merit of an organizational form depends on the curvature of the screening functions of the individual evaluation units. We find that for certain parameters organizations would want to implement asymmetric decision rules across screening units. This allows us to derive sufficient conditions for a polyarchy to dominate a hierarchy. We also find conditions for which the cost curves associated with the two organizational forms cross each other. In this case the optimal organizational form will depend on product market conditions and on the "lumpiness" of cost-reducing R&D.

JEL Classification: D23, D83, L22

Key Words: Organisations, screening, information aggregation, hierarchies, polyarchies

---

We would like to thank Siegfried Berninghaus, Hans Gersbach, Kai-Uwe Kühn, Meg Meyer, Armin Schmutzler and Nicholas Vettas, as well as participants of the Winter Meeting of the Econometric Society in Washington, the Annual Meeting of the Industrieökonomischer Ausschuss in Vienna, the CEPR-ECARE-conference on Information Processing Organizations in Brussels and seminar participants at Rice University, the University of Padova and the University of Zurich. We are particularly grateful for the comments and suggestions of Martin Hellwig and three anonymous referees. Gehrig gratefully acknowledges financial support of the Schweizerischer Nationalfonds and the hospitality of the Institut d'Analisi Economica and Rice University. Regibeau and Rockett gratefully acknowledge support from the Spanish Ministry of Education under a DGICYT grant. Regibeau also acknowledges support of the EU under a TMR-program.

1 Introduction

When firms search for new products or ideas they need to develop judgements about the likelihood of success. If these judgements are not perfectly accurate it may be desirable to ask different individuals to evaluate the idea and provide independent assessments. These evaluations can then be used to decide whether, or not, to pursue the product or idea in question. If all assessments resemble each other, an overall decision will be easily reached. If there is disagreement, however, the overall decision will depend on the nature of the aggregation rule used by the organization.

In this paper we focus on the case where firms must evaluate (potentially) cost-reducing R&D projects. Following Sah and Stiglitz (1986,1988), we assume that individual reviewers cannot communicate perfectly their evaluation of a given project. They can only express whether or not they believe that the project exceeds a pre-specified measure of quality. We will refer to these minimum quality standards as "thresholds". An organization can then be seen as a set of review units capped by a "strategic" unit which sets the thresholds and decides how to aggregate the assessments of the reviewers. Two such aggregation rules are considered. In a hierarchy, unanimous approval by the review units is required for the R&D project to be approved and carried out. On the contrary, a polyarchy would pursue any project approved by at least one of its units.[1]

The assumption of limited communication seems to be reasonable. Individual reviewers may well develop sophisticated assessments of the project at hand but the sheer complexity of the task combined with differences in the skills and backgrounds of reviewing and strategic unit may hamper the effective communication of such detailed appraisals.[2] Also it may be difficult to articulate "gut feelings" about the profitability of a project. Alternatively, incentive reasons may obscure public statements by researchers who may feel uneasy about revealing areas in which their knowledge is rather imprecise. In order to concentrate on the informational differences associated with different decision rules we abstract from any explicit consideration of incentive effects. Instead, our review units behave rather mechanically as truthful information generating devices.[3]

For the type of cost-reducing R&D projects that we consider we show that the performance of an organization can be summarized by a "cost function" $C(q)$, where q is the joint probability of accepting a project and this project being successful. $C(q)$ then is the minimum expected cost of actually carrying out a successful project with probability q. This reflects the cost of carrying out all approved projects, whether or not they turn out to be successful. We can then compare polyarchies and hierarchies by ranking their corresponding cost func-

[1] It should be stressed that the terms "polyarchy" and "hierarchy" only refer to specific aggregation rules. They do not refer to any further aspects of hierarchical decision making. Hence decision problems of the type analyzed in Radner (1993) are not considered .

[2] See, however, Quian, Roland and Xu (1999) for a different approach to modelling imperfect communication.

[3] See, for example, Melumad, Mookherjee and Reichelstein (1995)for an analysis of incentives in organizations when communication is limited.

tions. To achieve this, we depart from Sah and Stiglitz (1986, 1988) by allowing the strategic unit to set different thresholds for different review units. This extra flexibility allows the organization to affect the quality of an observation communicated to the strategic unit. Typically, the quality of an individual observation differs across organizational forms. We find that the polyarchy always uses its two observations, i.e. the thresholds chosen are such that, for each review unit, there are values of the signal for which a project must be rejected. On the other hand, there are situations where the hierarchy optimally chooses to let one of the review units accept all projects, irrespective of the signal received. In such cases, the hierarchy effectively uses only one of its two observations. This striking result is explained by the differential informational value of an additional observation under the two organizational forms. When additional observations are possible, a hierarchy always loosens the thresholds assigned to its decision units, thereby reducing the quality of the communicated signals.[4] This means that a hierarchy must trade off the costs of a higher probability of erroneously accepting bad projects against the benefit from additional observations. In contrast, a polyarchy always tightens the threshold of its review units when it uses more of them, thus reducing the likelihood of falsely accepting bad projects. Therefore the polyarchy always prefers to use more observations.

We show that, whenever a hierarchy chooses to only use one of its review units the cost function of a polyarchy lies everywhere below the cost function of a hierarchy. Such a situation occurs when the distribution of signals received by the review units has a decreasing likelihood ratio and signals are not too informative. Whenever the hierarchy prefers to use their two observations, polyarchies and hierarchies would both choose the same threshold for all review units so that Sah and Stiglitz's assumption is actually verified. Still we can extend their results by showing that, for our cost-reducing R&D projects, the cost functions associated with hierarchy and polyarchy must cross at least once. This suggests that the optimal organizational form depends on the desired level of q and thus on market conditions. Moreover, a polyarchy must be more efficient than a hierarchy for high levels of q while the opposite must be true for low levels of q.

The paper is organized as follows. In Sect. 2 we present the market environment, the screening processes, and the stochastic environment faced by the firm. In Sect. 3 we obtain conditions under which polyarchies and hierarchies choose interior or corner solutions for their thresholds. We use this result in Sect. 4 to rank hierarchies and polyarchies according to their cost functions and discuss how the choice of organizational form might depend on the firm's external environment. Section 5 presents parametric examples. Section 6 provides conditions for optimality of the threshold decision rule and discusses further extensions. Section 7 concludes.

[4] I.e., the probability of erroneous acceptances rises.

2 The Model

We will first describe the market environment in which the firm operates. We will then turn to the internal organization of the firm and to a precise specification of the stochastic environment.

Consider a single firm which has the option of conducting cost-reducing research. The outcome of the research effort is uncertain. However, the firm may hire experts, who will develop some imperfect judgement about the project's likelihood of success. If the project is successful the firm can reduce marginal costs of production to zero. If the project is unsuccessful, production continues at the current marginal costs $c > 0$. The cost of carrying out the project is assumed to be fixed and is equal to $F > 0$.

Architecture of the Firm and the Screening Process

The firm is viewed as consisting of a strategic (policy-setting) unit and two screening units. Screening units $i = 1, 2$ have to evaluate potential research projects. The result of their screening activities are two imperfect signals $\tilde{y}_i, i = 1, 2$ of a project's quality. Based on its own signal each unit decides whether or not to recommend the project for adoption. The recommendation is the only information passed on to the strategic unit. The decision to recommend a project is based on a decision rule A_i.

Different organizational forms are identified with rules that aggregate individual decisions. When unanimity is required to implement the project we refer to the organisation as a *hierarchy*. When the project requires only one vote of approval we shall call the organization a *polyarchy* (see Fig. 1).

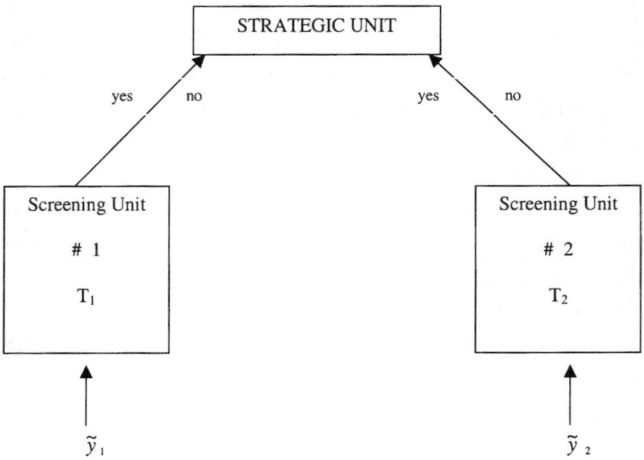

Fig. 1. Information aggregation within the firm

It should be noted that our definition of a polyarchy implies some form of coordination, which excludes the duplication of projects. In our setting the project will be adopted by the organization only after individual decisions are aggregated.[5]

The strategic unit selects an organizational form and determines the decision rules $A_i, i = 1, 2$ to maximize the firm's expected profits. While a general screening rule would specify precisely the set of signals for which adoption is recommended, we concentrate on *threshold decision rules*. A screening unit will vote for adoption, whenever $A_i = \{\tilde{y}_i \mid \tilde{y}_i \leq T_i\}$, where T_i is the decision threshold. The main appeal of these rules is their simplicity. They also correspond to some widely observed screening rules such as the internal "hurdle rates" used by most US firms. Finally, since the initial analysis of Sah and Stiglitz was conducted for such rules, it seems appropriate to use them as well in order to isolate the effect of allowing the two organizational forms to differ in the strictness of the screening criterion that they impose on the individual units. Still, we will show in Sect. 6, that "threshold" rules are actually optimal, when the signals satisfy the monotone likelihood property as defined below.

Screening units are assumed to do their prescribed tasks rather mechanically: they observe their signals and only report whether or not they meet the assigned thresholds. We abstract from incentive issues by assuming that the quality of the signal $\tilde{y}_i$ does not depend on the effort exerted by the screening agent and that the welfare of the screening unit is independent of its report.

In order to compare our two organizational forms, some kind of normalization is necessary. One could explicitly introduce reviewing costs and let each organization decide how many of the available projects to review. Rather, we decided to ignore review costs and to normalize the number of projects reviewed by each organization to one. This approach has the advantage that each organizational form gets the same number of independent signals. In Sect. 6 we discuss the implications of our findings for alternative normalizations.

The Stochastic Environment

We take the signal received by a single review unit to be a one-dimensional random variable distributed on the interval $[0, 1]$. Let $p \in]0, 1[$ be the a priori success probability of the project.

The conditional densities $h_0(y)$ of observing $\tilde{y} = y$ given that the project is really good, and $h_c(y)$ of observing $\tilde{y} = y$ when the project is actually bad, are of particular interest. Denote their respective cumulative distributions by $H_0(y)$ and $H_c(y)$ and assume $h_0(y) > 0$ and $h_c(y) > 0$ for all $y \in [0, 1]$.

Given a decision rule A_i, the probability $\hat{q}_i$ that a single unit i accepts the project is equal to the probability that the project is good times the conditional probability of acceptance given that the project is good plus the probability of

[5] Such a view seems reasonable, when the organization is interpreted as a firm or a committee. When economic systems are compared, as in Sah and Stiglitz (1986), presumably one would interpret each screening unit as a firm that could adopt the project on its own. In this case duplication of projects will occur in the case of polyarchies. This cost would not apply to hierarchies. Surprisingly, Sah and Stiglitz (1986) simply assume that duplication does not occur.

acceptance given that the project is bad, i.e.

$$\begin{aligned}\hat{q}_i &= p \int_{y \in A_i} h_0(y)\,dy + (1-p) \int_{y \in A_i} h_c(y)\,dy \\ &= pH_0(T_i) + (1-p)H_c(T_i)\,.\end{aligned}$$

The joint probability of unit i accepting a project and this project being successful, q_i, is determined by:

$$\begin{aligned}q_i &= p \int_{y \in A_i} h_0(y)\,dy \\ &= p\,H_0(T_i)\,.\end{aligned}$$

Accordingly, q_i is the probability that the final outcome of the application of the decision rule is acceptance of a project that turns out to be successful.

We consider the case where observation errors across screening units are conditionally independent.

Assumption: Independent observation errors The joint conditional distributions of $(\tilde{y}_1, \tilde{y}_2)$ given $\tilde{c}$ can be written
$H_0(y_1, y_2) = H_0(y_1)H_0(y_2)$ and $H_c(y_1, y_2) = H_c(y_1)H_c(y_2)$.

Finally we discuss the meaning of signals. We assume that low realizations of $\tilde{y}$ can be taken as an indication of low costs, and hence constitute good news, while high realizations are bad news. This is formalized as:

Definition: Monotone likelihood ratio property (MLRP) Let $h_0(y) > 0$ and $h_c(y) > 0$ and let $h_0(y)$ and $h_c(y)$ be differentiable for $0 < y < 1$. Furthermore, let $H_0(0) = H_c(0) = 0$. The monotone likelihood property (MLRP) is satisfied when[6]

$$\frac{h_0'(y)}{h_0(y)} \leq \frac{h_c'(y)}{h_c(y)} \quad , \; 0 < y < 1 \quad . \tag{1}$$

3 Optimal Organizational Structures

After the project has been evaluated and, possibly, carried out, the firm competes in a market game. Denote its market payoff $R(\tilde{c}) \geq 0$, where $\tilde{c} = 0$ if the project was approved and successful, and $\tilde{c} = c > 0$ if the project was rejected or if it was approved but it was not successful.

Recall that q was defined as the probability of the event that "the project is good and accepted". Therefore the expected profit can be written as

$$qR(0) + (1-q)R(c) - C(q) \tag{2}$$

where $C(q)$ is the expected cost of carrying out the project conditional on actually approving and developping a project that is successful with probability q. In

[6] The likelihood ratio is defined as $\frac{h_0(y)}{h_c(y)}$. Our definition of MLRP implies a declining likelihood ratio.

other words, $C(q)$ is equal to the cost F of developing the project multiplied by the probability that the project, good or bad, is approved by the organization. To compare the expected profits of the hierarchy and the polyarchy we must therefore rank their corresponding cost functions $C^H(q)$ and $C^P(q)$. These functions will not usually be the same.[7] For identical thresholds, the polyarchy will accept both good and bad projects with a higher probability than the hierarchy since it only needs one "yes" to go ahead with a project while the hierarchy requires unanimity (Sah and Stiglitz 1986). The polyarchy's thresholds could of course be lowered to yield the same probability of acceptance of a good project as the hierarchy but then the two organizations would still differ in the probability of acceptance of a bad project.

In the case of the hierarchy, for given thresholds T_i , $i = 1, 2$, the probability that the organization actually reduces its cost from c to 0 is the probability that the given project is good, p, times the probability that the organization accepts the project, conditional on the project being good.

$$\begin{aligned} q^H(T_1, T_2) &= p \int_{y \le T_1} h_0(y)dy \int_{y \le T_2} h_0(y)dy \\ &= p\, H_0(T_1)\, H_0(T_2) \end{aligned}$$

Likewise in the case of the polyarchy, we have:[8]

$$\begin{aligned} q^P(T_1, T_2) &= p\left(1 - \left(1 - \int_{y \le T_1} h_0(y)dy\right)\left(1 - \int_{y \le T_2} h_0(y)dy\right)\right) \\ &= p\ (1 - (1 - H_0(T_1))\ (1 - H_0(T_2))) \end{aligned}$$

The probability $\hat{q}^H(T_1, T_2)$ that the organization accepts the project also includes the possibility of erroneously accepting a bad project, i.e. $\hat{q}^H(T_1, T_2) =$

[7] While we choose to compare cost functions for different aggregation rules one could also analyze expected returns under the different aggregation rules in an alternative framework as suggested by a referee. As in much of Sah and Stiglitz (1986) assume that the expected return of a project x can have two values $x_s > 0$ and $x_u < 0$, with prior probabilities p_s and p_u, respectively. Consider the H-aggregation procedure. Given thresholds T_1 and T_2, let $r_s^H(T_1, T_2)$ (resp. $r_u^H(T_1, T_2)$) denote the probability of accepting the project conditional on $x = x_s$ (resp. on $x = x_u$). Then the expected payoff given T_1 and T_2 is $\pi^H(T_1, T_2) = p_s r_s^H(T_1, T_2) x_s + p_u r_u^H(T_1, T_2) x_u$.

Define r_s^L, r_u^L and π^L analogously for the L-aggregation rule. The problem then is to solve $\max_{T_1, T_2} \pi^k(T_1, T_2)$ for $k \in \{H, L\}$, and to compare the maximised values.

[8] Notice that the threshold assigned to one unit does not depend on the decision taken by the other unit. One interpretation is that the units conduct their review simultaneously. However, under our assumption of independent observation errors, such simultaneity is not essential: allowing for sequential screening, where the threshold of the second unit could differ according to the message received from the first unit does not affect our results. The intuition behind this result is that, in the simultaneous setting, the coarseness of communication between the two units effectively makes the second unit's threshold conditional on the message obtained from the first unit. In the case of a hierarchy, the threshold of the second unit is only relevant when the first unit accepts. Hence, the unconditional threshold used here can be thought of as a threshold that only applies if the first unit communicates a "yes". The threshold used after a "no" message is received is irrelevant since the project will be rejected anyway. In the case of a polyarchy, the threshold used in earlier sections corresponds to the threshold that applies following a "no" from the first unit. The threshold applying when a "yes" is received is irrelevant since the project will be adopted, regardless of the message sent by the second unit.

$q^H(T_1,T_2)+(1-p)H_c(T_1)H_c(T_2)$. This reads in the case of a hierarchy as:

$$\hat{q}^H(T_1,T_2) = p\,H_0(T_1)\,H_0(T_2) + (1-p)\,H_c(T_1)\,H_c(T_2)$$

Likewise in the case of the polyarchy the project is accepted, if either screening unit accepts, or alternatively, if both units do not reject. So the probability of acceptance is:

$$\begin{aligned}\hat{q}^P(T_1,T_2) &= p\left(1-(1-H_0(T_1))\,(1-H_0(T_2))\right)\\ &\quad +(1-p)\left(1-(1-H_c(T_1))\,(1-H_c(T_2))\right)\end{aligned}$$

We are now in a position to derive the cost functions $C^H(q)$ and $C^P(q)$. Suppose the strategic unit would like the organization to accept a good project with probability q. The cost associated with this requirement consists of the erroneous acceptance of bad projects. The probability of an erroneous acceptance will depend on the choice of T_1 and T_2. The cost of achieving success probability q is defined by the choice of (T_1,T_2) that minimizes erroneous adoptions. So a firm organized as a hierarchy solves

$$C^H(q) := \min_{T_1,T_2}\left[\hat{q}^H(T_1,T_2)F \mid q^H(T_1,T_2)=q\right] \qquad \textbf{(H)}$$

while a firm organized as a polyarchy solves

$$C^P(q) := \min_{T_1,T_2}\left[\hat{q}^P(T_1,T_2)F \mid q^P(T_1,T_2)=q\right] \qquad \textbf{(P)}$$

The concept of *log concavity* and *log convexity* will prove useful in characterizing different regimes for organizational form.

Definition. *A function $h : D \to IR$, where $D \subset IR$ is called log concave if and only if $\ln h(x)$ is concave for all $x \in D$. The function $h : D \to IR$ is called log convex if and only if $\ln h(x)$ is convex.*

Clearly a concave function is also log concave while a log convex fucntion is also convex. Lemma 1 summarizes some useful properties of log concave and log convex functions.

Lemma 1. *Let $h(x) \geq 0$ be differentiable for $x \in D$, where $D \subset IR$ is compact.*

a) The function $h(.)$ is (strictly) log concave if and only if $h''(x)h(x)-(h'(x))^2 < 0$ for all $x \in D$. It is (strictly) log convex if and only if $h''(x)h(x)-(h'(x))^2 > 0$ for all $x \in D$.

b) When $h''(x)h(x)-(h'(x))^2 = 0$ for all $x \in D$, the function $h(x)$ is necessarily of the form $h(x) = exp(Ax+B)+C$, where A, B and C are parameters.

c) Let $k : D \to D$ and $k(x) = h(1-x)$. Then $h(.)$ is log concave (log convex) if and only if $k(.)$ is log concave (log convex).

d) Let $x \leq 0$. Then $h(exp(x)) : [-\infty, 0] \to [0,1]$ is log concave if $h''(x)h(x) - (h'(x))^2 < \frac{-1}{x}h(x)h'(x)$ for all x and $k(1-exp(x)) : IR_{\leq 0} \to [0,1]$ is log concave if $k''(x)k(x) - (k'(x))^2 < \frac{1}{1-x}k(x)k'(x)$ for all x.

$h(exp(x)) : [-\infty, 0] \to [0,1]$ is log convex if $h''(x)h(x) - (h'(x))^2 > \frac{-1}{x}h(x)h'(x)$ for all x and $k(1-exp(x)) : [\infty, 0] \to [0,1]$ is log convex if $k''(x)k(x) - (k'(x))^2 > \frac{1}{1-x}k(x)k'(x)$ for all x.

Proof. See appendix.

Define the conditional success probability for a good project $z := \frac{q}{p}$. We are now in a position to discuss the solutions to the optimization problems **(H)** and **(P)**.

Result 1: Hierarchy
a. If $H_c(H_0^{-1}(e^x))$ is log concave in x, the solution to **(H)** *is a corner solution with $(1-T_1)(1-T_2) = 0$.*
b. If $H_c(H_0^{-1}(e^x))$ is log convex in x, the solution to **(H)** *is uniquely determined and symmetric, i.e. $T_1 = T_2$.*
c. If $H_c(H_0^{-1}(e^x))$ is both log concave and log convex in x, organizational form is indeterminate.

Proof. The proof translates the cost minimization problem into an equivalent problem, which makes transparent the conditions for (global) convexity and concavity of the objective function. Other than that standard arguments are used.

With $z = \frac{q}{p} \in [0,1]$, the hierarchy's planning problem is equivalent to

$$\min_{T_1,T_2}\Big[(1-p)\,H_c(T_1)\,H_c(T_2)\,F \mid H_0(T_1)\,H_0(T_2) = z\Big]$$

Equivalently,

$$\min_{T_1,T_2}\Big[\quad H_c(H_0^{-1}(exp(lnH_0(T_1))))\,H_c(H_0^{-1}(exp(lnH_0(T_2)))) \mid \\ lnH_0(T_1) + lnH_0(T_2) = ln(z)\Big]$$

or

$$\min_{S_1,S_2}\Big[H_c(H_0^{-1}(exp(S_1)))\,H_c(H_0^{-1}(exp(S_2))) \mid S_1 + S_2 = ln(z)\Big] \qquad (3)$$

where $S_i = lnH_0(T_i))$. Substitute $A(S_i) := H_c(H_0^{-1}(exp(S_i)))$. A is increasing in S_i. The first order conditions imply

$$\frac{A'(S_1)}{A(S_1)} = \frac{A'(S_2)}{A(S_2)}$$

Moreover, the determinant of the bordered Hessian of the cost minimization problem implies a cost minimum if

$$A''(S_1)A(S_2) + A(S_1)A''(S_2) - 2A'(S_1)A'(S_2) > 0$$

With the help of the first order condition this inequality can be rewritten as

$$\left(\frac{A''(S_1)A(S_1)}{A'(S_1)} - A'(S_1)\right)A'(S_2) + \left(\frac{A(S_2)A''(S_2)}{A'(S_2)} - A'(S_2)\right)A'(S_1) > 0$$

Accordingly, $A''(S_i)A(S_i) > (A'(S_i))^2$ for $i = 1, 2$ implies global convexity of the objective function and, hence, an interior solution, while $A''(S_i)A(S_i) < (A'(S_i))^2$ implies global concavity and, hence, a corner solution. So, by Lemma 1.a, the optimization problem **(H)** attains a corner solution with $(1-T_1)(1-T_2) = 0$ when $A(S)$ is log concave and **(H)** attains a unique interior solution when $A(S)$ is log convex. Because the optimization problem is symmetric in $T_i, i = 1, 2$ the unique equilibrium is characterized by symmetric thresholds $T_1^H = T_2^H$.

Finally, under the condition of c., by virtue of Lemma 1.b, $H_c(H_0^{-1}(e^x)) = \exp(Ax + B) + C$ for some parameters A, B, C. Hence, in this case the cost minimization problem (3) is equivalent to

$$\min_{S_1,S_2}\left[B^2 e^{AS_1} e^{AS_2} + C \mid S_1 + S_2 = ln(z)\right]$$

which again is equivalent to

$$\min_{T_1,T_2}\left[(T_1 T_2)^A \mid T_1 T_2 = z\right]$$

and thus proves the claim. Q.E.D.

The proof generalizes easily to the case of $N > 2$ reviewing units with a hierarchical decision rule. The same applies the the case of the polyarchical decision rule.

Result 2: Polyarchy

a. If $1 - H_c(H_0^{-1}(1 - e^x))$ *is log concave in x, the solution to* **(P)** *is uniquely determined and symmetric, i.e.,* $T_1 = T_2$.

b. If $1 - H_c(H_0^{-1}(1 - e^x))$ *is log convex in x, the solution to* **(P)** *is a corner solution with* $(1 - T_1)(1 - T_2) = 0$.

c. If $1 - H_c(H_0^{-1}(1 - e^x))$ *is both log concave and log convex in x, organizational form is indeterminate.*

Proof. The logic of the proof is the same as for Result 1. The firm's minimization problem is

$$\min_{T_1,T_2}\left[1 - (1 - H_c(T_1))\,(1 - H_c(T_2)) \mid ((1 - H_0(T_1))\,(1 - H_0(T_2))) = 1 - z\right]$$

Or, equivalently,

$$\max_{S_1,S_2}\Big[\ \left(1 - H_c(H_0^{-1}(1 - exp(S_1)))\right)\,\left(1 - H_c(H_0^{-1}(1 - exp(S_2)))\right) \mid S_1 + S_2 = ln(1 - z)\Big]$$

where the monotonic transformation $S_i = ln(1 - H_0(T_i))$ has been made. Defining $A(S_i) = 1 - H_c(H_0^{-1}(1 - \exp(S_i)))$ we have

$$\max_{S_1,S_2}\left[A(S_1)A(S_2) \mid S_1 + S_2 = ln(1-z)\right]$$

This problem has an interior solution if the isocost curve is convex to the origin, and corner solutions if the isocost curve is concave to the origin. As for Result 1, it is easily shown that convexity of the isocost curve obtains if $A(S_i)$ is log concave while concavity obtains if $A(S_i)$ is log convex. As in Proposition 1.c under the condition c. organizational form does not matter. Q.E.D.

The conditions for the potential selection of corner solutions under the two organizational forms are of particular interest, since this phenomenon was ruled out in the analysis of Sah and Stiglitz (1986). First note that the monotone likelihood ratio property implies the convexity of $H_c(H_0^{-1}(x))$.

Lemma 2. *Under the monotone likelihood ratio property the function* $H_c(H_0^{-1}(x))$ *is convex in* $x \in [0,1]$.

Proof. See appendix.

An immediate consequence of Lemmas 1 and 2 is that the function $1 - H_c(H_0^{-1}(1-exp(x)))$ is log concave. Therefore, we find that under the monotone likelihood ratio property the polyarchy will always select an interior solution. With a hierarchy, however, corner solutions can still obtain.

Result 3
Under the monotone likelihood ratio property the polyarchy will always select an interior solution $T_1^P = T_2^P$, *but the hierarchy will choose a corner solution when signals are not very informative.*

Proof. The proof proceeds in three steps. First conditions for corner solutions are derived for the two organizational forms. Then it is shown that the conditions for corner solutions for a polyarchy cannot arise under MLRP. Finally, it is shown that corner solutions can arise for a hierarchy even under MLRP.
1. Let $H_0(0) = H_c(0)$ and $h_0(y) > 0$, $h_c(y) > 0$ and differentiable for $0 < y < 1$. Then

a) **(H)** attains corner solutions when

$$\frac{h_c(y)}{H_c(y)} - \frac{h_c'(y)}{h_c(y)} > \frac{h_0(y)}{H_0(y)} - \frac{h_0'(y)}{h_0(y)}, \quad 0 < y < 1$$

b) and **(P)** attains corner solutions when

$$\frac{-h_c(y)}{1-H_c(y)} - \frac{h_c'(y)}{h_c(y)} > \frac{-h_0(y)}{1-H_0(y)} - \frac{h_0'(y)}{h_0(y)}, \quad 0 < y < 1$$

The proof a statements a) and b) is by direct evaluation of the conditions for corner solutions in Result 1.a and Result 2.b. Observe that the (first) derivative of $\ln(H_c(y))$ with respect to x, where $y = H_0^{-1}(e^x)$, can be written as $\frac{H_0(y)}{H_c(y)}\frac{h_c(y)}{h_0(y)}$. Likewise the (first) derivative of $\ln(1 - H_c(y))$ where $y = H_0^{-1}(1-e^x)$ can be written as $\frac{1-H_0(y)}{1-H_c(y)}\frac{h_c(y)}{h_0(y)}$. The conditions of a) and b) then follow by differentiating again and checking for concavity and convexity respectively.

2. Polyarchy

We show that under MLRP the function $1-H_c(H_0^{-1}(1-e^x))$ is log concave. According to Result 2 this implies that the polyarchy selects an interior solution.

To do this define the auxiliary function $l(x) := H_c(H_0^{-1}(x))$. This function has the following derivatives

$$l'(x) = \frac{h_c(H_0^{-1}(x))}{h_0(H_0^{-1}(x))}$$

$$l''(x) = \frac{h_c(H_0^{-1}(x))}{h_0^2(H_0^{-1}(x))}\left(\frac{h_c'(H_0^{-1}(x))}{h_c(H_0^{-1}(x))} - \frac{h_0'(H_0^{-1}(x))}{h_0(H_0^{-1}(x))}\right)$$

Accordingly, $\frac{l''(x)}{l'(x)} - \frac{l'(x)}{l(x)} > \frac{1}{1-x}$ is equivalent to

$$\frac{h'_c(H_0^{-1}(x))}{h_c(H_0^{-1}(x)} - \frac{h'_0(H_0^{-1}(x))}{h_0(H_0^{-1}(x))} > \frac{h_c(H_0^{-1}(x))}{H_c(H_0^{-1}(x))} + \frac{h_0(H_0^{-1}(x))}{1-H_0^{-1}(x)}$$

Since the right hand side is always positive, according to Lemma 2, MLRP implies this relation. Hence, by application of Lemmas 1.d and 1.c we find that $1-l(1-e^x)$ is log concave.

3. Hierarchy

We show: $H_c(H_0^{-1}(exp(x)))$ is log convex in x if and only if:

$$\frac{h_c'(H_0^{-1}(x))}{h_c(H_0^{-1}(x))} - \frac{h_0'(H_0^{-1}(x))}{h_0(H_0^{-1}(x))} > \frac{h_c(H_0^{-1}(x))}{H_c(H_0^{-1}(x))} - \frac{h_0(H_0^{-1}(x))}{H_0^{-1}(x)} \quad , 0 < x < 1 \,. \tag{4}$$

Using the auxiliary function $l(x)$ of step 2 one finds that $\frac{l''(x)}{l'(x)} - \frac{l'(x)}{l(x)} > \frac{-1}{x}$ for all x iff

$$\frac{h'_c(H_0^{-1}(x))}{h_c(H_0^{-1}(x))} - \frac{h'_0(H_0^{-1}(x))}{h_0(H_0^{-1}(x))} > \frac{h_c(H_0^{-1}(x))}{H_c(H_0^{-1}(x))} - \frac{h_0(H_0^{-1}(x))}{H_0^{-1}(x)} \quad , 0 < x < 1$$

According to Lemma 1.d this implies that $l(e^x)$ is log convex.

The condition for an interior solution of the hierarchy's problem is restrictive, since (4) is not implied by MLRP. As a result, with MLRP two possibilities arise. Either the hierarchy chooses a corner solution, in which case it is (weakly) dominated by the polyarchy, or it chooses an interior solution. Section 5 provides an example, in which the hierarchy (weakly) selects a corner solution while the polyarchy selects an interior solution. Q.E.D.

The hierarchy attains corner solutions when the relative slopes $\frac{h_c'(y)}{h_c(y)}$ and $\frac{h_0'(y)}{h_0(y)}$ are sufficiently close. In other words, corner solutions occur when the conditional

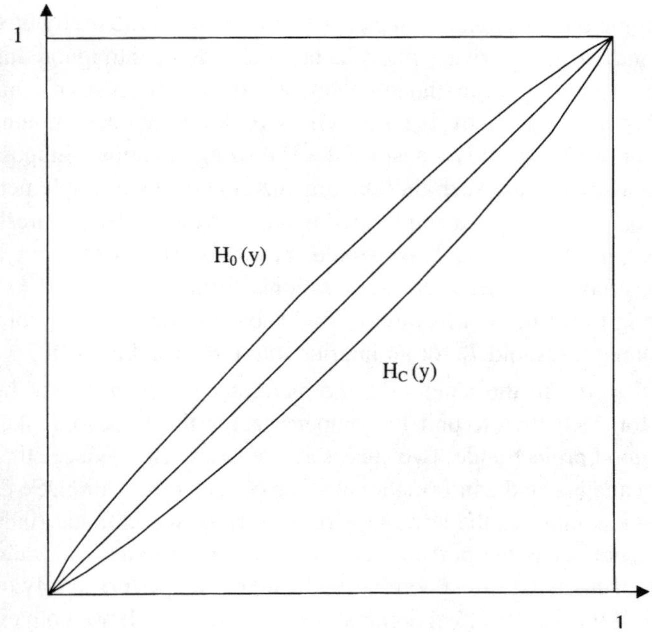

Fig. 2. Observations with little informational content

distribution functions $H_0(y)$ and $H_c(y)$ are relatively close, or signals are not very informative (see Fig. 2).

The condition for corner solutions for a polyarchically structured organization imply that $\frac{h_c'(y)}{h_c(y)} - \frac{h_0'(y)}{h_0(y)} < 0$ whenever the signal $\tilde{y}$ is informative. Hence, under MLRP, the polyarchical firm will never select corner solutions.

4 Properties of Optimal Organizations

We are now in a position to compare the performance of the two organizational forms.

Result 4: Dominance of the polyarchy
When $H_c(H_0^{-1}(exp(x)))$ is log concave in $x \in IR_{\leq 0}$ under the monotone likelihood ratio property the polyarchy has lower costs than the hierarchy, i.e. $C^P(q) \leq C^H(q)$.

Proof. According to Result 3 the polyarchy selects an interior solution, while the hierarchy selects a corner solution. This corner solution is a feasible choice for the polyarchy. Q.E.D.

Interestingly, the polyarchy is the dominant form of organization whenever MLRP is satisfied, and when $H_c(H_0^{-1}(exp(x)))$ is log concave in x. This condition is more likely to be met, when signals are not very informative. In such situations a hierarchy may be too conservative. Therefore, the organization has to restrict itself to a single observation, in order to meet the desired probability of success q. This result may seem surprising since the hierarchy chooses to ignore information from further sampling even though there is no explicit cost of sampling. In fact, this result is driven by the coarseness of information communicated by the review units. To see this, consider the following situation. Suppose product market characteristics are such that the organization wants to implement success probability q_0. If there was a single review unit it would select a threshold level T_0 such that $pH_0(T_0) = q_0$. How would an additional review unit affect the organizations payoffs under each organizational form?

According to Result 4, a hierarchy would select either a corner solution with a corresponding threshold T_0, or an interior solution with threshold T^H such that $p\left(H_0(T^H)\right)^2 = q_0$. In the latter case the hierarchy has to relax the hurdle rate $T^H > T_0$ for each review unit to compensate for the increased likelihood of rejecting a good project under two successive reviews. This reduces the tightness of screening and hence the informational value of each observation. So employing a second review unit has the advantage of acquiring an additional (independent) observation, but it has the potential cost of a lower informational value of each observation. If the latter effect dominates the hierarchy prefers to rely on a single observation. If the former effect dominates the hierarchy will use both evaluations and implement a symmetric solution. The loss in information value of a single observation is particularly damaging, when signals are not very informative. This can be seen for example for the case when $h_c(y) = 1$ and $\frac{h'_0(y)}{h_0(y)}$ is close to zero (Fig. 3a). When $\frac{h'_0(y)}{h_0(y)}$ is rather negative the signal may still remain fairly informative even when its tightness is reduced (Fig. 3b).

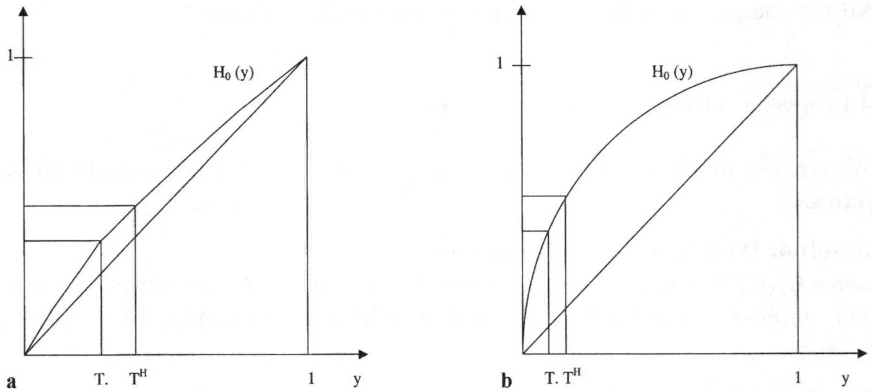

Fig. 3a,b. Threshold adjustment for additional observations. **a** Low informational content; **b** High informational content

In the case of the polyarchy thresholds T^P are chosen symmetrically such that $p(1 - H_0(T^P))^2 = q_0$. This implies $T^P < T_0$. Hence, the polyarchy increases the tightness of filtering when additional observations are used. This actually means that the polyarchy gains both, from additional observations *and* from a higher value of each observation. This explains why the polyarchy will never select a corner solution.

Result 4 demonstrates that organizational choice will typically depend on the curvature of the likelihood ratio. When signals are more informative, i.e. $H_c(H_0^{-1}(exp(x)))$ is log convex, both organizations will select (symmetric) interior solutions. This means that the assumption of symmetric thresholds made by Sah and Stiglitz (1986) is justified. Even in this case, however, we can extend their analysis by presenting conditions under which $C^P(q)$ and $C^H(q)$ must cross at least once. We also obtain some limit results on the ranking of the two cost functions for sufficiently high and sufficiently low values of q.

Result 5: Crossing cost curves
When $H_c(H_0^{-1}(exp(x)))$ is log convex and when the MLRP is satisfied, both organizations select interior solutions and the cost functions associated with the two organizations cross at least once. Furthermore, the hierarchy is more efficient for low levels of q while the polyarchy is more efficient for very high levels of q, i.e. there are $\underline{q} > 0$ and $\bar{q} < p$ such that $C^H(q) < C^P(q)$ for $0 < q \leq \underline{q}$ and $C^H(q) > C^P(q)$ for $p > q \geq \bar{q}$.

Proof. See appendix.

The intuition for this result relies on the fact that the value of "tight" screening rules changes as one moves from low to high values of q. Let us first remember that, for any given q, the threshold chosen by the hierarchy, T^H, will be less tight (i.e. higher) than the threshold chosen by the polyarchy, T^P. For small q, T^H and T^P are fairly close to zero, i.e. both organizational forms must be very strict in their evaluations in order to achieve the desired probability of success q. In short, the gains obtained through tighter evaluations (as in the polyarchy) are small compared to the advantages of additional evaluations with looser rules so that a hierarchy is preferred. For high values of q, on the other hand, T^P and T^H are both close to one so that both organizational forms do little filtering. In this case the value of somewhat tighter rules is great so that the polyarchy is the most efficient organizational form.

According to Result 5, for bounded and strictly decreasing signal densities we can find situations, in which firms may prefer the hierarchical organization, when they choose a (rather) low conditional success probability $\frac{q}{p}$, while they will prefer a polyarchical organization, when their desired conditional success probability $\frac{q}{p}$ is (rather) large. A direct consequence of this is that the optimal organizational form depends on the level of q that the strategic unit wants to achieve. This in turn will depend on the precise shape of the payoff function $R(c)$, i.e., on *market conditions*. This means that, in contrast to Sah and Stiglitz (1986), the prior probability p is not essential: for a given p, the same firm might choose different forms of organization depending on its market environment.

At the points of intersection of $C^H(q)$ and $C^P(q)$, typically, the firm's cost curve $\min(C^H(q), C^P(q))$ is not differentiable (see Fig. 4). Under MLRP, for example, both $C^H(q)$ and $C^P(q)$ are convex functions. This implies that the set of acceptance probabilities that is potentially chosen by the firm may be non-convex. This non-differentiability also implies potentially drastic reactions in the optimal organizational form with respect to small changes in product market conditions (see Fig. 5). Indeed, if one were to consider a two-stage game where firms choose their organizational forms and threshold before reviewing a project, observe their new marginal cost, and compete in prices, one would easily obtain multiple equilibria, some of them asymmetric.[9]

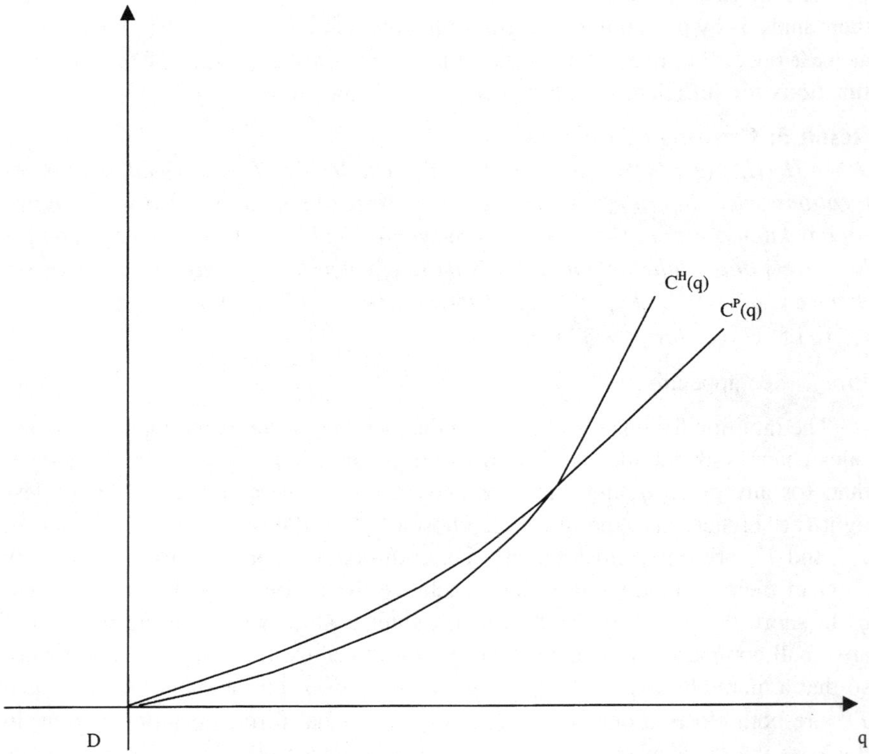

Fig. 4. Typical cost curves

5 Parametric Examples

Results 4 and 5 provide sufficient conditions for dominance of the polyarchical form or for crossing cost curves. In this section we shall provide examples, demonstrating that these possibilites do actually occur for some classes of distribution functions. First we provide a class of distribution functions, for which the polyarchy is the dominant organizational form and then we provide an example with crossing cost curves.

Example 1. Consider the following distribution functions:

[9] Our analysis also suggests that reaction functions $q_i(q_j), i \neq j$ of such a game are typically lacking continuity.

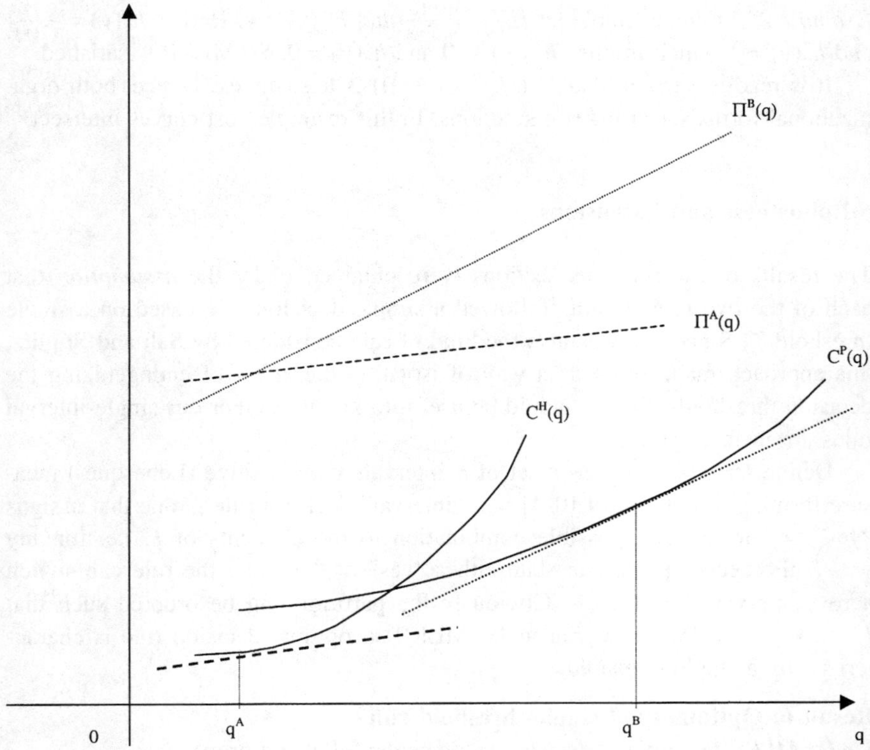

Fig. 5. Organizational choice and market conditions

$$H_0(y) := y^a,\ a < 1$$

$$H_c(y) := y^b,\ b > 1$$

In this case, $h_0(y) = ay^{a-1}$ is declining and $h_c(y) = by^{b-1}$ is increasing in y. So MLRP is satisfied. Moreover,

$$H_c(H_0^{-1}(\exp(x))) = \exp(x)^{\frac{b}{a}}$$

$$1 - H_c(H_0^{-1}(1 - \exp(x)) = 1 - (1 - \exp(x))^{\frac{b}{a}}$$

Since $lnH_c(H_0^{-1}(exp(x))) = \frac{b}{a}$ this function is both log convex and log concave. Consequently, according to Result 1.c the hierarchy is indifferent in the number of screens it uses. The corresponding cost function turns out to be:

$$C^H(q) = \min_{T_1,T_2}\left[(1-p)\,(T_1\,T_2)^{\frac{b}{a}}\ F \mid pT_1T_2 = q\right] = (1-p)\left(\frac{q}{p}\right)^{\frac{b}{a}} F$$

On the other hand $1 - H_c(H_0^{-1}(1 - \exp(x))$ is log concave. In accordance with Result 2 the polyarchy will employ both filters. This example shows, that the possibility described in Result 3 can actually occur.

Example 2. In this example let $H_0(y) = \frac{4y}{3+y}$ and $H_c(y) = y$. Hence, $h_0(y) = \frac{12}{(3+y)^2}$ and $h_c(y) = 1$ which implies $h'_0(y) < 0$ and $h_c(y) = 0$. So MLRP is satisfied.

It is readily verified that $H_c(H_0^{-1}(exp(z)))$ is log convex. Hence, both organizational forms select interior solutions. In this example cost curves intersect.[10]

6 Robustness and Extensions

The results of the previous sections were obtained under the *assumption* that each of the two review units followed a simple decision rule based on a single threshold T. Since this was also the kind of rule considered by Sah and Stiglitz, this approach made sense as a way of isolating the effect of endogenizing the decision thresholds. Still, it would be useful to know whether our single-interval rules are in fact optimal.

Define $I = \{I_1, ..., I_n\}$ as a set of n intervals with positive (Lebesque-) measure forming a partition of $[0, 1]$. An "interval" decision rule is one that assigns "yes" or "no" in any possible combination to the elements of I, i.e. for any $y \in I_l$ the decision rule can state either "yes" or "no" and the rule can switch across intervals $l = 1, ..., n$. Obviously the partition can be ordered such that $I_1 < I_2 < ...I_n$. We show that under MLRP an optimal decision rule is characterized by a single threshold.

Result 6: Optimality of single-threshold rules
Under MLRP the optimal decision rule has the following properties:
i) If $y \in I_k$ implies "yes", then $y \in I_l$ implies "yes" for all $l < k$.

ii) If $y \in I_{k'}$ implies "no", then $y \in I_{l'}$ implies "no" for all $l' > k'$.

Proof. See appendix.

In other words: Under MRLP the optimal decision rule implies the existence of a threshold $T \in [0, 1]$ such that for any interval decision rule $y < T$ implies acceptance of the project while $y > T$ implies rejection.

The optimality of our single-threshold rules comes from the combination of the MLRP-property, the conditional independence of signals and the coarseness of the information that can be transmitted. Because of independence and the fact that review units can only accept or reject, the behavior of one unit only appears as a multiplicative term in the maximization problem of the other unit so that, effectively, we need only to show that the decision rule of an isolated review unit must be monotonic. This, in turn, is guaranteed by our monotone likelihood ratio property.

In order to compare polyarchy and hierarchy some form of standardization is necessary. We have chosen to force both organizational forms to evaluate the same number of projects (i.e. one). This contrasts with the analysis of Gersbach and Wehrspohn (1998), who allow the organizational forms to evaluate different numbers of projects but constrain them to implement the same expected number

[10] One finds, for example, $C^H(.1) < C^P(.1)$ (and $C^H(.9) < C^P(.9)$), while $C^H(.99) > C^P(.99)$.

of projects. For exogenous and identical thresholds across review units and organizational forms they find that the hierarchy will screen projects more tightly (as pointed out by Sah and Stiglitz, 1986) and, consequently, that it will evaluate more projects. Accordingly, in their framework the hierarchy always performs better. Our analysis suggests that endogenizing thresholds might modify the results of Gersbach and Wehrspohn. This is especially likely for the cases, where we find that the polyarchy dominates the hierarchy because it can use a second signal about the value of the project without having to loosen decision rules of individual units. This effect would also arise with Gersbach and Wehrspohn's normalization. However, the strength of this effect would probably be less important than in our framework because the hierarchy could, to some extent, compensate the lesser ability to exploit a second signal about the same project by reviewing more projects than the polyarchy.

A more satisfying, but more complex, normalization would be to consider that the organization has a maximum budget B that it can spend on both the cost of carrying out projects (F per project) and the cost of project evaluation (say M per project reviewed). For large values of M (i.e. $M > \frac{B-F}{2}$), the organization can only evaluate a single project so that we are back to our own normalization. As M gets small we converge to a case where the two organizational forms will effectively carry out the same number of projects.[11] Under the conditions for which we find the polyarchy dominant we would expect the relative profitability of the hierarchy to improve as M decreases since the cost of "compensating" by evaluating more projects decreases.[12]

7 Concluding Comments

Firms must often decide whether or not to pursue projects of uncertain pay-offs. In making that decision, companies rely on the judgement of their own managers and/or of outside experts. We consider the case of cost-reducing R&D projects. Following Sah and Stiglitz (1986, 1988) we concentrate on the situation where the "review units" can only communicate whether or not the project should be undertaken according to a simple threshold rule such as a hurdle rate. We also assume that all units review the project simultaneoulsy. We compare polyarchic organizations, where the approval of one unit is enough for the project to proceed and hierarchical organizations, where unanimity is required. We also allow the threshold rule of each unit to be set optimally by a "strategic" unit.

We show that, when the signals received by the review units are not very informative, the hierarchy optimally chooses to disregard some of the signals received. In this case, the polyarchy unambiguously dominates the hierarchy.

[11] This is not quite the same normalization as in Gersbach and Wehrspohn where the two organizations have the same expected number of projects approved for development. With fixed and equal thresholds, however, their results would obtain with both organizations carrying out the same actual number of projects.

[12] In the extreme case of an infinite number of projects and $M = 0$, polyarchy, hierarchy and single units perform equally well as they optimally wait until a signal $y = 0$ is received.

If, on the other hand, both types of organization use all of the available signals, then their relative performance depends on market conditions and on the nature of R&D projects. For example, one would expect the polyarchy to be relatively more efficient when innovation is "lumpy" while the hierarchy would be preferable if innovation typically occurs in small increments. Our results can be readily extended to situations where the review units can use more complex decision rules and where the decision process is sequential.

Several unanswered questions remain. For example, although our results suggest that the choice of organizational form can crucially depend on product market conditions faced by the firm, we cannot shed much light on this relationship. There is clearly room for models that could investigate the interaction between product market competition and the choice of organizational form by the various competitors in more detail. Another question worth pursuing is the effect of the *degree* of coarseness of the message space on the relative performance of the two types of organizational form. Does one type of organization become relatively more efficient as one moves from our extreme case where review units can only transmit a binary signal to cases where they can communicate the signal that they perceive more precisely?

Appendix

Proof of Lemma 1. These results are based on standard techniques and straightforward differentiation. In case d) observe that $h(k(x)) = x$ implies $h'(k(x))k'(x) = 1$ and (by differentiating again) $h''(k(x))\,(k'(x))^2 + h'(k(x))k''(x) = 0$. Application of a) yields the result. Q.E.D.

Proof of Lemma 2. Observe that $\frac{\partial}{\partial x} H_c(H_0^{-1}(x)) = \frac{h_c(H_0^{-1}(x))}{h_0(H_0^{-1}(x))}$ and $\frac{\partial^2}{\partial x^2} H_c(H_0^{-1}(x)) \geq 0$ if and only if $\frac{h_0'(H_0^{-1}(x))}{h_0(H_0^{-1}(x)}) \leq \frac{h_c'(H_0^{-1}(x))}{h_c(H_0^{-1}(x))}$ Q.E.D.

Proof of Result 5. Under the conditions of Result 5 both organizations will choose interior solutions. These are uniquely determined and symmetric, i.e. $T_1^H = T_2^H$ and $T_1^P = T_2^P$. (This follows from the fact that log convexity/concavity are imposed globally and screening units are identical). So the cost functions can be written as:

$$C^H(q) = \left(q + (1-p)A^2(\sqrt{\frac{q}{p}})\right) F$$

$$C^P(q) = \left(q + (1-p)(1-(1-A(1-\sqrt{1-\frac{q}{p}}))^2)\right) F \quad ,$$

where $A(z) = H_c(H_0^{-1}(z))$.

Obviously, $C^H(0) = C^P(0) = 0$ and $C^H(p) = C^P(p) = F$. We shall demonstrate that the two organizational forms exhibit different marginal behaviour in the limits. Define

$$B^H(z) := A^2(\sqrt{z})$$

$$B^P(z) := 1 - (1 - A(1 - \sqrt{1-z}))^2$$

First consider the marginal behaviour for small z, i.e. $z \to 0$. By application of l'Hospital's rule one finds:

$$\begin{aligned}
\lim_{z\to 0} \frac{\partial}{\partial z} B^H(z) &= \lim_{z\to 0} \frac{1}{\sqrt{z}} \frac{\partial}{\partial z} A(\sqrt{z}) A(\sqrt{z}) \\
&= \frac{\partial}{\partial z} A(0) \frac{\lim_{z\to 0} \frac{\partial}{\partial z} A(\sqrt{z})}{\lim_{z\to 0} \frac{\partial}{\partial z} \sqrt{z}} \\
&= \left(\frac{\partial}{\partial z} A(0)\right)^2
\end{aligned}$$

$$\begin{aligned}
\lim_{z\to 0} \frac{\partial}{\partial z} B^P(z) &= \lim_{z\to 0} \left(1 - (1 - A(1 - \sqrt{1-z}))^2\right) \\
&= \lim_{z\to 0} \frac{1}{\sqrt{1-z}} (1 - A(1 - \sqrt{1-z}) \frac{\partial}{\partial z} A(1 - \sqrt{1-z}) \\
&= \frac{\partial}{\partial z} A(0)
\end{aligned}$$

According to Lemma 2 the function $A(z)$ is convex for $z \in [0,1]$. Therefore, $A : [0,1] \to [0,1]$ implies $\frac{\partial}{\partial z} A(0) < 1$. Hence, there is a $\underline{z} > 0$, such that $B^H(z) < B^P(z)$ for $0 < z \le \underline{z}$.

The reverse is true as $z \to 1$. In this case we find:

$$\lim_{z\to 1} \frac{\partial}{\partial z} B^H(z) = \frac{\partial}{\partial z} A(1)$$

$$\begin{aligned}
\lim_{z\to 1} \frac{\partial}{\partial z} B^P(z) &= \lim_{z\to 1} \left(1 - (1 - A(1 - \sqrt{1-z}))^2\right) \\
&= \frac{\partial}{\partial z} A(1) \lim_{z\to 1} \frac{1 - A(1 - \sqrt{1-z})}{\sqrt{1-z}} \\
&= \left(\frac{\partial}{\partial z} A(1)\right)^2
\end{aligned}$$

Again, because of convexity of $A(z)$ the derivative $\frac{\partial}{\partial z} A(1) > 1$. So the cost function has a steeper slope in case of the polyarchy. Therefore, in a sufficiently small neighbourhood of $q = p$ we find $C^P(q) < C^H(q)$. Q.E.D.

Proof of Result 6. For each review unit $i = 1, 2$ define Y_i as the set of intervals to which a "yes" has been assigned and N_i as the set of intervals to which a "no" has been assigned. We will prove the claim by contradiction. Take any possible interval rule for unit 1. Now assume that the optimal rule for unit 2 is *not* a single threshold rule. This means that there must be a "yes" interval that lies immediately to the right of a "no" interval. Let us define the "no" interval as $[T_1, T_2]$ and the corresponding "yes" interval as $[T_2, T_3]$. We are going to show that this cannot be an optimal rule because a reshuffling of these two intervals decreases the cost of obtaining a given q. The proof will be shown for the hierarchy. The case for the polyarchy is easily derived along similar lines.

Define $Y_{2-} := Y_2 - [T_1, T_2]$ and $N_{2-} := N_2 - [T_2, T_3]$. We have

$$q = p \int_{y_1 \in Y_1} h_0(y_1) dy_1 \left(\int_{y_2 \in Y_{2-}} h_0(y_2) dy_2 + H_0(T_3) - H_0(T_4) \right)$$

and

$$\begin{aligned} \hat{q} = \; & p \int_{y_1 \in Y_1} h_0(y_1) dy_1 \left(\int_{y_2 \in Y_{2-}} h_0(y_2) dy_2 + H_0(T_3) - H_0(T_2) \right) + \\ & (1-p) \int_{y_1 \in Y_1} h_c(y_1) dy_1 \left(\int_{y_2 \in Y_{2-}} h_c(y_2) dy_2 + H_c(T_3) - H_c(T_2) \right) \end{aligned}$$

Now let us decrease T_3 by an arbitrarily small $\epsilon > 0$ (i.e. expand the "no" interval to the right of our "yes" interval) and increase T_1 by ϵ' (i.e. expand the "yes" interval to the left of our "no" interval). Notice that such reshuffling is always possible. We can select ϵ and ϵ' such that $dq = 0$. This implies that:

$$H_0(T_3) - H_0(T_3 - \epsilon) = H_0(T_1 + \epsilon') - H_0(T_1)$$

We can now determine the effect of such a change on the cost of achieving q. After the reshuffling we have

$$\begin{aligned} \hat{q}_r = \; & p \int_{y_1 \in Y_1} h_0(y_1) dy_1 \\ & \left(\int_{y_2 \in Y_{2-}} h_0(y_2) dy_2 + H_0(T_3 - \epsilon) - H_0(T_2) + H_0(T_1 + \epsilon') - H_0(T_1) \right) + \\ & (1-p) \int_{y_1 \in Y_1} h_c(y_1) dy_1 \\ & \left(\int_{y_2 \in Y_{2-}} h_c(y_2) dy_2 + H_c(T_3 - \epsilon) - H_c(T_2) + H_c(T_1 + \epsilon') - H_c(T_3 - \epsilon) \right) \end{aligned}$$

where the subscript r refers to the values after reshuffling. Hence the change in $\hat{q}$ induced by reshuffling is:

$$\begin{aligned} \hat{q}_r - \hat{q} = \; & p \int_{y_1 \in Y_1} h_0(y_1) dy_1 \left(H_0(T_3 - \epsilon) - H_0(T_3) + H_0(T_1 + \epsilon') - H_0(T_1) \right) + \\ & (1- \; p) \int_{y_1 \in Y_1} h_c(y_1) dy_1 \left(H_c(T_3 - \epsilon) - H_c(T_3) + H_c(T_1 + \epsilon') - H_c(T_1) \right) \end{aligned}$$

Using the condition obtained from $dq = 0$, we get

$$\hat{q}_r - \hat{q} = (1-p) \int_{y_1 \in Y_1} h_c(y_1) dy_1 \left(H_c(T_1 + \epsilon') - H_c(T_1) - (H_c(T_3) - H_c(T_3 - \epsilon)) \right)$$

Hence, $\hat{q}_r < \hat{q}$ iff $H_c(T_1 + \epsilon') - H_c(T_1) < H_c(T_3) - H_c(T_3 - \epsilon)$.

Since the conditional density functions are assumed to be continuous[13] the term $\frac{H_0(T_1+\epsilon') - H_0(T_1)}{\epsilon'} \; \frac{\epsilon'}{H_c(T_1+\epsilon') - H_c(T_1)}$ approximates $\frac{h_0(T_1)}{h_c(T_1)}$ as $\max(\epsilon, \epsilon')$ becomes sufficiently small. Moreover, MLRP implies that $\frac{h_0(T)}{h_c(T)}$ is decreasing in T. Accordingly,

[13] Our definition of MLRP even requires differentiability.

$$\frac{H_0(T_1+\epsilon')-H_0(T_1)}{H_c(T_1+\epsilon')-H_c(T_1)} > \frac{H_0(T_3)-H_0(T_3-\epsilon)}{H_c(T_3)-H_c(T_3-\epsilon)}$$

Recall that ϵ and ϵ' were chosen such that $H_0(T_1+\epsilon') - H_0(T_1) = H_0(T_3) - H_0(T_3-\epsilon)$. Therefore, $H_c(T_1+\epsilon') - H_c(T_1) < H_c(T_3) - H_c(T_3-\epsilon)$, which again implies that the cost of achieving any given q are lower under reshuffling (since $\hat{q}_r < \hat{q}$).

Hence the only possible rules are "no" intervals everywhere, "yes" intervals everywhere, or "yes" intervals for all $y \leq T$ and "no" for all $y > T$. All these are special cases of the single-threshold rule. Q.E.D.

References

Gersbach, H., Wehrspohn, U. (1998) Organizational design with a budget constraint. *Review of Economic Design* 3(2): 149–157

Melumad, N., Mookherjee, D., Reichelstein, S. (1995) Hierarchical decentralization of incentive schemes. *Rand Journal of Economics* 26(4): 654–672

Quian, Y., Roland, G., Xu, C. (1999) Coordinating changes in M-form and U-form organizations. Working Paper, London School of Economics, August 1999

Radner, R. (1993) The organization of decentralized information processing. *Econometrica* 62: 1109–1146

Sah, R., Stiglitz, J. (1986) The architecture of economic systems: Hierarchies and polyarchies. *American Economic Review* 76: 716–727

Sah, R., Stiglitz, J. (1988) Committees, hierarchies and polyarchies. *The Economic Journal* 98: 451–470

References

[1] Dutta, B., Jackson, M.O. (2001) On the Formation of Networks and Groups

[2] Myerson, R. (1977) Graphs and Cooperation in Games. *Mathematical Operations Research* 2: 225–229

[3] Jackson, M.O., Wolinsky, A. (1996) A Strategic Model of Social and Economic Networks. *Journal of Economic Theory* 71: 44–74

[4] Johnson, C. Gilles, R.P. (2000) Spatial Social Networks. *Review of Economic Design* 5: 273–299

[5] Dutta, B., Mutuswami, S. (1997) Stable Networks. *Journal of Economic Theory* 76: 322–344

[6] Jackson, M.O. (2002) The Stability and Efficiency of Economic and Social Networks. In Murat Sertel (ed.) *Advances in Economic Design*, forthcoming. Springer-Verlag

[7] Bala, V., Goyal, S. (2000) A non-cooperative model of network formation. *Econometrica* 68: 1181–1229

[8] Dutta, B., Jackson, M.O. (2000) The Stability and Efficiency of Directed Communication Networks. *Review of Economic Design* 5: 251–272

[9] Aumann, R., Myerson, R. (1988) Endogenous Formation of Links Between Players and Coalitions: An Application of the Shapley Value. In A. Roth, *The Shapley Value*, Cambridge University Press, 175–191

[10] Dutta, B., van den Nouweland, A., Tijs, S. (1998) Link Formation in Cooperative Situations. *International Journal of Game Theory* 27: 245–256

[11] Slikker, M., van den Nouweland, A. (2000) Network Formation Models with Costs for Establishing Links. *Review of Economic Design* 5: 333–362

[12] Currarini, S., Morelli, M. (2000) Network Formation with Sequential Demands. *Review of Economic Design* 5: 229–249

[13] Gerber, A. (2000) Coalition Formation in General NTU Games. *Review of Economic Design* 5: 149–175

[14] Bala, V., Goyal, S. (2000) A Strategic Analysis of Network Reliability. *Review of Economic Design* 5: 205–228

[15] Watts, A. (2001) A Dynamic Model of Network Formation. *Games and Economic Behavior* 34: 331–341

[16] Kranton, R., Minehart, D. (2001) A Theory of Buyer-Seller Networks. *American Economic Review* 61: 485–508

[17] Kranton, R., Minehart, D. (2000) Competition for Goods in Buyer-Seller Networks. *Review of Economic Design* 5: 301–331

[18] Bloch, F., Ghosal, S. (2000) Buyers' and Sellers' Cartels on Markets with Indivisible Goods. *Review of Economic Design* 5: 129–147

[19] Bienenstock, E., Bonacich, P. (1997) Network Exchange as a Cooperative Game. *Rationality and Society* 9: 37–65

[20] Barbera, S., Dutta, B. (2000) Incentive Compatible Reward Schemes for Labour-Managed Firms. *Review of Economic Design* 5: 111–127

[21] Gehrig, T., Regibeau, P., Rockett, K. (2000) Project Evaluation and Organizational Form. *Review of Economic Design* 5: 177–199